90.00
6-92
GL
W9-CRB-727

THE
INTERNATIONAL SERIES
OF
MONOGRAPHS ON PHYSICS

GENERAL EDITORS

R. J. ELLIOTT J. A. KRUMHANSL
W. MARSHALL D. H. WILKINSON

# THEORY OF NEUTRON SCATTERING FROM CONDENSED MATTER

## Volume 1: Nuclear Scattering

STEPHEN W. LOVESEY

*Rutherford Appleton Laboratory*

CLARENDON PRESS · OXFORD
1984

Repl. ~~B000223042~~
PHYSICS
04688636

*Oxford University Press, Walton Street, Oxford OX2 6DP*
*London New York Toronto*
*Delhi Bombay Calcutta Madras Karachi*
*Kuala Lumpur Singapore Hong Kong Tokyo*
*Nairobi Dar es Salaam Cape Town*
*Melbourne Auckland*

*and associated companies in*
*Beirut Berlin Ibadan Mexico City Nicosia*

*Oxford is a trade mark of Oxford University Press*

*Published in the United States*
*by Oxford University Press, New York*

© *Stephen W. Lovesey, 1984*

*All rights reserved. No part of this publication may be reproduced, stored in a retrieval system, or transmitted, in any form or by any means, electronic, mechanical, photocopying, recording, or otherwise, without the prior permission of Oxford University Press*

*British Library Cataloguing in Publication Data.*
*Lovesey, S. W.*
*Theory of neutron scattering from condensed matter.—(International series of monographs on physics)*
*Vol. 1*
*1. Neutrons—Scattering*
*I. Title II. Series*
*539.7′213 QC793.5.N4628*
*ISBN 0-19-852015-8*

*Library of Congress Cataloging in Publication Data*
*Lovesey, S. W. (Stephen W.)*
*Theory of neutron scattering from condensed matter.*
*(The International series of monographs on physics)*
*Includes bibliographies and index.*
*1. Neutrons—Scattering. 2. Condensed matter.*
*I. Title. II. Series: International series of monographs on physics (Oxford, Oxfordshire)*
*QC793.5.N4628L68 1984 539.7′213 84-9684*
*ISBN 0-19-852015-8 (v. 1)*
*ISBN 0-19-852017-4 (v. 2)*

*Filmset and printed in Northern Ireland by The Universities Press (Belfast) Ltd.*

QC 793
.5
N4628
L68
1984
v.1
PHYS

# PREFACE

Neutron scattering measurements provide information at an atomic level on the chemical and physical properties of matter. By virtue of the weakness of the neutron–matter interaction, the information is undistorted by the neutron probe and, in some instances, the information is obtainable in no other way. Utilization of neutron diffraction and spectroscopy has continued apace in the past decade (Bromley 1983), rendering *Theory of thermal neutron scattering*, by Marshall and Lovesey (published in 1971) incomplete and partially obsolete. At first sight, a revision exercise is daunting because of the vast range of applications. For neutron scattering is now a standard experimental technique in many branches of physical chemistry and biophysics, in addition to more traditional applications in the physics of condensed matter and materials science. Yet it is just this burgeoning which provides much of the motivation to undertake a revision. Moreover, there is good reason to expect an acceleration of research, and new vistas exposed with the exploitation of advanced spallation neutron sources (Windsor 1981).

My aim is to provide a comprehensive account of the basic concepts and theoretical methods for the interpretation of neutron scattering experiments designed to elucidate properties of condensed matter. By the very nature of research, the reader will, more than likely, want results for a situation slightly different from the one described here. Consequently, I have reported most calculations in some detail, so that modifications and embellishments might be straightforward. By the same token, the reader should have ready access to specialist books and reviews. The availability of recent books and review articles has often been the deciding factor in determining the depth of treatment of various topics.

One of the great advantages of the neutron, referred to above, is that it is a very delicate probe. This manifests itself in the theoretical interpretation of neutron scattering experiments by the fact that we are concerned wholly with the chemical and physical properties of the target sample, with no distortion of its properties by the neutron probe. All the examples I discuss are systems that are close to thermal equilibrium, i.e. I exclude examples that involve chemical kinetics, turbulence, convective instabilities, etc. In consequence, linear response theory is omnipresent in the theoretical framework used here; indeed, most of the examples are case studies for the statistical physics of systems close to thermal equilibrium.

No attempt has been made to prepare a bibliography of relevant literature, since a source is already available (Sakamoto *et al.* 1979). However, I have listed recent books and review articles, and papers that report calculations germane to the developments I discuss. Nor is there any attempt to discuss experimental methods, multiple scattering, resolution corrections, instrumentation, and neutron sources.

The revision exercise has, not surprisingly, led to an increase in the size of the book, to an extent where it seemed prudent to separate the material into two volumes. The first volume, on the scattering from nuclei, contains much new material. Polarization phenomena and magnetic scattering are covered in the second volume. Because the topics now include hot and cold neutron scattering, as well as thermal neutron scattering, the title of the two-volume set has been suitably modified.

Choice of notation always poses problems for an author, and here the problem is exacerbated, to some extent, by the wide range of backgrounds of potential readers. By and large, I have opted for the notation in current usage, although this leads, inevitably, to conflicts where usage is not common between different scientific communities. For example, many physicists denote the neutron scattering wave vector by $\boldsymbol{\kappa}$ (used here) or $\mathbf{k}$, whereas chemists frequently use $\mathbf{Q}$. A list of important symbols is given in the front of the book. I should draw attention to the use of a circumflex to distinguish quantum mechanical operators from ordinary c-numbers; e.g. $\hat{b}$ is the scattering amplitude operator and $b$ is a scattering length. Time-dependent operators $\hat{\mathbf{R}}(t)$, say, that feature in the Van Hove representation of the cross-section are sometimes required in the formalism for time $t=0$ and we often use the shorthand $\hat{\mathbf{R}}(0)=\hat{\mathbf{R}}$ where it is unlikely to cause confusion. This is in line with an effort to use the least cumbersome notation. However, I have not succumbed to the use of reduced units with $\hbar=k_{\mathrm{B}}=1$, or the neutron mass $m=1$. The definitions of special mathematical functions follow Gradshteyn and Ryzhik (1980).

I am indebted to many colleagues who, over the years, have helped my understanding of the subject. None more so than the late Dr John Hubbard who also perused the original book and supplied many comments which I have heeded. My colleague of long standing Prof Ewald Balcar kindly pointed out several points of possible confusion in the text.

Invitations to lecture on neutron scattering in university courses and summer schools have provided me with the motivation to revise and up-date my general knowledge of the theory. Some material prepared for these occasions is incorporated in the new texts. I should like to acknowledge my gratitude to Dr John Dore, Prof Marco Fontana, Dr Julia Higgins and Prof George Trammell for inviting me to lecture in the courses and schools they have organized.

The secretarial talents of Miss Caroline Monypenny made light the arduous task of manuscript preparation. Margaret, my Achates and wife, and my children Jane and Kirke made the exercise entirely possible through their unflagging support and forebearance.

*Stanford-in-the-Vale, Oxfordshire*
*March 1984* S. W. L.

## REFERENCES

Bromley, D. A. (1983) Physics Today, December p. 31.
Gradshteyn, I. S. and Ryzhik, I. M. (1980). *Table of integrals series and products.* Academic Press, New York.
Sakamoto, M., Chihara, J., Kadotani, H., Sekiya, T., Gotoh, Y., Inoue, K., Kuriyama, K., and Yoda, O. (1979). *Bibliography for thermal neutron scattering* (6th edn). Japan Atomic Energy Research Institute 8417.
Windsor, C. G. (1981). *Pulsed neutron scattering.* Taylor and Francis, London.

# ACKNOWLEDGEMENTS

I am indebted to the following for permission to use data and diagrams as a basis for tables and figures in the text: H. Bilz, J. R. D. Copley, B. Dorner, J.-P. Hansen, W. A. Kamitakahara, L. Koester, W. Kress, J. Mayers, I. R. McDonald, J. Smith, O. Söderstrom, J.-B. Suck, M. Warner, W. B. Yelon, Academic Press (London), Institute of Physics (London), Springer-Verlag (Heidelberg), and American Institute of Physics and *The Physical Review*.

CONTENTS


**CONTENTS OF VOLUME 2** xiii

**LIST OF IMPORTANT SYMBOLS** xv

**1. PRINCIPAL FEATURES** **1**
- 1.1. Introduction 1
- 1.2. Neutron cross-section 4
- 1.3. Scattering from bound nuclei 10
- 1.4. Total single-atom cross-sections 16
- 1.5. Contrast factors 21
- 1.6. Scattering amplitude operator 23
- 1.7. Homonuclear diatomic molecule 24
- 1.8. Static approximation 29
- 1.9. Scattering by a single free nucleus 30

**2. ELASTIC NUCLEAR SCATTERING** **34**
- 2.1. Crystal lattices and reciprocal lattices 34
- 2.2. Coherent cross-sections 38
- 2.3. Crystal symmetry effects 39
- 2.4. Bragg scattering 43
- 2.5. Diffuse scattering 54

**3. CORRELATION AND RESPONSE FUNCTIONS** **59**
- 3.1. Response function 59
- 3.2. Coherent and incoherent cross-sections 63
- 3.3. Properties of correlation functions and $S(\boldsymbol{\kappa}, \omega)$ 67
- 3.4. Impulse approximation 73
- 3.5. Classical approximation 76
- 3.6. Simple model systems 78
  - 3.6.1. Perfect fluid 78
  - 3.6.2. Harmonic oscillator 86
- 3.7. Total cross-section for a perfect Boltzmann fluid 93
- 3.8. Proof of Bloch's identity 95

**4. LATTICE DYNAMICS** **98**
- 4.1. Harmonic lattice vibrations 98
- 4.2. Elastic scattering 107
- 4.3. Debye–Waller factor 112
- 4.4. Inelastic one-phonon scattering 114

4.5. Anharmonic phonon theory 124
4.6. Structural phase transitions 129
4.7. Mixed harmonic lattices 133
4.7.1. Particle hybridized with phonon environment 134
4.7.2. Mass defect 140
4.7.3. Small concentration of mass defects 150
4.8. Multi-phonon effects 160

**5. DENSE FLUIDS** **170**
5.1. Cross-section for unpolarized neutrons 171
5.2. Classical monatomic fluid: sum rules 177
5.3. Classical monatomic fluid: single-particle response 181
5.3.1. Gaussian approximation 182
5.3.2. Hydrodynamic limit 185
5.3.3. Intermediate $\kappa$ 189
5.4. Classical monatomic fluid: coherent response 191
5.4.1. Static approximation 195
5.4.2. Hydrodynamic theory 199
5.4.3. Viscoelastic theory 205
5.5. Classical multicomponent fluid 217
5.6. Quantum fluids 224

**6. PHYSICO-CHEMICAL APPLICATIONS** **239**
6.1. Diffraction by a molecular fluid 239
6.2. Quasi-elastic scattering from single particles 244
6.3. Gas of isotropic harmonic oscillators 253
6.4. Molecular spectroscopy 257
6.5. Scattering by molecular gases 264
6.5.1. Diatomic molecules 269
6.5.2. Polyatomic molecules 274
6.5.3. Quasi-classical approximation 279

**APPENDIX A** **283**
A.1. Scattering amplitude $f$ 283
A.2. Complex and energy-dependent scattering lengths 286

**APPENDIX B: LINEAR RESPONSE THEORY** **291**
B.1. Generalized susceptibility $\chi_{AB}[\omega]$ 293
B.2. Analytic properties of $\chi_{AB}[\omega]$ 294
B.3. Damped harmonic oscillator 296
B.4. Calculation of the response function $\phi_{AB}(t)$ and the fluctuation-dissipation theorem 299
B.5. Relaxation function $R_{AB}(t)$ 303

B.6. Isothermal susceptibility $\chi_{AB}$ 305
B.7. Correlation functions and symmetry relations 306
B.8. Green functions $G(\omega)$ 311
B.9. Generalized Langevin equation 314
B.9.1. Definitions 314
B.9.2. Anharmonic lattice vibrations 320

**INDEX** **327**

CONTENTS OF VOLUME 2


| | |
|---|---|
| **7. PRINCIPAL FEATURES OF MAGNETIC SCATTERING** | **1** |
| **8. MAGNETIC CORRELATION AND RESPONSE FUNCTIONS** | **31** |
| **9. SPIN WAVES** | **57** |
| **10. POLARIZATION ANALYSIS** | **147** |
| **11. ATOMIC ELECTRONS** | **195** |
| **12. ELASTIC MAGNETIC SCATTERING** | **250** |
| **13. PARAMAGNETIC AND CRITICAL SCATTERING** | **310** |
| **INDEX** | **340** |

# IMPORTANT SYMBOLS

$A = \{(i+1)b^{(+)} + ib^{(-)}\}/(2i+1)$;
  $= \bar{b}$ coherent scattering amplitude; § 1.6
$\mathbf{a}_1, \mathbf{a}_2, \mathbf{a}_3$ basic vectors of unit cell
$\hat{a}^+$, $\hat{a}$ Bose creation and annihilation operators
Å (angstrom) $= 10^{-8}$ cm

$B = 2\{b^{(+)} - b^{(-)}\}/(2i+1)$; § 1.6
$b^{(+)}$ scattering length for $i+\frac{1}{2}$ state } for bound nuclei
$b^{(-)}$ scattering length for $i-\frac{1}{2}$ state } for bound nuclei
$\hat{b}$ scattering amplitude operator; § 1.6
$\bar{b}$ coherent scattering length for bound nuclei; Table 1.1
bn (barn) $= 10^{-24}$ cm$^2$

$c$ concentration of impurities
$c_p$, $c_v$ specific heats per particle at constant pressure $c_p$ and constant volume $c_v$
$c_\xi$ concentration of $\xi$ isotope

$\mathbf{d}$ position vector
$D_s$ self-diffusion constant
$d\sigma/d\Omega$ differential cross-section; eqn (1.12)
$d^2\sigma/d\Omega\, dE'$ partial differential cross-section; eqns (1.23), (3.4), (A.8)

$E$, $E'$ initial and final energy of neutron, respectively
eV electron volt
$E_R$ recoil energy

$F(\boldsymbol{\kappa})$ form factor
$F_N(\boldsymbol{\tau})$ unit-cell structure factor

$G(\mathbf{r}, t)$ pair correlation function; eqn (3.10)
$G_s(\mathbf{r}, t)$ self pair correlation function; eqn (3.23)
$G'(\mathbf{r}, t)$; eqn (3.32)
$G_s'(\mathbf{r}, t)$; eqn (3.34)
$g(r)$ static pair-distribution function

$\hat{\mathcal{H}}$ quantum mechanical Hamiltonian
h(t) applied perturbation
$\hbar$ Planck's constant/$2\pi$

$\hat{\mathbf{i}}$ nuclear spin operator of magnitude $i$
$I$ total nuclear spin
$\mathrm{I}_n(y)$ modified Bessel function of the first kind
$I(\boldsymbol{\kappa}, t)$ spatial Fourier transform of pair correlation function
Im imaginary part

$\mathrm{j}_K(x)$ spherical Bessel function of order $K$

$\mathbf{k}, \mathbf{k}'$ initial and final wave vector of neutron, respectively
$k_\mathrm{B}$ Boltzmann's constant

$l$ cell indices
$\mathbf{l}$ position vector

$m$ mass of neutron
$M$ mass of nucleus
meV (millielectron volt) $= 10^{-3}$ eV

$N$ number of unit cells in crystal or particles
$n(\omega) = [\exp(\hbar\omega\beta) - 1]^{-1}$ Bose factor

$p_\mathrm{f}$ Fermi wave vector
$p_\sigma$ incident neutron spin probability
$p_\lambda$ probability distribution for initial target states
$\mathrm{P}\int$ principal part integral

$\mathbf{q}$ wave vector

$\mathbf{R}_{ld} = \mathbf{l} + \mathbf{d}$ position vector of nucleus in rigid lattice; eqn (2.3)
$R_\mathrm{AB}(t)$ relaxation function; Table B.1
$R_\mathrm{AB}(\omega)$ Fourier transform of relaxation function
$\tilde{R}_\mathrm{AB}(s)$ Laplace transform of relaxation function
Re real part

$S(\boldsymbol{\kappa}, \omega)$ response function; eqn (3.8)
$S_\mathrm{i}(\boldsymbol{\kappa}, \omega)$ incoherent response function; eqn (3.22)

$T$ absolute temperature $\beta = (k_\mathrm{B}T)^{-1}$
$t$ time variable

$\hat{\mathbf{u}}(l, t)$ displacement operator

$V$ volume of target sample

$v_0$ volume of unit cell
$\hat{V}(\boldsymbol{\kappa})$ Fourier transform of neutron–matter interaction potential; eqns (2.18), (3.2)

$W(\boldsymbol{\kappa})$ exponent in Debye–Waller factor; § 4.3

$Y_{jj'}(\boldsymbol{\kappa}, t)$ correlation function; eqn (3.6)

$Z = \mathrm{Tr}\exp(-\beta\hat{\mathscr{H}})$ partition function
$Z(\omega)$ normalized vibrational density of states; eqn (4.39)
$z(\tau)$ number of reciprocal lattice vectors with magnitude $|\boldsymbol{\tau}| = \tau$

$\alpha$ (with $\beta$) Cartesian component index

$\beta$ (with $\alpha$) Cartesian component index
$\beta = (k_{\mathrm{B}}T)^{-1}$

$\delta(x)$ Dirac delta function; eqns (1.19), (1.20), (1.87)
$\delta_{l,m}$ Kronecker delta function

$\epsilon_{\mathrm{f}}$ Fermi energy

$\eta$ viscosity

$\theta(t)$ unit step function
$\theta$ scattering angle $\mathbf{k}\cdot\mathbf{k}' = kk'\cos\theta$

$\boldsymbol{\kappa} = \mathbf{k} - \mathbf{k}'$ neutron wave vector change

$\lambda$ thermal conductivity

$\nu$ longitudinal viscosity

$\xi$ isotope label

$\rho$, $\rho_0$ particle density $= N/V$
$\hat{\rho}(\mathbf{r}, t) = \sum \delta\{\mathbf{r} - \hat{\mathbf{R}}_j(t)\}$ particle density operator
$\hat{\rho}_{\mathbf{q}}(t) = \sum \exp\{-i\mathbf{q}\cdot\hat{\mathbf{R}}_j(t)\}$ Fourier component of particle density operator

$\frac{1}{2}\hat{\boldsymbol{\sigma}}$ neutron spin operator
$\boldsymbol{\sigma}^j(\mathbf{q})$ phonon polarization vector for $j$th branch
$\sigma$, $\sigma'$ initial and final neutron spin quantum numbers in cross-section

$\sigma_a$ single (bound) nucleus absorption cross-section; Table 1.1
$\sigma_c$ single (bound) nucleus coherent cross-section
$\sigma_i = \sigma - \sigma_c$ single (bound) nucleus incoherent cross-section
$\sigma$ total single (bound) nucleus cross-section; Table 1.1

$\boldsymbol{\tau} = t_1\boldsymbol{\tau}_1 + t_2\boldsymbol{\tau}_2 + t_3\boldsymbol{\tau}_3$ reciprocal lattice vectors; eqn (2.5)

$\phi(t)$ linear response function; Table B.1
$\phi_{\mathbf{q}}(t)$ wave vector dependent linear response function

$\chi$ isothermal susceptibility $= -\chi[0]$
$\chi[\omega] = \chi'[\omega] + i\chi''[\omega]$ generalized susceptibility; Table B.1

$\omega_D$ Debye frequency; eqn (4.42)

$\hbar\omega = \hbar^2(k^2 - k'^2)/2m$ neutron energy change

$\omega_j(\mathbf{q})$ frequency function of $j$th phonon branch

$\Omega$ solid angle and particle volume

$\partial_t$ derivative with respect to $t$

# 1
# PRINCIPAL FEATURES

## 1.1. Introduction

Let us review briefly some intrinsic neutron properties and their ramifications in the use of neutrons as a probe of condensed matter (Foderaro 1971; Koester 1977).

The energy $E$ of a neutron with a wave vector $\mathbf{k}$ is

$$E = \hbar^2 k^2/2m, \tag{1.1}$$

where $m$ is the neutron mass. In neutron spectroscopy, energies are often given in units of $\text{meV} = 10^{-3}\,\text{eV}$; using $m = 9.383 \times 10^8\,\text{eV}$, we have

$$(\hbar^2/2m) = 2.08\,\text{meV}\,\text{Å}^2. \tag{1.2}$$

From this last value we see that a neutron with a wave vector $k \sim$ few $\text{Å}^{-1}$ possesses an energy $E \sim$ few meV. This energy is typical of the energy carried by lattice vibrations and by spin excitations in magnetic insulators, for example.

It is sometimes useful to rewrite the energy relation (1.1) in terms of the neutron wavelength $\lambda$ or alternatively the inverse relation,

$$\begin{aligned}\lambda &= (h^2/2mE)^{1/2}\\ &= 9.04E^{-1/2}\,\text{Å}\end{aligned} \tag{1.3}$$

where we have used (1.2) in arriving at the final result and $E$ is in units of meV. The result (1.3) shows that the mass of the neutron is such that a neutron with an energy $E = 25$ meV has a wavelength $\lambda = 1.81$ Å which is compatible with atomic spacings in condensed matter. In consequence, the diffraction of neutrons from condensed matter can display pronounced interference effects.

Thus far, we have quoted neutron energies in meV. Another energy unit frequently used in neutron spectroscopy is the terahertz, usually abbreviated THz, where

$$1\,\text{THz} = 10^{12}\,\text{s}^{-1} \equiv 4.14\,\text{meV}. \tag{1.4}$$

Users of other spectroscopic techniques often use wave numbers in units of $\text{cm}^{-1}$. We then have

$$1\,\text{meV} \equiv 0.24\,\text{THz} \equiv 8.07\,\text{cm}^{-1} \equiv 11.61\,\text{K} \tag{1.5}$$

and

$$1\,\text{THz} \equiv 33.35\,\text{cm}^{-1} \equiv 48.02\,\text{K},$$

where the conversion to temperature is included for completeness.

Low-energy neutron beams are often described as being cold, thermal, hot, or epithermal. There is no standard definition of these terms, although there is a consensus that thermal neutrons have an energy $\sim 300\,\mathrm{K} \doteq 25\,\mathrm{meV}$. For future reference we give here our interpretation of the thermal nomenclature for neutron energies.

| | $E$ (meV) |
|---|---|
| Cold | 0.1– 10 |
| Thermal | 10 –100 |
| Hot | 100 –500 |
| Epithermal | >500. |

In an actual scattering experiment the key variables are the *change* in the neutron energy and the concomitant *change* in wave vector. We shall denote the transfer of energy and wave vector to the target sample by $\hbar\omega$ and $\boldsymbol{\kappa}$, respectively. Thus, if the initial and final energies and wave vectors are $E$, $\mathbf{k}$ and $E'$, $\mathbf{k}'$, then

$$\hbar\omega = E - E' = \frac{\hbar^2}{2m}(k^2 - k'^2) \tag{1.6}$$

and the scattering vector

$$\boldsymbol{\kappa} = \mathbf{k} - \mathbf{k}'. \tag{1.7}$$

The spectrum of the scattered neutrons is a function of $\boldsymbol{\kappa}$, and not some other function of $\mathbf{k}$ and $\mathbf{k}'$, because the scattering of low-energy neutrons is a weak process. A perturbative calculation is therefore adequate, and this leads directly to a scattering amplitude that depends on the difference of the incident and scattered wave vectors (cf. § 1.2).

The motion of atoms and their spatial correlations are revealed in a neutron scattering experiment when $\hbar\omega$ and $\kappa = |\boldsymbol{\kappa}|$ match the corresponding energies and wave-vectors involved. The dependence of the scattered intensity on $\omega$ and $\boldsymbol{\kappa}$ can often be readily ascribed to the conservation of energy and wave vector in the scattering process; for example, scattering from a lattice vibration with a long lifetime occurs only when the neutron-energy and wave-vector changes coincide with the energy and wave vector of the vibration. The effect of energy and wave-vector conservation are pronounced in the example of scattering from a lattice vibration because the energy and wave vector of the vibration satisfy a dispersion relation. In contrast, there is a broad spectrum of final states for neutrons scattered from a free atom. Even so, the shape of the spectrum reflects the conservation laws since it is centred about the recoil energy of the atom.

The range of values of $\omega$ and $\kappa$ that are readily measured is large. With today's facilities the range is typically, $1\,\mu\mathrm{eV} < \hbar\omega < 1\,\mathrm{eV}$ and $0.01 < \kappa\ (\text{Å}^{-1}) < 30$ and the ranges will surely increase with advances in neutron production and spectrometer design. Various types of spectrometer are required to cover the indicated range of the variables, but a survey is well beyond the scope of this book. The variables can be measured with good accuracy, and a spread of a few per cent is not unusual.

It is evident from (1.6) and (1.7) that $\omega$ and $\kappa$ are related. The relation imposes constraints—kinematic constraints—on the scattering experiments. For example, in order to gain access to the domain of large $\omega$ and modest $\kappa$ it is necessary to use incident neutrons with very high energies and to measure at small scattering angles. In other words, not all of $(\kappa, \omega)$-space is accessible in a neutron scattering experiment. The 'kinematically allowed region' is a function of the incident neutron energy $E$ and lies between the $(\kappa, \omega)$-loci corresponding to forward and backward scattering. From this observation it follows that the characteristics of a neutron source to some extent limit the range of experiments that can be performed (OECD/NEA 1983).

The penetration depth of neutrons in matter is extremely long, largely because they are uncharged. Neutron scattering is therefore well suited to the study of bulk properties, and surface scattering effects can usually be neglected in the interpretation of data. The scattering provides information on the physical and chemical properties of the target sample which is undistorted by the experimental probe (setting aside resolution effects arising from the spectrometer). The interpretation of neutron scattering data is then relatively straightforward since we have only to unravel the properties of the sample and not the properties of the composite system comprised of the sample and neutrons. This feature of the scattering emerges in the cross-section as the factorization into a term that depends on the nature of the neutron–matter interaction and a response function which depends on the physical and chemical properties of the sample.

A corollary of the weak neutron–matter interaction is that neutron-scattering experiments are intensity limited. However, there have been significant developments in the design of neutron sources in the past few years, with the result that the intensity limitations are gradually being eased (OECD/NEA 1983).

Let us now discuss the neutron–matter interaction in more detail. The neutron–nucleus scattering varies in magnitude from one isotope to another, and there are no systematics as a function of their $Z$, $N$ values. This feature of the scattering can be very valuable since nuclear contrast factors can be varied by isotope substitution. A particularly important example in physical-chemistry studies (Bacon 1977) is the contrast between hydrogen and deuterium; the total scattering from hydrogen exceeds that from deuterium by an order of magnitude (cf §§ 1.4 and 1.5).

The magnetic scattering of neutrons can provide information on the properties of materials that is not obtainable by any other technique. Moreover, the ability to determine the state of polarization of a neutron beam can be of immense value. Unfortunately the polarization analysis methods currently available are not very efficient, except at low neutron energies, and the intensity penalty is often unacceptable in a technique that is intrinsically intensity limited.

In the present volume we are concerned solely with nuclear scattering. Magnetic neutron scattering and polarization phenomena are the subject of the second volume. Our immediate task is to derive an expression for the neutron cross-section. The development given in the following section is valid equally for nuclear and magnetic scattering.

## 1.2. Neutron cross-section

The geometry of the scattering experiment is shown in Fig. 1.1. A neutron specified by the wave vector $\mathbf{k}$ is scattered into a state with wave vector $\mathbf{k}'$; the transfer of momentum to the target sample is $\hbar\boldsymbol{\kappa}$ where the scattering vector, $\boldsymbol{\kappa}$, is defined in (1.7). The basic quantity that is measured is the partial differential cross-section (Foderaro 1971) which gives the fraction of neutrons of incident energy $E$ scattered into an element of solid angle $\mathrm{d}\Omega$ with an energy between $E'$ and $E'+\mathrm{d}E'$. The cross-section is denoted by

$$\mathrm{d}^2\sigma/\mathrm{d}\Omega\,\mathrm{d}E'$$

and has the dimension of (area/energy).

The only satisfactory way to derive an expression for the cross-section is to use formal scattering theory. However, in our case the appeal of such a derivation is offset to a large extent by the complexity of the formalism and the minimal need for scattering theory in the rest of the book. We stressed in the preceding section that the interpretation of neutron scattering data requires only an understanding of the properties of the target. This stems from the fact that neutron scattering is a very weak probe and it does not distort the intrinsic properties of the target sample. More precisely, the amplitude of a wave scattered from one atom of the sample becomes small even at the position of a neighbouring atom; this condition is satisfied if the cross-sections for individual atoms are small compared with the square of the mean separation of the atoms. When this condition holds, the total scattering amplitude for the sample is equal to the sum of those for individual atoms (Newton 1982).

We will not descant on the beauties of scattering theory, but merely note that while the derivation of the cross-section used here has the merit of simplicity it lacks nicety. One reason why the derivation is not really satisfactory is that we must artifically enclose the composite system

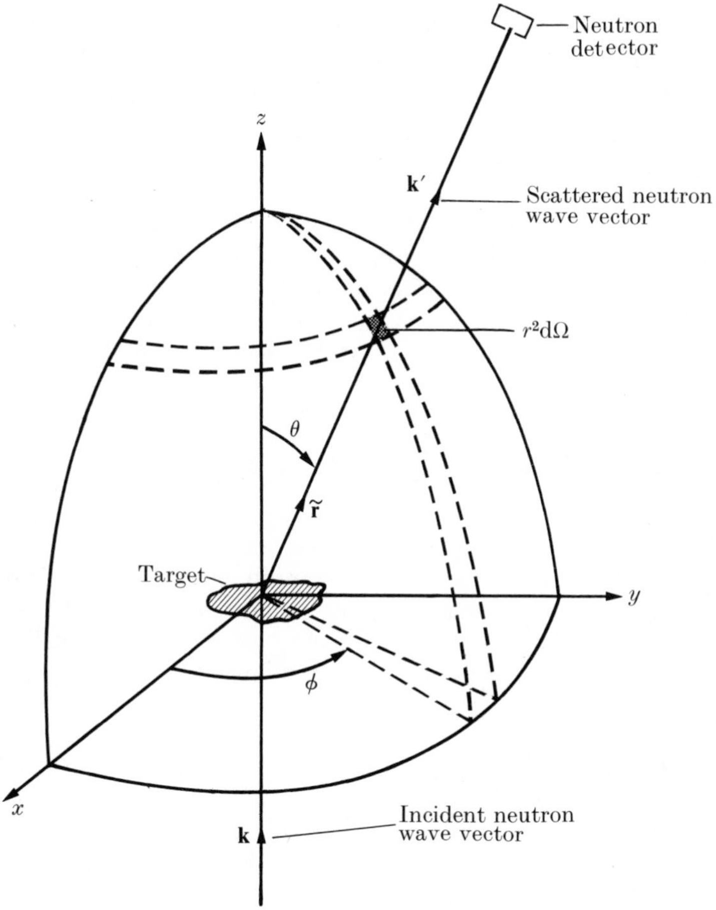


FIG. 1.1. Geometry of scattering problem.

of neutron and target in a large box in order to normalize the neutron wave functions. Because the density of final neutron states is proportional to the normalization volume, the cross-section is independent of the volume, as should be expected.

We begin by obtaining an expression for the differential cross-section which is appropriate for elastic scattering, i.e. no energy is transferred to the target sample. The expression is then generalized to the partial differential cross-section that includes inelastic events. The incident neutron has the state $\psi_{\mathbf{k}}$ and the scattered neutron the state $\psi_{\mathbf{k}'}$. The direction of propagation of the latter with respect to the incident neutron is defined by the polar angle $\theta$ and the azimuthal angle $\phi$. If the flux of incident neutrons, defined as the number per unit area per unit time, is $N$, then the number scattered per unit time into the element of solid angle $d\Omega = \sin\theta \, d\theta \, d\phi$ is

$$N\left(\frac{d\sigma}{d\Omega}\right) d\Omega$$

where $d\sigma/d\Omega$ is the differential cross-section. It follows that $d\sigma/d\Omega$ and the total cross-section $\sigma$ have the dimensions of area.

If the target sample has no internal structure, then clearly the energy of the scattered neutron is identical to that of the incident neutron, i.e. the scattering is elastic. To calculate the differential cross-section for this case we need to know the probability of a transition from the plane-wave state defined by the wave vector $\mathbf{k}$ to the plane-wave state defined by the wave vector $\mathbf{k}'$, both having the same energy $E = \hbar^2 k^2/2m$. This probability is given by Fermi's Golden rule, namely

$$W_{k \to k'} = \frac{2\pi}{\hbar} \left| \int d\mathbf{r} \, \psi_{\mathbf{k}'}^* \hat{V} \psi_{\mathbf{k}} \right|^2 \rho_{\mathbf{k}'}(E). \tag{1.8}$$

Here $\hat{V}$ is the interaction potential that causes the transition, in our case the interaction between the incident neutron and the target sample, and $\rho_{\mathbf{k}'}(E)$ is the density of final scattering states per unit energy range.

To evaluate the latter we must first decide upon the normalization of our wave functions. For a large box of volume $L^3$ the states $\psi_{\mathbf{k}}$ and $\psi_{\mathbf{k}'}$ are, respectively,

$$\psi_{\mathbf{k}} = \frac{1}{L^{3/2}} \exp(i\mathbf{k} \cdot \mathbf{r})$$

and

$$\psi_{\mathbf{k}'} = \frac{1}{L^{3/2}} \exp(i\mathbf{k}' \cdot \mathbf{r}).$$

It follows that our density of final states is

$$\rho_{\mathbf{k}'}(E) = \left(\frac{L}{2\pi}\right)^3 \frac{d\mathbf{k}'}{dE}.$$

Clearly,

$$d\mathbf{k}' = k'^2 \, d\Omega \, dk' = k^2 \, d\Omega \, dk$$

and

$$dE = \frac{\hbar^2 k}{m} dk,$$

so that

$$\rho_{\mathbf{k}'}(E) = \left(\frac{L}{2\pi}\right)^3 \frac{mk}{\hbar^2} d\Omega. \tag{1.9}$$

To obtain the cross-section we have now only to determine the incident flux of neutrons. This is

$$(\text{velocity of incident neutrons})/L^3 = \frac{\hbar k}{mL^3}. \tag{1.10}$$

Hence, from (1.8), (1.9), and (1.10),

$$\begin{aligned} d\sigma &= W_{k\to k'}/\text{incident flux} \\ &= L^6\left(\frac{m}{2\pi\hbar^2}\right)^2 \left|\int d\mathbf{r}\ \psi^*_{\mathbf{k}'}\hat{V}\psi_{\mathbf{k}}\right|^2 d\Omega. \end{aligned}$$

With the notation

$$(L^3 m/2\pi\hbar^2)\int d\mathbf{r}\ \psi^*_{\mathbf{k}'}\hat{V}\psi_{\mathbf{k}} = \langle\mathbf{k}'|\ \hat{V}\ |\mathbf{k}\rangle, \tag{1.11}$$

we have

$$\frac{d\sigma}{d\Omega} = |\langle\mathbf{k}'|\ \hat{V}\ |\mathbf{k}\rangle|^2, \tag{1.12}$$

which is the desired result.

A useful concept is that of the scattering amplitude $f(\boldsymbol{\kappa})$. This is defined so that

$$\frac{d\sigma}{d\Omega} = |f(\boldsymbol{\kappa})|^2. \tag{1.13a}$$

The phase of the scattering amplitude is not defined by this relationship, but has been chosen to agree with that obtained from a more complete analysis given in Appendix A. We can therefore take

$$f(\boldsymbol{\kappa}) = -\langle\mathbf{k}'|\ \hat{V}\ |\mathbf{k}\rangle. \tag{1.13b}$$

Notice that with the definition (1.11), the matrix element $\langle\mathbf{k}'|\ \hat{V}\ |\mathbf{k}\rangle$ has the dimension of length.

We turn now to the calculation of the partial differential cross-section that includes inelastic scattering events. For such events the neutron energy is changed and $\hbar\omega$, defined in (1.6), is finite. As defined here, $\omega$ is positive for neutron energy loss and negative for neutron energy gain. The change in the neutron energy is taken up in the response of the target sample through a rearrangement of its various states.

A state of the target is labelled by the index $\lambda$. In most cases this will have to be a composite label in order to specify a state completely. The corresponding eigenvector is $|\lambda\rangle$, so that the initial state of the system, composed of the incident neutron and target, is described by a product state function

$$|\mathbf{k}\rangle\,|\lambda\rangle \equiv |\mathbf{k}\lambda\rangle, \tag{1.14}$$

where the second form is used in the following discussion for ease of notation. The states $|\lambda\rangle$ form a complete set. We will assume a normalization for the eigenvectors such that the closure relation is,

$$\sum_{\lambda} |\lambda\rangle\langle\lambda| = 1. \tag{1.15}$$

The energy associated with the target state described by $|\lambda\rangle$ is denoted by $E_\lambda$. If the response of the target sample to the neutron interaction is to change from the state labelled $\lambda$ to the state labelled $\lambda'$, then, given our definition of $\omega$, conservation of energy requires that

$$\hbar\omega = E_{\lambda'} - E_\lambda. \tag{1.16}$$

With the notation given in (1.11) and (1.14), the associated cross-section is readily shown to be

$$\left(\frac{d\sigma}{d\Omega}\right)_{\lambda'}^{\lambda} = \frac{k'}{k} |\langle \mathbf{k}'\lambda'| \, \hat{V} \, |\mathbf{k}\lambda\rangle|^2 \tag{1.17}$$

where the factor $(k'/k)$ arises from the density of final neutron states divided by the incident neutron flux.

The partial differential cross-section is obtained from (1.17) by incorporating the energy conservation (1.16). This is accomplished with the aid of a delta function

$$\delta(E + E_\lambda - E' - E_{\lambda'}) = \delta(\hbar\omega + E_\lambda - E_{\lambda'}), \tag{1.18}$$

that vanishes unless the condition (1.16) is satisfied. We recall that, for any reasonably behaved function $g(x)$ of the variable $x$,

$$\int_{-\infty}^{\infty} dx' \, \delta(x - x') g(x') = g(x) \tag{1.19}$$

and also

$$\delta(ax) = \frac{1}{|a|} \delta(x), \tag{1.20}$$

where $a$ is independent of $x$. From (1.19) we deduce that, for the delta function (1.18),

$$\int_{-\infty}^{\infty} dE' \, \delta(E + E_\lambda - E' - E_{\lambda'}) = 1,$$

so the partial differential cross-section is

$$\left(\frac{d^2\sigma}{d\Omega \, dE'}\right)_{\lambda'}^{\lambda} = \frac{k'}{k} |\langle \mathbf{k}'\lambda'| \, \hat{V} \, |\mathbf{k}\lambda\rangle|^2 \, \delta(\hbar\omega + E_\lambda - E_{\lambda'}). \tag{1.21}$$

Using (1.20), we can verify that the cross-section (1.21) has the correct dimension. The finite energy resolution available in an experiment means that in observed spectra the delta function will be smeared to some extent, and might approximate to a Lorentz function whose width is a measure of the resolution.

The cross-section (1.21) relates to specific initial and final target states. In general there will be a range of accessible initial states. The weight for the state $|\lambda\rangle$ is denoted by $p_\lambda$, and we choose a normalization

$$\sum_\lambda p_\lambda = 1. \tag{1.22}$$

The weight $p_\lambda$ is, in general, the product of the thermodynamic factor $\exp(-E_\lambda/k_B T)$, where $T$ is the sample temperature, and appropriate degeneracy factors. Moreover, if the final states in (1.21) are not selected by some form of discrimination they will all be observed. In view of what has just been said, the basic quantity that is measured is the partial differential cross-section defined by the expression,

$$\frac{d^2\sigma}{d\Omega\, dE'} = \frac{k'}{k} \overline{\sum_{\lambda\lambda'} p_\lambda\, |\langle \mathbf{k}'\lambda'|\, \hat{V}\, |\mathbf{k}\lambda\rangle|^2\, \delta(\hbar\omega + E_\lambda - E_{\lambda'})}. \tag{1.23}$$

The horizontal bar in (1.23) stands for any relevant averages over and above those included in the weights $p_\lambda$. For example, we do not include in $p_\lambda$ the distribution of isotopes, the nuclear spin orientations, or the precise positions of the nuclei.

Since expression (1.8) for the transition probability is derived from perturbation theory, our expression for the neutron cross-section is itself only approximate. Expression (1.23) is the first Born approximation (or more colloquially, the Born approximation) to the cross-section. The nature of the approximation is discussed in more detail in Appendix A. One consequence of the approximation is that the nuclear scattering amplitude depends on $\boldsymbol{\kappa} = \mathbf{k} - \mathbf{k}'$ (rather than some other function of $\mathbf{k}$ and $\mathbf{k}'$) as we shall verify in the next section.

When it comes to calculating the scattering amplitude for many atoms, we will assume that it is given by the sum of the scattering amplitudes for the separate atoms. This approximation for the total scattering amplitude of the target sample is entirely consistent with the perturbative calculation of the separate scattering amplitudes.

The final topic in this section concerns the inclusion in the cross-section of the neutron spin states. Let the spin states be labelled by an index $\sigma$. A complete initial state of the neutron and target system is described by the product state function

$$|\mathbf{k}\lambda\rangle\, |\sigma\rangle \tag{1.24a}$$

and a complete final state is described by

$$|\mathbf{k}'\lambda'\rangle\,|\sigma'\rangle. \tag{1.24b}$$

If the initial spin state occurs with a (normalized) weight $p_\sigma$, then, when we include the neutron spin states in the cross-section, the squared matrix element in (1.23) is replaced as follows

$$|\langle\mathbf{k}'\lambda'|\,\hat{V}\,|\mathbf{k}\lambda\rangle|^2 \Rightarrow \sum_{\sigma\sigma'} p_\sigma\,|\langle\sigma'|\,\langle\mathbf{k}'\lambda'|\,\hat{V}\,|\mathbf{k}\lambda\rangle\,|\sigma\rangle|^2. \tag{1.25}$$

This is the only modification to the expression for the cross-section when the energies are independent of the neutron spin. For the special case when the interaction operator $\hat{V}$ is completely independent of the neutron spin,

$$\langle\sigma'|\,\hat{V}\,|\sigma\rangle = \langle\sigma'\,|\,\sigma\rangle\,\hat{V} = \delta_{\sigma,\sigma'}\hat{V}, \tag{1.26}$$

where the Kronecker delta function results from the orthogonality of different spin states. Now, when the result (1.26) is inserted in the right-hand side of (1.25) all dependence on the neutron spin disappears and we recover our original expression (1.23). Thus (1.23) is, in fact, correct when the interaction $\hat{V}$ is independent of the neutron spin.

For the general case in which $\hat{V}$ is a function of the neutron spin, it is useful to simplify the right-hand side of (1.25) by using the closure relation for the spin states, i.e.

$$\sum_\sigma |\sigma\rangle\langle\sigma| = 1. \tag{1.27}$$

For ease of notation in subsequent working let us represent the matrix element $\langle\mathbf{k}'\lambda'|\,\hat{V}\,|\mathbf{k}\lambda\rangle$ by $\hat{W}$, an operator with respect to the neutron spin. Using the following identity for the complex conjugate of a matrix element

$$\langle\sigma'|\,\hat{W}\,|\sigma\rangle^* = \langle\sigma|\,\hat{W}^+\,|\sigma'\rangle, \tag{1.28}$$

where $\hat{W}^+$ is the Hermitian conjugate of $\hat{W}$, the right-hand side of (1.25) becomes

$$\sum_{\sigma\sigma'} p_\sigma\langle\sigma|\,\hat{W}^+\,|\sigma'\rangle\,\langle\sigma'|\,\hat{W}\,|\sigma\rangle = \sum_\sigma p_\sigma\langle\sigma|\,\hat{W}^+\hat{W}\,|\sigma\rangle. \tag{1.29}$$

The final expression is simply the average value of the operator $\hat{W}^+\hat{W}$ with respect to the neutron spin states. For an unpolarized beam of neutrons, $p_\sigma$ is independent of $\sigma$ and equal to $\frac{1}{2}$.

## 1.3. Scattering from bound nuclei

A relatively simple, yet fundamental problem is the scattering of neutrons by independent, bound nuclei. In reality, of course, the nuclei in a solid are not strictly bound nor are they completely free. Whether or not they can be regarded as being in either of these two extreme states, or intermediate to them, depends, however, on the energy of the incident neutrons relative to the chemical binding. We set aside a discussion of these points for the present and concentrate on the concepts of coherent and incoherent nuclear scattering of unpolarized neutrons.

There is at present no complete theory of the nucleon-nucleon interaction (Koester 1977; Mughabghab *et al.* 1981) but we know from experimental results that it has a very short range (of the order $1.5 \times 10^{-13}$ cm). Because this is much less than the wavelength of low-energy neutrons and the nuclear radius is only about an order of magnitude greater, the neutron-nucleus scattering can contain only s-wave components, i.e. the scattering is isotropic, and, as is well known, can therefore be characterized by a single parameter $b$, called the scattering length. $b$ can be complex, and the real part may be either positive or negative depending on the energy of the incident neutron and the particular nucleus involved in the scattering. The imaginary part of $b$ represents absorption (principally radiative capture for thermal neutrons) and in most cases it is small.

In general we can have different scattering lengths not only for each atomic type but also for each isotope, and furthermore the scattering length depends on the relative orientation of the neutron spin and nuclear spin (if the latter exists). For the moment we set aside these complications by assuming the target is a single spinless isotope with scattering length $b$. In the following $b$ is taken as complex but independent of energy.

The only form of $\hat{V}(\mathbf{r})$ that, using the Born approximation, gives isotropic scattering is a delta function. If the nucleus is at the position defined by $\mathbf{R}$, the Fermi pseudo-potential is defined as

$$\hat{V}(\mathbf{r}) = \frac{2\pi\hbar^2}{m} b\,\delta(\mathbf{r}-\mathbf{R}). \tag{1.30}$$

Ths sign of the scattering length is defined such that $b$ is positive for the scattering by an impenetrable sphere of radius $b$, cf. eqn (1.31) and Appendix A.

We emphasize that (1.30) is not the true potential (Mughabghab *et al.* 1981). Indeed, perturbation theory is inapplicable to the scattering of a neutron by a nucleus because, although the interaction potential has a very short range, it is very strong. The Fermi pseudo-potential as given by eqn (1.30) is a formal artifice defined to give, in the Born approximation, what we know to be correct for simple s-wave scattering (Nowak 1982).

If we substitute (1.30) in (1.11), and set $\mathbf{R}=0$,

$$\langle \mathbf{k}' |\, \hat{V}\, | \mathbf{k} \rangle = \left(\frac{m}{2\pi\hbar^2}\right)\frac{2\pi\hbar^2}{m}\, b \int \mathrm{d}\mathbf{r}\, \exp(-\mathrm{i}\mathbf{k}'\cdot\mathbf{r})\,\delta(\mathbf{r})\exp(\mathrm{i}\mathbf{k}\cdot\mathbf{r}) = b. \tag{1.31}$$

From (1.12),

$$\frac{d\sigma}{d\Omega} = |b|^2 \tag{1.32}$$

and hence the total cross-section is

$$\sigma = 4\pi\ |b|^2. \tag{1.33}$$

Next consider the scattering from a rigid array of $N$ nuclei, the position vector of the $l$th nucleus being denoted by $\mathbf{R}_l$. We also remove the restriction to a single spinless isotope and describe the scattering length of the $l$th nucleus as $b_l$. The interaction potential in this instance is

$$\hat{V}(\mathbf{r}) = \frac{2\pi\hbar^2}{m} \sum_l b_l\, \delta(\mathbf{r} - \mathbf{R}_l). \tag{1.34}$$

Thus

$$\begin{aligned}\langle \mathbf{k}'|\ \hat{V}\ |\mathbf{k}\rangle &= \sum_l b_l \int \mathrm{d}\mathbf{r}\ \exp(-\mathrm{i}\mathbf{k}' \cdot \mathbf{r})\ \delta(\mathbf{r} - \mathbf{R}_l)\exp(\mathrm{i}\mathbf{k} \cdot \mathbf{r}) \\ &= \sum_l b_l\ \exp(\mathrm{i}\boldsymbol{\kappa} \cdot \mathbf{R}_l),\end{aligned}$$

with $\boldsymbol{\kappa} = \mathbf{k} - \mathbf{k}'$ and the cross-section (1.23) becomes

$$\frac{d^2\sigma}{d\Omega\, dE'} = \frac{k'}{k} \sum_{\lambda,\sigma} p_\lambda p_\sigma \sum_{\lambda',\sigma'} \overline{\left|\left\langle \sigma'\lambda' \right| \sum_l b_l\ \exp(\mathrm{i}\boldsymbol{\kappa} \cdot \mathbf{R}_l) \left| \sigma\lambda \right\rangle\right|^2}\ \delta(\hbar\omega + E_\lambda - E_{\lambda'}). \tag{1.35}$$

The matrix element in (1.35) describes a scattering process in which there is a sudden transfer of momentum, $\hbar\boldsymbol{\kappa}$, to the target sample. To verify this interpretation of the Born approximation we begin by considering the momentum of the initial target state. If the latter is described by a wave function $\phi_\lambda(\mathbf{R})$ that is normalized to unity, the momentum of the initial state is

$$\mathbf{p} = \int \mathrm{d}\mathbf{R}\ \phi^*_\lambda(\mathbf{R})(-\mathrm{i}\hbar\boldsymbol{\nabla})\phi_\lambda(\mathbf{R}).$$

The matrix element in (1.35) contains the function $\exp(\mathrm{i}\boldsymbol{\kappa} \cdot \mathbf{R})\phi_\lambda(\mathbf{R})$, and the momentum associated with it is

$$\int \mathrm{d}\mathbf{R}\phi^*_\lambda(\mathbf{R})\exp(-\mathrm{i}\boldsymbol{\kappa} \cdot \mathbf{R})(-\mathrm{i}\hbar\boldsymbol{\nabla})\exp(\mathrm{i}\boldsymbol{\kappa} \cdot \mathbf{R})\phi_\lambda(\mathbf{R}) = \mathbf{p} + \hbar\boldsymbol{\kappa}.$$

In view of this last result the matrix element is the overlap integral between a final target state and a state which differs from the initial state by the addition of momentum $\hbar\boldsymbol{\kappa}$. The scattering process therefore amounts to a transfer of momentum to the initial target state without otherwise changing it. Moreover, the process is independent of the chemical binding forces although they determine the target wave functions. These observations are consistent with a sudden, or impulsive, process in which target states do not adjust apart from taking up the change in the wave vector of the neutron.

Since the nuclei are rigidly bound $\exp(\mathrm{i}\boldsymbol{\kappa}\cdot\mathbf{R}_l)$ in (1.35) is not an operator, so it follows that the matrix element is proportional

$$\langle\lambda\mid\lambda'\rangle=\delta_{\lambda,\lambda'} \tag{1.36}$$

which means that there is just one term in the sum over $\lambda'=\lambda$. The sum of $p_\lambda$ is unity by definition (cf. eqn (1.22)). The result (1.36) implies that, in (1.35), $E_\lambda=E_{\lambda'}$ and the scattering is therefore elastic. In view of these various results, the expression (1.35) for scattering by a rigid array of nuclei reduces to

$$\frac{\mathrm{d}\sigma}{\mathrm{d}\Omega}=\sum_\sigma p_\sigma\sum_{ll'}\exp\{\mathrm{i}\boldsymbol{\kappa}\cdot(\mathbf{R}_l-\mathbf{R}_{l'})\}\,\langle\sigma|\,\overline{b_{l'}^*b_l}\,|\sigma\rangle, \tag{1.37}$$

where the sum over $\sigma'$ has been done by closure as in the development at the end of the previous section. The quantity $\overline{b_{l'}^*b_l}$ is the value of $b_{l'}^*b_l$ averaged over random nuclear spin orientations and random isotope distributions. Having performed this averaging, all dependence on $\sigma$ must disappear (see, for example, eqn (1.76)) and since $\sum p_\sigma$ is unity by definition (1.37) becomes

$$\frac{\mathrm{d}\sigma}{\mathrm{d}\Omega}=\sum_{ll'}\exp\{\mathrm{i}\boldsymbol{\kappa}\cdot(\mathbf{R}_l-\mathbf{R}_{l'})\}\overline{b_{l'}^*b_l}. \tag{1.38}$$

In general $b_l$ will depend on which isotope is at the site $\mathbf{R}_l$ and what nuclear spin is associated with the isotope. Clearly there is no correlation between values of $b_l$ and $b_{l'}$ if $l$ and $l'$ refer to different sites, i.e.

$$\overline{b_{l'}^*b_l}=\overline{b_{l'}^*}\,\overline{b_l}=|\bar{b}|^2 \quad \text{if} \quad l\neq l'.$$

But if $l=l'$,

$$\overline{b_{l'}^*b_l}=\overline{|b_l|^2}=\overline{|b|^2},$$

so that in general

$$\overline{b_{l'}^*b_l}=|\bar{b}|^2+\delta_{l,l'}(\overline{|b|^2}-|\bar{b}|^2). \tag{1.39}$$

Substituting (1.39) into (1.38), the cross-section can be written as the sum of two parts,

$$\frac{\mathrm{d}\sigma}{\mathrm{d}\Omega}=\left(\frac{\mathrm{d}\sigma}{\mathrm{d}\Omega}\right)_{\mathrm{coh}}+\left(\frac{\mathrm{d}\sigma}{\mathrm{d}\Omega}\right)_{\mathrm{incoh}}, \tag{1.40}$$

where the *coherent* cross-section is

$$\left(\frac{\mathrm{d}\sigma}{\mathrm{d}\Omega}\right)_{\mathrm{coh}} = |\bar{b}|^2 \left|\sum_l \exp(\mathrm{i}\boldsymbol{\kappa}\cdot\mathbf{R}_l)\right|^2 \tag{1.41}$$

and the *incoherent* cross-section

$$\left(\frac{\mathrm{d}\sigma}{\mathrm{d}\Omega}\right)_{\mathrm{incoh}} = N\{\overline{|b|^2} - |\bar{b}|^2\} = N\,\overline{|b-\bar{b}|^2}. \tag{1.42}$$

It is obvious from these formulae that the coherent and the incoherent scattering are profoundly different. In the coherent scattering there is a strong interference between the waves scattered from each nucleus; so strong that we find coherent scattering from a crystal only if strict geometrical conditions are satisfied. In the incoherent scattering there is no interference at all, and the cross-section is completely isotropic. This last comment is not strictly true for a real lattice, i.e. a lattice in which the nuclei can move; see Chapter 4. Physically we can think of the coherent and incoherent scattering as follows: because the scattering lengths vary from one isotope to another and also depend on the nuclear spin orientation (relative to the neutron), the neutron does not see a crystal of uniform scattering potential but one in which the scattering varies from one point to the next. It is only the average scattering potential that can give interference effects and thus coherent scattering; this average scattering potential is proportional to $\bar{b}$ and hence the coherent scattering cross-section is proportional to $|\bar{b}|^2$. The deviations from the average potential are randomly distributed and therefore cannot give interference effects; they therefore give incoherent scattering proportional to the mean-square deviation, i.e. to $\overline{|b-\bar{b}|^2} = \overline{|b|^2} - |\bar{b}|^2$.

We now seek the values of $\bar{b}$ and $\overline{|b|^2}$. As a first example we assume the scattering lengths to be independent of spin, so that only the variation with isotope need be considered. We label the isotopes with the subscript $\xi$, so that $b_\xi$ is the scattering length and $c_\xi$ the fractional concentration of the $\xi$ isotope. Then

$$\bar{b} = \sum_\xi c_\xi b_\xi \tag{1.43}$$

and

$$\overline{|b|^2} = \sum_\xi c_\xi\,|b_\xi|^2. \tag{1.44}$$

As a second example we assume a single isotope but recall that the interaction between two nucleons depends on their spins (Foderaro 1971). Thus, if the nuclei have a spin $i$ we must make allowance for this in our scattering cross-section. Physically, the basis of this dependence on spin orientation can be seen as follows: when the neutron, with spin $\frac{1}{2}$, interacts with a nucleus of spin $i$ it can do so in states of total spin $i+\frac{1}{2}$ or $i-\frac{1}{2}$, and, accordingly, there are scattering lengths $b^{(+)}$ and $b^{(-)}$ associated with the two states. (Recall that only s-wave components of the neutron wave function are involved.) We remember that there are $2t+1$ states of total angular momentum $t$, there are therefore $2(i+\frac{1}{2})+1$ states of spin $i+\frac{1}{2}$ and $2(i-\frac{1}{2})+1$ states of spin $i-\frac{1}{2}$. Each of these has the same intrinsic probability, so we conclude that the probability of interaction in the $i+\frac{1}{2}$ state is

$$\frac{2i+2}{(2i+2)+2i}=\frac{i+1}{2i+1},$$

while the probability of interaction in the $i-\frac{1}{2}$ state is

$$\frac{2i}{(2i+2)+2i}=\frac{i}{2i+1}.$$

Hence

$$\bar{b}=\left(\frac{i+1}{2i+1}\right)b^{(+)}+\left(\frac{i}{2i+1}\right)b^{(-)} \tag{1.45}$$

and

$$\overline{|b|^2}=\left(\frac{i+1}{2i+1}\right)|b^{(+)}|^2+\left(\frac{i}{2i+1}\right)|b^{(-)}|^2. \tag{1.46}$$

From these two examples we can now combine (1.43)–(1.46) to give the general expressions for the case where there is more than one isotope present and some of these have non-zero spins

$$\bar{b}=\sum_{\xi} c_{\xi}\frac{1}{2i_{\xi}+1}\{(i_{\xi}+1)b_{\xi}^{(+)}+i_{\xi}b_{\xi}^{(-)}\} \tag{1.47}$$

and

$$\overline{|b|^2}=\sum_{\xi} c_{\xi}\frac{1}{2i_{\xi}+1}\{(i_{\xi}+1)\,|b_{\xi}^{(+)}|^2+i_{\xi}\,|b_{\xi}^{(-)}|^2\}. \tag{1.48}$$

These expressions have to be inserted into (1.41) and (1.42) to give the coherent and incoherent cross-sections.

The division of the cross-section into coherent and incoherent terms is to some extent arbitrary, and often a source of confusion. For one thing the division depends on the state of the nuclei, and on whether or not exchange forces are important. If quantum mechanical exchange forces are significant, the spin and spatial states of identical nuclei are correlated and the division of the cross-section into coherent and incoherent contributions, as defined here, does not occur, as we will see in § 1.7

and Chapter 3. Moreover, a sample with spatial disorder generates diffuse scattering that is defined to be the difference between the total coherent scattering and the scattering which is proportional to the square of the average scattering length (Bragg scattering for a crystalline sample). An example of diffuse scattering for a dilute solution of macromolecules is given in § 1.6. Strictly speaking, coherent scattering arises from the averaged scattering amplitude, all relevant averages (isotope distribution, nuclear spin orientations, composition and configuration distributions, etc.) being included, and such scattering is elastic. Having sounded this note of caution we employ, for the most part, the loose terminology of coherent and incoherent contributions since its usage is widespread.

## 1.4. Total single-atom cross-sections

To illustrate how the preceding formulae work out in practice we discuss a few examples from Table 1.1. Cross-sections are given in units of barn $\equiv$ bn $= 10^{-24}\,\mathrm{cm}^2$. The various quantities tabulated are the coherent bound scattering length $\bar{b}$, the total scattering cross-section $\sigma$, and the absorption cross-section $\sigma_a$. Thus $\sigma$ is

$$\sigma = 4\pi\,\overline{|b|^2} \tag{1.49}$$

and the coherent cross-section is

$$\sigma_c = 4\pi\,|\bar{b}|^2. \tag{1.50}$$

The absorption cross-section is that for incident neutrons with a velocity $2200\ \mathrm{m\,s^{-1}}$.

The scattering of neutrons by protons affords a simple example of the use of the above formulae. The relevant scattering lengths are

$$b^{(+)} = 1.04\times 10^{-12}\ \mathrm{cm}\quad \text{(triplet)}$$

and

$$b^{(-)} = -4.74\times 10^{-12}\ \mathrm{cm}\quad \text{(singlet)}$$

and hence

$$\bar{b} = \tfrac{3}{4}b^{(+)} + \tfrac{1}{4}b^{(-)} = -0.38\times 10^{-12}\ \mathrm{cm}$$

and

$$\overline{|b|^2} = \tfrac{3}{4}\,|b^{(+)}|^2 + \tfrac{1}{4}\,|b^{(-)}|^2 = 6.49\ \mathrm{bn}$$

which give

$$\left.\begin{aligned}\sigma_c &= 1.8\ \mathrm{bn}\\ \sigma &= 81.7\ \mathrm{bn}\end{aligned}\right\}{}^{1}\mathrm{H}$$

as given in Table 1.1.

## Table 1.1

*Summary of low-energy neutron scattering lengths and cross-sections. (Compilation by L. Koester and W. B. Yelon)*

| Element | A | Per cent Abundance | $b$ | $\sigma$ | $\sigma_a$ |
|---|---|---|---|---|---|
| H | 1 | 100 | −03741 | 81.67 | 0.3326 (t) |
| | 2 | 0.0149 | 0.6674 | 7.63 | 0.000519 |
| | 3 | | 0.494 | 3.03 | <0.000006 |
| He | * | | 0.326 | 1.21 | <0.001 |
| | 3 | 0.00014 | 0.574 | 5.6 | 5333. (t) |
| | 4 | 100. | 0.326 | 1.21 | ~0 |
| Li | * | | −0.203 | 1.40 | 70.5 |
| | 6 | 7.5 | +0.187 −0.026i | 0.98 | 940. |
| | 7 | 92.5 | −0.220 | 1.44 | 0.0454 |
| Be | 9 | 100. | 0.779 | 7.61 | 0.0076 |
| B | * | | 0.535 −0.021i | 5.01 | 767. |
| | 10 | 20. | 0.0 −0.11i | 0.98 | 3837. |
| | 11 | 80. | 0.666 | 5.8 | 0.006 |
| C | * | | 0.6648 | 5.564 | 0.00350 (t) |
| | 12 | 98.89 | 0.6653 | 5.564 | 0.00353 |
| | 13 | 1.11 | 0.62 | 5.5 | 0.00137 |
| | 14 | | | | $<10^{-6}$ |
| N | * | | 0.930 | 11.5 | 1.90 |
| | 14 | 99.63 | 0.937 | 11.5 | 1.90 |
| | 15 | 0.37 | 0.644 | 5.21 | 0.000024 |
| O | * | | 0.5805 | 4.234 | 0.00019 |
| | 16 | 99.762 | 0.5805 | 4.234 | 0.00019 |
| | 17 | 0.038 | 0.578 | 3.96 | 0.24 |
| | 18 | 0.200 | 0.584 | 4.3 | 0.00016 |
| F | 19 | 100. | 0.565 | 4.02 | 0.0096 |
| Ne | * | | 0.455 | 2.66 | 0.030 (t) |
| | 20 | 90.51 | 0.463 | 2.71 | 0.037 (t) |
| | 21 | 0.27 | 0.666 | 5.60 | <1.5 |
| | 22 | 9.22 | 0.387 | 1.86 | 0.045 |
| Na | 23 | 100. | 0.363 | 3.23 | 0.530 (t) |
| Mg | * | | 0.5375 | 3.681 | 0.063 (t) |
| | 24 | 78.99 | 0.549 | 4.1 | 0.051 (t) |
| | 25 | 10.00 | 0.362 | 1.8 | 0.19 |
| | 26 | 11.01 | 0.492 | 3.2 | 0.0382 (t) |
| Al | 27 | 100. | 0.3449 | 1.506 | 0.231 (t) |
| Si | * | | 0.4149 | 2.173 | 0.171‡ (t) |
| | 28 | 92.23 | 0.4106 | 2.119 | 0.177‡ |
| | 29 | 4.67 | 0.47 | 2.9 | 0.10 |
| | 30 | 3.10 | 0.458 | 2.65 | 0.107 (t) |
| P | 31 | 100. | 0.513 | 3.314 | 0.172 (t) |
| S | * | | 0.2847 | 1.023 | 0.53 |
| | 32 | 95.02 | 0.2804 | 1.00 | 0.53 |
| | 33 | 0.75 | 0.47 | 3.0 | 0.54 |
| | 34 | 4.21 | 0.348 | 1.54 | 0.227 |
| Cl | * | | 0.9579 | 16.63 | 33.5‡ |
| | 35 | 75.77 | 1.17 | 21.62 | 44.1‡ |
| | 37 | 24.43 | 0.308 | 1.2 | 0.433 |
| Ar | * | | 0.1884 | 0.663 | 0.675 |
| | 36 | 0.337 | 2.42 | 77.8 | 5.2 |
| | 38 | 0.063 | | | 0.8 |
| | 40 | 99.60 | 0.183 | 0.42 | 0.66 |
| K | * | | 0.367 | 210 | 2.1 |
| | 39 | 93.258 | 0.379 | 2.19 | 2.1 |
| | 40 | 0.012 | | | 34. |
| | 41 | 6.73 | 0.258 | 0.83 | 1.46 |
| Ca | * | | 0.490 | 3.00 | 0.43 (t) |
| | 40 | 96.94 | 0.49 | 3.03 | 0.41 (t) |
| | 42 | 0.647 | | | 0.68 |
| | 43 | 0.135 | | | 6.2 |
| | 44 | 2.09 | 0.18 | 0.38 | 0.88 (t) |
| | 46 | 0.0035 | | | 0.74 |
| | 48 | 0.187 | | | 1.1 |
| Sc | 45 | 100. | 1.23 | 23.6 | 27.2‡ (t) |
| Ti | * | | −0.3438 | 4.06 | 6.1 (t) |
| | 46 | 8.2 | 0.473 | 2.85 | 0.6 |
| | 47 | 7.4 | 0.349 | 3.1 | 1.7 |
| | 48 | 73.8 | −0.584 | 4.24 | 7.8 |
| | 49 | 5.4 | 0.100 | 3.5 | 2.2 |
| | 50 | 5.2 | 0.593 | 4.4 | 0.179 (t) |
| V | * | | −0.0382 | 4.953 | 5.08 (t) |
| | 50 | 0.25 | | | 60. (r) |
| | 51 | 99.75 | −0.0414 | 4.946 | 4.9 (t) |
| Cr | * | | 0.3635 | 3.47 | 3.07 (t) |
| | 50 | 4.35 | −0.450 | 2.47 | 15.9 |
| | 52 | 83.79 | 0.491 | 3.04 | 0.76 |
| | 53 | 9.50 | −0.420 | 8.04 | 18. |
| | 54 | 2.36 | 0.455 | 2.6 | 0.36 |
| Mn | 55 | 100. | −0.373 | 2.10 | 13.3 (t) |
| Fe | * | | 0.954 | 11.66 | 2.56 |
| | 54 | 5.8 | 0.42 | 2.2 | 2.3 |
| | 56 | 91.72 | 1.01 | 12.8 | 2.6 |
| | 57 | 2.20 | 0.23 | <1. | 2.5 |
| | 58 | 0.28 | 1.5 | 28. | 1.28 |
| Co | 59 | 100. | 0.253 | 6.15 | 37.18 (t) |
| Ni | * | | 1.03 | 17.56 | 4.5 |
| | 58 | 68.27 | 1.44 | 25.87 | 4.6 |
| | 60 | 26.10 | 0.28 | 0.96 | 2.9 |
| | 61 | 1.13 | 0.76 | 7.23 | 2.5 |
| | 62 | 3.59 | −0.87 | 9.6 | 14.5 |
| | 64 | 0.91 | −0.38 | 0.02 | 1.52 (t) |
| Cu | * | | 0.7718 | 7.93 | 3.78 (t) |
| | 63 | 69.17 | 0.67 | 5.54 | 4.50 (t) |
| | 65 | 30.83 | 1.00 | 13.1 | 2.17 (t) |
| Zn | * | | 0.5680 | 4.115 | 1.11 (t) |
| | 64 | 48.6 | 0.52 | 3.44 | 0.76 |
| | 66 | 27.9 | 0.60 | 4.54 | 0.9 |
| | 67 | 4.1 | 0.76 | 7.32 | 6.8 |
| | 68 | 18.8 | 0.60 | 4.61 | 1.1 |
| | 70 | 0.6 | | | 0.09 |
| Ga | * | | 0.729 | 7.07 | 2.9 (t) |
| | 69 | 60.1 | | | 1.68 (t) |
| | 71 | 39.9 | | | 4.71 (t) |

**Table 1.1 (ctd.)**

| Element | A | Per cent Abundance | $b$ | $\sigma$ | $\sigma_a$ |
|---|---|---|---|---|---|
| Ge | * | | 0.8193 | 8.75 | 2.3 (r) |
| | 70 | 20.5 | 0.84 | 9.0 | 3.43 |
| | 72 | 27.4 | 0.78 | 7.7 | 0.98 (r) |
| | 73 | 7.8 | 0.28 | 1.0 | 15. (r) |
| | 74 | 36.5 | 0.7 | 6.3 | 0.51 |
| | 76 | 7.8 | | | 0.15 |
| As | 75 | 100. | 0.658 | 5.51 | 4.5 (t) |
| Se | * | | 0.797 | 8.34 | 11.7 (t) |
| | 74 | 0.9 | 0.08 | ~0 | 52. |
| | 76 | 9.0 | 1.22 | 18.8 | 85. |
| | 77 | 7.6 | 0.825 | 8.9 | 42. (r) |
| | 78 | 23.5 | 0.824 | 8.5 | 0.43 |
| | 80 | 49.6 | 0.748 | 7.04 | 0.61 |
| | 82 | 9.4 | 0.634 | 5.0 | 0.44 |
| Br | * | | 0.679 | 5.92 | 6.9 (t) |
| | 79 | 50.69 | 0.679 | 6.12 | 11.0 (t) |
| | 81 | 49.31 | 0.678 | 5.8 | 2.7 (t) |
| Kr | * | | 0.785 | 7.70 | 25.‡ |
| | 78 | 0.35 | | | 6.2 |
| | 80 | 2.25 | | | 11.5 |
| | 82 | 11.6 | | | 28. |
| | 83 | 11.5 | | | 180.‡ |
| | 84 | 57.0 | | | 0.11 |
| | 86 | 17.4 | | | 0.003 |
| Rb | * | | 0.708 | 6.30 | 0.38‡ |
| | 85 | 72.17 | 0.707 | 6.26 | 0.48‡ |
| | 87 | 27.83 | 0.727 | 6.66 | 0.12 |
| Sr | * | | 0.702 | 6.22 | 1.28 |
| | 84 | 0.56 | | | 0.87 |
| | 86 | 9.86 | 0.568 | 4.0 | 1.04 |
| | 87 | 7.00 | 0.741 | 6.9 | 16. |
| | 88 | 82.58 | 0.716 | 6.4 | 0.058 |
| Y | 89 | 100 | 0.775 | 7.70 | 1.28 |
| Zr | * | | 0.716 | 6.56 | 0.185 |
| | 90 | 51.45 | 0.65 | 5.2 | 0.011 |
| | 91 | 11.32 | 0.88 | 9.78 | 1.2 |
| | 92 | 17.19 | 0.75 | 6.9 | 0.22 |
| | 94 | 17.28 | 0.83 | 8.6 | 0.050 |
| | 96 | 2.76 | 0.55 | 3.7 | 0.023 |
| Nb | 93 | 100. | 0.7054 | 6.25 | 1.15 (r) |
| Mo | * | | 0.695 | 5.9 | 2.55 |
| | 92 | 14.48 | | | 0.019 (c) |
| | 94 | 9.25 | | | 0.015 (c) |
| | 95 | 15.92 | | | 14.0 |
| | 96 | 16.68 | | | 0.5 |
| | 97 | 9.55 | | | 2.1 |
| | 98 | 24.13 | | | 0.130 |
| | 100 | 9.63 | | | 0.199 |
| Tc | 98 | | | | 0.9 |
| | 98 | | 0.68 | | 20. (t) |
| Ru | * | | 0.721 | 6.6 | 2.6 |
| | 96 | 5.52 | | | 0.29 (t) |
| | 98 | 1.86 | | | <8. (r) |
| | 99 | 12.7 | | | 7. (t) |
| | 100 | 12.6 | | | 5.0 (t) |
| | 101 | 17.0 | | | 3.4 (t) |
| | 102 | 31.6 | | | 1.21 (t) |
| | 104 | 18.7 | | | 0.32 (t) |
| Rh | 103 | 100. | 0.593 | 4.2 | 145. (t) |
| Pd | * | | 0.591 | 5.1 | 6.9 |
| | 102 | 1.02 | | | 3.4 (r) |
| | 104 | 11.14 | | | 0.6 (c) |
| | 105 | 22.33 | | | 20. (c) |
| | 106 | 27.33 | | | 0.30 (t) |
| | 108 | 26.46 | | | 8.5 |
| | 110 | 11.72 | | | 0.23 (r) |
| Ag | * | | 0.597 | 5.09 | 63.3 (t) |
| | 107 | 51.83 | 0.764 | | 38. (t) |
| | 109 | 48.17 | 0.419 | | 91. (t) |
| Cd | * | | 0.50 −0.16i | 5.7 | 2520.‡ (t) |
| | 106 | 1.25 | | | ~1. (r) |
| | 108 | 0.89 | | | 1.1 (r) |
| | 110 | 12.51 | | | 11. (r) |
| | 111 | 12.81 | | | 24. (r) |
| | 112 | 24.13 | 0.74 | 7.0 | 2.2 (r) |
| | 113 | 12.22 | −0.8 −0.75i | | 20600.‡ (t) |
| | 114 | 28.72 | 0.64 | 5.3 | 0.34 |
| | 116 | 7.47 | 0.71 | 6.5 | 0.07 |
| In | * | | 0.406 −0.054i | 2.75 | 194.‡ |
| | 113 | 4.28 | 0.539 | 3.8 | 12. (t) |
| | 115 | 95.72 | 0.400 −0.056i | 2.6 | 202.‡ (t) |
| Sn | * | | 0.6228 | 4.893 | 0.626 |
| | 112 | 1.0 | | | 1.0 |
| | 114 | 0.7 | | | 0.12 |
| | 115 | 0.4 | | | 30. (r) |
| | 116 | 14.7 | 0.58 | 4.2 | 0.14 (r) |
| | 117 | 7.7 | 0.64 | | 2.3 (r) |
| | 118 | 24.3 | 0.58 | 4.2 | 0.22 (r) |
| | 119 | 8.6 | 0.60 | | 2.2 (r) |
| | 120 | 32.4 | 0.64 | 5.1 | 0.14 |
| | 122 | 4.6 | 0.55 | 3.8 | 0.18 |
| | 124 | 5.6 | 0.59 | 4.4 | 0.13 |
| Sb | * | | 0.564 | 4.2 | 5.1 |
| | 121 | 57.25 | | | 5.9 |
| | 123 | 42.75 | | | 4.1 |
| Te | * | | 0.580 | 4.33 | 4.7 |
| | 120 | 0.091 | 0.53 | 3.5 | 2.3 (r) |
| | 122 | 2.6 | | | 3.4 (r) |
| | 123 | 0.908 | | | 418.‡ |
| | 124 | 4.816 | 0.56 | 4.0 | 6.8 |
| | 125 | 7.14 | 0.56 | 4.0 | 1.6 |
| | 126 | 18.95 | | | 1.0 |
| | 128 | 31.69 | | | 0.215 |
| | 130 | 33.80 | 0.57 | 4.1 | 0.29 |
| I | 127 | 100. | 0.528 | 4.52 | 6.2 (r) |
| Xe | * | | 0.489 | 4.3 | 23.9 (r) |
| | 124 | 0.1 | | | 165. (t) |
| | 126 | 0.09 | | | 3.5 (t) |
| | 128 | 1.91 | | | <8. (r) |
| | 129 | 26.4 | | | 21. (t) |
| | 130 | 4.1 | | | <26. (r) |
| | 131 | 21.2 | | | 85. (t) |
| | 132 | 26.9 | | | 0.45 |
| | 134 | 10.4 | | | 0.26 (t) |

**Table 1.1 (ctd.)**

| Element | A | Per cent Abundance | $b$ | $\sigma$ | $\sigma_a$ |
|---|---|---|---|---|---|
| | 136 | 8.9 | | | 0.26 (t) |
| Cs | 133 | 100. | 0.542 | 3.96 | 29. (t) |
| Ba | * | | 0.525 | 3.37 | 1.2 |
| | 130 | 0.106 | | | 11.3 |
| | 132 | 0.101 | | | 7.0 (r) |
| | 134 | 2.417 | | | 2. (r) |
| | 135 | 6.592 | | | 5.8 |
| | 136 | 7.854 | | | 0.4 |
| | 137 | 11.23 | | | 5.1 |
| | 138 | 71.70 | | | 0.36 |
| La | * | | 0.827 | 10.2 | 8.97 |
| | 138 | 0.09 | | | 8.93 |
| | 139 | 99.91 | 0.827 | 10.2 | 57. (t) |
| Ce | * | | 0.484 | 2.9 | 0.63 |
| | 136 | 0.19 | | | 7. |
| | 138 | 0.25 | | | 1.1 |
| | 140 | 88.48 | 0.47 | 2.87 | 0.57 |
| | 142 | 11.05 | 0.45 | 2.55 | 0.95 (r) |
| Pr | 141 | 100. | 0.445 | 2.58 | 11.5 (t) |
| Nd | * | | 0.769 | 16. | 51. |
| | 142 | 27.13 | 0.77 | 7.7 | 18.7 |
| | 143 | 12.18 | | | 325. (t) |
| | 144 | 23.80 | 0.24 | 1.0 | 3.6 |
| | 145 | 8.30 | | | 42. (t) |
| | 146 | 17.19 | 0.87 | 9.6 | 1.4 (t) |
| | 148 | 5.76 | | | 2.5 (t) |
| | 150 | 5.64 | | | 1.2 (t) |
| Pm | 146 | | | | 8400. |
| | 147 | | 1.26 | | 168.‡ (t) |
| Sm | * | | | | 5670.‡ (t) |
| | 144 | 3.1 | | | 0.7 |
| | 147 | 15.1 | | | 57.‡ (t) |
| | 148 | 11.3 | | | 2.4 |
| | 149 | 13.9 | | | 40140.‡ (t) |
| | 150 | 7.4 | | | 104. (t) |
| | 152 | 26.6 | −0.50 | 3.1 | 206. (t) |
| | 154 | 22.6 | 0.80 | 8.0 | 8.4 |
| Eu | * | | 0.60 | 8. | 4600.‡ (t) |
| | 151 | 47.86 | | | 9200.‡ (t) |
| | 153 | 52.14 | 0.82 | 8.0 | 307. (t) |
| Gd | * | | 0.95 | | 29400.‡ (t) |
| | 152 | 0.2 | | | 1100. (t) |
| | 154 | 2.1 | | | 85. (t) |
| | 155 | 14.8 | | | 60900.‡(t) |
| | 156 | 20.6 | | | 1.5 |
| | 157 | 15.7 | | | 25400.‡ (t) |
| | 158 | 24.8 | | | 2.0 (t) |
| | 160 | 21.8 | 0.915 | | 0.77 |
| Tb | 159 | 100. | 0.738 | | 23.0‡ (t) |
| Dy | * | | 1.69 | 82.6 | 940.‡ (t) |
| | 156 | 0.057 | | | 33. (t) |
| | 158 | 0.10 | | | 43. (t) |
| | 160 | 2.34 | 0.67 | 5.6 | 56. (t) |
| | 161 | 19.0 | 1.03 | 22.0 | 570.‡ (t) |
| | 162 | 25.5 | −0.14 | 0.25 | 194. (t) |
| | 163 | 24.9 | 0.50 | 9.7 | 130.‡ (t) |
| | 164 | 28.1 | 4.94 | 307. | 2650.‡ (t) |

**Table 1.1 (ctd.)**

| Element | A | Per cent Abundance | $b$ | $\sigma$ | $\sigma_a$ |
|---|---|---|---|---|---|
| Ho | 165 | 100. | 0.808 | 9.5 | 65. (t) |
| Er | * | | 0.803 | 11. | 159.‡ (t) |
| | 162 | 0.14 | | | 19. |
| | 164 | 1.56 | 0.98 | 12.1 | 13. |
| | 166 | 33.4 | 1.23 | 19.1 | 20. |
| | 167 | 22.9 | | | 659.‡ (t) |
| | 168 | 27.1 | 1.0 | 13.1 | 2.74 (t) |
| | 170 | 14.9 | 1.1 | 15.9 | 5.8 |
| Tm | 169 | 100. | 0.705 | 12. | 105.‡ |
| Yb | * | | 1.24 | 25. | 35. (t) |
| | 168 | 0.127 | | | 2300.‡ (t) |
| | 170 | 3.04 | | | 11.4 (t) |
| | 171 | 14.28 | | | 49. (t) |
| | 172 | 21.83 | | | 0.8 (t) |
| | 173 | 16.13 | | | 17. (t) |
| | 174 | 31.83 | | | 67. (t) |
| | 176 | 12.76 | | | 4.4 (r) |
| Lu | * | | 0.73 | 7. | 76.‡ (t) |
| | 175 | 97.41 | | | 23. (t) |
| | 176 | 2.59 | | | 2090.‡ (t) |
| Hf | * | | 0.77 | 8. | 103.‡ (t) |
| | 174 | 0.163 | | | 510. (t) |
| | 176 | 5.21 | | | 24. (t) |
| | 177 | 18.56 | | | 370.‡ (t) |
| | 178 | 27.10 | | | 84. (t) |
| | 179 | 13.75 | | | 45. (t) |
| | 180 | 35.22 | | | 13.04 (t) |
| Ta | * | | 0.691 | 6.0 | 20.5 (t) |
| | 180 | 0.012 | | | 563. (t) |
| | 181 | 99.988 | 0.691 | 6.0 | 20.5 (t) |
| W | * | | 0.477 | 5.7 | 18.4 (t) |
| | 180 | 0.126 | | | ~30. |
| | 182 | 26.31 | 0.83 | 8.6 | 20.7 (t) |
| | 183 | 14.28 | 0.43 | | 10.1 (t) |
| | 184 | 30.64 | 0.76 | 7.2 | 1.7 (t) |
| | 186 | 28.64 | −0.119 | 0.18 | 37.9 (t) |
| Re | * | | 0.92 | 11.3 | 91. (t) |
| | 185 | 37.4 | | | 112. (t) |
| | 187 | 62.6 | | | 78. (t) |
| Os | * | | 1.07 | 14.9 | 16.0 (r) |
| | 184 | 0.02 | | | 3000. |
| | 186 | 1.58 | | | 80. (t) |
| | 187 | 1.6 | | | 320. (t) |
| | 188 | 13.3 | 0.78 | 7.6 | 5. (t) |
| | 189 | 16.1 | 1.10 | 14. +10 | 25. (t) |
| | 190 | 26.4 | 1.14 | 16. | 13.1 (t) |
| | 192 | 41.0 | 1.19 | 18. | 2.0 (t) |
| Ir | * | | 1.06 | 14. | 425.‡ (t) |
| | 191 | 37.3 | | | 955.‡ (t) |
| | 193 | 62.7 | | | 111.‡ (t) |
| Pt | * | | 0.95 | 11.2 | 10.3‡ (t) |
| | 190 | 0.01 | 0.9 | ~10. | ~800. (t) |
| | 192 | 0.79 | 0.99 | 12. | 10. (t) |
| | 194 | 32.9 | 1.06 | 14. | 1.2 (t) |
| | 195 | 33.8 | | | 27.‡ (t) |

**Table 1.1 (ctd.)**

| Element | A | Per cent Abundance | $b$ | $\sigma$ | $\sigma_a$ |
|---|---|---|---|---|---|
| | 196 | 25.3 | 0.89 | 12.3 | 0.72 (t) |
| | 198 | 7.2 | 0.78 | 7.7 | 3.8 (t) |
| Au | 197 | 100. | 0.763 | 7.81 | 98.65 (t) |
| Hg | * | | 1.266 | 26.3 | 372. (t) |
| | 196 | 0.15 | | | 3080. |
| | 198 | 10.1 | | | 1.9 (c) |
| | 199 | 16.9 | | | 2162. (t) |
| | 200 | 23.1 | | | <60. |
| | 201 | 13.2 | | | 8. (t) |
| | 202 | 29.7 | | | 4.9 (r) |
| | 204 | 6.9 | | | 0.4 |
| Tl | * | | 0.879 | 9.86 | 3.4 |
| | 203 | 29.5 | | | 11.0 |
| | 205 | 70.5 | | | 0.10 |
| Pb | * | | 0.9401 | 11.11 | 0.17 |
| | 203 | 1.4 | | | 0.66 |
| | 206 | 24.1 | | | 0.0305 |
| | 207 | 22.1 | | | 0.709 |
| | 208 | 52.4 | | | 0.00049 |
| Bi | 209 | 100. | 0.8533 | 9.156 | 0.033 |

**Table 1.1 (ctd.)**

| Element | A | Per cent Abundance | $b$ | $\sigma$ | $\sigma_a$ |
|---|---|---|---|---|---|
| Po | 210 | | | | <0.03 |
| Rn | 222 | | | | 0.72 |
| Ra | 226 | | 1.0 | | 11.‡ |
| | 228 | | | | 36. |
| Ac | 227 | | | | 515. |
| Th | 232 | 100. | 0.984 | 12.2 | 7.40 |
| Pa | 231 | | 0.91 | | 210.‡ |
| U | * | | 0.842 | 8.9 | 7.5 |
| | 234 | 0.0055 | | | 100. |
| | 235 | 0.71 | 0.98 | | 680.‡ |
| | 238 | 99.27 | 0.855 | 9.20 | 2.7 |
| Np | 237 | | 1.06 | | 169.‡ |
| Pu | 238 | | | | 564. |
| | 239 | | 0.77 | | 1011.‡ |
| | 240 | | 0.35 | | 289. |
| | 242 | | 0.81 | | 18.5 |
| Am | 243 | | 0.83 | | 79.‡ |
| Cm | 244 | | 0.95 | | 15. |

$\bar{b}(10^{-12}\,\text{cm})$: Coherent scattering length for bound atoms. Complex values correspond to a neutron wavelength of 1 Å. For references and complete data see Koester *et al.* (1981).
$\sigma(10^{-24}\,\text{cm}^2)$: Total scattering cross-sections of bound atoms for 'thermal' neutrons.
$\sigma_a(10^{-24}\,\text{cm}^2)$: Absorption cross-section for thermal neutrons ($2200\,\text{m s}^{-1}$) compiled from Mughabghab *et al.* (1981) and from Mughabghab and Garber (1973).
(t): Absorption cross-section measured with 2200 m/s neutrons.
(r): Absorption cross-section measured with reactor neutron spectrum.
(c): Calculated cross-section.
All other cross-sections measured with Maxwellian neturon spectrum.
* Natural isotopic mixture.
‡ Check BNL-325 for full data on absorption cross-sections in the resonance region.
Note: Values are uncertain in the given digit. For nuclei giving resonance scattering in the low energy region the tabulated $\bar{b}$ and $\sigma$ are valid only for zero neutron energy.

In contrast to the large incoherent cross-section for hydrogen,

$$\sigma_i = \sigma - \sigma_c = 79.8\ \text{bn}$$

the situation for deuterium is very different. Using the values

$$b^{(+)} = 0.95 \times 10^{-12}\ \text{cm},$$
$$b^{(-)} = 0.10 \times 10^{-12}\ \text{cm},$$

and $i = 1$, we find

$$\bar{b} = \tfrac{2}{3} b^{(+)} + \tfrac{1}{3} b^{(-)} = 0.67 \times 10^{-12}\ \text{cm}$$

and

$$\overline{|b|^2} = \tfrac{2}{3}\,|b^{(+)}|^2 + \tfrac{1}{3}\,|b^{(-)}|^2 = 0.61\ \text{bn}$$

which give

$$\left.\begin{aligned}\sigma_c &= 5.6\ \text{bn}\\ \sigma &= 7.6\ \text{bn}\end{aligned}\right\}{}^2\text{H} = \text{D}.$$

Thus for deuterium only a relatively small part of the scattering is incoherent. The incoherent scattering is, in the case of hydrogen, purely spin-dependent. Vanadium is another example of an element that scatters almost entirely incoherently, since $\sigma_c = 0.03$ bn and $\sigma = 5.1$ bn. This comes about because, although there is only one stable V isotope, the scattering lengths $(+)$ and $(-)$ have opposite signs and are almost exactly in the ratio $(i+1)$: $i$, i.e. almost $9/2:7/2$, so that $\bar{b}$ is very close to zero. In contrast to $^1H$ and V, iron has a very small incoherent cross-section, due to the high abundance of an isotope with $i = 0$. On the other hand nickel has a relatively high incoherent cross-section, even though its three principal isotopes have $i = 0$, because one of these has a large negative scattering length.

## 1.5. Contrast factors (Kostorz 1983; Jacrot 1976; Ivin 1976)

In order to illustrate the value of isotope substitution in the study of the sizes and shapes of macromolecules we derive an expression for the scattering by macromolecules in a solution. The generic form of the expression is the cornerstone of the method of small-angle scattering in which the diffuse scattering around the direct beam is measured. The diffuse scattering is caused by the variation of scattering length density over distances which exceed the normal interatomic spacings obtained in condensed matter. Aggregates of small particles, macromolecules in liquid or solid solution (e.g. polymers and alloys), and samples with smoothly varying scattering length density profiles are all readily investigated by the small-angle method (cf. § 6.1).

To be concrete we consider a dilute solution of macromolecules. If the solution is sufficiently dilute, the coherent scattering from *different* macromolecules can be neglected since it will be confined to extremely small scattering vectors $\kappa \sim \pi/l$ where $l$ is a measure of the mean separation of the macromolecules. The coherent scattering from a single macromolecule and the solvent is, from (1.41) and omitting the subscript,

$$\frac{d\sigma}{d\Omega} = \left| \sum_j \bar{b}_j \exp(i\boldsymbol{\kappa} \cdot \mathbf{R}_j) \right|^2. \qquad (1.51)$$

Here, $\bar{b}_j$ is the coherent scattering length of the chemical species at the position defined by the vector $\mathbf{R}_j$.

The sum in (1.51) is separated into contributions from the macromolecule and solvent. For the latter we replace the scattering length by a locally averaged scattering length density $\rho_s$; the dimension of $\rho_s$ is $(\text{length})^{-2}$. The solvent contribution to the sum in (1.51) is then

$$\rho_s \left\{ \int d\mathbf{R} \exp(i\boldsymbol{\kappa} \cdot \mathbf{R}) - \int_\Omega d\mathbf{R} \exp(i\boldsymbol{\kappa} \cdot \mathbf{R}) \right\}$$

where the first integral is over the volume of the whole target sample and the second integral is restricted to the volume $\Omega$ of the macromolecule under consideration. In the limit of a bulk sample the first integral approximates a delta function $\delta(\boldsymbol{\kappa})$ which is zero for $\boldsymbol{\kappa} \neq 0$. Hence this term corresponds to no scattering. Let $\rho(\mathbf{R})$ be the scattering length density within the macromolecule. We then have for (1.51) the result

$$\frac{d\sigma}{d\Omega} = \left| \int_{\Omega} d\mathbf{R} \{\rho(\mathbf{R}) - \rho_s\} \right|^2. \tag{1.52}$$

The scattering is seen to be due to the excess of the molecule scattering density over the displaced solvent scattering density. The magnitude of the excess scattering may often be manipulated with advantage by isotope enrichment.

For some applications it suffices to replace $\rho(\mathbf{R})$ by a suitably averaged scattering density. In this case,

$$\frac{d\sigma}{d\Omega} = K^2 \, |F(\boldsymbol{\kappa})|^2 \tag{1.53}$$

where the contrast factor

$$K = \Omega \, |\rho - \rho_s| \tag{1.54}$$

and

$$F(\boldsymbol{\kappa}) = \frac{1}{\Omega} \int_{\Omega} d\mathbf{R} \exp(i\boldsymbol{\kappa} \cdot \mathbf{R}) \tag{1.55}$$

is usually referred to as the molecular form factor. From the definition of the form factor it follows that $F(0) = 1$.

To illustrate the variation in contrast factors that can be achieved by deuteration we consider the example of dilute polystyrene in various solvents. The ratio of the contrast factors for deuterated and hydrogenous polystyrene in the solvent $CS_2$ is 28.7, while for the solvent $C_6D_6$ the ratio is 0.3 and for $C_6H_6$ it is 19.5.

The form factor for a spherical molecule is easily calculated. The result for a molecule of radius $R = x/\kappa$ is,

$$F(\kappa) = \left(\frac{3}{x}\right) j_1(x) \tag{1.56}$$

where $j_1(x)$ is a spherical Bessel function of order 1. The corresponding $|F|^2$ is peaked sharply about $\kappa = 0$; the first zero occurs for $x = 4.49$, and the second maximum occurs for $x = 5.77$ at which $|F|^2 = 7.42 \times 10^{-3}$. The effective radius of the molecule can be obtained from a measurement of the position of the first minimum in the structure factor.

## 1.6. Scattering amplitude operator

In some circumstances (for example in § 1.7), it is useful to allow for the spin dependence of the neutron-nucleus interaction by introducing the concept of a scattering amplitude operator $\hat{b}$. It is convenient to discuss this concept here and reproduce the results of § 1.3 by this alternative technique. For a single nucleus we define

$$\hat{b} = A + \tfrac{1}{2}B\hat{\boldsymbol{\sigma}} \cdot \hat{\mathbf{i}}, \tag{1.57}$$

where $\frac{1}{2}\hat{\boldsymbol{\sigma}}$ is the spin of the neutron and $\hat{\mathbf{i}}$ that of the nucleus. The coefficients $A$ and $B$ are determined from the requirement that $\hat{b}$ should have eigenvalues $b^{(\pm)}$ for the two possible states of total spin, $t = i \pm \frac{1}{2}$. Since

$$\hat{\mathbf{t}}^2 = (\tfrac{1}{2}\hat{\boldsymbol{\sigma}} + \hat{\mathbf{i}})^2 = \tfrac{1}{4}\hat{\boldsymbol{\sigma}}^2 + \hat{\mathbf{i}}^2 + \hat{\boldsymbol{\sigma}} \cdot \hat{\mathbf{i}}$$

and $\hat{\mathbf{t}}^2 = t(t+1)$, $\hat{\boldsymbol{\sigma}}^2 = 3$, $\hat{\mathbf{i}}^2 = i(i+1)$, then

$$\hat{\boldsymbol{\sigma}} \cdot \hat{\mathbf{i}} = t(t+1) - \tfrac{3}{4} - i(i+1),$$

and hence we have, for total spin $t = i + \frac{1}{2}$,

$$b^{(+)} = A + B\tfrac{1}{2}\{(i+\tfrac{1}{2})(i+\tfrac{3}{2}) - i(i+1) - \tfrac{3}{4}\},$$

and for $t = i - \frac{1}{2}$,

$$b^{(-)} = A + B\tfrac{1}{2}\{(i-\tfrac{1}{2})(i+\tfrac{1}{2}) - i(i+1) - \tfrac{3}{4}\}.$$

These two equations give

$$A = \frac{1}{2i+1}\{(i+1)b^{(+)} + ib^{(-)}\} \tag{1.58}$$

and

$$B = \frac{2}{2i+1}\{b^{(+)} - b^{(-)}\}. \tag{1.59}$$

We have now to calculate the average values of the operators $\hat{b}$ and $|\hat{b}|^2$ for unpolarized incident neutrons. The directions of the spins of the neutrons and the nuclei are completely random, and the averaging of the spins of the neutrons and of the nuclei are independent, each spin averaging to zero. We therefore have

$$\overline{\hat{b}} = A. \tag{1.60}$$

To calculate $\overline{|\hat{b}|^2}$ we use the identity†

$$(\hat{\mathbf{i}} \cdot \hat{\boldsymbol{\sigma}})^2 = \hat{\mathbf{i}}^2 + \mathrm{i}\hat{\boldsymbol{\sigma}} \cdot (\hat{\mathbf{i}} \times \hat{\mathbf{i}}) = \hat{\mathbf{i}}^2 + \mathrm{i}\hat{\boldsymbol{\sigma}} \cdot (\mathrm{i}\hat{\mathbf{i}}) = \hat{\mathbf{i}}^2 - \hat{\boldsymbol{\sigma}} \cdot \hat{\mathbf{i}}, \tag{1.61}$$

† If $\hat{\mathbf{A}}$ and $\hat{\mathbf{B}}$ commute with $\hat{\boldsymbol{\sigma}}$ then $[\mathrm{i} = \sqrt{(-1)}]$

$$(\hat{\mathbf{A}} \cdot \hat{\boldsymbol{\sigma}})(\hat{\mathbf{B}} \cdot \hat{\boldsymbol{\sigma}}) = \hat{\mathbf{A}} \cdot \hat{\mathbf{B}} + \mathrm{i}\hat{\boldsymbol{\sigma}} \cdot \hat{\mathbf{A}} \times \hat{\mathbf{B}}.$$

which leads to the result

$$\overline{|\hat{b}|^2} = |A|^2 + |B|^2 \tfrac{1}{4} i(i+1) = \frac{1}{2i+1}\{(i+1)\,|b^{(+)}|^2 + i\,|b^{(-)}|^2\}. \qquad (1.62)$$

Eqns (1.60) and (1.62) are in agreement with (1.45) and (1.46).

## 1.7. Homonuclear diatomic molecule

There are three main reasons for discussing the scattering of neutrons from diatomic molecules at this juncture. First, it is clearly an example of considerable interest in physical-chemistry studies, and we will give a more comprehensive account of the cross-section in Chapter 6. We choose to give a preliminary account here because the problem demonstrates a striking quantum effect that stems from the exchange interaction of the nuclei, and we encounter another illustration of the use of scattering amplitude operator discussed in the preceding section. In the course of the discussion given at the end of the section we obtain an expression for the cross-section that is appropriate for scattering from a molecule with an arbitrary number of *different* nuclei (see also § 6.5.2).

If the molecule is assumed to be non-magnetic, practically all the scattering takes place at the nuclei.† We label the two nuclei by $\xi = 1, 2$, and the position vectors $\mathbf{R}_\xi = \frac{1}{2}\mathbf{r}(-1)^\xi$. The partial cross-section (1.17) is then

$$\left(\frac{d\sigma}{d\Omega}\right)^{\lambda}_{\lambda'} = \frac{k'}{k}\left|\langle\lambda'|\sum_\xi \hat{b}_\xi \exp(\mathrm{i}\boldsymbol{\kappa}\cdot\hat{\mathbf{R}}_\xi)\,|\lambda\rangle\right|^2$$

$$= \frac{k'}{k}\,|\langle\lambda'|\,\{(\hat{b}_1+\hat{b}_2)\cos(\tfrac{1}{2}\boldsymbol{\kappa}\cdot\hat{\mathbf{r}}) + \mathrm{i}(\hat{b}_2-\hat{b}_1)\sin(\tfrac{1}{2}\boldsymbol{\kappa}\cdot\hat{\mathbf{r}})\}\,|\lambda\rangle|^2. \qquad (1.63)$$

This expression can be developed if we relate the matrix elements of $\exp(\mathrm{i}\boldsymbol{\kappa}\cdot\hat{\mathbf{r}}/2)$ and $\exp(-\mathrm{i}\boldsymbol{\kappa}\cdot\hat{\mathbf{r}}/2)$. In the subsequent discussion we shall, for the moment, neglect the translational motion of the centre of mass of molecule.

The composite labels $\lambda$ and $\lambda'$ in (1.63) specify the allowed states of the nuclei. The Hamiltonian of a homonuclear diatomic molecule is invariant with respect to an interchange of the coordinates of the two nuclei. A state is said to be symmetric with respect to the nuclei if its wave function changes sign. If both $|\lambda\rangle$ and $|\lambda'\rangle$ are symmetric or antisymmetric, a situation we refer to as case (a),

$$\langle\lambda'|\exp\left(\frac{\mathrm{i}}{2}\boldsymbol{\kappa}\cdot\hat{\mathbf{r}}\right)|\lambda\rangle = \langle\lambda'|\exp\left(-\frac{\mathrm{i}}{2}\boldsymbol{\kappa}\cdot\hat{\mathbf{r}}\right)|\lambda\rangle,$$

† The neutron–electron scattering length $\sim 10^{-16}$ cm (Koester 1977).

whereas, if one is symmetric and the other is antisymmetric (case (b)),

$$\langle\lambda'|\exp\left(\frac{i}{2}\boldsymbol{\kappa}\cdot\hat{\mathbf{r}}\right)|\lambda\rangle = -\langle\lambda'|\exp\left(-\frac{i}{2}\boldsymbol{\kappa}\cdot\hat{\mathbf{r}}\right)|\lambda\rangle.$$

Hence, in (1.63),

$$\langle\lambda'|\sin(\tfrac{1}{2}\boldsymbol{\kappa}\cdot\hat{\mathbf{r}})|\lambda\rangle = 0 \quad \text{case (a)}$$
$$\langle\lambda'|\cos(\tfrac{1}{2}\boldsymbol{\kappa}\cdot\hat{\mathbf{r}})|\lambda\rangle = 0 \quad \text{case(b)}.$$

The coordinate wave function of a system of two identical particles is symmetric when the total spin of the system $I$ is even, and antisymmetric when it is odd. We conclude that for case (a) both states are either even- or odd-$I$ states, and for case (b) one must be even-$I$ and the other odd. Alternatively we can say that $\Delta I = I' - I$ is an even integer for case (a) and odd for case (b).

The symmetry with respect to an interchange of the coordinates of the two nuclei depends also on the symmetry of the total wave function of the electrons and nuclei. The majority of chemically stable molecules have an electronic ground state that is invariant with respect to all symmetry transformations in the molecule. Moreover, the total spin is zero too. Because the electronic ground state is separated from the first excited state by several eV, we will assume that the molecule remains in its electronic ground state in the scattering process. Energy transferred from the neutron beam to the molecule is taken up by the relative motion of the two nuclei. For a molecule with a totally symmetric electronic state, the motion of the nuclei is equivalent to that of a single particle, of orbital angular momentum $K = 0, 1, 2, \ldots,$ in a centrally symmetric potential. When the sign of the coordinates is changed, the nuclear wave function is multiplied by a factor $(-1)^K$. From this it follows that a state is symmetric with respect to the nuclei for even $K$, and antisymmetric for odd $K$. In consequence, the values of $K$ and $I$ are correlated; namely,

symmetric state, $I$ and $K$ even

antisymmetric state, $I$ and $K$ odd.

Let us now apply this knowledge to the evaluation of the cross-section. First we expand our labelling of the states, and replace $\lambda$ by $K$, $\nu$, $I$ where $\nu$ denumerates the levels with a given $K$. Furthermore,

$$|\lambda\rangle = |K\nu\rangle\,|I\rangle \tag{1.64}$$

where $|I\rangle$ is a nuclear spin state. For case (a) we require,

$$\langle\lambda'|(\hat{b}_1+\hat{b}_2)\cos(\tfrac{1}{2}\boldsymbol{\kappa}\cdot\hat{\mathbf{r}})|\lambda\rangle = \langle I'|\,\hat{b}_1+\hat{b}_2\,|I\rangle\,\langle K'\nu'|\cos(\tfrac{1}{2}\boldsymbol{\kappa}\cdot\hat{\mathbf{r}})\,|K\nu\rangle \tag{1.65}$$

and $\Delta I$ and $\Delta K = K' - K$ are even integers. From the definition of the scattering amplitude operator (1.57),

$$\langle I'|\, \hat{b}_1 + \hat{b}_2 \,|I\rangle = \langle I'|\, 2A + \tfrac{1}{2}B\hat{\boldsymbol{\sigma}} \cdot \hat{\mathbf{I}} \,|I\rangle \tag{1.66}$$

where $\hat{\mathbf{I}} = \hat{\mathbf{i}}_1 + \hat{\mathbf{i}}_2$ is the total nuclear spin.

If the initial and final nuclear states in (1.66) have different total nuclear spins, then, obviously, the term containing $A$ is zero. In addition, the matrix elements of $\hat{\mathbf{I}}$ between states of different spin are zero. We conclude that for case (a) the states that contribute to the cross-section satisfy

$$\Delta I = 0, \qquad |\Delta K| = 0, 2, \ldots, \quad \text{case (a)}.$$

The energy levels for the nuclear motion are independent of the nuclear spin. This means that in scattering we do not discriminate between the various values of $I'$, and therefore all allowed values must be included in the cross-section. In view of this, the partial cross-section for case (a) is

$$\left(\frac{d\sigma}{d\Omega}\right)_a = \frac{k'}{k} \sum_{I'} |\langle I'|\, \hat{b}_1 + \hat{b}_2 \,|I\rangle|^2 \quad |\langle K'\nu'|\cos(\tfrac{1}{2}\boldsymbol{\kappa} \cdot \hat{\mathbf{r}})\,|K\nu\rangle|^2. \tag{1.67}$$

Following the same line of reasoning as that used at the end of § (1.2), we obtain

$$\begin{aligned}\sum_{I'} |\langle I'|\, \hat{b}_1 + \hat{b}_2 \,|I\rangle|^2 &= \langle I|\, (\hat{b}_1 + \hat{b}_2)^+ (\hat{b}_1 + \hat{b}_2)\,|I\rangle \\ &= 4A^2 + \langle I|\, 2AB\hat{\boldsymbol{\sigma}} \cdot \hat{\mathbf{I}} + \tfrac{1}{4}B^2(\hat{\boldsymbol{\sigma}} \cdot \hat{\mathbf{I}})^2 \,|\mathrm{I}\rangle \\ &= 4A^2 + \langle I|\, (2AB - \tfrac{1}{4}B^2)\hat{\boldsymbol{\sigma}} \cdot \hat{\mathbf{I}} + \tfrac{1}{4}B^2\hat{\mathbf{I}}^2 \,|I\rangle,\end{aligned} \tag{1.68}$$

where in the last line we have used the identity (1.61) and we assume that $A$ and $B$ are real. If the incident neutron beam is unpolarized, the average value of the neutron spin $\frac{1}{2}\hat{\boldsymbol{\sigma}}$ in (1.68) is zero. The final expression for the partial cross-section for the scattering of unpolarized neutrons by a homonuclear diatomic molecule for case (a) is,

$$\begin{aligned}\left(\frac{d\sigma}{d\Omega}\right)_a &= \frac{k'}{k}\{4A^2 + \tfrac{1}{4}B^2 I(I+1)\} \\ &\quad \times |\langle K'\nu'|\cos(\tfrac{1}{2}\boldsymbol{\kappa} \cdot \hat{\mathbf{r}})\,|K\nu\rangle|^2.\end{aligned} \tag{1.69}$$

For case (b) the explicit dependence of the cross-section on the nuclear spin appears in the term

$$\langle I'|\, \hat{b}_1 - \hat{b}_2 \,|I\rangle = \tfrac{1}{2}B\langle I'|\, \hat{\boldsymbol{\sigma}} \cdot (\hat{\mathbf{i}}_1 - \hat{\mathbf{i}}_2)\,|I\rangle, \tag{1.70}$$

and it has only matrix elements which are non-diagonal with respect to the nuclear spin; in fact the matrix elements are non-zero for $\Delta I = \pm 1$. A case of particular interest is molecular hydrogen. The two nuclear spin states are parahydrogen, $I=0$, and orthohydrogen, $I=1$. We denote the respective nuclear spin state functions by $|0\rangle$, and $|1, M\rangle$ where $M=0, \pm 1$. Writing $\hat{\mathbf{Q}}=\hat{\mathbf{i}}_1-\hat{\mathbf{i}}_2$ it is straightforward to show that

$$\langle 0|\, \hat{Q}^x\, |1, 1\rangle = -\langle 0|\, \hat{Q}^x\, |1, -1\rangle = -\frac{1}{\sqrt{2}}$$

$$\langle 0|\, \hat{Q}^y\, |1, 1\rangle = \langle 0|\, \hat{Q}^y\, |1, -1\rangle = -\frac{\mathrm{i}}{\sqrt{2}}$$

and

$$\langle 0|\, \hat{Q}^z\, |1, 0\rangle = 1,$$

and all other matrix elements are zero. These results confirm that for case (b), $\Delta I = \pm 1$. We conclude that,

$$|\Delta I| = 1, \quad \text{and} \quad |\Delta K| = 1, 3, \ldots, \quad \text{case (b).}$$

This means that transitions between states that have different symmetries occur in those processes that involve the spin incoherent scattering length $B$. In addition, the only states which contribute to the neutron cross-section are those with $\Delta I = 0$ as in case (a), or $\Delta I = \pm 1$ as in case (b). Thus, for molecular deuterium, for example, the transition $I=0 \leftrightarrow I=2$ does not contribute to the cross-section.

The next step in the calculation of the cross-section for case (b) is the evaluation of

$$\begin{aligned}\sum_{I'} |\langle I'|\, \hat{b}_1-\hat{b}_2\, |I\rangle|^2 &= \langle I|\, (\hat{b}_1-\hat{b}_2)^+(\hat{b}_1-\hat{b}_2)\, |I\rangle \\ &= \tfrac{1}{4}B^2\langle I|\, (\hat{\mathbf{i}}_1-\hat{\mathbf{i}}_1)^2 - \hat{\boldsymbol{\sigma}}\cdot(\hat{\mathbf{i}}_1-\hat{\mathbf{i}}_2)\, |I\rangle \\ &= \tfrac{1}{4}B^2\langle I|\, 2\hat{\mathbf{i}}_1^2+2\hat{\mathbf{i}}_2^2-\hat{\mathbf{I}}^2\, |I\rangle \\ &= \tfrac{1}{4}B^2\{4i(i+1)-I(I+1)\}. \end{aligned} \tag{1.71}$$

We then arrive at the following expression for the partial cross-section for case (b),

$$\left(\frac{d\sigma}{d\Omega}\right)_{\mathrm{b}} = \frac{k'}{k}\frac{\sigma_{\mathrm{i}}}{\pi}\left\{1-\frac{I(I+1)}{4i(i+1)}\right\}|\langle K'\nu'|\, \sin(\tfrac{1}{2}\boldsymbol{\kappa}\cdot\hat{\mathbf{r}})\, |K\nu\rangle|^2, \tag{1.72}$$

where the incoherent cross-section,

$$\sigma_{\mathrm{i}} = \pi i(i+1)B^2. \tag{1.73}$$

Notice that for case (b) the cross-section vanishes in the limit $\boldsymbol{\kappa} \to 0$.

The evaluation of the remaining matrix elements in (1.69) and (1.72) involves considerable algebra. We will postpone the study for consideration in § 6.5.1 and conclude the present discussion by answering the question, what is the cross-section for a molecule when the correlation between the spatial and spin wave functions is neglected? The error involved in neglecting the quantum correlation is expected to be small for heavy nuclei.

The key step in the approximate treatment of the cross-section is to take the scattering amplitude operators from inside the matrix elements in (1.63). Our starting point is therefore,

$$\left(\frac{\mathrm{d}\sigma}{\mathrm{d}\Omega}\right)_{\lambda'}^{\lambda}=\frac{k'}{k}\sum_{\xi\xi'}\hat{b}_{\xi'}^{+}\hat{b}_{\xi}\,\langle K\nu|\exp(-\mathrm{i}\boldsymbol{\kappa}\cdot\hat{\mathbf{R}}_{\xi'})\,|K'\nu'\rangle\,\langle K'\nu'|\exp(\mathrm{i}\boldsymbol{\kappa}\cdot\hat{\mathbf{R}}_{\xi})\,|K\nu\rangle. \tag{1.74}$$

The next step is to average the product of the scattering amplitude operators over the nuclear spin states. From the definition (1.57),

$$\hat{b}_{\xi'}^{+}\hat{b}_{\xi}=A_{\xi'}A_{\xi}+\tfrac{1}{2}A_{\xi'}B_{\xi}\hat{\boldsymbol{\sigma}}\cdot\hat{\mathbf{i}}_{\xi}+\tfrac{1}{2}A_{\xi}B_{\xi'}\hat{\boldsymbol{\sigma}}\cdot\hat{\mathbf{i}}_{\xi'}+\tfrac{1}{4}B_{\xi'}B_{\xi}(\hat{\boldsymbol{\sigma}}\cdot\hat{\mathbf{i}}_{\xi'})(\hat{\boldsymbol{\sigma}}\cdot\hat{\mathbf{i}}_{\xi}). \tag{1.75}$$

We have attached labels to $A$ and $B$ for completeness. If the nuclear spins are randomly orientated, the average value of $\hat{\mathbf{i}}$ in (1.75) is zero; thus the middle two terms vanish on taking the average. The average of the product of two spin operators will vanish if they belong to different nuclei. When the operators belong to the same nucleus we can make the average required in (1.74) by using the identity (1.61). Denoting the average over the spin orientations by a bar, we find for randomly oriented nuclear spins the result

$$\overline{\hat{b}_{\xi'}^{+}\hat{b}_{\xi}}=A_{\xi'}A_{\xi}+\tfrac{1}{4}\delta_{\xi\xi'}i_{\xi}(i_{\xi}+1)B_{\xi}^{2}. \tag{1.76}$$

From this result we conclude that the cross-section is independent of the neutron spin when the nuclear spins are randomly oriented. This is what physical intuition would lead us to expect given that the molecule does not possess a preferred axis when the nuclear spins are randomly oriented.

Returning to the cross-section (1.74), and performing the average over the nuclear spin orientations with the aid of (1.76), we obtain

$$\left(\frac{\mathrm{d}\sigma}{\mathrm{d}\Omega}\right)_{\lambda'}^{\lambda}=\frac{k'}{k}\Big\{\sum_{\xi}[A_{\xi}^{2}+\tfrac{1}{4}i_{\xi}(i_{\xi}+1)B_{\xi}^{2}]\,|\langle K'\nu'|\exp(\mathrm{i}\boldsymbol{\kappa}\cdot\hat{\mathbf{R}}_{\xi})\,|K\nu\rangle|^{2}+{\sum_{\xi\xi'}}'\,A_{\xi'}A_{\xi}\langle K\nu|\exp(-\mathrm{i}\boldsymbol{\kappa}\cdot\hat{\mathbf{R}}_{\xi'})\,|K'\nu'\rangle\langle K'\nu'|\exp(\mathrm{i}\boldsymbol{\kappa}\cdot\hat{\mathbf{R}}_{\xi})\,|K\nu\rangle\Big\} \tag{1.77}$$

where terms with $\xi = \xi'$ are excluded in the double sum. While the result (1.77) is an approximation when the nuclei are identical, it is exact when they are different species (including different isotopes of the same element).

Let us apply (1.77) to a homonuclear diatomic molecule. When $\Delta K$ is even we find

$$\left(\frac{d\sigma}{d\Omega}\right)_a \doteq \frac{k'}{k}\{4A^2 + \tfrac{1}{2}i(i+1)B^2\}\,|\langle K'\nu'|\cos(\tfrac{1}{2}\boldsymbol{\kappa}\cdot\hat{\mathbf{r}})\,|K\nu\rangle|^2 \tag{1.78}$$

and for $\Delta K$ odd,

$$\left(\frac{d\sigma}{d\Omega}\right)_b \doteq \frac{k'}{k}\frac{\sigma_i}{2\pi}\,|\langle K'\nu'|\sin(\tfrac{1}{2}\boldsymbol{\kappa}\cdot\hat{\mathbf{r}})\,|K\nu\rangle|^2. \tag{1.79}$$

These approximate results should be compared with the corresponding exact expressions (1.69) and (1.72).

## 1.8. Static approximation

The rationale for this approximation is based on two observations. First, the respose of the target sample to the incident neutrons is, in many instances, derived from states whose energy spectrum is bounded. For example, the number of states in the energy spectrum of phonons in a crystal is minimal at energies that exceed the Debye energy. Moreover, the response to an energetic beam of neutrons will entail essentially all the available states and, in the absence of fine energy resolution, the discrimination of the energy-conserving delta function in the cross-section is unimportant. From these observations we conclude that, with relaxed energy resolution and an incident neutron energy very much in excess of the maximum energy of the responding states, the conservation of energy in (1.23) can be neglected. Since the change in the energy of the neutrons must be relatively small, the scattering appears to be elastic to a good approximation.

For the conditions described in the preceding paragraph, the cross-section (1.23) reduces to

$$\frac{d^2\sigma}{d\Omega\, dE'} \doteq \overline{\sum_{\lambda\lambda'} p_\lambda\,|\langle \mathbf{k}'\lambda'|\,\hat{V}\,|\mathbf{k}\lambda\rangle|^2\delta(\hbar\omega)}. \tag{1.80}$$

Result (1.80) is called the static approximation. Note that the factor $(k'/k)$ in (1.23) is unity since the scattering is elastic. Having removed the energy discrimination on the final states in (1.80), we can use the closure relation (1.15) and the manipulations described at the end of § 1.2 to reduce (1.80) to the final form

$$\frac{d\sigma}{d\Omega} \doteq \overline{\langle|\langle\mathbf{k}'|\ \hat{V}\ |\mathbf{k}\rangle|^2\rangle}. \tag{1.81}$$

Here we have introduced a notation that will be used throughout the book, namely,

$$\langle(\cdot\cdot\cdot)\rangle = \sum_{\lambda} p_{\lambda}\langle\lambda|\,(\cdot\cdot\cdot)\,|\lambda\rangle. \tag{1.82}$$

Thus, the angular brackets denote the thermal average of the enclosed quantity.

Although the static approximation reduces the cross-section to elastic scattering it must be emphasized that the approximation is distinct from purely elastic scattering. In the latter case, the final states included in the cross-section are identical with the initial states; in (1.23) the elastic cross-section is given by the contributions for which $\lambda' = \lambda$. In contrast, the static approximation, by its very nature, includes all possible final states.

One use of the static approximation is to estimate the maximum scattering which can be observed. The approximation can be applied to a subset of the available states whose energy spread is small compared to the incident neutron energy. For example, in scattering thermal neutrons from a heavy molecule, for which the rotational energy constant is a fraction of an meV, it would be valid to sum over all the rotational states.

The leading-order correction to the static approximation results in the so-called impulse approximation, which is the subject of § 3.4. In the following section we calculate the cross-section for a free nucleus, an example for which the static approximation is never valid.

## 1.9. Scattering by a single free nucleus

In § 1.3 we calculated the cross-section for the scattering by a single bound nucleus of zero spin. We now consider the problem in the opposite limit, namely when the nucleus is completely free. The state of the latter is completely determined by specifying its wave vector $\mathbf{q}$. The wave function is simply

$$|\mathbf{q}\rangle = V^{-1/2}\exp(\mathrm{i}\mathbf{q}\cdot\mathbf{R}),$$

where $V$ is the volume of the box in which the motion of the nucleus is quantized and, as before, $\mathbf{R}$ is its position vector. (1.23) now becomes with the potential (1.30),

$$\frac{\mathrm{d}^2\sigma}{\mathrm{d}\Omega\,\mathrm{d}E'} = |b|^2 \frac{k'}{k} \sum_{\mathbf{q}',\mathbf{q}} p_{\mathbf{q}} \left| \frac{1}{V} \int \mathrm{d}\mathbf{R} \exp\{\mathrm{i}\mathbf{R}\cdot(\boldsymbol{\kappa}+\mathbf{q}-\mathbf{q}')\} \right|^2 \times \delta\left\{\hbar\omega + \frac{\hbar^2}{2M}(q^2-q'^2)\right\}, \tag{1.83}$$

where $\boldsymbol{\kappa} = \mathbf{k} - \mathbf{k}'$ and $M$ is the mass of the nucleus.

The integral in (1.83) is zero unless the wave vectors satisfy

$$\boldsymbol{\kappa} + \mathbf{q} - \mathbf{q}' = 0.$$

Given that the wave vectors have a density $V/(2\pi)^3$, we find

$$\left| \frac{1}{V} \int \mathrm{d}\mathbf{R} \exp\{\mathrm{i}\mathbf{R}\cdot(\boldsymbol{\kappa}+\mathbf{q}-\mathbf{q}')\} \right|^2 = \delta_{\boldsymbol{\kappa},\mathbf{q}'-\mathbf{q}} \tag{1.84}$$

where the Kronecker delta function is unity for $\boldsymbol{\kappa} = \mathbf{q}' - \mathbf{q}$ and zero otherwise. Suppose that the target nucleus is at rest initially, so that

$$p_{\mathbf{q}} = \delta_{\mathbf{q},0}; \tag{1.85}$$

then the cross-section (1.83) reduces to

$$\frac{\mathrm{d}^2\sigma}{\mathrm{d}\Omega\,\mathrm{d}E'} = \overline{|b|^2} \frac{k'}{k} \delta\left(\hbar\omega - \frac{\hbar^2\kappa^2}{2M}\right) = \overline{|b|^2}\left(\frac{\xi}{k^2}\right)\frac{2m}{\hbar^2}\,\delta\{(1-\gamma) - \xi^2(1+\gamma) + 2\gamma\mu\xi\}. \tag{1.86}$$

Here we have used (1.20) and the dimensionless variables $\xi = k'/k$, and $\gamma = m/M$, and $\mu = \cos\theta$ where $\theta$ is the angle between $\mathbf{k}$ and $\mathbf{k}'$. The partial cross-section is obtained from (1.86) by integrating over $\mathrm{d}E' = \hbar^2 k'\,\mathrm{d}k'/m = \hbar^2 k^2 \xi\,\mathrm{d}\xi/m$. Since,

$$\delta\{f(\xi)\} = \delta(\xi - \xi_0)/|f'(\xi_0)|, \tag{1.87}$$

where $f(\xi_0) = 0$, and $f'(\xi)$ is the derivative of $f(\xi)$, we find

$$\frac{\mathrm{d}\sigma}{\mathrm{d}\Omega} = \frac{2\overline{|b|^2}}{(1+\gamma)}\{\xi_0^2/|f'(\xi_0)|\}$$

where

$$f(\xi) = \xi^2 - \frac{2\gamma\mu}{(1+\gamma)}\xi - \left(\frac{1-\gamma}{1+\gamma}\right)$$

$$(1+\gamma)\xi_0 = \gamma\mu + \{1-\gamma^2+(\gamma\mu)^2\}^{1/2}$$

and

$$(1+\gamma)f'(\xi_0) = 2\{1-\gamma^2+(\gamma\mu)^2\}^{1/2}.$$

One of the two solutions of $f(\xi_0)=0$ is discarded because it does not satisfy $\xi_0>0$.

From these results we obtain the desired expression for the partial cross-section for scattering by a single free nucleus

$$\frac{d\sigma}{d\Omega}=\frac{\overline{|b|^2}}{(1+\gamma)^2}\left\{2\gamma\mu+\frac{(1-\gamma^2+2(\gamma\mu)^2)}{(1-\gamma^2+(\gamma\mu)^2)^{1/2}}\right\}. \tag{1.88}$$

The difference between the cross-sections for forward- and backward-scattering is $4\gamma\overline{|b|^2}/(1+\gamma)^2$ and the difference vanishes for a fixed nucleus since, in this instance, $\gamma=m/M\to 0$. In fact, the scattering for arbitrary $\gamma$ is weighted toward the forward hemisphere ($\theta>\pi/2$) as we shall see shortly.

The total cross-section is obtained from (1.88) by using the relation

$$\sigma=2\pi\int_{-1}^{1}d\mu\left(\frac{d\sigma}{d\Omega}\right)$$

and the result, for a free nucleus, is

$$\sigma=4\pi\,\overline{|b|^2}/(1+\gamma)^2. \tag{1.89}$$

Two comments are appropriate here. First, use of the static approximation for scattering from a free particle yields the result $4\pi\,\overline{|b|^2}$. The error produced by the approximation is therefore significant for light mass particles, but then the use of the static approximation for a free particle clearly violates the conditions for its validity established in § 1.8. However, note that the static approximation yields an upper bound for the cross-section. The second comment concerns coherent scattering from a crystal; this is called Bragg scattering and it is a main topic of Chapter 2. In this instance the appropriate mass for the scatterer is the mass of the crystal, so the use of the bound scattering length is always correct. An alternative viewpoint is that recoil of the nuclei destroys the coherence in the scattered neutron beam, and hence purely coherent scattering can depend only on the bound scattering amplitude.

To learn more about the scattering from a free nucleus we return to (1.88) and calculate the angular distribution of the scattered neutrons. The average value of $\mu=\cos\theta$, $\bar{\mu}$, is

$$\bar{\mu}=\frac{2\pi}{\sigma}\int_{-1}^{1}d\mu\,\mu\left(\frac{d\sigma}{d\Omega}\right)=\tfrac{2}{3}\gamma. \tag{1.90}$$

Because $\bar{\mu}\geqslant 0$, the angular distribution is weighted toward the forward hemisphere, except for a fixed nucleus ($\gamma=0$) when the scattering is isotropic. The value $\gamma=1$, achieved to a very good approximation by a proton target, is a special case since there is no scattering in the backward hemisphere. Setting $\gamma = 1$ in (1.88), we find

$$\frac{d\sigma}{d\Omega} = \tfrac{1}{2}\,\overline{|b|^2}\,(\mu + |\mu|) = \overline{|b|^2}\cos\theta; \qquad \theta \leqslant \pi/2$$

$$= 0; \qquad \theta > \pi/2. \tag{1.91}$$

This result reflects the fact that, for $\theta > \pi/2$, all the energy is transferred to the target nucleus and the neutron is brought to rest.

## REFERENCES


Bacon, G. E. (1977). *Neutron scattering in chemistry*. Butterworths, London.

Foderaro, A. (1971). *The elements of neutron interaction theory*. MIT Press, Cambridge, Massachusetts.

Ivin, K. J. (1976). *Structural studies of macromolecules by spectroscopic methods*. John Wiley; New York.

Jacrot, B. (1976). *Rep. Prog. Phys.* **39,** 911.

Koester, L. (1977). *Springer tracts in modern physics*, Vol. 80. Springer-Verlag, Berlin.

——, Rauch, H., Herkens, M., and Schröder, K. (1981). Summary of neutron scattering lengths. KFA-Report, Jül-1755. Kernforschungsanlage, Jülich, GmbH.

Kostorz, G. (1983). Physical Metallury Ch. 12 R. W. Cahn and P. Haasen (eds) Elsevier Science Publishers, New York.

Mughabghab, S. F., Divadeenam, M., and Holden, N. E. (1981). *Neutron cross-Sections*, Vol. 1. Academic Press, New York.

——, and Garber, D. E. (1973). BNL-325 (3rd edn), Vol. 1. Brookhaven National Laboratory, Long Island, NY.

Newton, R. G. (1982). *Scattering theory of waves and particles*. Texts and Monographs in Physics, Springer-Verlag, Berlin.

Nowak, E. (1982). *Z. Phys.* **B45,** 265.

OECD/NEA (1983). *Neutron sources*, an OECD/NEA Report. Pergamon Press, Oxford.

# 2
# ELASTIC NUCLEAR SCATTERING

A main topic of this chapter is a detailed discussion of the coherent cross-section given by eqn (1.41) in the case of a rigid perfect crystal. First, however, we digress from this task to consider a general description of crystal lattices and their associated reciprocal lattices. This description will serve to define a notation that we use throughout this book. A second topic is the diffuse scattering from a real crystal.

## 2.1. Crystal lattices and reciprocal lattices

A perfect crystal lattice may be constructed by repeating a unit cell periodically in space. The unit cell is defined in terms of three non-coplanar *basic vectors* $\mathbf{a}_1$, $\mathbf{a}_2$, and $\mathbf{a}_3$ and has a volume $v_0 = \mathbf{a}_1 \cdot (\mathbf{a}_2 \times \mathbf{a}_3)$. In general the unit cell will contain more than one lattice site. The vector leading from a point in one unit cell to the corresponding point in another cell is called a *lattice vector*. We denote the lattice vectors by $\mathbf{l}$. By definition

$$\mathbf{l} = l_1\mathbf{a}_1 + l_2\mathbf{a}_2 + l_3\mathbf{a}_3, \tag{2.1}$$

where the integers $l_1$, $l_2$, $l_3$, known as cell indices and frequently denoted for brevity by the single letter $l$, take all values.

If the unit cell contains just one lattice site, so that the lattice vectors $\mathbf{l}$ give every lattice site in the crystal ($\mathbf{R}_l = \mathbf{l}$), then we have a Bravais lattice. We note that in a Bravais lattice each lattice point is a centre of inversion symmetry, as is evident from its definition in terms of the lattice vectors $\mathbf{l}$. There are in total 14 Bravais lattices, (Landau and Lifshitz 1980) but we shall be concerned in the main with the three cubic and the hexagonal Bravais lattices. One simple choice of the basic vectors of these four Bravias lattices is given in Table 2.1 in Cartesian components in terms of the lattice constant $a$. The basic vectors for the body-centred and face-centred cubic and the hexagonal lattices are illustrated in Fig. 2.1.

In the fourth example in Table 2.1, the hexagonal Bravais lattice, the vectors $\mathbf{a}_1$ and $\mathbf{a}_2$ are seen to generate a plane of hexagonal network, and $\mathbf{a}_3$ displaces sites in a direction perpendicular to this plane, with a separation $\gamma a$ between neighbouring planes.

In general, crystal lattices have more than one atom per unit cell and are constructed from several ($r$) interpenetrating *identical* Bravais lattices (i.e. Bravais lattices with identical basic vectors), each lattice not necessarily containing atoms of the same type.

**Table 2.1**
*Basic vectors of four Bravais lattices*

| | $\mathbf{a}_1$ | $\mathbf{a}_2$ | $\mathbf{a}_3$ | $v_0$ |
|---|---|---|---|---|
| Simple cubic (s.c.) | $a(1, 0, 0)$ | $a(0, 1, 0)$ | $a(0, 0, 1)$ | $a^3$ |
| Body-centred cubic (b.c.c.) | $\frac{a}{2}(-1, 1, 1)$ | $\frac{a}{2}(1, -1, 1)$ | $\frac{a}{2}(1, 1, -1)$ | $\frac{1}{2}a^3$ |
| Faced-centred cubic (f.c.c.) | $\frac{a}{2}(0, 1, 1)$ | $\frac{a}{2}(1, 0, 1)$ | $\frac{a}{2}(1, 1, 0)$ | $\frac{1}{4}a^3$ |
| Hexagonal | $a(1, 0, 0)$ | $a(\frac{1}{2}, \frac{1}{2}\sqrt{3}, 0)$ | $a(0, 0, \gamma)$ | $\gamma\frac{\sqrt{3}}{2}a^3$ |

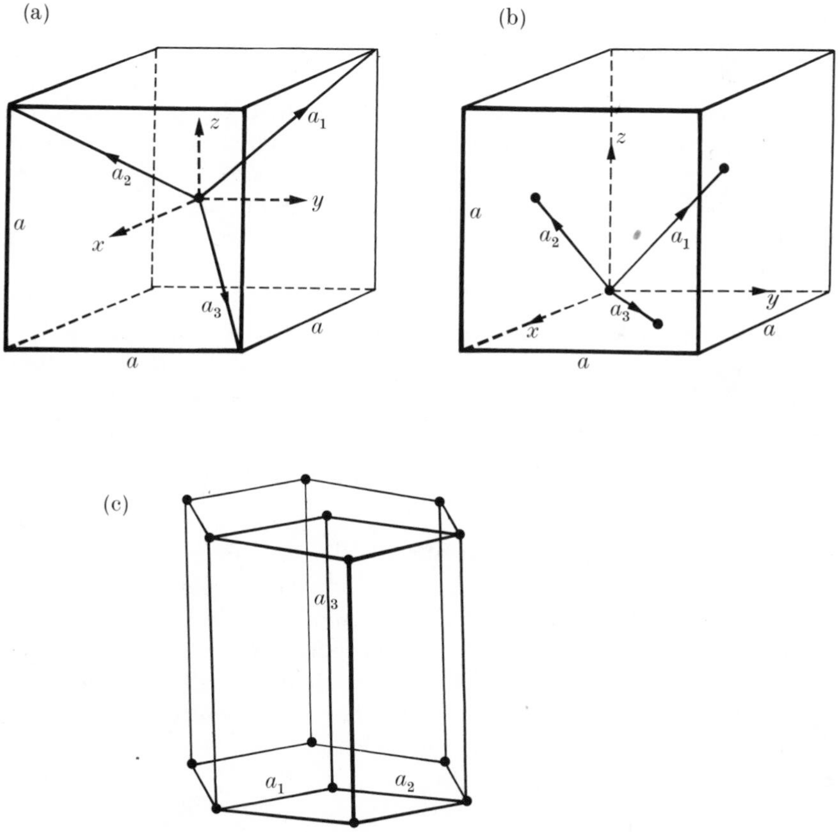


FIG. 2.1. The basic vectors of three Bravais lattices: (a) body-centred cubic; (b) face-centred cubic; and (c) hexagonal. In (a) and (b), $a$ is the lattice constant.

The positions of the $r$ atoms within a unit cell are denoted by the vectors $\mathbf{d}$, and these can be expressed in terms of the basic vectors $\mathbf{a}_1$, $\mathbf{a}_2$, and $\mathbf{a}_3$ that define the unit cell. The $r-1$ non-null vectors $\mathbf{d}$ are

$$\mathbf{d} = d_1\mathbf{a}_1 + d_2\mathbf{a}_2 + d_3\mathbf{a}_3, \tag{2.2}$$

with $0 \leqslant d_i \leqslant 1$, $i = 1, 2, 3$, the site $\mathbf{d} = 0$ coinciding with the corner of the unit cell. The position vector $\mathbf{R}_{ld}$ of an atom in the crystal is now given by

$$\mathbf{R}_{ld} = \mathbf{l} + \mathbf{d}. \tag{2.3}$$

As a first example of a lattice with more than one atom per unit cell, consider a body-centred crystal with different atoms at the two sets of sites

$$a(n_1, n_2, n_3)$$

and

$$a\{(n_1+\tfrac{1}{2}), (n_2+\tfrac{1}{2}), (n_3+\tfrac{1}{2})\},$$

where the integers $n_1$, $n_2$, and $n_3$ take all possible values. Each set generates a simple cubic lattice, so that this crystal structure consists of two interpenetrating simple cubic lattices with a lattice constant $a$, the unit cell being identical with the unit cell of one of the single cubic lattices and the $r = 2$ atoms within the cell located at the sites

$$\mathbf{d} = (0, 0, 0) \quad \text{and} \quad \mathbf{d} = a(\tfrac{1}{2}, \tfrac{1}{2}, \tfrac{1}{2}).$$

Another example of a crystal with two atoms per unit cell is the hexagonal close-packed lattice (h.c.p.). This type of lattice is constructed from a hexagonal Bravais lattice, with basic vectors $\mathbf{a}_1$, $\mathbf{a}_2$, $\mathbf{a}_3$, by adding to each lattice site another at a distance

$$\boldsymbol{\rho} = \tfrac{1}{3}\mathbf{a}_1 + \tfrac{1}{3}\mathbf{a}_2 + \tfrac{1}{2}\mathbf{a}_3.$$

Thus the lattice consists of two identical hexagonal Bravais lattices, the six nearest-neighbour sites on the other lattice to a given site being at a distance $|\boldsymbol{\rho}| = a(\tfrac{1}{3} + \tfrac{1}{4}\gamma^2)^{1/2}$. There are also six nearest-neighbour sites on the same lattice. These are at a distance $a$ and lie in the basal plane through the atom (perpendicular to the hexagonal axis). The close-packed structure is formed when all these 12 neighbouring sites, six on the second lattice and six on the same lattice, are at the same distance away, i.e. $|\boldsymbol{\rho}| = a$, which requires $\gamma = (8/3)^{1/2}$. This structure is illustrated in Fig. 2.2.

It is to be noted that the vector $-\boldsymbol{\rho}$ does not coincide with the position of a lattice site in the h.c.p. crystal structure, i.e. there is not inversion symmetry between neighbouring sites on different sublattices though, of course, there is within each sublattice because they are

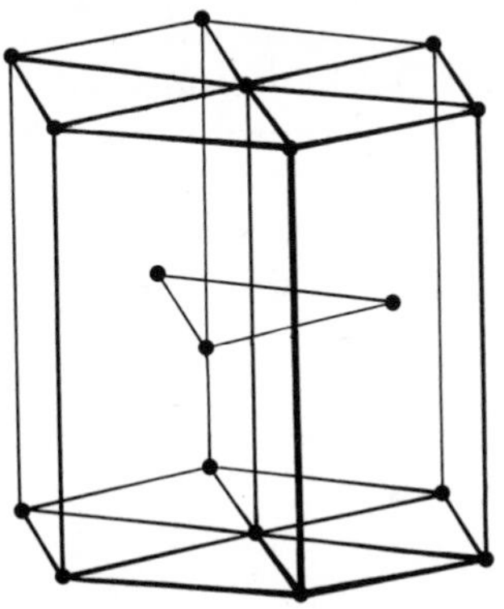

FIG. 2.2. The ideal hexagonal close-packed lattice, i.e. with $\gamma = (8/3)^{1/2}$.

Bravais. This simply means that the h.c.p. structure is not a Bravais lattice, in contrast to our first example of a crystal lattice with two atoms per unit cell, in which a b.c.c. lattice was generated from two s.c. sublattices. It is worthwhile to note in this connection that the f.c.c. lattice can be generated from four s.c. sublattices, and a s.c. lattice from two f.c.c. lattices. The most general lattice point for a f.c.c. lattice can be expressed as

$$\tfrac{1}{2}a(n_1, n_2, n_3),$$

with $n_1+n_2+n_3$ an even integer. Thus the two sets of vectors,

$$\mathbf{l} = \tfrac{1}{2}a(n_1, n_2, n_3), \quad n_1+n_2+n_3 \text{ even},$$

and

$$\mathbf{l} = \tfrac{1}{2}a(n_1, n_2, n_3), \quad n_1+n_2+n_3 \text{ odd},$$

together generate a simple cubic lattice of side $\tfrac{1}{2}a$.

For every crystal lattice defined by vectors $\mathbf{l}$ as in (2.1), we can also define a reciprocal lattice with vectors $\boldsymbol{\tau}$ such that

$$\exp(\mathrm{i}\boldsymbol{\tau}\cdot\mathbf{l}) = 1 \quad \text{for all } \mathbf{l}. \tag{2.4}$$

If we write, by analogy with (2.1),

$$\boldsymbol{\tau} = t_1\boldsymbol{\tau}_1 + t_2\boldsymbol{\tau}_2 + t_3\boldsymbol{\tau}_3, \tag{2.5}$$

where the integers $t_1$, $t_2$, and $t_3$ take all possible values, the basic vectors of the reciprocal lattice are given by

$$\boldsymbol{\tau}_1 = \frac{2\pi}{v_0}\mathbf{a}_2\times\mathbf{a}_3, \qquad \boldsymbol{\tau}_2 = \frac{2\pi}{v_0}\mathbf{a}_3\times\mathbf{a}_1, \tag{2.6}$$

**Table 2.2**
*Reciprocal lattices*

| | $\boldsymbol{\tau}_1$ | $\boldsymbol{\tau}_2$ | $\boldsymbol{\tau}_3$ |
|---|---|---|---|
| s.c. | $\frac{2\pi}{a}(1, 0, 0)$ | $\frac{2\pi}{a}(0, 1, 0)$ | $\frac{2\pi}{a}(0, 0, 1)$ |
| b.c.c. | $\frac{2\pi}{a}(0, 1, 1)$ | $\frac{2\pi}{a}(1, 0, 1)$ | $\frac{2\pi}{a}(1, 1, 0)$ |
| f.c.c. | $\frac{2\pi}{a}(-1, 1, 1)$ | $\frac{2\pi}{a}(1, -1, 1)$ | $\frac{2\pi}{a}(1, 1, -1)$ |
| hexagonal | $\frac{2\pi}{a}\left(1, -\frac{1}{\sqrt{3}}, 0\right)$ | $\frac{2\pi}{a}\left(0, \frac{2}{\sqrt{3}}, 0\right)$ | $\frac{2\pi}{a}\left(0, 0, \frac{1}{\gamma}\right)$ |

and

$$\boldsymbol{\tau}_3 = \frac{2\pi}{v_0}\,\mathbf{a}_1 \times \mathbf{a}_2,$$

as is easily verified. The cartesian components of $\boldsymbol{\tau}_1$, $\boldsymbol{\tau}_2$, and $\boldsymbol{\tau}_3$ for tne s.c., b.c.c., f.c.c., and hexagonal lattices are given in Table 2.2.

On comparing the results given in Table 2.1 with those given in Table 2.2 we note that the reciprocal lattice of a s.c. crystal corresponds to a s.c. lattice, while the reciprocal lattice of a b.c.c. crystal corresponds to a f.c.c. lattice and vice versa.

From (2.6) we notice that the volume of the unit cell of the reciprocal lattice

$$\boldsymbol{\tau}_1 \cdot (\boldsymbol{\tau}_2 \times \boldsymbol{\tau}_3) = (2\pi)^3/v_0. \tag{2.7}$$

## 2.2. Coherent cross-sections

We have now defined the notation we need immediately and therefore return to an evaluation of (1.41). We suppose first we have a rigid Bravais lattice of $N$ atoms so that (1.41) becomes simply

$$\left(\frac{\mathrm{d}\sigma}{\mathrm{d}\Omega}\right)_{\text{coh}} = |\bar{b}|^2 \left|\sum_l \exp(\mathrm{i}\boldsymbol{\kappa}\cdot\mathbf{l})\right|^2. \tag{2.8}$$

Obviously when the scattering vector $\boldsymbol{\kappa}$ is equal to zero, or indeed to any reciprocal lattice vector, the right-hand side is large because all the terms in the sum over $\mathbf{l}$ add up in phase. However, as $\boldsymbol{\kappa}$ moves away from a reciprocal lattice vector, the terms rapidly come out of phase and the right-hand side drops to a negligible value. For a large crystal we can show that

$$\left|\sum_l \exp(\mathrm{i}\boldsymbol{\kappa}\cdot\mathbf{l})\right|^2 = N\frac{(2\pi)^3}{v_0}\sum_{\tau}\delta(\boldsymbol{\kappa}-\boldsymbol{\tau}) \tag{2.9}$$

and (2.8) becomes

$$\left(\frac{d\sigma}{d\Omega}\right)_{coh} = \frac{N(2\pi)^3}{v_0} |\bar{b}|^2 \sum_{\tau} \delta(\boldsymbol{\kappa} - \boldsymbol{\tau}). \tag{2.10}$$

The generalization of this formula to the case with more than one atom per unit cell is immediately clear as

$$\left(\frac{d\sigma}{d\Omega}\right)_{coh} = N\frac{(2\pi)^3}{v_0} \sum_{\tau} \delta(\boldsymbol{\kappa} - \boldsymbol{\tau})\, |F_N(\boldsymbol{\tau})|^2, \tag{2.11}$$

where the nuclear *unit-cell structure factor* $F_N(\boldsymbol{\tau})$ is defined as

$$F_N(\boldsymbol{\tau}) = \sum_{d} \exp(i\boldsymbol{\tau} \cdot \mathbf{d})\bar{b}_d. \tag{2.12}$$

We emphasize that coherent scattering determines the bound scattering length. The generalization of (1.42) is simply

$$\left(\frac{d\sigma}{d\Omega}\right)_{incoh} = N \sum_{d} \{\overline{|b_d|^2} - |\bar{b}_d|^2\}. \tag{2.13}$$

For a real crystal the Debye–Waller factors must be included in (2.12) and (2.13) (cf. Chapter 4).

## 2.3. Crystal symmetry effects

The form of the results (2.10) and (2.11) is noteworthy because they demonstrate that no coherent elastic scattering occurs from a perfect crystal unless $\boldsymbol{\kappa}$ coincides with a reciprocal lattice vector. This result is a direct consequence of the symmetry of the crystal lattice and does not, for example, depend upon the nature of the neutron—lattice interaction; we later derive similar formulae for the magnetic scattering of neutrons. To illustrate this point, it is convenient in this section to consider in a very general way the scattering from a periodic structure. We shall then verify that the general formulae agree with those of § 2.2 for the special case of coherent nuclear scattering.

The interaction potential between the neutron and the crystal, $\hat{V}(\mathbf{r})$, does not, in general, have the periodicity of the lattice, because the potential will depend upon the nuclear spin orientation, electron spin orientation, particular isotope, etc., in the neighbourhood of $\mathbf{r}$. Nevertheless, the major property of a solid is that it has, on average, a periodic structure. We therefore define the mean potential seen by the neutron as

$$Q(\mathbf{r}) = \overline{\sum_{\lambda} p_\lambda \langle\lambda|\, \hat{V}(\mathbf{r})\, |\lambda\rangle}, \tag{2.14}$$

where $Q(\mathbf{r})$ is now a c-number so far as the target is concerned. It could itself be an operator in so far as $Q(\mathbf{r})$ could depend on the neutron spin, but we set this aside for ease of notation. The deviation in the potential is then $\delta\hat{V}(\mathbf{r})$, where

$$\hat{V}(\mathbf{r}) = Q(\mathbf{r}) + \delta\hat{V}(\mathbf{r}). \tag{2.15}$$

Note that $Q(\mathbf{r})$ can, in general, be temperature dependent and that

$$\overline{\sum_{\lambda} p_{\lambda} \langle\lambda|\, \delta\hat{V}(\mathbf{r})\, |\lambda\rangle = 0} \tag{2.16}$$

by definition. Now, in the expression (1.23) we need to calculate

$$\left(\frac{m}{2\pi\hbar^2}\right)^2 \sum_{\lambda\lambda'} p_{\lambda}\, |\langle\lambda'|\, \hat{V}(\boldsymbol{\kappa})\, |\lambda\rangle|^2\, \delta(\hbar\omega + E_{\lambda} - E_{\lambda'}), \tag{2.17}$$

where, for convenience in the subsequent development, we have introduced the notation

$$\hat{V}(\boldsymbol{\kappa}) = \frac{2\pi\hbar^2}{m} \langle\mathbf{k}'|\, \hat{V}\, |\mathbf{k}\rangle = \int d\mathbf{r}\, \exp(-i\mathbf{k}' \cdot \mathbf{r})\, \hat{V}(\mathbf{r}) \exp(i\mathbf{k} \cdot \mathbf{r}). \tag{2.18}$$

Because $Q(\mathbf{r})$ is a c-number,

$$\langle\lambda|\, Q(\mathbf{r})\, |\lambda'\rangle = \delta_{\lambda\lambda'} Q(\mathbf{r}),$$

and substituting (2.15) into (2.17) gives

$$\left(\frac{m}{2\pi\hbar^2}\right)^2 \left\{|Q(\boldsymbol{\kappa})|^2\, \delta(\hbar\omega) + \sum_{\lambda\lambda'} p_{\lambda}\, |\langle\lambda'|\, \delta\hat{V}(\boldsymbol{\kappa})\, |\lambda\rangle|^2\, \delta(\hbar\omega + E_{\lambda} - E_{\lambda'})\right\}. \tag{2.19}$$

The first term of (2.19) gives rise to Bragg scattering. The remainder of the expression gives either inelastic scattering or elastic diffuse or incoherent scattering. For the moment we concentrate on Bragg scattering. Later in this chapter we discuss elastic diffuse scattering, and many examples of inelastic scattering are discussed in subsequent chapters.

The mean potential $Q(\mathbf{r})$ seen by the neutrons must have the full periodicity of the lattice. In consequence, it can be expanded in a Fourier series in the reciprocal lattice vectors $\boldsymbol{\tau}$. We shall take

$$Q(\mathbf{r}) = (2\pi\hbar^2/mv_0) \sum_{\boldsymbol{\tau}} \exp(-i\boldsymbol{\tau} \cdot \mathbf{r}) F(\boldsymbol{\tau}), \tag{2.20}$$

and the choice of constants means that the Fourier components $F(\boldsymbol{\tau})$, have the dimension of length. From (2.20) it follows that $F(\boldsymbol{\tau})$ is proportional to the Fourier transform of $Q(\mathbf{r})$ over a unit cell,

$$F(\boldsymbol{\tau}) = (m/2\pi\hbar^2) \int_{\text{cell}} d\mathbf{r}\, \exp(i\boldsymbol{\tau} \cdot \mathbf{r}) Q(\mathbf{r}). \tag{2.21}$$

This expression is a general definition of our unit-cell structure factor.

To obtain $|Q(\boldsymbol{\kappa})|^2$ in (2.19) in terms of $F(\boldsymbol{\tau})$ we first note that, from (2.20),

$$\begin{aligned} Q(\boldsymbol{\kappa}) &= \int \mathrm{d}\mathbf{r}\, \exp(\mathrm{i}\boldsymbol{\kappa}\cdot\mathbf{r})Q(\mathbf{r}) \\ &= (2\pi\hbar^2/mv_0)\sum_{\boldsymbol{\tau}} \int \mathrm{d}\mathbf{r}\, \exp\{\mathrm{i}\mathbf{r}\cdot(\boldsymbol{\kappa}-\boldsymbol{\tau})\}F(\boldsymbol{\tau}) \\ &= (2\pi\hbar^2/mv_0)(2\pi)^3\sum_{\boldsymbol{\tau}} \delta(\boldsymbol{\kappa}-\boldsymbol{\tau})F(\boldsymbol{\tau}). \end{aligned} \tag{2.22}$$

However, to obtain $|Q(\boldsymbol{\kappa})|^2$ we cannot simply square (2.22) because this would give the square of a Dirac delta-function. Instead we notice that $|Q(\boldsymbol{\kappa})|^2$ must have the form

$$|(m/2\pi\hbar^2)Q(\boldsymbol{\kappa})|^2 = \sum_{\boldsymbol{\tau}} x_{\boldsymbol{\tau}}\, \delta(\boldsymbol{\kappa}-\boldsymbol{\tau}), \tag{2.23}$$

and to determine $x_{\boldsymbol{\tau}}$ we multiply by $\exp(\mathrm{i}\boldsymbol{\kappa}\cdot\mathbf{y})$ and integrate over all $\boldsymbol{\kappa}$ to give

$$\begin{aligned} \sum_{\tau} x_{\tau} \exp(\mathrm{i}\boldsymbol{\tau}\cdot\mathbf{y}) &= \int \mathrm{d}\boldsymbol{\kappa} \int \mathrm{d}\mathbf{r} \int \mathrm{d}\mathbf{r}'\, \exp\{\mathrm{i}\boldsymbol{\kappa}\cdot(\mathbf{r}-\mathbf{r}'+\mathbf{y})\} \\ &\quad \times \frac{N^2}{V^2}\sum_{\tau,\tau'} \exp(-\mathrm{i}\boldsymbol{\tau}\cdot\mathbf{r}+\mathrm{i}\boldsymbol{\tau}'\cdot\mathbf{r}')F(\boldsymbol{\tau})F(\boldsymbol{\tau}')^* \\ &= N^2\frac{(2\pi)^3}{V^2}\int \mathrm{d}\mathbf{r} \sum_{\tau,\tau'} \exp\{\mathrm{i}(\boldsymbol{\tau}'-\boldsymbol{\tau})\cdot\mathbf{r}+\mathrm{i}\boldsymbol{\tau}'\cdot\mathbf{y}\}F(\boldsymbol{\tau})F(\boldsymbol{\tau}')^* \\ &= N^2\frac{(2\pi)^3}{V}\sum_{\tau,\tau'} \delta_{\boldsymbol{\tau}',\boldsymbol{\tau}} \exp(\mathrm{i}\boldsymbol{\tau}'\cdot\mathbf{y})F(\boldsymbol{\tau})F(\boldsymbol{\tau}')^* \\ &= N^2\frac{(2\pi)^3}{V}\sum_{\tau} \exp(\mathrm{i}\boldsymbol{\tau}\cdot\mathbf{y})\, |F(\boldsymbol{\tau})|^2. \end{aligned}$$

Hence

$$x_{\boldsymbol{\tau}} = N^2\frac{(2\pi)^3}{V}|F(\boldsymbol{\tau})|^2, \tag{2.24}$$

and we obtain

$$|(m/2\pi\hbar^2)Q(\boldsymbol{\kappa})|^2 = N\frac{(2\pi)^3}{v_0}\sum_{\boldsymbol{\tau}} \delta(\boldsymbol{\kappa}-\boldsymbol{\tau})\, |F(\boldsymbol{\tau})|^2. \tag{2.25}$$

This last formula shows that, irrespective of the scattering potential, the coherent elastic cross-section must contain the characteristic Bragg peaks at the reciprocal lattice positions.

Returning for the moment to (2.19), it is worthwhile to give a brief discussion of the second term, which involves the deviations $\delta\hat{V}$ from the average scattering potential. The deviations are of two types; static variations from the overall crystal periodicity, and dynamic fluctuations due to the movement of atoms or electrons. We shall find that the static variations give rise to scattering that is strictly elastic, i.e. the partial differential cross-section contains $\delta(\hbar\omega)$, whereas the dynamic fluctuations give rise to inelastic scattering. One good example of the effect of static variations is given by the isotope disorder scattering we have considered earlier. Another example, which is discussed in § 2.5, is the scattering from a distorted alloy where the interaction potential varies from one type of atom to the next. The scattering from crystal vibrations (Chapter 4) is an example of the effects of dynamic fluctuations. Sometimes the distinction between static variations and dynamic fluctuations is, to some extent, arbitrary. An example of this is the nuclear spin orientation effects we discussed in Chapter 1. Clearly the orientation of a nuclear spin the target must change with time, and therefore, formally, the spin-dependent scattering potential can be classified as a dynamic fluctuation and the corresponding scattering as 'inelastic'. In a formal sense this is true, but the energies connected with different nuclear spin orientations are typically $10^{-4}$ K, and therefore the frequencies at which they change orientation are extremely low compared to those the neutron can usually observe. Therefore, for most practical purposes, the nuclear spin orientation effects can be treated as static and the corresponding scattering as strictly elastic.

Throughout this book we use the following nomenclature. The scattering due to random isotope distributions and random nuclear spin orientations we call *incoherent.* Scattering due to other random effects in the crystal we call *diffuse.* Both give rise to elastic and inelastic scattering; the elastic scattering comes from the static variations directly and the inelastic scattering from the dynamic fluctuations superimposed on the random static variations. Similarly we get coherent elastic scattering from the average potential $Q(\mathbf{r})$ and coherent inelastic scattering from the dynamic fluctuations not associated with some intrinsic inhomogeneity in the sample.

We now examine the form of $\hat{V}(\mathbf{r})$ in more detail. In the simple case examined in Chapter 1, $\hat{V}(\mathbf{r})$ was the sum of interaction potentials with each atom, see (1.34) for example. In general this simple superposition of potentials is a good, but not rigorous, approximation. (It is rigorous to express $\hat{V}(\mathbf{r})$ as a superposition of potentials from each nucleus and each electron; the problem is that, when we assemble this into a sum over atoms, we cannot rigorously identify any part of the electron density with a specific atom, without introducing a model of the target sample.) But, whether or not the superposition of potentials is a good enough approximation, we know by definition that the average potential is periodic.

When it is satisfactory to regard $\hat{V}(\mathbf{r})$ as a superposition of potentials from each atom, we put

$$\hat{V}(\mathbf{r}) = \sum_{l,d} \hat{v}_{l,d}(\mathbf{r}-\mathbf{l}-\mathbf{d}) \tag{2.26}$$

and

$$Q(\mathbf{r}) = \sum_{l,d} v_d(\mathbf{r}-\mathbf{l}-\mathbf{d}), \tag{2.27}$$

where $v_d$ is now the average potential given by atoms of type $d$. Then the unit-cell structure factor is written

$$F(\boldsymbol{\tau}) = \left(\frac{m}{2\pi\hbar^2}\right) \sum_d \int_{\text{cell}} \mathrm{d}\mathbf{r} \exp(\mathrm{i}\boldsymbol{\tau}\cdot\mathbf{r}) v_d(\mathbf{r}). \tag{2.28}$$

For nuclear scattering it is rigorous to express $\hat{V}(\mathbf{r})$ as a superposition of potentials, and $v_d(\mathbf{r})$ is simply $(2\pi\hbar^2/m)\bar{b}_d\,\delta(\mathbf{r}-\mathbf{d})$. The unit-cell structure factor is then

$$F_{\mathrm{N}}(\boldsymbol{\tau}) = \sum_d \exp(\mathrm{i}\boldsymbol{\tau}\cdot\mathbf{d})\bar{b}_d, \tag{2.29}$$

and this reproduces the results of (2.11) and (2.12).

## 2.4. Bragg scattering (Dachs 1978; Brown 1979)

From (2.11) we see that there is no *Bragg* scattering unless the condition

$$\boldsymbol{\kappa} = \mathbf{k} - \mathbf{k}' = \boldsymbol{\tau} \tag{2.30}$$

is satisfied. We can discuss this condition with the aid of Fig. 2.3. We first mark the reciprocal lattice points, and then, to represent the incident neutron, we draw a vector $\mathbf{k}$ from some point $A$ chosen so that $\mathbf{k}$ ends at the origin $O$. Now because the scattering is elastic the final wave vector $\mathbf{k}'$ must have the same magnitude as $\mathbf{k}$ and therefore if drawn from $A$ must end on the surface of a sphere drawn with centre $A$ and radius $|\mathbf{k}| = |\mathbf{k}'| = k$. But in general the surface of this sphere does not pass through any reciprocal lattice points other than the origin, so in general (2.30) is not satisfied except for $\boldsymbol{\tau} = 0$, which corresponds to no scattering. Therefore in general a single crystal gives no elastic coherent scattering and, in fact, if the incoherent scattering happens to be zero then it gives no elastic scattering at all.

But for certain special orientations of the crystal, i.e. of the reciprocal lattice, relative to the incident beam, or with fixed orientation for certain special values of the wave vector $\mathbf{k}$, the point $A$ is such that the

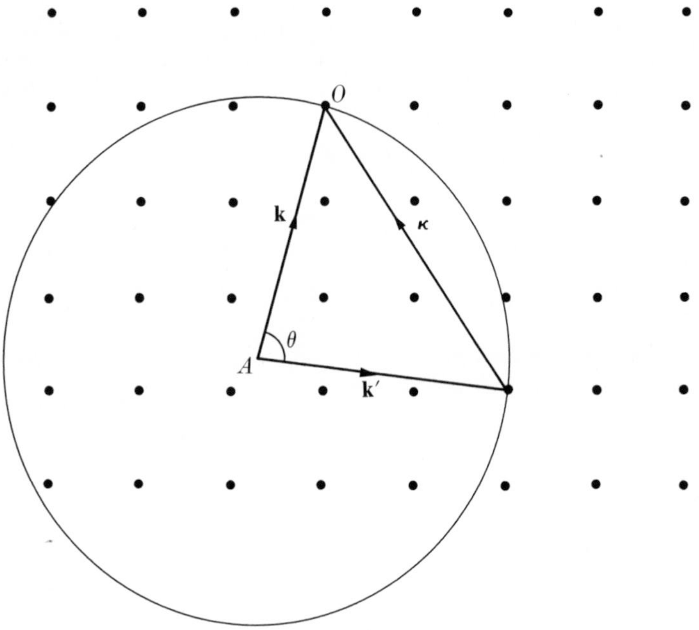


FIG. 2.3. Condition on the scattering vector $\boldsymbol{\kappa} = \mathbf{k} - \mathbf{k}'$ for Bragg scattering. $|\boldsymbol{\kappa}| = 2\,|\mathbf{k}|\sin\frac{1}{2}\theta$.

sphere drawn around it does pass through a reciprocal lattice point, so that (2.30) can be satisfied. Then we get scattering in a direction given by $\mathbf{k}'$. If $\theta$ is the angle between $\mathbf{k}$ and $\mathbf{k}'$,

$$|\boldsymbol{\tau}| = |\boldsymbol{\kappa}| = 2k\sin\tfrac{1}{2}\theta. \tag{2.31}$$

Hence scattering occurs only at angles $\theta$ given by ($|\boldsymbol{\tau}| = \tau$)

$$\sin\tfrac{1}{2}\theta = \frac{\tau}{2k}, \tag{2.32}$$

and then only if the crystal is correctly orientated. To show that (2.32) is the familiar Bragg law we remember that each vector $\boldsymbol{\tau}$ is perpendicular to a set of planes in the original lattice and has a magnitude that is an integral multiple of $2\pi$ times the inverse of the plane spacing. Thus

$$\tau = \frac{2\pi}{d}n \quad \text{and} \quad k = \frac{2\pi}{\lambda},$$

so

$$2d\sin\tfrac{1}{2}\theta = n\lambda, \tag{2.33}$$

which is the more familiar form of Bragg's law.

Notice that (2.32) cannot possibly be satisfied if

$$k < \tfrac{1}{2}\tau_{\min}, \tag{2.34}$$

where $\tau_{\min}$ is the smallest $\tau$ (other than zero). Hence if (2.34) is satisfied, i.e. if the neutron wavelength is long enough, there is no Bragg scattering, whatever the crystal orientation.

We obtain the total cross-section from (2.11) by integrating over $\mathrm{d}\Omega$, i.e. over $\mathrm{d}\tilde{\mathbf{k}}'$, where the tilde ~ denotes a unit vector. Hence

$$\sigma_{\mathrm{c}} = N\frac{(2\pi)^3}{v_0}\sum_{\boldsymbol{\tau}} |F_N(\boldsymbol{\tau})|^2 \int \mathrm{d}\tilde{\mathbf{k}}'\,\delta(\mathbf{k}-\mathbf{k}'-\boldsymbol{\tau}).$$

To perform this integral we write $\tilde{\mathbf{k}}' = \tilde{\mathbf{s}}$ and multiply by

$$(2/k')\int_0^\infty \mathrm{d}s\, s^2\,\delta(s^2-k'^2),$$

which is unity; we then rearrange the order of integration, make use of the vector $\mathbf{s} = s\tilde{\mathbf{s}}$, so $\mathrm{d}\mathbf{s} = s^2\,\mathrm{d}s\,\mathrm{d}\tilde{\mathbf{s}}$, and finally recall that $k' = k$ for elastic scattering. Thus

$$\begin{aligned}\int \mathrm{d}\tilde{\mathbf{k}}'\,\delta(\mathbf{k}-\mathbf{k}'-\boldsymbol{\tau}) &= (2/k')\int \mathrm{d}\tilde{\mathbf{s}}\,\delta(\mathbf{k}-k'\tilde{\mathbf{s}}-\boldsymbol{\tau})\int_0^\infty \mathrm{d}s\,s^2\,\delta(s^2-k'^2),\\ &= (2/k')\int \mathrm{d}\mathbf{s}\,\delta(\mathbf{k}-\mathbf{s}-\boldsymbol{\tau})\,\delta(s^2-k'^2)\\ &= (2/k')\delta\{(\mathbf{k}-\boldsymbol{\tau})^2-k'^2\} = (2/k)\,\delta(\tau^2-2k\tau\sin\tfrac{1}{2}\theta).\end{aligned}$$

Hence

$$\sigma_{\mathrm{c}} = N\frac{(2\pi)^3}{v_0}\frac{2}{k}\sum_{\boldsymbol{\tau}} |F_N(\boldsymbol{\tau})|^2\,\delta(\tau^2-2k\tau\sin\tfrac{1}{2}\theta). \tag{2.35}$$

The fact that the total cross-section involves a $\delta$-function tells us that there is no scattering unless very special conditions are satisfied.

There are three methods of observing Bragg scattering and we will discuss these in turn. Of course, they correspond exactly to the three methods of observing the Bragg scattering of X-rays.

*Method* 1. The Laue method utilizes a fixed single crystal and a 'white beam', i.e. an incident beam which contains all wavelengths. In Fig. 2.3 this corresponds to keeping the reciprocal lattice fixed and the orientation of $\mathbf{k}$ fixed but varying the length $k$. For some length $k$ the sphere drawn about $A$ will intersect a reciprocal lattice point and these neutrons are those suitable for scattering. If the incident flux is such that $N(\lambda)\,\mathrm{d}\lambda$ is the flux of neutrons between $\lambda$ and $\lambda+\mathrm{d}\lambda$, then the scattered intensity in one of the scattered peaks is

$$P_{\mathrm{s}} = \int \mathrm{d}\lambda\, N(\lambda)\sigma_{\mathrm{c}} = VQ^{\lambda}N(\lambda), \tag{2.36}$$

where

$$Q^\lambda = \frac{2\pi\,|F_N(\boldsymbol{\tau})|^2\,\lambda^3}{v_0^2\tau\sin\frac{1}{2}\theta} = \frac{\lambda^4\,|F_N(\boldsymbol{\tau})|^2}{2v_0^2\sin^2\frac{1}{2}\theta}. \tag{2.37}$$

This formula is obtained by inserting (2.35) into (2.36), writing $k = 2\pi/\lambda$, and integrating. Here $\lambda$ is the particular wavelength that is suitable for scattering and $V$ is, as uual, the volume of the crystal.

Notice that in this method every reciprocal lattice vector $\boldsymbol{\tau}$ gives rise to some scattering. Each one selects out from the incident beam those neutrons that have suitable wavelengths and scatters them out in a direction characteristic of that particular $\boldsymbol{\tau}$ vector. This Laue method is clearly a very convenient way of producing a monochromatic beam from reactor beams, because if we look at those neutrons scattered in a particular Bragg peak $\boldsymbol{\tau}$ they will all have the same wavelength. Unfortunately there is a complication. We notice that, if the crystal is scattering neutrons of wavelength $\lambda$ in a certain direction via the reciprocal lattice vector $\boldsymbol{\tau}$, then it is also scattering neutrons of wavelength $\frac{1}{2}\lambda$ in the same direction via the reciprocal lattice vector $2\boldsymbol{\tau}$, and those of $\frac{1}{3}\lambda$ via $3\boldsymbol{\tau}$, etc. The scattered beam is therefore contaminated by these higher-order reflections. From (2.36) and (2.37) the ratio of the second-order contaminant to the primary scattered beam is

$$\frac{P_s(\frac{1}{2}\lambda)}{P_s(\lambda)} = \frac{N(\frac{1}{2}\lambda)}{N(\lambda)}\cdot\frac{1}{2^4}\frac{|F_N(2\boldsymbol{\tau})|^2}{|F_N(\boldsymbol{\tau})|^2}.$$

The ratio of the structure factors is fixed by the structure of the crystal: for simple crystals the ratio is unity. Except when working at long wavelengths, the factor $2^{-4}$ usually ensures that the second-order contaminant is small compared to the primary beam, but sometimes (e.g. when examining antiferromagnets) it is particularly important to reduce this second-order contaminant to a very small quantity. A convenient way of doing this is to reduce the ratio $N(\frac{1}{2}\lambda)/N(\lambda)$ by choosing $\lambda$ suitably. We notice from (2.35) that by varying the orientation of the crystal relative to the incident beam we can change the wavelength that a vector $\boldsymbol{\tau}$ will scatter. Now if $N(\lambda)$ is a function roughly of the form shown in Fig. 2.4, one would first guess it is best to orientate the crystal so that neutrons with wavelengths corresponding to the peak, point $A$, are those that are scattered. But then we get a second-order contaminant with an intensity corresponding to the height of the curve at the point $A'$, i.e. $\frac{1}{2}\lambda$. It is preferable to reorient the crystal so as to scatter neutrons at wavelengths $B$, thus sacrificing some intensity, so that the second-order contaminant, represented by the height at $B'$, is smaller proportionally.

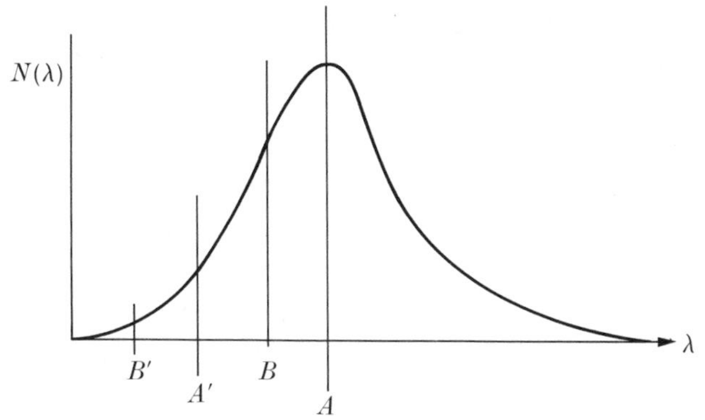


FIG. 2.4. Selecting Bragg conditions to reduce second-order contaminant.

If this way is not convenient to reduce the second-order contaminant it is possible to place a filter in the beam that absorbs neutrons of wavelength $\frac{1}{2}\lambda$ much more strongly than those of $\lambda$. Of course this also reduces the primary intensity, but the second-order contaminant is reduced even more so. Yet another idea is to use as monochromator a crystal such as germanium, which has a structure such that, working with the (1, 1, 1) reflection, the second-order contaminant is zero.

The analysis we have presented so far has presupposed a perfect crystal and a perfectly collimated incident beam. Of course neither condition is realized in practice, and it is of importance to inquire how monochromatic a beam is produced by the Laue method in practice. We will not have time to go into this question in great detail but we will give it a short discussion.

But first of all we notice that (2.35) and (2.36) are certainly wrong in one respect, they are both linear in the size of the crystal (i.e. in $N$ and $V = Nv_0$). This suggests that the scattered flux can be increased by increasing $V$ indefinitely. This of course is not true; in using the Born approximation we assumed the crystal was uniformly bathed in the incident beam. This is only a good approximation if the scattering is small, i.e. the formulae we have are the 'small crystal' limit. Just how small a crystal has to be to satisfy the 'small crystal' criterion varies considerably. For scattering from a weak reflection the critical volume may be quite large, so the same crystal may be 'large' for some reflections (the strong ones) and 'small' for others (the weak ones). The criterion is also very sensitive to the value of the mosaic spread in the crystal. All crystals, no matter how perfect they may seem, are made up of mosaic blocks of linear dimensions of the order of thousands of ångströms, each separated from one another by dislocations. Each small block may be regarded as a truly perfect crystal, but they are slightly displaced and slightly misorientated relative to one another so that the scattering from them is not coherent, i.e. there is no interference between the scattered waves coming from different mosaic blocks. The degree of misorientation $\eta$ is usually several minutes but may occasionally be as large as a degree.

An important consequence of the mosaic spread is that it is impossible to align the whole volume of the crystal in the same orientation. The crystal may nominally be aligned in a certain direction but in reality each mosaic block is slightly misaligned from this direction. This is an extremely fortunate circumstance, for if all the blocks were exactly aligned the first few layers of mosaic blocks would scatter all the 'suitable' neutrons, leaving none 'suitable' to be scattered by the more remote layers of the crystal. As it is, in a real crystal each block selects its own group of 'suitable neutrons' and there is only competition between them for the same neutrons if the crystal is very thick. When the crystal is 'thick'—and what this means depends very sensitively on the magnitude of the mosaic spread—we say 'secondary extinction' is present. ('Primary extinction' would occur in the less likely circumstance that the mosaic blocks themselves were so thick that the incident beam was appreciably attenuated in traversing each one.)

Because of this mosaic structure it is clear that even if the incident beam were perfectly collimated the scattered beam would not be so (it would have an angular divergence of the order of the mosaic spread) and neither would it be monochromatic, because each mosaic block selects a slightly different wavelength neutron to scatter and does so through slightly different angles. But of course in practice the incident beam is not perfectly collimated and this leads to a further angular spread and wavelength variation in the scattered beam. These effects can be discussed with the aid of Figs. 2.5 and 2.6.

We suppose that the collimator is of such a length and width that it permits passage of neutrons at angles $\pm\alpha$ relative to the central path, which we suppose is scattered at an angle $\theta_c$ by the crystal. Now from Fig.

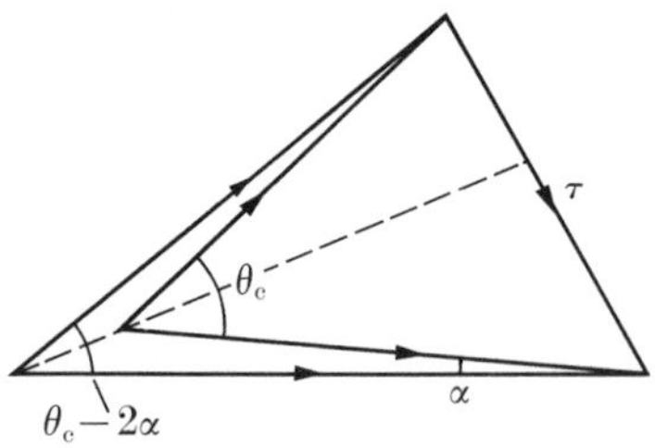


FIG. 2.5. The effect of collimation angle on Laue scattering with relative angle of emergence $+\alpha$. This is shown schematically in Fig. 2.6.

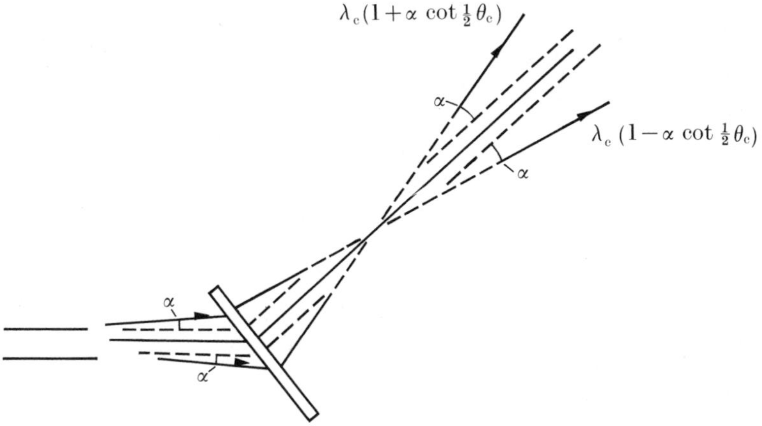


FIG. 2.6. Schematic diagram of Laue scattering.

2.5 we notice that neutrons incident at an angle $+\alpha$ relative to the central path are scattered through an angle $\theta_c - 2\alpha$ and therefore emerge at an angle $-\alpha$ relative to the central path. Similarly those incident at relative angle $-\alpha$ are scattered through $\theta_c + 2\alpha$. If we ignore the effect of the mosaic spread for the moment the lines drawn in Fig. 2.6 represent the extreme paths of the neutrons. Now from (2.32) we have

$$\sin \tfrac{1}{2}\theta = \frac{\tau}{4\pi}\lambda,$$

where $\theta$ is the scattering angle. If scattering through $\theta_c$ corresponds to a wavelength $\lambda_c$, scattering through $\theta_c - 2\alpha$ corresponds to

$$\lambda_c - \frac{4\pi}{\tau}\alpha \cos(\tfrac{1}{2}\theta_c) = \lambda_c - \lambda_c \alpha \cot(\tfrac{1}{2}\theta_c).$$

Hence the neutrons emerging at relative angle $-\alpha$ have wavelengths $\lambda_c(1-\alpha \cot \frac{1}{2}\theta_c)$, while those emerging at relative angle $\alpha$ have wavelengths $\lambda_c(1+\alpha \cot \frac{1}{2}\theta_c)$. This is indicated in Fig. 2.6. The total wavelength spread is

$$2\lambda_c \alpha \cot(\tfrac{1}{2}\theta_c).$$

Actually the mosaic spread of the crystal increases this wavelength spread somewhat but we will not discuss this.

*Method* 2. This method utilizes a monochromatic beam (produced for example by a Laue method) and a single crystal that is free to rotate. Referring to Fig. 2.3 we can think of this as keeping **k** fixed and rotating the reciprocal lattice so that points cross the surface of the sphere. It is customary to measure the orientation of the crystal by the value of the angle $\frac{1}{2}\theta$ and by rocking the crystal obtain a plot of scattered power against $\frac{1}{2}\theta$. According to (2.35) this plot is a $\delta$-function, but of course in practice the peak has a non-zero width determined by the mosaic spread, the degree of collimation, and the wavelength spread in the incident beam, which, of course, is only nominally monochromatic. Whatever the width of the peak may be, the area is as determined from (2.35) and this is the significant parameter that is measured experimentally. The intensity integrated over the rocking curve is

$$\int d(\tfrac{1}{2}\theta)N(\lambda)\sigma_c = N(\lambda)VQ^\theta, \tag{2.38}$$

where

$$Q^\theta = \frac{\lambda^3\,|F_N(\boldsymbol{\tau})|^2}{v_0^2 \sin\theta}. \tag{2.39}$$

These formulae are obtained by substituting for $\sigma_c$ and performing the integral.

It is rather interesting to consider the width of these rocking curves. In Fig. 2.6 we showed a schematic diagram of the production of a 'monochromatic' beam by a Laue type of method. In the figure the neutrons are shown as being scattered to the left. If the 'monochromatic' beam is now used for an experiment with a second crystal, it is preferable to arrange this second crystal so that the second scattering takes place to the right, because the rocking curves are then narrower than they would be if the second scattering were also to go to the left. The reason for this asymmetry between left and right is that the beam produced by the first crystal is not really monochromatic nor perfectly collimated; those neutrons incident at relative angle $+\alpha$ (compared to the central path) have longer wavelengths than those incident at relative angle $-\alpha$. Now the longer wavelengths are scattered through wider angles than the shorter wavelengths, and hence if the neutrons are again scattered to the left the angular divergence between the extreme wavelengths will increase while if this second scattering takes place to the right the angular divergence will be decreased. Similarly the rocking curves for scattering to the right will be narrower than for scattering to the left. This is illustrated in Fig. 2.7. Of course, in either case the areas under the rocking curves are the same, but when the rocking curve is narrow it is easier to subtract off the background scattering and the peaks are easier to resolve.

*Method* 3. The powder method utilizes a monochromatic beam and a powder sample. This corresponds to keeping **k** fixed and averaging over all orientations of the reciprocal lattice. To obtain the cross-section for scattering from a powder sample we have to average (2.11) over all

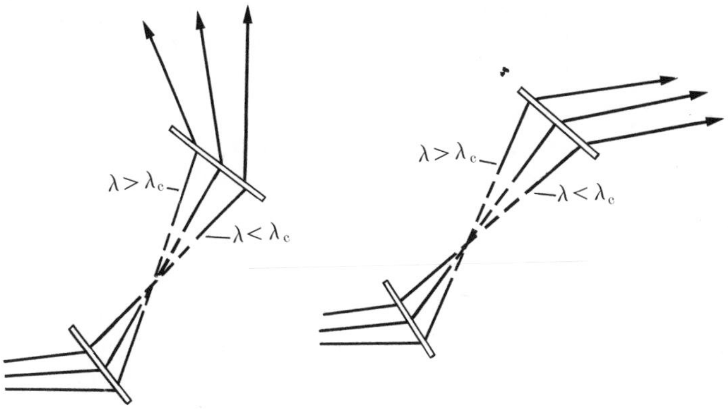


FIG. 2.7. Bragg scattering for 'left–left' and 'left–right' arrangements.

orientations of $\boldsymbol{\tau}$. This is equivalent to averaging over orientations of $\boldsymbol{\kappa}$ and this is easier to do. The answer is

$$\left(\frac{d\sigma}{d\Omega}\right)_{\text{coh}} = \frac{2\pi^2 N}{k^2 v_0} \sum_{\boldsymbol{\tau}} \frac{1}{\tau} |F_N(\boldsymbol{\tau})|^2 \, \delta\left(1 - \frac{\tau^2}{2k^2} - \cos\theta\right). \tag{2.40}$$

(The derivation is similar to that of (2.35).)

In this formula all reference to the aximuthal scattering angle has disappeared. Hence the scattering takes place in cones, called Debye–Scherrer cones (Fig. 2.8), which have the direction $\mathbf{k}$ as axis and have semi-angles $\theta$ defined by

$$\cos\theta = 1 - \tau^2/2k^2.$$

We recognize this last condition to be the same as (2.32).

According to (2.40) each cone is infinitely thin; of course in practice each cone has a width determined by the degree of collimation and the wavelength spread in the incident beam. The total cross-section associated with each cone is

$$\sigma_c(\tau) = \frac{4\pi^3 N}{k^2 v_0} \frac{z(\tau)}{\tau} |F_\tau|^2, \tag{2.41}$$

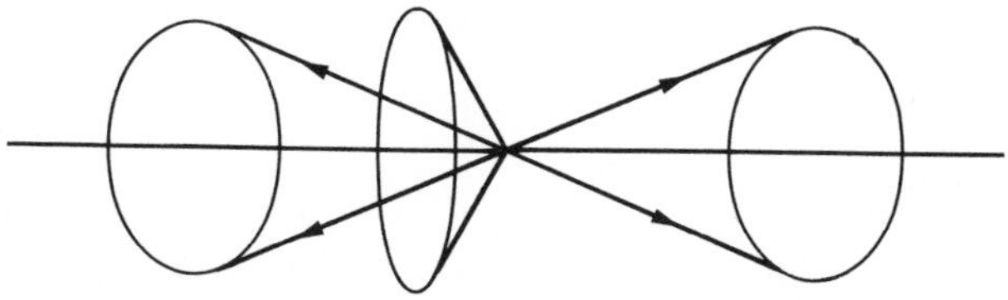

FIG. 2.8. Debye–Scherrer cones for Bragg scattering from a powder.

where $z(\tau)$ is the number of $\boldsymbol{\tau}$ vectors with magnitude $\tau$ and $|F_\tau|^2$ is the mean value of $|F_N(\boldsymbol{\tau})|^2$ for just these vectors. When using this expression one must take care to allow for the fact that the counter can observe only a fraction of the complete cones. This fraction is $l/2\pi r \sin\theta$, where $l$ is the height of the counter and $r$ is the distance from the sample to the counter.

It is instructive to watch how the positions of these cones vary with increasing $k$. For very small $k$, (2.32) cannot be satisfied and there is no scattering. When $k$ is equal to the critical value $\frac{1}{2}\tau_{\text{min}}$, scattering occurs for $\theta = \pi$, i.e. in the backward direction. As $k$ slowly increases above this value this backward scattering opens out into a cone, which moves steadily towards the forward direction. Eventually, when $k$ reaches a second critical value, backward scattering again appears and then in turn this opens out into a cone, and so on. When $k$ is large there are many, many cones all closely spaced and, since they have an appreciable width in practice, they overlap and become indistinguishable. From (2.41) the total cross-section is

$$\sigma_c = \frac{4\pi^3 N}{k^2 v_0} \sum_{\tau}^{\tau<2k} \frac{1}{\tau} |F_\tau|^2. \tag{2.42}$$

Let us evaluate this for a simple lattice with one atom per unit cell, so that the form factor is simply $\bar{b}$, and in the limit when $k$ is large. In this limit the sum can be replaced by an integral. We remember

$$\sum_{\boldsymbol{\tau}} \rightarrow \frac{v_0}{(2\pi)^3} \int \mathrm{d}\boldsymbol{\tau}$$

to get

$$\sigma_c \rightarrow \frac{4\pi^3 N}{k^2 v_0} |\bar{b}|^2 \frac{v_0}{(2\pi)^3} \int_0^{2k} \mathrm{d}\boldsymbol{\tau} \frac{1}{\tau},$$

i.e.

$$\sigma_c \rightarrow 4\pi N |\bar{b}|^2 \quad \text{as} \quad k \rightarrow \infty.$$

Similarly, for a more complex lattice it is possible to show that

$$\sigma_c \rightarrow 4\pi N \sum_d |\bar{b}_d|^2 \quad \text{as} \quad k \rightarrow \infty. \tag{2.43}$$

Hence, using (2.13), we get for the total cross-section

$$\sigma = \sigma_c + \sigma_i \rightarrow 4\pi N \sum_d \overline{|b_d|^2}, \tag{2.44}$$

i.e. all interference effects vanish for high incident momenta. This is just what we expect. Actually, it is not true that elastic cross-sections approach a non-zero limit as $k$ increases. We have obtained answers that

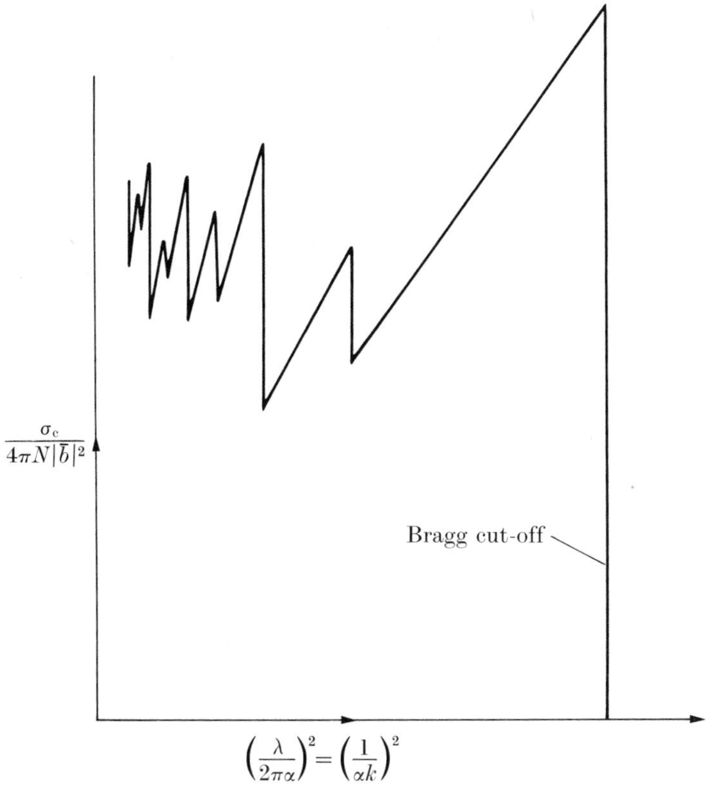


FIG. 2.9. The total coherent scattering from a body-centred cubic *rigid* lattice.

say so only because we have assumed our lattice is rigid. When we take account of lattice vibrations we will find all elastic cross-sections approach zero as $k$ increases. It is true, however, that all interference effects vanish as $k$ increases (cf. Chapter 4).

This powder method is often used because it is usually much easier to obtain a sample in powder form than as a single crystal and because there are no serious extinction problems with a powder. But, of course, powder intensities are much lower than single crystal intensities, so it is always preferable to use a single crystal if one is available.

In Fig. 2.9 we show a plot of the total coherent cross-section (2.42) for a body-centred cubic lattice of cube side $a$. Notice how this cross-section is zero for wavelengths beyond the Bragg cut-off and how the cross-section changes discontinuously as $\lambda$ is decreased and new Debye–Scherrer cones appear. This curve is seriously modified for short wavelengths when proper account of lattice vibrations is made; instead of approaching a non-zero limit as $\lambda \to 0$, as it does in this graph, it should approach zero.

In practice nuclear Bragg scattering is used to investigate only those materials where it is impossible or inconvenient to use X-rays. For example, it is hard to locate light atoms in the presence of heavy atoms using X-rays because the X-ray scattering length is proportional to $Z$, the number of electrons in the atom. But for neutrons the scattering lengths for light and heavy nuclei are comparable and therefore the light nuclei can be located. Neutron Bragg scattering is specially valuable for the location of hydrogen atoms, i.e. protons. Also, using X-rays it is not easy to distinguish between atoms of similar $Z$ values (though it can be done by using wavelengths close to an absorption edge of one of the atoms), and therefore it is often convenient to use neutrons because the neutron scattering lengths vary irregularly from one nucleus to another and therefore are likely to be quite different for the nuclei concerned (Bacon 1977).

## 2.5. Diffuse scattering (Krivoglaz 1969; Bauer 1979; Schultz 1982)

Consider a binary alloy that forms a rigid Bravais crystal lattice. The coherent and incoherent cross-sections for elastic scattering are

$$\left(\frac{d\sigma}{d\Omega}\right)_{coh} = \overline{\left|\sum_l \bar{b}_l \exp(i\boldsymbol{\kappa}\cdot\mathbf{l})\right|^2} \tag{2.45}$$

and

$$\left(\frac{d\sigma}{d\Omega}\right)_{incoh} = \frac{N}{4\pi}\overline{\sigma_i} \tag{2.46}$$

where $N$ is the total number of atoms in the alloy. The horizontal bars in these two expressions denote an average over the spatial configurations of the two components of the alloy.

We label the components 1, and 2. Let $c$ be the fractional concentration of atoms of type-2. The configurational average of the scattering length is clearly

$$\overline{(\bar{b})} = (1-c)\bar{b}_1 + c\bar{b}_2 = b, \tag{2.47}$$

where the second equality defines $b$. Also,

$$\overline{\sigma_i} = (1-c)\sigma_i^{(1)} + c\sigma_i^{(2)} \tag{2.48}$$

which when inserted in (2.46) gives the spin incoherent cross-section for the alloy.

In order to perform the configurational average of the coherent cross-section it is useful to introduce a site occupation function $p_l$ with the property

$$\begin{aligned} p_l &= 1, \quad \text{if } l \text{ labels an atom of type-2} \\ &= 0, \quad \text{otherwise} \end{aligned} \tag{2.49}$$

Note that, by definition,

$$\overline{p_l} = c \tag{2.50}$$

and

$$\sum_l p_l = cN. \tag{2.51}$$

From the definition of $p_l$ we deduce that

$$\begin{aligned} \bar{b}_l &= \bar{b}_1 + p_l(\bar{b}_2 - \bar{b}_1) \\ &= b + (\bar{b}_2 - \bar{b}_1)(p_l - c), \end{aligned} \tag{2.52}$$

and so, in view of (2.50),

$$\overline{\bar{b}_{l'}\bar{b}_l} = b^2 + (\bar{b}_2 - \bar{b}_1)^2\overline{(p_{l'} - c)(p_l - c)}. \tag{2.53}$$

To evaluate the second term in (2.53) we observe that

$$\begin{aligned} \overline{p_{l'}p_l} &= c, \quad \text{if} \quad l = l' \\ &= c^2, \quad \text{if} \quad l \neq l' \end{aligned}$$

or

$$\overline{p_{l'}p_l} = c^2 + c(1-c)\,\delta_{l,l'}. \tag{2.54}$$

Substituting (2.54) in (2.53) we find for the coherent cross-section (2.45) the result,

$$\left(\frac{\mathrm{d}\sigma}{\mathrm{d}\Omega}\right)_{\text{coh}} = \left(\frac{\mathrm{d}\sigma}{\mathrm{d}\Omega}\right)_{\text{Bragg}} + N(\bar{b}_2 - \bar{b}_1)^2 c(1-c), \tag{2.55}$$

where the Bragg cross-section is

$$\left(\frac{\mathrm{d}\sigma}{\mathrm{d}\Omega}\right)_{\text{Bragg}} = b^2 \left|\sum_l \exp(\mathrm{i}\boldsymbol{\kappa}\cdot\mathbf{l})\right|^2 \tag{2.56}$$

and the second term represents diffuse scattering. We see that, for an alloy, the Bragg scattering is just a part of the coherent elastic scattering.

Notice that the amplitude of the Bragg scattering depends on the average scattering length. This behaviour is to be expected in view of the fact that Bragg scattering results from perfect lattice periodicity, and for an alloy this is found in the configurationally averaged sample. The result (2.56) applies to an arbitrary lattice structure if we interpret $b$ as the configurationally averaged unit-cell structure factor.

While Bragg scattering is observed at special values of the scattering vector, the diffuse scattering in (2.55) is independent of $\boldsymbol{\kappa}$.† The expression given in (2.55) is usually called the Laue monotonic diffuse scattering cross-section. The scattering vanishes for a monatomic lattice, for which $c=0$ or 1. It depends on the difference of the scattering lengths for the two types of atom, and the quantity $|\bar{b}_2-\bar{b}_1|$ is analogous to a contrast factor.

Laue monotonic scattering derives from compositional effects, and makes no allowance for the fact that alloying often produces lattice distortions. To see the effect of static lattice distortions in the coherent elastic cross-section we shall analyse a model which is appropriate for weak distortions in a dilute alloy.

Local distortions of the lattice are described by vectors $\mathbf{h}_l$, and the distortions are weak if the average value of these vectors is small compared to the lattice spacing. The average lattice possess perfect periodicity so it is defined by a set of lattice vectors $\mathbf{l}$. The reciprocal lattice vectors satisfy $\exp(\mathrm{i}\boldsymbol{\tau}\cdot\mathbf{l})=1$ for all $\mathbf{l}$. Because the sample is not macroscopically deformed, the distortions obey the constraint,

$$\sum_l \mathbf{h}_l = 0. \tag{2.57}$$

The coherent elastic cross-section for our model binary alloy with static distortions is

$$\left(\frac{\mathrm{d}\sigma}{\mathrm{d}\Omega}\right)_{\mathrm{coh}} = \left|\overline{\sum_l \bar{b}_l \exp\{\mathrm{i}\boldsymbol{\kappa}\cdot(\mathbf{l}+\mathbf{h}_l)\}}\right|^2 \tag{2.58}$$

Because the concentration of type-2 atoms is very small the correlation between the distributions of the scattering lengths and distortion vectors is negligible, in which case the average scattering amplitude in (2.58) is

$$\overline{(\bar{b}_l)}\ \overline{\exp(\mathrm{i}\boldsymbol{\kappa}\cdot\mathbf{h}_l)} = b\exp\{-H(\boldsymbol{\kappa})\}, \tag{2.59}$$

where the equality defines a static Debye–Waller factor with an exponent $H(\boldsymbol{\kappa})$. If the distortions follow a Gaussian distribution with $\bar{\mathbf{h}}=0$, so that (2.57) is satisfied, on the average

$$H(\boldsymbol{\kappa}) = \tfrac{1}{2}\overline{(\boldsymbol{\kappa}\cdot\mathbf{h})^2} = \tfrac{1}{6}\kappa^2\overline{h^2} \tag{2.60}$$

where $\overline{h^2}$ is the mean-square distortion. The Bragg cross-section follows directly from (2.59),

$$\left(\frac{\mathrm{d}\sigma}{\mathrm{d}\Omega}\right)_{\mathrm{Bragg}} = b^2\exp\{-2H(\boldsymbol{\kappa})\}\left|\sum_l \exp(\mathrm{i}\boldsymbol{\kappa}\cdot\mathbf{l})\right|^2. \tag{2.61}$$

† This statement is correct for a rigid lattice. Thermal motion introduces a Debye–Waller factor (cf. Chapter 4), which is accounted for by the replacement, $\bar{b}_j \to \bar{b}_j \exp(-W_j)$ and in most cases $W_j \propto \kappa^2$.

When we subtract the Bragg cross-section from (2.58) the remaining terms correspond to diffuse scattering. To develop the expression for diffuse scattering we use (2.52), and the approximation

$$\exp(i\boldsymbol{\kappa}\cdot\mathbf{h}_l) \doteq 1 + i\boldsymbol{\kappa}\cdot\mathbf{h}_l \tag{2.62}$$

which is appropriate for weak distortions. We then obtain the following expression for the diffuse cross-section,

$$\left(\frac{d\sigma}{d\Omega}\right)_{\text{diff}} = \left(\frac{d\sigma}{d\Omega}\right)_{\text{coh}} - \left(\frac{d\sigma}{d\Omega}\right)_{\text{Bragg}}$$
$$\doteq N(\bar{b}_2 - \bar{b}_1)^2 c(1-c) + b^2 \sum_{ll'} \overline{(\boldsymbol{\kappa}\cdot\mathbf{h}_{l'})(\boldsymbol{\kappa}\cdot\mathbf{h}_l)} \exp\{i\boldsymbol{\kappa}\cdot(\mathbf{l}-\mathbf{l}')\} \tag{2.63}$$

The first contribution on the right-hand side is identical with the expression (2.55) for Laue monotonic scattering.

The contribution in (2.63) which arises from the distortions can be simplified by noting that for a centrosymmetric lattice,

$$\overline{(\boldsymbol{\kappa}\cdot\mathbf{h}_{l'})(\boldsymbol{\kappa}\cdot\mathbf{h}_l)} = \tfrac{1}{3}\kappa^2\,\overline{\mathbf{h}_{l'}\cdot\mathbf{h}_l}. \tag{2.64}$$

In order to pursue the discussion further we introduce a specific model for the distortion vectors. Given that the alloy is dilute, we will assume a linear superposition of distortions with

$$\mathbf{h}_l = \sum_R \mathbf{e}(\mathbf{R})(p_{l+R} - c), \tag{2.65}$$

where $\mathbf{e}(\mathbf{R})$ is the distortion produced at a lattice position a distance $R$ away from a type-2 atom and $p_l$ is defined by (2.49). By construction $\mathbf{e}(0) = 0$, and it is reasonable to assume that $\mathbf{e}(\mathbf{R}) \to 0$ as $R \to \infty$. For the approximation (2.65) the constraint (2.57) is satisfied on the average.

The calculation of the averaged quantity on the right-hand side of (2.64) for the model (2.65) follows directly from the relation (2.54),

$$\overline{\mathbf{h}_{l'}\cdot\mathbf{h}_l} = c(1-c)\sum_R \mathbf{e}(\mathbf{R})\cdot\mathbf{e}(\mathbf{l}-\mathbf{l}'+\mathbf{R}) \tag{2.66}$$

and then

$$\sum_{ll'} \overline{\mathbf{h}_{l'}\cdot\mathbf{h}_l} \exp\{i\boldsymbol{\kappa}\cdot(\mathbf{l}-\mathbf{l}')\} = Nc(1-c)\left|\sum_R \mathbf{e}(\mathbf{R})\exp(i\boldsymbol{\kappa}\cdot\mathbf{R})\right|^2. \tag{2.67}$$

Assembling the various results we find that the diffuse cross-section (2.63) is

$$\left(\frac{d\sigma}{d\Omega}\right)_{\text{diff}} \doteq Nc(1-c)\left\{(\bar{b}_2 - \bar{b}_1)^2 + \tfrac{1}{3}\kappa^2 b^2 \left|\sum_R \mathbf{e}(\mathbf{R})\cos(\boldsymbol{\kappa}\cdot\mathbf{R})\right|^2\right\}. \tag{2.68}$$

Because $\cos(\boldsymbol{\tau} \cdot \mathbf{R}) = 1$, the Fourier transform of the distortion field is a maximum at the Bragg positions. The behaviour of the Fourier transform in the vicinity of a Bragg position is controlled by the large-distance behaviour of the distortion field. A calculation for an elastic continuum model leads to the result $|\mathbf{e}(\mathbf{R})| \propto R^{-2}$ for large $R$, in which case the square of the Fourier transform in (2.68) behaves like $|\boldsymbol{\kappa} - \boldsymbol{\tau}|^{-2}$ as $\boldsymbol{\kappa}$ approaches the reciprocal lattice vector $\boldsymbol{\tau}$. For the case of a centrosymmetric lattice, where the distortions have inversion symmetry with respect to the impurity site, the Fourier transform of the distortion field is zero for $\boldsymbol{\kappa} = \boldsymbol{\tau}/2$, and so the diffuse scattering is equal to the Laue monotonic scattering. To prove this we note that, since $\mathbf{e}(\mathbf{R}) = -\mathbf{e}(-\mathbf{R})$, its Fourier transform must be an odd function of $\boldsymbol{\kappa}$. However, the Fourier transform is unchanged under the transformation $\boldsymbol{\kappa} \rightarrow \boldsymbol{\kappa} + \boldsymbol{\tau}$. These two properties are satisfied simultaneously for $\boldsymbol{\kappa} = \boldsymbol{\tau}/2$ if the Fourier transform is zero.

## REFERENCES

Bacon, G. E. (1977). *Neutron scattering in chemistry*. Butterworths, London.

Bauer, G. S. (1979). In G. Kostorz (ed.) *Treatise on materials science and technology*, Vol. 15. Academic Press, New York.

Brown, P. J. (1979). In G. Kostorz (ed.) *Treatise on materials science and technology*, Vol. 15. Academic Press, New York.

Dachs, H. (ed.) (1978). *Topics in current physics*, Vol. 6. Springer-Verlag.

Krivoglaz, M. A. (1969). *The theory of X-ray and thermal neutron scattering by real crystals*. Plenum, London.

Landau, L. D. and Lifschitz, E. M. (1980). *Statistical physics*, Part 1. Pergamon Press, Oxford.

Schultz, J. M. (1982). *Diffraction for material scientists*. Prentice-Hall, Englewood Cliffs, N.J.

# 3
# CORRELATION AND RESPONSE FUNCTIONS

The master expression we developed in Chapter 1 for the neutron cross-section can be expressed in terms of a correlation function that is the thermal average of a combination of operators belonging to the target sample, so that the scattering problem is reduced essentially to a problem in statistical mechanics. The techniques for handling the latter are manifold (Lovesey 1980; Reichl 1980; Rickayzen 1980; Inkson 1984). Hence, the reduction of the calculation of the neutron cross-section to a study of an appropriate correlation function is an exceedingly important one. In the following we pursue this problem and indicate the manipulations involved to calculate these functions for a number of cases of interest.

## 3.1. Response function (Van Hove 1954; Lovesey 1980)

The interaction potential between the neutrons and the target for purely nuclear scattering is of the general form

$$\left(\frac{m}{2\pi\hbar^2}\right)\hat{V}(\mathbf{r}) = \sum_j \hat{V}_j(\mathbf{r}-\mathbf{R}_j), \tag{3.1}$$

where $\mathbf{R}_j$ is the position vector of the $j$th scattering nucleus. If we substitute (3.1) into (1.23), then the cross-section for the scattering of unpolarized neutrons is easily shown to be

$$\frac{d^2\sigma}{d\Omega\, dE'} = \frac{k'}{k}\sum_{\lambda,\lambda'} p_\lambda \overline{\left|\langle\lambda'|\sum_j \hat{V}_j(\boldsymbol{\kappa})\exp(i\boldsymbol{\kappa}\cdot\mathbf{R}_j)\,|\lambda\rangle\right|^2}\,\delta(\hbar\omega + E_\lambda - E_{\lambda'}), \tag{3.2}$$

where

$$\hat{V}_j(\boldsymbol{\kappa}) = \int d\mathbf{r}\,\exp(i\boldsymbol{\kappa}\cdot\mathbf{r})\,\hat{V}_j(\mathbf{r}).$$

so that

$$\langle\mathbf{k}'|\,\hat{V}\,|\mathbf{k}\rangle = \sum_j \hat{V}_j(\boldsymbol{\kappa})\exp(i\boldsymbol{\kappa}\cdot\mathbf{R}_j).$$

We can write the $\delta$-function in (3.2) in the form

$$\delta(\hbar\omega + E_\lambda - E_{\lambda'}) = \frac{1}{2\pi\hbar}\int_{-\infty}^{\infty} dt\,\exp\{-it(\hbar\omega + E_\lambda - E_{\lambda'})/\hbar\},$$

where $t$ has the dimension of time, and thus

$$\sum_{\lambda,\lambda'} p_\lambda \left|\langle\lambda'|\sum_j \hat{V}_j(\boldsymbol{\kappa})\exp(\mathrm{i}\boldsymbol{\kappa}\cdot\mathbf{R}_j)\,|\lambda\rangle\right|^2 \delta(\hbar\omega + E_\lambda - E_{\lambda'})$$
$$= \frac{1}{2\pi\hbar}\int_{-\infty}^{\infty} \mathrm{d}t\,\exp(-\mathrm{i}\omega t)\sum_{\lambda,\lambda'} p_\lambda\,\langle\lambda|\sum_j \exp(-\mathrm{i}\boldsymbol{\kappa}\cdot\mathbf{R}_j)\hat{V}_j^+(\boldsymbol{\kappa})\,|\lambda'\rangle$$
$$\times\langle\lambda'|\exp(\mathrm{i}tE_{\lambda'}/\hbar)\sum_{j'}\hat{V}_{j'}(\boldsymbol{\kappa})\exp(\mathrm{i}\boldsymbol{\kappa}\cdot\mathbf{R}_{j'})\exp(-\mathrm{i}tE_\lambda/\hbar)\,|\lambda\rangle$$
$$= \frac{1}{2\pi\hbar}\int_{-\infty}^{\infty} \mathrm{d}t\,\exp(-\mathrm{i}\omega t)\sum_{\lambda,\lambda'} p_\lambda\,\langle\lambda|\sum_j \exp(-\mathrm{i}\boldsymbol{\kappa}\cdot\mathbf{R}_j)\hat{V}_j^+(\boldsymbol{\kappa})\,|\lambda'\rangle$$
$$\times\langle\lambda'|\sum_{j'}\exp(\mathrm{i}t\hat{\mathcal{H}}/\hbar)\hat{V}_{j'}(\boldsymbol{\kappa})\exp(\mathrm{i}\boldsymbol{\kappa}\cdot\mathbf{R}_{j'})\exp(-\mathrm{i}t\hat{\mathcal{H}}/\hbar)\,|\lambda\rangle,$$

where $\hat{\mathcal{H}}$ is the Hamiltonian that describes the target system. The sum over the final states can be performed using (1.15) and in terms of Heisenberg operators

$$\hat{V}_j(\boldsymbol{\kappa}, t)\exp\{\mathrm{i}\boldsymbol{\kappa}\cdot\hat{\mathbf{R}}_j(t)\} = \exp(\mathrm{i}t\hat{\mathcal{H}}/\hbar)\,\hat{V}_j(\boldsymbol{\kappa})\exp(-\mathrm{i}t\hat{\mathcal{H}}/\hbar)$$
$$\times\exp(\mathrm{i}t\hat{\mathcal{H}}/\hbar)\exp(\mathrm{i}\boldsymbol{\kappa}\cdot\mathbf{R}_j)\exp(-\mathrm{i}t\hat{\mathcal{H}}/\hbar), \quad (3.3)$$

with $\hat{\mathbf{R}}(0) \equiv \hat{\mathbf{R}}$, for example, the cross-section (1.23) can be written in the preponderant form,

$$\frac{\mathrm{d}^2\sigma}{\mathrm{d}\Omega\,\mathrm{d}E'} = \frac{k'}{k}\frac{1}{2\pi\hbar}\int_{-\infty}^{\infty}\mathrm{d}t\,\exp(-\mathrm{i}\omega t)$$
$$\times\sum_{jj'}\sum_\lambda p_\lambda\,\langle\lambda|\exp(-\mathrm{i}\boldsymbol{\kappa}\cdot\hat{\mathbf{R}}_j)\hat{V}_j^+(\boldsymbol{\kappa})\hat{V}_{j'}(\boldsymbol{\kappa}, t)\exp\{\mathrm{i}\boldsymbol{\kappa}\cdot\hat{\mathbf{R}}_{j'}(t)\}\,|\lambda\rangle$$
$$= \frac{k'}{k}\frac{1}{2\pi\hbar}\int_{-\infty}^{\infty}\mathrm{d}t\,\exp(-\mathrm{i}\omega t)$$
$$\times\sum_{jj'}\langle\exp(-\mathrm{i}\boldsymbol{\kappa}\cdot\hat{\mathbf{R}}_j)\hat{V}_j^+(\boldsymbol{\kappa})\hat{V}_{j'}(\boldsymbol{\kappa}, t)\exp\{\mathrm{i}\boldsymbol{\kappa}\cdot\hat{\mathbf{R}}_{j'}(t)\}\rangle. \quad (3.4)$$

The final form is written with the notation of Chapter 1 for thermal averages (cf. (1.82)) and the quantity $\langle\cdots\rangle$ is usually referred to as a *correlation function.*

Expression (3.4) allows for coupling between the target sample states and the interaction potentials which exists, for example, when quantum mechanical exchange forces are significant (see § 1.7). This is the exception rather than the rule, so we develop (3.4) for the case when the averaging over nuclear spin orientations and distributions, denoted by a horizontal bar, is independent of the thermal averaging. In this instance,

$$\frac{d^2\sigma}{d\Omega\, dE'} = \frac{k'}{k} \frac{1}{2\pi\hbar} \int_{-\infty}^{\infty} dt \exp(-i\omega t) \sum_{jj'} \overline{\hat{V}_j^+(\boldsymbol{\kappa}) \hat{V}_{j'}(\boldsymbol{\kappa})}\; Y_{jj'}(\boldsymbol{\kappa}, t) \tag{3.5}$$

where the correlation function

$$Y_{jj'}(\boldsymbol{\kappa}, t) = \langle \exp(-i\boldsymbol{\kappa} \cdot \hat{\mathbf{R}}_j) \exp\{i\boldsymbol{\kappa} \cdot \hat{\mathbf{R}}_{j'}(t)\}\rangle. \tag{3.6}$$

When the average of the product of the potentials in (3.5) is independent of the type of the nuclei, i.e. independent of $j$ and $j'$, as in coherent scattering, it is customary to write the cross-section for scattering from $N$ particles in the form

$$\frac{d^2\sigma}{d\Omega\, dE'} = N \frac{k'}{k} \overline{|\hat{V}_j(\boldsymbol{\kappa})|^2}\; S(\boldsymbol{\kappa}, \omega) \tag{3.7}$$

where the *response function*

$$S(\boldsymbol{\kappa}, \omega) = \frac{1}{2\pi\hbar N} \int_{-\infty}^{\infty} dt \exp(-i\omega t) \sum_{jj'} Y_{jj'}(\boldsymbol{\kappa}, t), \tag{3.8}$$

has the dimension of $(\text{energy})^{-1}$. The response function is often called the *scattering law* or *dynamic structure factor.*

The terminology used for $S(\boldsymbol{\kappa}, \omega)$ merits some comment since while we call it a response function it is determined by spontaneous fluctuations in the target sample (eqn (3.8)). The intimate relation between the response and the spectrum of spontaneous fluctuations stems from the fact that neutron scattering is a weak process and the sample response is linear, i.e. describable by first-order perturbation theory. It follows that, the cross-section, which clearly is a measure of the response of the sample, is determined by the spectrum of spontaneous fluctuations as we have just seen. In other words, the neutron cross-section measures the undistorted properties of the sample. The formal theory that relates linear response to spontaneous fluctuations is embodied in the fluctuation–dissipation theorem (see eqn (3.54) and Appendix B).

An alternative expression to (3.8) for $S(\boldsymbol{\kappa}, \omega)$ can be obtained in terms of the microscopic particle density operator

$$\hat{\rho}(\mathbf{r}, t) = \sum_j \delta\{\mathbf{r} - \hat{\mathbf{R}}_j(t)\}. \tag{3.9}$$

To see this, consider the *pair correlation function*

$$G(\mathbf{r}, t) = \left(\frac{1}{2\pi}\right)^3 \int d\boldsymbol{\kappa} \exp(-i\boldsymbol{\kappa} \cdot \mathbf{r}) \frac{1}{N} \sum_{j,j'} \langle \exp(-i\boldsymbol{\kappa} \cdot \hat{\mathbf{R}}_j) \exp\{i\boldsymbol{\kappa} \cdot \hat{\mathbf{R}}_{j'}(t)\}\rangle. \tag{3.10}$$

We must be careful in evaluating the right-hand side of (3.10) because, except when $t=0$, the two operators $\exp\{-i\boldsymbol{\kappa}\cdot\hat{\mathbf{R}}_j(0)\}$ and $\exp\{i\boldsymbol{\kappa}\cdot\hat{\mathbf{R}}_{j'}(t)\}$ do not commute. Hence we cannot replace their product by

$$\exp[-i\boldsymbol{\kappa}\cdot\{\hat{\mathbf{R}}_j-\hat{\mathbf{R}}_{j'}(t)\}]$$

and, indeed, must be careful in all algebraic manipulations to keep $\hat{\mathbf{R}}_{j'}(t)$ to the right of $\hat{\mathbf{R}}_j$. Hence we write

$$\begin{aligned} G(\mathbf{r},t) &= \left(\frac{1}{2\pi}\right)^3 \int d\boldsymbol{\kappa}\, \exp(-i\boldsymbol{\kappa}\cdot\mathbf{r})\frac{1}{N}\sum_{j,j'}\int d\mathbf{r}' \\ &\quad\times\langle\exp\{-i\boldsymbol{\kappa}\cdot\hat{\mathbf{R}}_j+i\boldsymbol{\kappa}\cdot\mathbf{r}'\}\,\delta\{\mathbf{r}'-\hat{\mathbf{R}}_{j'}(t)\}\rangle \\ &= \frac{1}{N}\sum_{j,j'}\int d\mathbf{r}'\langle\delta\{\mathbf{r}-\mathbf{r}'+\hat{\mathbf{R}}_j\}\,\delta\{\mathbf{r}'-\hat{\mathbf{R}}_{j'}(t)\}\rangle, \end{aligned} \tag{3.11}$$

and from (3.9) it follows that

$$G(\mathbf{r},t)=\frac{1}{N}\int d\mathbf{r}'\langle\hat{\rho}(\mathbf{r}'-\mathbf{r})\hat{\rho}(\mathbf{r}',t)\rangle, \tag{3.12}$$

with

$$\int d\mathbf{r}\, G(\mathbf{r},t)=N. \tag{3.13}$$

Thus

$$\begin{aligned} S(\boldsymbol{\kappa},\omega) &= \frac{1}{2\pi\hbar}\int_{-\infty}^{\infty} dt\, \exp(-i\omega t)\int d\mathbf{r}\, \exp(i\boldsymbol{\kappa}\cdot\mathbf{r})G(\mathbf{r},t) \\ &= \frac{1}{2\pi\hbar N}\int_{-\infty}^{\infty} dt\, \exp(-i\omega t)\int d\mathbf{r}\int d\mathbf{r}'\, \exp(i\boldsymbol{\kappa}\cdot\mathbf{r}) \\ &\quad\times\langle\hat{\rho}(\mathbf{r}'-\mathbf{r})\hat{\rho}(\mathbf{r}',t)\rangle. \end{aligned} \tag{3.14}$$

In some instances it is convenient to introduce the Fourier components of $\hat{\rho}(\mathbf{r})$, i.e. we write

$$\hat{\rho}(\mathbf{r})=\frac{1}{V}\sum_{\mathbf{q}}\hat{\rho}_{\mathbf{q}}\exp(i\mathbf{q}\cdot\mathbf{r}). \tag{3.15}$$

The wave vectors $\mathbf{q}$ are defined by imposing cyclic boundary conditions on the target sample and have a density $V/(2\pi)^3$, where $V$ is the volume of the sample. Therefore, from (3.15),

$$\begin{aligned} \sum_j \exp(-i\mathbf{q}'\cdot\mathbf{R}_j) &= \int d\mathbf{r}\, \exp(-i\mathbf{q}'\cdot\mathbf{r})\frac{1}{V}\sum_{\mathbf{q}}\hat{\rho}_{\mathbf{q}}\exp(i\mathbf{q}\cdot\mathbf{r}) \\ &= \frac{(2\pi)^3}{V}\sum_{\mathbf{q}}\hat{\rho}_{\mathbf{q}}\,\delta(\mathbf{q}-\mathbf{q}')=\frac{(2\pi)^3}{V}\frac{V}{(2\pi)^3}\int d\mathbf{q}\,\hat{\rho}_{\mathbf{q}}\,\delta(\mathbf{q}-\mathbf{q}')=\hat{\rho}_{\mathbf{q}'}, \end{aligned}$$

i.e.

$$\hat{\rho}_{\boldsymbol{\kappa}}(t) = \sum_j \exp\{-\mathrm{i}\boldsymbol{\kappa} \cdot \hat{\mathbf{R}}_j(t)\}. \tag{3.16}$$

From (3.14) (or directly from (3.8)) it follows that

$$S(\boldsymbol{\kappa}, \omega) = \frac{1}{2\pi\hbar N} \int_{-\infty}^{\infty} \mathrm{d}t \exp(-\mathrm{i}\omega t) \langle \hat{\rho}_{\boldsymbol{\kappa}} \hat{\rho}_{-\boldsymbol{\kappa}}(t) \rangle. \tag{3.17}$$

For many complex problems it is more convenient to deal with the spatial transform of $G(\mathbf{r}, t)$ rather than $G(\mathbf{r}, t)$ itself. The spatial transform is denoted as $I(\boldsymbol{\kappa}, t)$, where

$$I(\boldsymbol{\kappa}, t) = \int \mathrm{d}\mathbf{r} \exp(\mathrm{i}\boldsymbol{\kappa} \cdot \mathbf{r}) G(\mathbf{r}, t) = \frac{1}{N} \langle \hat{\rho}_{\boldsymbol{\kappa}} \hat{\rho}_{-\boldsymbol{\kappa}}(t) \rangle, \tag{3.18}$$

and hence

$$S(\boldsymbol{\kappa}, \omega) = \frac{1}{2\pi\hbar} \int_{-\infty}^{\infty} \mathrm{d}t \exp(-\mathrm{i}\omega t) I(\boldsymbol{\kappa}, t). \tag{3.19}$$

We draw attention to the form of $S(\boldsymbol{\kappa}, \omega)$ given by (3.14), which shows that $S(\boldsymbol{\kappa}, \omega)$ can be expressed in terms of a density–density correlation function. The reason for this is simply that, so far as coherent nuclear scattering is concerned, the interaction potential couples the neutron to the density of the target system. Indeed, using (3.9), the interaction potential (3.1), averaged over nuclear spins and isotopes, can be written in the form

$$\hat{V}(\mathbf{r}) = \int \mathrm{d}\mathbf{r}' \, \hat{V}(\mathbf{r} - \mathbf{r}') \hat{\rho}(\mathbf{r}').$$

The cross-section as expressed by eqn (3.7) is the product of two factors, the first of which depends only on the properties of the individual scatterers in the target, while the second $S(\boldsymbol{\kappa}, \omega)$ is, for given energy and momentum transfers, determined solely by the structure and dynamical behaviour of the target sample. Since it is $S(\boldsymbol{\kappa}, \omega)$ that is measured in an experiment, we obtain directly from an analysis of experimental data information about the structure and dynamics of the target.

## 3.2. Coherent and incoherent cross-sections

From Chapter 1 it follows that, for a monatomic target sample, in (3.5)

$$\overline{\hat{V}_j^+(\boldsymbol{\kappa}) \hat{V}_{j'}(\boldsymbol{\kappa})} = \{|\bar{b}|^2 + \delta_{j,j'}[\overline{|b|^2} - |\bar{b}|^2]\} = \left\{\frac{\sigma_\mathrm{c}}{4\pi} + \delta_{j,j'} \frac{\sigma_\mathrm{i}}{4\pi}\right\}.$$

Whence

$$\left(\frac{\mathrm{d}^2\sigma}{\mathrm{d}\Omega \, \mathrm{d}E'}\right)_{\mathrm{coh}} = N \frac{k'}{k} \frac{\sigma_\mathrm{c}}{4\pi} S(\boldsymbol{\kappa}, \omega) \tag{3.20}$$

and

$$\left(\frac{d^2\sigma}{d\Omega\, dE'}\right)_{\text{incoh}} = N\frac{k'}{k}\frac{\sigma_i}{4\pi} S_i(\boldsymbol{\kappa}, \omega). \tag{3.21}$$

Here

$$S_i(\boldsymbol{\kappa}, \omega) = \frac{1}{2\pi\hbar N}\int_{-\infty}^{\infty} dt \exp(-i\omega t) \sum_j Y_{jj}(\boldsymbol{\kappa}, t), \tag{3.22}$$

and, in analogy to the function $G(\mathbf{r}, t)$, of which, we recall, $S(\boldsymbol{\kappa}, \omega)$ is, apart from a constant, the space and time Fourier transform, we define the *self pair correlation function*,

$$G_s(\mathbf{r}, t) = \frac{1}{N}\sum_j \int d\mathbf{r}' \langle \delta\{\mathbf{r} - \mathbf{r}' + \hat{\mathbf{R}}_j\}\, \delta\{\mathbf{r}' - \hat{\mathbf{R}}_j(t)\}\rangle, \tag{3.23}$$

with

$$\int d\mathbf{r} G_s(\mathbf{r}, t) = 1. \tag{3.24}$$

Note that $S(\boldsymbol{\kappa}, \omega)$ contains $S_i(\boldsymbol{\kappa}, \omega)$. Moreover, in the limit of large $\boldsymbol{\kappa}$, which corresponds to short wavelengths, coherent processes are minimal so that $S(\boldsymbol{\kappa}, \omega) \doteq S_i(\boldsymbol{\kappa}, \omega)$.

At this juncture in our discussion we ask what physical significance can be attached to the two functions $G_s(\mathbf{r}, t)$ and $G(\mathbf{r}, t)$ that have been introduced in this and the previous section. For this we examine the functions for special, simple conditions, since $G_s$ and $G$ are in general complicated functions on account of the fact that the operators involved in their definitions do not as a rule commute.

If the system under discussion is classical and all nuclei are equivalent (this enables us to drop the averaging $(1/N)\sum_j$ in both $G_s$ and $G$), then, clearly,

$$G_s^{cl}(\mathbf{r}, t) = \langle \delta\{\mathbf{r} - \mathbf{R}_0(t) + \mathbf{R}_0\}\rangle, \tag{3.25}$$

and this is the probability that, given a particle at the origin at time $t = 0$, the same particle is at $\mathbf{r}$ at time $t$. Also, from (3.11),

$$G^{cl}(\mathbf{r}, t) = \sum_j \langle \delta\{\mathbf{r} - \mathbf{R}_j(t) + \mathbf{R}_0\}\rangle, \tag{3.26}$$

and this is just the probability that, given a particle at the origin at time $t = 0$, any particle (including the original particle) is to be found at the position $\mathbf{r}$ at time $t$.

Returning to the general discussion, the operators involved in the definitions of $G_s$ and $G$ commute at equal times. Thus,

$$G_s(\mathbf{r}, 0) = \delta(\mathbf{r}) \tag{3.27}$$

and

$$G(\mathbf{r}, 0) = \delta(\mathbf{r}) + \sum_{j \neq 0} \langle \delta(\mathbf{r} - \mathbf{R}_j + \mathbf{R}_0) \rangle = \delta(\mathbf{r}) + \rho g(\mathbf{r}), \tag{3.28}$$

where $g(\mathbf{r})$ is the *static pair-distribution function,* and gives the average particle-density about a given particle.

Consider now the limit $t \to \infty$ for a bulk sample. In this instance there is, clearly, no correlation between the two terms on the right-hand sides of (3.11) and (3.23). From (3.12) it is clear that $G$ in the limit $t \to \infty$ can be written

$$G(\mathbf{r}, \infty) = \frac{1}{N} \int d\mathbf{r}' \langle \rho(\mathbf{r}' - \mathbf{r}) \rangle \langle \rho(\mathbf{r}') \rangle. \tag{3.29}$$

In the case of a liquid the density is uniform, i.e. $\langle \rho(\mathbf{r}) \rangle = \rho$. For this particular case (3.29) reduces to

$$G(\mathbf{r}, \infty) = \frac{1}{N} \int d\mathbf{r}' \, \rho^2 = \rho. \tag{3.30}$$

Also for this liquid example

$$G_s(\mathbf{r}, \infty) = 0. \tag{3.31}$$

If we compare (3.27) and (3.31), and (3.28) and (3.30) then $G_s(\mathbf{r}, t)$ and $G(\mathbf{r}, t)$ are seen to possess a pronounced local structure for $t = 0$, but as $t$ increases this dies away, and in the limit $t \to \infty$ the functions are independent of $\mathbf{r}$.

The results (3.30) and (3.31) are applicable only to the example of liquids. Thus, for example, for a rigid crystal $G_s(\mathbf{r}, \infty)$ would be the same as $G_s(\mathbf{r}, 0)$, i.e. $\delta(\mathbf{r})$, and for a real crystal, as we shall show in Chapter 4, $G_s(\mathbf{r}, \infty)$ is a Gaussian centred around the parent lattice site.

In general we can write

$$G(\mathbf{r}, t) = G(\mathbf{r}, \infty) + G'(\mathbf{r}, t), \tag{3.32}$$

so that

$$\lim_{t \to \infty} G'(\mathbf{r}, t) = 0. \tag{3.33}$$

Similarly, we define

$$G_s(\mathbf{r}, t) = G_s(\mathbf{r}, \infty) + G'_s(\mathbf{r}, t), \tag{3.34}$$

where again

$$\lim_{t \to \infty} G'_s(\mathbf{r}, t) = 0. \tag{3.35}$$

The cross-sections $(d^2\sigma/d\Omega\, dE')_{\text{coh}}$ and $(d^2\sigma/d\Omega\, dE')_{\text{incoh}}$ can now be written as the sum of elastic and inelastic parts

$$\left(\frac{d^2\sigma}{d\Omega\, dE'}\right)_{\text{coh}} = \left(\frac{d^2\sigma}{d\Omega\, dE'}\right)^{\text{el}}_{\text{coh}} + \left(\frac{d^2\sigma}{d\Omega\, dE'}\right)^{\text{inel}}_{\text{coh}} \tag{3.36}$$

and

$$\left(\frac{\mathrm{d}^2\sigma}{\mathrm{d}\Omega\,\mathrm{d}E'}\right)_{\text{incoh}} = \left(\frac{\mathrm{d}^2\sigma}{\mathrm{d}\Omega\,\mathrm{d}E'}\right)^{\text{el}}_{\text{incoh}} + \left(\frac{\mathrm{d}^2\sigma}{\mathrm{d}\Omega\,\mathrm{d}E'}\right)^{\text{inel}}_{\text{incoh}} \tag{3.37}$$

where

$$\left(\frac{\mathrm{d}^2\sigma}{\mathrm{d}\Omega\,\mathrm{d}E'}\right)^{\text{el}}_{\text{coh}} = N\frac{\sigma_{\text{c}}}{4\pi}\,\delta(\hbar\omega)\int \mathrm{d}\mathbf{r}\,\exp(\mathrm{i}\boldsymbol{\kappa}\cdot\mathbf{r})G(\mathbf{r},\infty)$$

$$= \frac{\sigma_{\text{c}}}{4\pi}\,\delta(\hbar\omega)\left|\int \mathrm{d}\mathbf{r}\,\exp(\mathrm{i}\boldsymbol{\kappa}\cdot\mathbf{r})\langle\rho(\mathbf{r})\rangle\right|^2,$$

where we have used (3.29) which is valid for bulk samples. Thus

$$\left(\frac{\mathrm{d}\sigma}{\mathrm{d}\Omega}\right)^{\text{el}}_{\text{coh}} = \frac{\sigma_{\text{c}}}{4\pi}\left|\int \mathrm{d}\mathbf{r}\,\exp(\mathrm{i}\boldsymbol{\kappa}\cdot\mathbf{r})\langle\rho(\mathbf{r})\rangle\right|^2 \tag{3.38}$$

and

$$\left(\frac{\mathrm{d}^2\sigma}{\mathrm{d}\Omega\,\mathrm{d}E'}\right)^{\text{inel}}_{\text{coh}} = \frac{N\sigma_{\text{c}}}{4\pi}\frac{k'}{k}\frac{1}{2\pi\hbar}\int_{-\infty}^{\infty}\mathrm{d}t\,\exp(-\mathrm{i}\omega t)\int \mathrm{d}\mathbf{r}\,\exp(\mathrm{i}\boldsymbol{\kappa}\cdot\mathbf{r})G'(\mathbf{r},t). \tag{3.39}$$

Similarly

$$\left(\frac{\mathrm{d}^2\sigma}{\mathrm{d}\Omega\,\mathrm{d}E'}\right)^{\text{el}}_{\text{incoh}} = N\frac{\sigma_{\text{i}}}{4\pi}\,\delta(\hbar\omega)\int \mathrm{d}\mathbf{r}\,\exp(\mathrm{i}\boldsymbol{\kappa}\cdot\mathbf{r})G_{\text{s}}(\mathbf{r},\infty)$$

$$= \frac{\sigma_{\text{i}}}{4\pi}\,\delta(\hbar\omega)\int \mathrm{d}\mathbf{r}\,\exp(\mathrm{i}\boldsymbol{\kappa}\cdot\mathbf{r})$$

$$\times\sum_j\int \mathrm{d}\mathbf{r}'\langle\delta(\mathbf{r}-\mathbf{r}'+\mathbf{R}_j)\rangle\langle\delta(\mathbf{r}'-\mathbf{R}_j)\rangle.$$

Hence, for a bulk sample,

$$\left(\frac{\mathrm{d}\sigma}{\mathrm{d}\Omega}\right)^{\text{el}}_{\text{incoh}} = \frac{\sigma_{\text{i}}}{4\pi}\sum_j|\langle\exp(\mathrm{i}\boldsymbol{\kappa}\cdot\mathbf{R}_j)\rangle|^2. \tag{3.40}$$

Finally we have

$$\left(\frac{\mathrm{d}^2\sigma}{\mathrm{d}\Omega\,\mathrm{d}E'}\right)^{\text{inel}}_{\text{incoh}} = N\frac{k'}{k}\frac{\sigma_{\text{i}}}{4\pi}\frac{1}{2\pi\hbar}\int_{-\infty}^{\infty}\mathrm{d}t\,\exp(-\mathrm{i}\omega t)\int \mathrm{d}\mathbf{r}\,\exp(\mathrm{i}\boldsymbol{\kappa}\cdot\mathbf{r})G'_{\text{s}}(\mathbf{r},t). \tag{3.41}$$

The formulae (3.38)–(3.41) are valid for any system with just one chemical species present. They are easily generalized to mixtures.

As a very simple example, consider the scattering from an array of bound nuclei, a problem already discussed in Chapters 1 and 2. In this instance $\mathbf{R}_j \equiv \mathbf{l}$ (we assume a Bravais lattice) and both $G'(\mathbf{r}, t)$ and $G'_{\text{s}}(\mathbf{r}, t)$ are zero. The number density is simply

$$\rho(\mathbf{r}) = \sum_l \delta(\mathbf{r}-\mathbf{l}).$$

Hence (3.38) becomes

$$\left(\frac{\mathrm{d}\sigma}{\mathrm{d}\Omega}\right)^{\mathrm{el}}_{\mathrm{coh}} = \frac{\sigma_{\mathrm{c}}}{4\pi}\left|\int \mathrm{d}\mathbf{r}\,\exp(\mathrm{i}\boldsymbol{\kappa}\cdot\mathbf{r})\sum_{l}\delta(\mathbf{r}-\mathbf{l})\right|^2 = \frac{\sigma_{\mathrm{c}}}{4\pi}\left|\sum_{l}\exp(\mathrm{i}\boldsymbol{\kappa}\cdot\mathbf{l})\right|^2.$$

This is just the result derived in Chapter 1. (3.40) is simply

$$\left(\frac{\mathrm{d}\sigma}{\mathrm{d}\Omega}\right)^{\mathrm{el}}_{\mathrm{incoh}} = N\frac{\sigma_{\mathrm{i}}}{4\pi}.$$

## 3.3. Properties of correlation functions and $S(\boldsymbol{\kappa}, \omega)$

The aim of this section is to verify the basic properties of $S(\boldsymbol{\kappa}, \omega)$, and to derive some results which will be useful in subsequent developments. For the most part, the discussion will be couched in terms of the properties of the correlation function that appears in the definition of $S(\boldsymbol{\kappa}, \omega)$ as a time Fourier transform. We shall begin by recording the pertinent properties of the correlation function. For example, the correlation function must possess a property that ensures that its time Fourier transform, $S(\boldsymbol{\kappa}, \omega)$, is a purely real quantity. The emergence of this requirement is not obvious at first sight, perhaps, given that the Fourier integral involves complex quantities.

The relationship (3.17) forms the starting point of the discussion. This shows that $S(\boldsymbol{\kappa}, \omega)$ is the time Fourier transform of a correlation function formed with the operator $\hat{\rho}_{\boldsymbol{\kappa}}$ (eqn (3.16)) and its Hermitian conjugate $\hat{\rho}^{+}_{\boldsymbol{\kappa}} = \hat{\rho}_{-\boldsymbol{\kappa}}$. For ease of notation we will often use the shorthand $\hat{\rho}_{\boldsymbol{\kappa}}(0) = \hat{\rho}_{\boldsymbol{\kappa}}$.

In all the properties of matter discussed in this book it is assumed that the system being studied obeys the fundamental condition of stationarity. For the correlation function this means that the origin of time is arbitrary, or

$$\langle \hat{\rho}_{\boldsymbol{\kappa}}(t_0)\hat{\rho}^{+}_{\boldsymbol{\kappa}}(t+t_0)\rangle = \langle \hat{\rho}_{\boldsymbol{\kappa}}(0)\hat{\rho}^{+}_{\boldsymbol{\kappa}}(t)\rangle, \tag{3.42}$$

for any value of $t_0$. With the choice $t_0 = -t$ in (3.42) we obtain a particularly useful identity

$$\langle \hat{\rho}_{\boldsymbol{\kappa}}(-t)\hat{\rho}^{+}_{\boldsymbol{\kappa}}\rangle = \langle \hat{\rho}_{\boldsymbol{\kappa}}\hat{\rho}^{+}_{\boldsymbol{\kappa}}(t)\rangle. \tag{3.43}$$

We shall now record two more identities. First, the complex conjugate of the correlation function satisfies

$$\begin{aligned}\langle \hat{\rho}_{\boldsymbol{\kappa}}\hat{\rho}^{+}_{\boldsymbol{\kappa}}(t)\rangle^{*} &= \langle \{\hat{\rho}^{+}_{\boldsymbol{\kappa}}(t)\}^{+}\hat{\rho}^{+}_{\boldsymbol{\kappa}}\rangle \\ &= \langle \hat{\rho}_{\boldsymbol{\kappa}}(t)\hat{\rho}^{+}_{\boldsymbol{\kappa}}\rangle. \end{aligned} \tag{3.44}$$

This identity is readily proved by expressing the correlation function in terms of matrix elements. The identities (3.43) and (3.44) enable us to verify that $S(\boldsymbol{\kappa}, \omega)$ is purely real. The last identity of immediate interest is

$$\langle\hat{\rho}_{\boldsymbol{\kappa}}(t)\hat{\rho}^{+}_{\boldsymbol{\kappa}}\rangle = \langle\hat{\rho}^{+}_{\boldsymbol{\kappa}}\hat{\rho}_{\boldsymbol{\kappa}}(t + \mathrm{i}\hbar\beta)\rangle, \quad \beta = 1/k_{\mathrm{B}}T, \tag{3.45}$$

which, like (3.44), is readily proved by expressing the correlation function in terms of matrix elements.

Two properties of the correlation function for a classical system are readily deduced from eqns (3.43)–(3.45). Taking the limit $\hbar \to 0$ in (3.45) shows that the correlation function is unaltered by the interchange of the operators, which do not remain operators in the sense of quantum mechanics after the limit is taken. Applied to (3.43) this means that, with $\hat{\rho} = \hat{\rho}^{+}$, the classical correlation function is an even function of time and, from (3.44), that it is purely real. A corollary of these findings is that the complex character of the correlation function is a quantum mechanical effect.

As a consequence of the reality of $S(\boldsymbol{\kappa}, \omega)$ it follows from (3.14) that $G(\mathbf{r}, t)$ satisfies

$$G(\mathbf{r}, t) = G^{*}(-\mathbf{r}, -t). \tag{3.46}$$

We also note at this point that the fact that $G(\mathbf{r}, t)$ is a complex function is directly related to the quantum mechanical aspect of the problem. This is made clear by (3.12), from which

$$\begin{aligned} \operatorname{Im} G(\mathbf{r}, t) &= \frac{1}{2\mathrm{i}N}\int \mathrm{d}\mathbf{r}'\langle\hat{\rho}(\mathbf{r}' - \mathbf{r})\hat{\rho}(\mathbf{r}', t) - \hat{\rho}(\mathbf{r}', t)\hat{\rho}(\mathbf{r}' - \mathbf{r})\rangle \\ &= \frac{\mathrm{i}}{2N}\int \mathrm{d}\mathbf{r}'\langle[\hat{\rho}(\mathbf{r}', t), \hat{\rho}(\mathbf{r}' - \mathbf{r})]\rangle. \end{aligned} \tag{3.47}$$

In the classical limit the commutator in (3.47) would vanish, and hence the classical correlation functions would be real, as is evident directly from eqns (3.25) and (3.26). Since $\hat{\rho}(\mathbf{r})$ is both real and Hermitian (cf. Appendix B, eqns (B.70) and (B.71)), and

$$\langle\hat{\rho}(\mathbf{r}, t)\hat{\rho}(\mathbf{r}')\rangle = \langle\hat{\rho}(\mathbf{r}', t)\hat{\rho}(\mathbf{r})\rangle, \tag{3.48}$$

and it therefore follows directly from (3.47) that $\operatorname{Im} G(\mathbf{r}, t)$ is an odd function of $t$. Similarly it can be shown that $\operatorname{Re} G(\mathbf{r}, t)$ is an even function of $t$.

We also note that because $S(\boldsymbol{\kappa}, \omega)$ is essentially the cross-section it must be positive definite. The analytical consequences of this show up most clearly in terms of the function $I(\boldsymbol{\kappa}, t)$ defined by (3.18). Inverting (3.19) gives

$$I(\boldsymbol{\kappa}, t) = \frac{1}{N}\langle\hat{\rho}_{\boldsymbol{\kappa}}\hat{\rho}_{-\boldsymbol{\kappa}}(t)\rangle = \hbar\int \mathrm{d}\omega \exp(\mathrm{i}\omega t)S(\boldsymbol{\kappa}, \omega),$$

$$-\left(\frac{\partial}{\partial t}\right)^{2} I(\boldsymbol{\kappa}, t) = \hbar\int \mathrm{d}\omega \exp(\mathrm{i}\omega t)\omega^{2}S(\boldsymbol{\kappa}, \omega),$$

and

$$\left(\frac{\partial}{\partial t}\right)^4 I(\boldsymbol{\kappa}, t) = \hbar \int d\omega \exp(i\omega t)\omega^4 S(\boldsymbol{\kappa}, \omega), \quad \text{etc.}$$

Putting $t=0$ in this series of equations we note that the integrals on the right-hand side are all positive definite and non-zero (unless the scattering is strictly elastic). It follows that either $I(\boldsymbol{\kappa}, t)$ is strictly independent of $t$, i.e. the target is rigidly fixed and the scattering is strictly elastic, or $I(\boldsymbol{\kappa}, t)$ at small times has a negative non-zero second derivative and a positive non-zero fourth derivative, with respect to $t$.

From the representation of $S(\boldsymbol{\kappa}, \omega)$ given by eqn (3.17) it is a simple matter to extract the so-called *condition of detailed balance.*

We find from (3.45)

$$S(\boldsymbol{\kappa}, \omega) = \exp(\hbar\omega\beta) S(-\boldsymbol{\kappa}, -\omega). \tag{3.49}$$

The physical significance of (3.49) is easily understood. When the incident neutrons come into thermal equilibrium with the target after many collisions with the nuclei, the number of neutrons that gain a given energy must equal the number that lose the same energy. Equation (3.49) is a mathematical statement of this fact, i.e. the probability that a neutron loses an energy $\hbar\omega$ is equal to $\exp(\hbar\omega\beta)$, where $\beta$ is the inverse temperature, times the probability that a neutron gains an energy $\hbar\omega$.

Note that if we define

$$\tilde{S}(\boldsymbol{\kappa}, \omega) = \exp(-\tfrac{1}{2}\hbar\omega\beta) S(\boldsymbol{\kappa}, \omega) \tag{3.50}$$

then it follows from (3.49) that $\tilde{S}(\boldsymbol{\kappa}, \omega)$ is an even function in both $\boldsymbol{\kappa}$ and $\omega$. Thus, if in analogy with (3.14) we define $\tilde{G}(\mathbf{r}, t)$ such that

$$\tilde{G}(\mathbf{r}, t) = \frac{\hbar}{(2\pi)^3} \int d\boldsymbol{\kappa} \int d\omega \, \exp(-i\boldsymbol{\kappa} \cdot \mathbf{r} + i\omega t) \tilde{S}(\boldsymbol{\kappa}, \omega) = G(\mathbf{r}, t + \tfrac{1}{2}i\hbar\beta), \tag{3.51}$$

then $\tilde{G}(\mathbf{r}, t)$ is purely real, and even in both $\mathbf{r}$ and $t$.

Finally we ask for the relation for $G(\mathbf{r}, t)$ that is equivalent to (3.49) for $S(\boldsymbol{\kappa}, \omega)$. It is straightforward to show with the aid of the identity (3.45) (or directly from (3.49)) that

$$G(\mathbf{r}, t) = G(-\mathbf{r}, -t + i\hbar\beta). \tag{3.52}$$

The next topic is the relation of $S(\boldsymbol{\kappa}, \omega)$ to the linear response function

$$\phi_{\boldsymbol{\kappa}}(t) = \frac{i}{\hbar N} \langle [\hat{\rho}_{\boldsymbol{\kappa}}(t), \hat{\rho}^+_{\boldsymbol{\kappa}}] \rangle. \tag{3.53}$$

It is established in Appendix B that $\phi_{\boldsymbol{\kappa}}(t)$ determines the linear response of the system, as measured by the change in $\hat{\rho}_{\boldsymbol{\kappa}}$, to a probe that couples to the system through the operator $\hat{\rho}^+_{\boldsymbol{\kappa}}$. The relation

$$S(\boldsymbol{\kappa}, \omega) = \{1 - \exp(-\hbar\omega\beta)\}^{-1} \frac{1}{2\pi \mathrm{i}} \int_{-\infty}^{\infty} \mathrm{d}t \exp(\mathrm{i}\omega t) \phi_{\boldsymbol{\kappa}}(t) \quad (3.54)$$

can be verified by using the identities (3.43) and (3.45). The expression (3.54) is often called the fluctuation–dissipation theorem since it relates the spectrum of spontaneous fluctuations $S(\boldsymbol{\kappa}, \omega)$ to the dissipative part of the response function. For neutron scattering, the response and spontaneous fluctuation spectrum are one and the same and we will continue to call $S(\boldsymbol{\kappa}, \omega)$ the response function.

Using the fact that $S(\boldsymbol{\kappa}, \omega)$ in (3.54) is real, we can deduce that

$$\phi^*_{\boldsymbol{\kappa}}(t) = -\phi_{\boldsymbol{\kappa}}(-t).$$

From the definition of $\phi_{\boldsymbol{\kappa}}(t)$ and the identity (3.44) it follows that $\phi_{\boldsymbol{\kappa}}(t)$ is purely real. We conclude that $\phi_{\boldsymbol{\kappa}}(t)$ is also an odd function of time,

$$\phi_{\boldsymbol{\kappa}}(t) = -\phi_{\boldsymbol{\kappa}}(-t). \quad (3.55)$$

From this it follows that the Fourier integral on the right-hand side of (3.54) is an odd function of $\omega$, and this behaviour is consistent with the condition (3.49). In this context, the factor $\{1 - \exp(-\hbar\omega\beta)\}^{-1} = 1 + n(\omega)$ on the right-hand side of (3.54) is often referred to as the detailed balance factor.

One virtue of the relation (3.54) is that it enables us to obtain useful results for the moments of $S(\boldsymbol{\kappa}, \omega)$. Inversion of the Fourier transform in (3.54) produces the result

$$\mathrm{i} \int_{-\infty}^{\infty} \mathrm{d}\omega \exp(-\mathrm{i}\omega t)\{1 - \exp(-\hbar\omega\beta)\} S(\boldsymbol{\kappa}, \omega) = \phi_{\boldsymbol{\kappa}}(t) \quad (3.56)$$

and therefore

$$\int_{-\infty}^{\infty} \mathrm{d}\omega\, \omega \{1 - \exp(-\hbar\omega\beta)\} S(\boldsymbol{\kappa}, \omega) = \partial_t \phi_{\boldsymbol{\kappa}}(t)|_{t=0}. \quad (3.57)$$

Consider the left-hand side of this relation. Provided that $S(\boldsymbol{\kappa}, \omega) = S(-\boldsymbol{\kappa}, \omega)$, the condition of detailed balance (3.49) leads to the result

$$2 \int_{-\infty}^{\infty} \mathrm{d}\omega\, \omega S(\boldsymbol{\kappa}, \omega).$$

A similar reduction is possible for any odd power of $\omega$. This means that the odd moments of $S(\boldsymbol{\kappa}, \omega)$ are related directly to time derivatives of the response function evaluated at $t = 0$. Turning to the evaluation of the first derivative of $\phi_{\boldsymbol{\kappa}}(t)$, which occurs on the right-hand side of (3.57), we note that the equation of motion for a Heisenberg operator is

$$\mathrm{i}\hbar\, \partial_t \hat{\rho}_{\boldsymbol{\kappa}}(t) = [\hat{\rho}_{\boldsymbol{\kappa}}(t), \hat{\mathcal{H}}] \quad (3.58)$$

where $\hat{\mathcal{H}}$ is the Hamiltonian. We then have

$$\partial_t \phi_{\boldsymbol{\kappa}}(t)|_{t=0} = \frac{1}{N\hbar^2}\langle[[\hat{\rho}_{\boldsymbol{\kappa}}, \hat{\mathcal{H}}], \hat{\rho}^{+}_{\boldsymbol{\kappa}}]\rangle$$

$$= -\frac{1}{N\hbar^2}\langle[\hat{\rho}_{\boldsymbol{\kappa}}, [\hat{\rho}^{+}_{\boldsymbol{\kappa}}, \hat{\mathcal{H}}]]\rangle \qquad (3.59)$$

and the second equality demonstrates a useful relation for nested commutators that involve the Hamiltonian. On combining the various results that have just been obtained, the first moment of $S(\boldsymbol{\kappa}, \omega)$ obeys the relation

$$\int_{-\infty}^{\infty} d\omega\, \omega S(\boldsymbol{\kappa}, \omega) = -\frac{1}{2\hbar^2 N}\langle[\hat{\rho}_{\boldsymbol{\kappa}}, [\hat{\rho}_{-\boldsymbol{\kappa}}, \hat{\mathcal{H}}]]\rangle. \qquad (3.60)$$

To discuss this it is convenient to evaluate (3.60) for non-interacting nuclei of mass $M$, i.e. for

$$\hat{\mathcal{H}} = \sum_{j=1}^{N} \frac{1}{2M}\hat{\mathbf{p}}_j^2.$$

First we need

$$[\hat{\rho}_{-\boldsymbol{\kappa}}, \hat{\mathcal{H}}] = \sum_{j,j'} \frac{1}{2M}[\exp(\mathrm{i}\boldsymbol{\kappa}\cdot\mathbf{R}_j), \hat{\mathbf{p}}_{j'}^2].$$

This commutator is evaluated by writing it as

$$[\exp(\mathrm{i}\boldsymbol{\kappa}\cdot\mathbf{R}_j), \hat{\mathbf{p}}_{j'}^2] = [\exp(\mathrm{i}\boldsymbol{\kappa}\cdot\mathbf{R}_j), \hat{\mathbf{p}}_{j'}]\hat{\mathbf{p}}_{j'} + \hat{\mathbf{p}}_{j'}[\exp(\mathrm{i}\boldsymbol{\kappa}\cdot\mathbf{R}_j), \hat{\mathbf{p}}_{j'}]$$

and making use of the result that, if $f(\mathbf{r})$ is an arbitrary function of position,

$$[f(\mathbf{r}), \hat{\mathbf{p}}] = \mathrm{i}\hbar \operatorname{grad} f(\mathbf{r}). \qquad (3.61)$$

Hence

$$[\exp(\mathrm{i}\boldsymbol{\kappa}\cdot\mathbf{R}_j), \hat{\mathbf{p}}_{j'}^2] = \delta_{j,j'}\{-\hbar\boldsymbol{\kappa}\exp(\mathrm{i}\boldsymbol{\kappa}\cdot\mathbf{R}_j)\hat{\mathbf{p}}_j - \hat{\mathbf{p}}_j\hbar\boldsymbol{\kappa}\exp(\mathrm{i}\boldsymbol{\kappa}\cdot\mathbf{R}_j)\}$$
$$= -\delta_{j,j'}\exp(\mathrm{i}\boldsymbol{\kappa}\cdot\mathbf{R}_j)\{2\hbar\boldsymbol{\kappa}\cdot\hat{\mathbf{p}}_j + \hbar^2\kappa^2\}.$$

So

$$[\hat{\rho}_{-\boldsymbol{\kappa}}, \hat{\mathcal{H}}] = -\frac{1}{2M}\sum_j \exp(\mathrm{i}\boldsymbol{\kappa}\cdot\mathbf{R}_j)\{2\hbar\boldsymbol{\kappa}\cdot\hat{\mathbf{p}}_j + \hbar^2\kappa^2\}$$

and

$$[\hat{\rho}_{\boldsymbol{\kappa}}, [\hat{\rho}_{-\boldsymbol{\kappa}}, \hat{\mathcal{H}}]] = -\frac{1}{2M}\sum_{j,j'} \delta_{j,j'}\exp(\mathrm{i}\boldsymbol{\kappa}\cdot\mathbf{R}_j)2\hbar\boldsymbol{\kappa}\cdot[\exp(-\mathrm{i}\boldsymbol{\kappa}\cdot\mathbf{R}_j), \hat{\mathbf{p}}_j]$$
$$= -\frac{1}{2M}\sum_j 2\hbar^2\kappa^2 = -N\frac{\hbar^2\kappa^2}{M}.$$

Hence

$$\tfrac{1}{2}\langle[\hat{\rho}_{\boldsymbol{\kappa}}, [\hat{\rho}_{-\boldsymbol{\kappa}}, \mathscr{H}]]\rangle = -N\frac{\hbar^2\kappa^2}{2M}, \tag{3.62}$$

and thus

$$\int_{-\infty}^{\infty} \mathrm{d}\omega\, \omega S(\boldsymbol{\kappa}, \omega) = \frac{\kappa^2}{2M}. \tag{3.63}$$

If we assume the nuclei of the target to be at rest prior to the scattering of a neutron, then the conservation of momentum demands for each nucleus

$$\hbar\mathbf{k} = \hbar\mathbf{k}' + M\mathbf{v}', \tag{3.64}$$

where $\mathbf{v}'$ is the velocity of the nucleus after the scattering process. The kinetic energy imparted to the nucleus, its recoil energy, is

$$\frac{M}{2}v'^2 = \frac{\hbar^2\kappa^2}{2M}. \tag{3.65}$$

(3.62) is the mean energy lost by the target, which, in the light of (3.64), is equal to minus $N$ times the recoil energy of a single nucleus. This is just what we should expect; the nuclei being free and initially at rest can only gain energy from the incident neutron, this accounts for the negative sign in (3.62), and since they are independent of one another the total energy gained is $N$ times that for a single nucleus.

This simple argument gives a neat physical interpretation of (3.63), but two further points are worth noting. First, that (3.63) is a result independent of temperature; hence for scattering from a gas the first moment of the scattering law is always $\kappa^2/2M$. Second, we have only used the Hamiltonian once in the commutation leading to (3.63); hence, if we added terms representing the interactions between the nuclei in the target system, we would obtain an identical result provided those interactions were momentum independent and varied only with the nuclear positions $\mathbf{R}_j$. In practice this last point is very substantially true, and therefore we also conclude that the first moment of the response function is given to an excellent approximation by (3.63) even for a dense gas or liquid.

The calculation of higher-order moments of $S(\boldsymbol{\kappa}, \omega)$ follows in a similar manner. However, because they involve multiply nested commutations they are very tedious to evaluate. Results for some higher-order moments are given in Chapter 5.

The classical analogue of the result (3.63) can be deduced from

(3.57) and the result

$$\partial_t \phi_{\boldsymbol{\kappa}}(t)|_{t=0} = (\kappa^2/M) \tag{3.66}$$

which follows from (3.59) and (3.62). We find that

$$\lim_{\hbar \to 0} \hbar \int_{-\infty}^{\infty} \mathrm{d}\omega\, \omega^2 S(\boldsymbol{\kappa}, \omega) = (\kappa^2/M\beta). \tag{3.67}$$

The reason that we obtain a result for an even moment stems from the fact that the correlation function for a classical system is an even function of time, thus $S(\boldsymbol{\kappa}, \omega)$ is an even function of $\omega$.

## 3.4. Impulse approximation

When the incident neutron energy is very much in excess of the maximum energy available within the response spectrum of the target, it is reasonable to approximate the correlation function by its behaviour at short times. In other words, if the time of passage of the neutrons through the sample is short compared to the characteristic time scale of the dynamic response of the sample, only the short-time properties of the sample are relevant. The approximation here in deriving the impulse approximation is akin to that for the static approximation described in § 1.8. We will find that the impulse approximation includes the first correction to the static approximation, and that this correction describes inelastic processes.

Starting with the correlation function (3.6), the key approximation involved is

$$\hat{\mathbf{R}}_{j'}(t) \doteq \hat{\mathbf{R}}_{j'} + \frac{t}{M_{j'}} \hat{\mathbf{p}}_{j'} \tag{3.68}$$

where $\hat{\mathbf{p}}_{j'}$ is the momentum conjugate to $\hat{\mathbf{R}}_{j'}$, and $M_{j'}$ is the mass of the $j'$th particle. The approximation (3.68) is valid for $t \to 0$. The corresponding approximation for the correlation function is

$$Y_{jj'}(\boldsymbol{\kappa}, t) \doteq \left\langle \exp(-\mathrm{i}\boldsymbol{\kappa} \cdot \hat{\mathbf{R}}_j) \exp\left(\mathrm{i}\boldsymbol{\kappa} \cdot \hat{\mathbf{R}}_{j'} + \frac{\mathrm{i}t}{M_{j'}} \boldsymbol{\kappa} \cdot \hat{\mathbf{p}}_{j'}\right)\right\rangle. \tag{3.69}$$

To make further progress we combine the two exponentials in (3.69). In doing so we recognize that the exponents contain operators. Because the commutator of the position and momentum operators is a c-number, namely,

$$[\hat{\mathbf{R}}_{\alpha j}, \hat{\mathbf{p}}_{\beta j'}] = \mathrm{i}\hbar\, \delta_{jj'}\, \delta_{\alpha\beta} \tag{3.70}$$

which is obtained as a special case of (3.61), the following relation holds for the exponentials,

$$\exp(-i\boldsymbol{\kappa}\cdot\hat{\mathbf{R}}_j)\exp\left(i\boldsymbol{\kappa}\cdot\hat{\mathbf{R}}_{j'}+\frac{it}{M_{j'}}\boldsymbol{\kappa}\cdot\hat{\mathbf{p}}_{j'}\right)$$
$$=\exp\left(-i\boldsymbol{\kappa}\cdot\hat{\mathbf{R}}_j+i\boldsymbol{\kappa}\cdot\hat{\mathbf{R}}_{j'}+\frac{it}{M_{j'}}\boldsymbol{\kappa}\cdot\hat{\mathbf{p}}_{j'}+\frac{1}{2}\left[\boldsymbol{\kappa}\cdot\hat{\mathbf{R}}_j,\left(\boldsymbol{\kappa}\cdot\hat{\mathbf{R}}_{j'}+\frac{t}{M_{j'}}\boldsymbol{\kappa}\cdot\hat{\mathbf{p}}_{j'}\right)\right]\right). \tag{3.71}$$

The identity employed here is obtained from the more general one given in the following section. The commutator in (3.71) is straightforward, since the position operators (at the same time $t=0$) commute, and the remaining commutator involves (3.70). For the correlation function (3.69) we obtain

$$Y_{jj'}(\boldsymbol{\kappa},t)\doteq\exp\left(\frac{i\hbar t\kappa^2}{2M_j}\delta_{jj'}\right)\left\langle\exp\left(i\boldsymbol{\kappa}\cdot(\hat{\mathbf{R}}_{j'}-\hat{\mathbf{R}}_j)+\frac{it}{M_{j'}}\boldsymbol{\kappa}\cdot\hat{\mathbf{p}}_{j'}\right)\right\rangle. \tag{3.72}$$

The cross-section obtained with the approximation $Y_{jj'}(\boldsymbol{\kappa},t)=Y_{jj'}(\boldsymbol{\kappa},0)$ is identical to the static approximation described in § 1.8.

The impulse approximation is derived from the term in (3.72) with $j=j'$. The reasoning behind this additional approximation is based on the assumption that large $\boldsymbol{\kappa}$ values are observed, for in this case coherent effects, arising from terms $j\neq j'$, are negligible. Thus, for energetic incident neutrons and large $\boldsymbol{\kappa}$, the dynamic structure factor is approximated by

$$S(\boldsymbol{\kappa},\omega)\doteq\frac{1}{2\pi\hbar N}\sum_j\int_{-\infty}^{\infty}dt\,\exp\left(-i\omega t+\frac{i\hbar t\kappa^2}{2M_j}\right)\left\langle\exp\left(\frac{it}{M_j}\boldsymbol{\kappa}\cdot\hat{\mathbf{p}}_j\right)\right\rangle. \tag{3.73}$$

By turning to § 3.6.1 we find that the impulse approximation (3.73) has the form obtained for free particles. This feature of the approximation reflects the fact that particle correlations are not apparent at short times. Alternatively, we recognize that, in a high-energy collision, the sample particle will behave as if it were free. In evaluating (3.73), however, the correlation function is evaluated for the conditions appropriate to the sample particle and not merely replaced by the correlation function for a free particle derived in § 3.6.1.

We consider two special applications of this result. First, when the motion of a particle approximates to that of a particle in an isotropic harmonic potential,

$$\left\langle \exp\left(\frac{it}{M_j}\boldsymbol{\kappa}\cdot\hat{\mathbf{p}}_j\right)\right\rangle = \exp\left\{\frac{-t^2}{2M_j^2}\langle(\boldsymbol{\kappa}\cdot\hat{\mathbf{p}}_j)^2\rangle\right\}. \tag{3.74}$$

This result is proved at the end of the chapter. The calculation of (3.73) is reduced now to a single integral. This integral amounts to the Fourier transform of a Gaussian function and the answer is a Gaussian; the result required is

$$\int_{-\infty}^{\infty} \mathrm{d}x \exp(-ax^2 + bx) = \sqrt{(\pi/a)}\exp(b^2/4a). \tag{3.75}$$

Hence, for a harmonic system the impulse approximation gives the result

$$S(\boldsymbol{\kappa}, \omega) \doteq \frac{1}{\hbar N}\sum_j \{2\pi\Delta_j^2\}^{-1/2} \exp\left\{-\left(\omega - \frac{\hbar\kappa^2}{2M_j}\right)^2 \Big/ 2\Delta_j^2\right\} \tag{3.76}$$

where the energy width of the Gaussian function is

$$\hbar\Delta_j = \frac{\hbar}{M_j}\{\langle(\boldsymbol{\kappa}\cdot\hat{\mathbf{p}}_j)^2\rangle\}^{1/2}. \tag{3.77}$$

From this result it follows that at high energies the dynamic structure factor is formed by Gaussians centred at the particle recoil energy $\hbar^2\kappa^2/2M_j$. The energy width increases with $\kappa$ and is related to the mean-square velocity of the particles.

The second application of (3.73) amounts to observing its form when the correlation function is evaluated in terms of momentum states. Labelling these states by wave vectors $\mathbf{q}$,

$$\hat{\mathbf{p}}\,|\mathbf{q}\rangle = \hbar\mathbf{q}\,|\mathbf{q}\rangle$$

and therefore

$$\left\langle \exp\left(\frac{it}{M}\boldsymbol{\kappa}\cdot\hat{\mathbf{p}}\right)\right\rangle = \sum_{\mathbf{q}} n_{\mathbf{q}} \exp\left(\frac{it\hbar}{M}\boldsymbol{\kappa}\cdot\mathbf{q}\right), \tag{3.78}$$

where $n_{\mathbf{q}}$ is the thermal distribution function. Inserting (3.78) into (3.73) yields the impulse approximation in the momentum representation

$$\hbar S(\boldsymbol{\kappa}, \omega) \doteq \sum_{\mathbf{q}} n_{\mathbf{q}}\, \delta\left(\omega - \frac{\hbar\kappa^2}{2M} - \frac{\hbar}{M}\boldsymbol{\kappa}\cdot\mathbf{q}\right), \tag{3.79}$$

where it has been assumed that the particles are identical so that the sum on $j$ in (3.73) gives a factor $N$. From the result (3.79) we deduce that, under the appropriate conditions, the dynamic structure factor is related to the momentum distribution of the particles. Corrections to the impulse approximation can be generated by including additional terms in the Taylor series expansion of $\hat{\mathbf{R}}(t)$ (eqn (3.68)). The ensuing manipulations are straightforward but tedious.

## 3.5. Classical approximation

In some instances the quantum mechanical nature of $S(\boldsymbol{\kappa}, \omega)$ can justifiably be neglected, and this greatly simplifies the calculation of the cross-section. In this section we seek a rigorous prescription for effecting this approximation, it being clearly incorrect merely to replace $G(\mathbf{r}, t)$ by the corresponding classical function.

In our discussion of the physical significance of $G(\mathbf{r}, t)$ and $G_s(\mathbf{r}, t)$ we noted that if we treat the operators involved as c-numbers, the resulting expressions, eqns (3.25) and (3.26), admit a very simple interpretation, but that in general it is not possible to give $G$ and $G_s$ such simple meanings. Bearing this in mind we re-write the operators involved in the definition of $S(\boldsymbol{\kappa}, \omega)$ to bring out their non-commuting nature, namely we write

$$\exp(-\mathrm{i}\boldsymbol{\kappa}\cdot\hat{\mathbf{R}}_j)\exp\{\mathrm{i}\boldsymbol{\kappa}\cdot\hat{\mathbf{R}}_{j'}(t)\} = \exp\{-\mathrm{i}\boldsymbol{\kappa}\cdot\{\hat{\boldsymbol{R}}_j - \hat{\mathbf{R}}_{j'}(t)\} + \mathrm{i}\hbar\hat{\Gamma}_{jj'}(t)\}. \tag{3.80}$$

To determine $\hat{\Gamma}$ we use the identity

$$\exp \hat{A} \exp \hat{B} = \exp(\hat{A} + \hat{B} + \hat{C}),$$

where

$$\hat{C} = \tfrac{1}{2}[\hat{A}, \hat{B}] + \tfrac{1}{12}[[\hat{A}, \hat{B}], \hat{B}] + \tfrac{1}{12}[[\hat{B}, \hat{A}], \hat{A}] + \tfrac{1}{24}[[[\hat{B}, \hat{A}], \hat{A}], \hat{B}] + \cdots, \tag{3.81}$$

and set

$$\hat{A} = -\mathrm{i}\boldsymbol{\kappa}\cdot\hat{\mathbf{R}}_j$$

and

$$\hat{B} = \mathrm{i}\boldsymbol{\kappa}\cdot\hat{\mathbf{R}}_{j'}(t).$$

Hence

$$\begin{aligned} \mathrm{i}\hbar\hat{\Gamma}_{jj'}(t) = {} & \tfrac{1}{2}[\boldsymbol{\kappa}\cdot\hat{\mathbf{R}}_j, \boldsymbol{\kappa}\cdot\hat{\mathbf{R}}_{j'}(t)] \\ & + \frac{\mathrm{i}}{12}[[\boldsymbol{\kappa}\cdot\hat{\mathbf{R}}_j, \boldsymbol{\kappa}\cdot\hat{\mathbf{R}}_{j'}(t)], \boldsymbol{\kappa}\cdot\hat{\mathbf{R}}_j] \\ & + \frac{\mathrm{i}}{12}[[\boldsymbol{\kappa}\cdot\hat{\mathbf{R}}_j, \boldsymbol{\kappa}\cdot\hat{\mathbf{R}}_{j'}(t)], \boldsymbol{\kappa}\cdot\hat{\mathbf{R}}_{j'}(t)] \\ & + \cdots. \end{aligned} \tag{3.82}$$

Using (3.80) we obtain for $S(\boldsymbol{\kappa}, \omega)$

$$S(\boldsymbol{\kappa}, \omega)=\frac{1}{2\pi\hbar}\int_{-\infty}^{\infty} \mathrm{d}t \exp(-\mathrm{i}\omega t)\frac{1}{N}\sum_{j,j'}\langle \exp\{-\mathrm{i}\boldsymbol{\kappa}\cdot\{\hat{\mathbf{R}}_j-\hat{\mathbf{R}}_{j'}(t)\}+\mathrm{i}\hbar\hat{\Gamma}_{jj'}(t)\}\rangle. \tag{3.83}$$

By writing $S(\boldsymbol{\kappa}, \omega)$ in this form we have in effect isolated the purely quantum mechanical features in $\hat{\Gamma}_{jj'}(t)$. The *classical approximation* to $S(\boldsymbol{\kappa}, \omega)$ is now obtained by (1) treating the operator

$$\exp[-\mathrm{i}\boldsymbol{\kappa}\cdot\{\hat{\mathbf{R}}_j-\hat{\mathbf{R}}_{j'}(t)\}]$$

as a c-number and (2) replacing the commutators in the expansion for $\hat{\Gamma}$, eqn (3.82), by $\mathrm{i}\hbar\{\ ,\ \}$, where the Poisson bracket $\{\ ,\ \}$ is defined by

$$\{A, B\}=\frac{\partial A}{\partial x}\frac{\partial B}{\partial p}-\frac{\partial A}{\partial p}\frac{\partial B}{\partial x}, \tag{3.84}$$

$p$ being the momentum variable conjugate to coordinate $x$.

If the commutator $[\boldsymbol{\kappa}\cdot\hat{\mathbf{R}}_j, \boldsymbol{\kappa}\cdot\hat{\mathbf{R}}_{j'}(t)]$ is a c-number, as is often the case, $\hat{\Gamma}$ is itself a c-number. If in addition the particles move independently of one another, $\hat{\Gamma}$ takes the form

$$\Gamma_{jj'}(t)=\delta_{j,j'}\gamma(t), \tag{3.85}$$

where, from (3.82),

$$\gamma(t)=\frac{1}{2\mathrm{i}\hbar}[\boldsymbol{\kappa}\cdot\hat{\mathbf{R}}_j, \boldsymbol{\kappa}\cdot\hat{\mathbf{R}}_j(t)]. \tag{3.86}$$

Inserting (3.85) into (3.83)

$$S(\boldsymbol{\kappa}, \omega)=\frac{1}{2\pi\hbar}\int_{-\infty}^{\infty} \mathrm{d}t \exp(-\mathrm{i}\omega t)\int \mathrm{d}\mathbf{r} \exp(\mathrm{i}\boldsymbol{\kappa}\cdot\mathbf{r})$$
$$\times[\exp\{\mathrm{i}\hbar\gamma(t)\}\bar{G}_s(\mathbf{r}, t)+\bar{G}_d(\mathbf{r}, t)], \tag{3.87}$$

where

$$\bar{G}_s(\mathbf{r}, t)=\frac{1}{N}\sum_j\langle\delta\{\mathbf{r}+\hat{\mathbf{R}}_j-\hat{\mathbf{R}}_j(t)\}\rangle \tag{3.88}$$

and

$$\bar{G}_d(\mathbf{r}, t)=\frac{1}{N}\sum_{\substack{j,j'\\ j\neq j'}}\langle\delta\{\mathbf{r}+\hat{\mathbf{R}}_j-\hat{\mathbf{R}}_{j'}(t)\}\rangle. \tag{3.89}$$

Note that $\bar{G}_s$ and $\bar{G}_d$ apparently have the same form as the classical functions $G_s^{cl}$ and $G^{cl}$ that could be defined from (3.25) and (3.26). Nevertheless the functions are not the same because (3.88) and (3.89) include quantum effects coming from the use of $\hat{\mathbf{R}}_j$ etc., as operators. Also

$$S_i(\boldsymbol{\kappa}, \omega) = \frac{1}{2\pi\hbar}\int_{-\infty}^{\infty} dt \exp\{-i\omega t + i\hbar\gamma(t)\}\int d\mathbf{r} \exp(i\boldsymbol{\kappa}\cdot\mathbf{r})\bar{G}_s(\mathbf{r}, t). \quad (3.90)$$

The two functions $\bar{G}_s(\mathbf{r}, t)$ and $\bar{G}_d(\mathbf{r}, t)$ are purely real and have a simple physical significance: $\bar{G}_s(\mathbf{r}, t)$ can be interpreted in exactly the same way as $G_s^{cl}(\mathbf{r}, t)$, eqn (3.25), while $\bar{G}_d(\mathbf{r}, t)$ is the probability of finding an atom at time $t=0$ and another at time $t$ separated by a distance $\mathbf{r}$.

In the following section we shall contrast the behaviour of $\bar{G}_s$ and $\bar{G}_d$ with the corresponding functions introduced in the previous section by evaluating them explicitly for simple models.

Before proceeding to this task it is worth while to note that $\tilde{G}(\mathbf{r}, t)$, eqn (3.51), possesses the same analytical properties as $G^{cl}(\mathbf{r}, t)$. This suggests that in some instances the cross-section may be well approximated by replacing $\tilde{G}$ by $G^{cl}$. Indeed, for the two examples of a monatomic gas and an isotropic oscillator this procedure gives a cross-section that is formally correct to first order in $\hbar$. Also note that the resulting cross-sections automatically fulfil the condition of detailed balance.

## 3.6. Simple model systems

We demonstrate how to use the formulae of the previous sections by examining in detail the scattering from two very simple model systems: (a) a *perfect fluid* and (b) a single nucleus bound in an isotropic, harmonic potential. In the course of discussing these systems we shall introduce several results that are used throughout the remainder of the book.

### 3.6.1. *Perfect fluid*

For a single free nucleus we can set $\mathbf{R}_j = \mathbf{r}$ and so from (3.8) the response function is

$$S(\boldsymbol{\kappa}, \omega) = \frac{1}{2\pi\hbar}\int_{-\infty}^{\infty} dt \exp(-i\omega t)\langle\exp(-i\boldsymbol{\kappa}\cdot\hat{\mathbf{r}})\exp\{i\boldsymbol{\kappa}\cdot\hat{\mathbf{r}}(t)\}\rangle. \quad (3.91)$$

The Hamiltonian of the target is simply

$$\mathcal{H} = \frac{M}{2}\dot{\hat{\mathbf{r}}}^2 = \frac{1}{2M}\hat{\mathbf{p}}^2,$$

where $M$ is the mass of the nucleus. The most direct way of determining the correlation function in (3.91) is to utilize an alternative expression that is useful when the Hamiltonian has a simple form, as in the present case.

Starting with the explicit form for the Heisenberg operator in the correlation function $Y_{jj'}(\boldsymbol{\kappa}, t)$ (eqn (3.6)) with $j = j'$,

$$Y_{jj}(\boldsymbol{\kappa}, t) = \langle \exp(-\mathrm{i}\boldsymbol{\kappa} \cdot \hat{\mathbf{R}}_j)\exp(\mathrm{i}\mathcal{H}t/\hbar)\exp(\mathrm{i}\boldsymbol{\kappa} \cdot \hat{\mathbf{R}}_j)\exp(-\mathrm{i}\mathcal{H}t/\hbar)\rangle.$$

The operator $\exp(\mathrm{i}\boldsymbol{\kappa} \cdot \hat{\mathbf{R}}_j)$ shifts the momentum of the $j$th particle so that

$$\exp(-\mathrm{i}\boldsymbol{\kappa} \cdot \hat{\mathbf{R}})\hat{\mathbf{p}} \exp(\mathrm{i}\boldsymbol{\kappa} \cdot \hat{\mathbf{R}}) = \hat{\mathbf{p}} + \hbar\boldsymbol{\kappa}.$$

In view of this result $Y_{jj}(\boldsymbol{\kappa}, t)$ can be written in the form

$$Y_{jj}(\boldsymbol{\kappa}, t) = \langle \exp(\mathrm{i}t\hat{\mathcal{H}}'_j/\hbar)\exp(-\mathrm{i}t\hat{\mathcal{H}}/\hbar)\rangle, \tag{3.92}$$

where the momentum of the $j$th particle in $\hat{\mathcal{H}}'_j$ is shifted by an amount $\hbar\boldsymbol{\kappa}$. Note that this representation applies to the self-correlation function only.

For the case of a single free particle,

$$\hat{\mathcal{H}}' = \frac{1}{2M}(\hat{\mathbf{p}} + \hbar\boldsymbol{\kappa})^2,$$

and from (3.92) it then follows that

$$\langle \exp(-\mathrm{i}\boldsymbol{\kappa} \cdot \hat{\mathbf{r}})\exp\{\mathrm{i}\boldsymbol{\kappa} \cdot \hat{\mathbf{r}}(t)\}\rangle = \exp(\mathrm{i}t\hbar\kappa^2/2M)\langle \exp(\mathrm{i}t\boldsymbol{\kappa} \cdot \hat{\mathbf{p}}/M)\rangle. \tag{3.93}$$

By definition of a Boltzmann particle,

$$\begin{aligned}
&\langle \exp(\mathrm{i}t\boldsymbol{\kappa} \cdot \hat{\mathbf{p}}/M)\rangle \\
&= \left\{\int \mathrm{d}\mathbf{p} \exp(-\beta p^2/2M)\exp(\mathrm{i}t\boldsymbol{\kappa} \cdot \mathbf{p}/M)\right\}\Big/\left\{\int \mathrm{d}\mathbf{p} \exp(-\beta p^2/2M\right\} \\
&= \exp(-t^2\kappa^2/2M\beta).
\end{aligned} \tag{3.94}$$

Here we have used result (3.75), and $\beta$ is the inverse temperature.

From (3.93) and (3.94) we have, finally,

$$\langle \exp(-\mathrm{i}\boldsymbol{\kappa} \cdot \hat{\mathbf{r}})\exp\{\mathrm{i}\boldsymbol{\kappa} \cdot \hat{\mathbf{r}}(t)\}\rangle = \exp\left[-\frac{\kappa^2}{2M}(t^2/\beta - \mathrm{i}\hbar t)\right]. \tag{3.95}$$

Let us now calculate both $G(\mathbf{r}, t)$ and $S(\boldsymbol{\kappa}, \omega)$. First

$$\begin{aligned}
G(\mathbf{r}, t) &= \left(\frac{1}{2\pi}\right)^3 \int \mathrm{d}\boldsymbol{\kappa} \exp(-\mathrm{i}\boldsymbol{\kappa} \cdot \mathbf{r})\langle \exp(-\mathrm{i}\boldsymbol{\kappa} \cdot \hat{\mathbf{r}})\exp\{\mathrm{i}\boldsymbol{\kappa} \cdot \hat{\mathbf{r}}(t)\}\rangle \\
&= \left(\frac{\beta M}{2\pi t(t - \mathrm{i}\hbar\beta)}\right)^{3/2} \exp\left[\frac{-\beta M r^2}{2t(t - \mathrm{i}\hbar\beta)}\right]
\end{aligned} \tag{3.96}$$

on using (3.75) and (3.95). Similarly we obtain for $S(\boldsymbol{\kappa}, \omega)$, eqn (3.91),

$$S(\boldsymbol{\kappa}, \omega) = \left(\frac{\beta M}{2\pi\hbar^2\kappa^2}\right)^{1/2} \exp\left[-\frac{\beta M}{2\hbar^2\kappa^2}\left(\hbar\omega - \frac{\hbar^2\kappa^2}{2M}\right)^2\right], \tag{3.97}$$

and thus the partial differential cross-section for the scattering by a single free nucleus at temperature $T = 1/k_B\beta$ is

$$\frac{d^2\sigma}{d\Omega\, dE'} = \frac{k'}{k}\overline{|b|^2}\left(\frac{\beta M}{2\pi\hbar^2\kappa^2}\right)^{1/2} \exp\left[-\frac{\beta M}{2\hbar^2\kappa^2}\left(\hbar\omega - \frac{\hbar^2\kappa^2}{2M}\right)^2\right]. \quad (3.98)$$

If we take the limit $\beta \to \infty$ in (3.98), we regain the result derived in Chapter 1 for the scattering by a single free nucleus at zero temperature, since

$$\delta(x) = \lim_{\epsilon \to 0^+} \frac{1}{\sqrt{(\pi\epsilon)}} \exp(-x^2/\epsilon).$$

Also, if $M \to \infty$ then (3.98) reduces to the scattering from a single bound nucleus, as expected.

Because there are no forces between the particles in a perfect fluid the coherent and single-particle response functions, $S(\kappa, \omega)$ and $S_i(\kappa, \omega)$, are identical.

At low temperatures, where the mean occupation numbers of the various quantum states are not small, identical particles are subject to exchange forces. Even so, the coherent and single-particle response functions are identical for a perfect quantum fluid, and (3.93) is the appropriate correlation function.

Before calculating the correlation function we amplify the comment concerning the role of exchange forces in a quantum fluid. When exchange forces are significant, the cross-section is derived from (3.4) with $\hat{V}_j(\boldsymbol{\kappa})$ equal to the scattering amplitude operator $\hat{b}_j$ defined in § 1.6; for identical nuclei,

$$\hat{V}_j(\boldsymbol{\kappa}) = \hat{b}_j = A + \tfrac{1}{2}B\hat{\boldsymbol{\sigma}} \cdot \hat{\mathbf{i}}_j, \quad (3.99)$$

where $\hat{\boldsymbol{\sigma}}/2$ is the spin of the neutron, and $\hat{\mathbf{i}}_j$ is the spin operator of the $j$th particle. When the neutrons are unpolarized, the correlation function in (3.4) is, apart from a factor $1/N$,

$$\frac{1}{N}\overline{\sum_{jj'} \langle \exp(-i\boldsymbol{\kappa} \cdot \hat{\mathbf{R}}_j)\hat{b}_j^+ \hat{b}_{j'}(t)\exp\{i\boldsymbol{\kappa} \cdot \hat{\mathbf{R}}_{j'}(t)\}\rangle}$$

$$= A^2 I(\boldsymbol{\kappa}, t) + \tfrac{1}{4}B^2 \frac{1}{N}\sum_{jj'} \langle \exp(-i\boldsymbol{\kappa} \cdot \hat{\mathbf{R}}_j)\{\hat{\mathbf{i}}_j \cdot \hat{\mathbf{i}}_{j'}(t)\}\exp\{i\boldsymbol{\kappa} \cdot \hat{\mathbf{R}}_{j'}(t)\}\rangle. \quad (3.100)$$

In this expression, $I(\boldsymbol{\kappa}, t)$ is defined in (3.16) and (3.18), and the second correlation function contains the Fourier transform of the nuclear spin density, namely

$$\sum_j \hat{\mathbf{i}}_j\, \delta(\mathbf{r} - \mathbf{R}_j).$$

Hence, for a quantum fluid, the cross-section is related to the weighted sum of the number particle and spin density correlation functions. For a perfect fluid, the spin correlation function reduces to

$$\sum_{jj'} \langle \hat{\mathbf{i}}_j \cdot \hat{\mathbf{i}}_{j'} \rangle Y_{jj'}(\boldsymbol{\kappa}, t) = i(i+1) \sum_{jj'} \delta_{jj'} Y_{jj'}(\boldsymbol{\kappa}, t)$$
$$= Ni(i+1) I(\boldsymbol{\kappa}, t).$$

The second equality follows because the single-particle and coherent correlation functions are identical in a perfect fluid.

The free particle states in a perfect quantum fluid are labelled by wave vectors $\mathbf{q}$ that are defined by imposing cyclic boundary conditions on the sample. The particles obey either Bose or Fermi statistics, and the occupation functions are denoted by $n_{\mathbf{q}}$ and $f_{\mathbf{q}}$, respectively. If $\epsilon_{\mathbf{q}} = \hbar^2 q^2/2M$ is the free-particle energy, the occupation functions are

$$n_{\mathbf{q}} = (\exp\{(\epsilon_{\mathbf{q}} - \mu)\beta\} - 1)^{-1}; \quad \text{Bose} \tag{3.101}$$

and

$$f_{\mathbf{q}} = (\exp\{(\epsilon_{\mathbf{q}} - \mu)\beta\} + 1)^{-1}; \quad \text{Fermi} \tag{3.102}$$

where $\mu$ is the chemical potential and $\beta = 1/k_B T$. From (3.93) it follows that

$$S(\boldsymbol{\kappa}, \omega) = (g/N) \sum_{\mathbf{q}} \delta\{\hbar\omega + \epsilon_{\mathbf{q}} - \epsilon_{\boldsymbol{\kappa}+\mathbf{q}}\} \begin{cases} n_{\mathbf{q}}; & \text{Bose} \\ f_{\mathbf{q}}; & \text{Fermi} \end{cases} \tag{3.103}$$

where $g = (2i+1)$ is the spin degeneracy. Boltzmann statistics are valid when the chemical potential satisfies $\exp(\mu\beta) \ll 1$, in which case the Bose and Fermi particle distributions are the same and (3.100) reduces to (3.97), using the identity (3.75), apart from the spin degeneracy factor that is set to unity for classical particles. Equivalent forms of (3.103), which are sometimes more convenient to work with, are

$$S(\kappa, \omega) = (g/N) \sum_{\mathbf{q}} \delta\{\hbar\omega + \epsilon_{\mathbf{q}} - \epsilon_{\boldsymbol{\kappa}+\mathbf{q}}\} \begin{cases} n_{\mathbf{q}}(1 + n_{\boldsymbol{\kappa}+\mathbf{q}}) \\ f_{\mathbf{q}}(1 - f_{\boldsymbol{\kappa}+\mathbf{q}}) \end{cases} \tag{3.104}$$

$$= \{1 - \exp(-\hbar\omega\beta)\}^{-1} (g/N) \sum_{\mathbf{q}} \delta\{\hbar\omega + \epsilon_{\mathbf{q}} - \epsilon_{\boldsymbol{\kappa}+\mathbf{q}}\} \begin{cases} (n_{\mathbf{q}} - n_{\boldsymbol{\kappa}+\mathbf{q}}) \\ (f_{\mathbf{q}} - f_{\boldsymbol{\kappa}+\mathbf{q}}) \end{cases}$$

$$= \{1 - \exp(-\hbar\omega\beta)\}^{-1} (g/N) \sum_{\mathbf{q}} [\delta\{\hbar\omega + \epsilon_{\mathbf{q}} - \epsilon_{\boldsymbol{\kappa}+\mathbf{q}}\}$$
$$- \delta\{-\hbar\omega + \epsilon_{\mathbf{q}} - \epsilon_{\boldsymbol{\kappa}+\mathbf{q}}\}] \begin{cases} n_{\mathbf{q}} \\ f_{\mathbf{q}} \end{cases}. \tag{3.105}$$

The equivalence of these expressions can be verified by using the condition of detailed balance, and a shift in the wave vector $\mathbf{q} \to \mathbf{q} - \boldsymbol{\kappa}$. We shall evaluate expression (3.105) for bosons and fermions.

The effects of quantum statistics are very apparent in a degenerate Fermi fluid, obtained in the limit $T \to 0$. In this instance the occupation function $f_{\mathbf{q}}$ vanishes if $\mathbf{q}$ exceeds the Fermi wave vector $p_f$, and $(1-f_{\boldsymbol{\kappa}+\mathbf{q}})$ vanishes if $|\boldsymbol{\kappa}+\mathbf{q}| < p_f$. Hence, the response function (3.104) for a degenerate Fermi fluid involves the promotion of particles from occupied $(q < p_f)$ to unoccupied $(q > p_f)$ states, usually referred to as particle and hole states; this behaviour is a manifestation of the Pauli exclusion principle for particles subject to Fermi quantum statistics. Particles subject to Bose quantum statistics display the phenomenon of (Bose–Einstein) condensation when a macroscopic number of particles occupy the state of zero momentum. Expression (3.105) displays the detailed balance factor explicitly. Note that for a degenerate quantum fluid the detailed balance factor approximates to a step function; $S(\kappa, \omega)$ vanishes for $\omega < 0$ and $T \to 0$ because in this limit there are no thermally excited states to participate in the response. The radically different behaviour of the Bose and Fermi distribution functions at low temperatures leads us to treat the two degenerate quantum fluids separately from now on.

The chemical potential for a Bose fluid is negative for all temperatures and densities, and it is implicitly determined as a function of these variables by the relation

$$N = g \sum_{\mathbf{q}} n_{\mathbf{q}} = g\{V/(2\pi)^3\} \int \mathrm{d}\mathbf{q}\, n_{\mathbf{q}}$$

or

$$\rho = (N/V) = \frac{gM^{3/2}}{2^{1/2}\pi^2\hbar^3} \int_0^\infty \frac{\mathrm{d}\epsilon\, \sqrt{\epsilon}}{\{\exp \beta(\epsilon-\mu)-1\}}. \tag{3.106}$$

An acceptable solution of this relation is obtained, for a fixed particle density $\rho$, for all temperatures down to $T_0 \propto \rho^{2/3}$ at which $\mu = 0$. For temperatures $T < T_0$ a fraction of the Bose particles are in the lowest state, i.e. have energy $\epsilon = 0$; this is the phenomenon of Bose–Einstein condensation.

The calculation of $S(\kappa, \omega)$ from the expression (3.105) entails the evaluation of the integral

$$(g/N)\{V/(2\pi)^3\} \int \mathrm{d}\mathbf{q}\, n_{\mathbf{q}}\, \delta\{\hbar\omega + \epsilon_{\mathbf{q}} - \epsilon_{\boldsymbol{\kappa}+\mathbf{q}}\}$$

$$= (g/N)\{MV/\hbar^2\kappa(2\pi)^3\} \int \mathrm{d}\mathbf{q}\, n_{\mathbf{q}}\, \delta(Q - \tilde{\boldsymbol{\kappa}} \cdot \mathbf{q}), \tag{3.107}$$

where $\tilde{\boldsymbol{\kappa}} = \boldsymbol{\kappa}/|\boldsymbol{\kappa}|$ and the wave vector

$$Q = (M/\hbar^2\kappa)(\hbar\omega - E_R) \tag{3.108}$$

in which the recoil energy

$$E_R = (\hbar^2\kappa^2/2M). \tag{3.109}$$

The integral (3.107) is readily performed in a coordinate system in which $\tilde{\boldsymbol{\kappa}}$ defines the $z$-axis, say, and $\rho^2 = q_x^2 + q_y^2 = q^2 - q_z^2$, for then

$$\int \mathrm{d}\mathbf{q}\, n_{\mathbf{q}}\, \delta(Q - \tilde{\boldsymbol{\kappa}} \cdot \mathbf{q}) = \int_{-\infty}^{\infty} \mathrm{d}q_z \int_0^{\infty} 2\pi\rho\, \mathrm{d}\rho\, n_{\mathbf{q}}\, \delta(Q - q_z)$$
$$= \pi \int_0^{\infty} \mathrm{d}\rho^2 \left( \exp\left\{ \frac{\hbar^2 \beta}{2M} (\rho^2 + Q^2) - \mu\beta \right\} - 1 \right)^{-1}. \tag{3.110}$$

The remaining integral in (3.110) can be evaluated using elementary methods. The final result for $S(\kappa, \omega)$, for an arbitrary temperature, is

$$S(\kappa, \omega) = \{1 - \exp(-\hbar\omega\beta)\}^{-1} \left( \frac{gM^2}{4\pi^2 \hbar^4 \kappa \beta \rho} \right)$$
$$\times \ln\{1 + [\{1 - \exp(-\hbar\omega\beta)\}/(\exp(\xi - \mu\beta) - 1)]\}; \quad \text{Bose} \tag{3.111}$$

where the dimensionless variable

$$\xi = \tfrac{1}{4}\beta E_{\mathrm{R}}(1 - \hbar\omega/E_{\mathrm{R}})^2. \tag{3.112}$$

Several points are worth making about result (3.111). The deviations of a perfect quantum fluid from classical properties are negligible when $\exp(\mu\beta) \ll 1$. In this instance Boltzmann statistics are valid and (3.111) coincides with (3.97). To verify this statement by an explicit calculation we note first that in the limit $\exp(\mu\beta) \ll 1$ expression (3.106) reduces to

$$\rho = \left( \frac{gM^{3/2}}{2^{1/2} \pi^2 \hbar^3} \right) \exp(\mu\beta) \int_0^{\infty} \mathrm{d}\epsilon\, \sqrt{\epsilon}\, \exp(-\epsilon\beta)$$

or

$$\exp(\mu\beta) = (\rho/g)(2\pi\hbar^2\beta/M)^{3/2}. \tag{3.113}$$

Turning to $S(\kappa, \omega)$ we find that, in the same limit,

$$S(\kappa, \omega) = \{1 - \exp(-\hbar\omega\beta)\}^{-1} \left( \frac{gM^2}{4\pi^2 \hbar^4 \kappa \beta \rho} \right)$$
$$\times \ln\{1 + \{1 - \exp(-\hbar\omega\beta)\} \exp(\mu\beta - \xi)\}.$$

Here, the logarithm can be replaced by

$$\{1 - \exp(-\hbar\omega\beta)\} \exp(\mu\beta - \xi)$$

with $\exp(\mu\beta)$ given by (3.113); the resulting expression for $S(\kappa, \omega)$ is identical with (3.97), as anticipated. The classical result, regarded as a function of $\omega$ for fixed $\boldsymbol{\kappa}$, has a maximum for $\xi = 0$ at which $\hbar\omega = E_{\mathrm{R}}$. A study of $S(\kappa, \omega)$ for a Bose fluid shows that it possesses the same property. Moreover, for $\beta\,|\omega| \to \infty$, $S(\kappa, \omega)$ shares the same frequency dependence as a classical fluid. Quantum effects are most apparent as the temperature is decreased until the onset of condensation. For $\mu = 0$ and $\xi \ll 1$, the result (3.111) contains $\ln(1/\xi)$, and hence the response of a perfect Bose fluid with a condensate diverges as the neutron energy transfer approaches the recoil energy.

A degenerate Fermi fluid is characterized by the fact that all states up to an energy $\mu = \epsilon_f = \hbar^2 p_f^2/2M$ are uniformly occupied. In view of this,

$$N = g\sum_{\mathbf{q}} f_{\mathbf{q}} = g\sum_{q<p_f} = g\{V/(2\pi)^3\}\int_{q<p_f} d\mathbf{q} = g\{V/(2\pi)^3\}\tfrac{4}{3}\pi p_f^3, \quad (3.114)$$

from which the Fermi wave vector $p_f$ is determined as a function of the particle density.

The calculation of $S(\kappa, \omega)$ for a degenerate Fermi fluid entails the evaluation of the integral,

$$(g/N)\{V/(2\pi)^3\}\int_{q<p_f} d\mathbf{q}\ \delta\{\hbar\omega + \epsilon_{\mathbf{q}} - \epsilon_{\boldsymbol{\kappa}+\mathbf{q}}\}$$

$$= (g/N)\{MV/\hbar^2\kappa(2\pi)^3\}\int_{-p_f}^{p_f} dq_z \int_0^{\rho_0} 2\pi\rho\ d\rho\ \delta(Q - q_z), \quad (3.115)$$

where $\rho_0^2 = p_f^2 - q_z^2$. For $|Q| < p_f$ the integral in (3.115) is simply

$$\pi(p_f^2 - Q^2),$$

and the complete expression is

$$(M/\hbar^2\kappa)(3/4p_f^3)(p_f^2 - Q^2),$$

where we have used (3.114) to eliminate the particle density and spin degeneracy in terms of $p_f^3$. Intriducing the reduced variables,

$$x = \hbar\omega/\epsilon_f \quad \text{and} \quad y = \kappa/p_f, \quad (3.116)$$

we find that the condition $|Q| < p_f$ is equivalent to

$$(y^2 - 2y) \leq x \leq (y^2 + 2y). \quad (3.117)$$

The complete expression for $S(\kappa, \omega)$ is obtained from the difference of (3.115) and the same function evaluated with $\omega \rightarrow -\omega$; for this second case the condition on $x$ is simply

$$x \leq (2y - y^2). \quad (3.118)$$

For a degenerate fluid it is evident that $S(\kappa, \omega)$ is zero for $\omega < 0$ since there are no thermally excited states in the sample. Hence, in addition to the conditions (3.117) and (3.118), we require $x > 0$. The net result is that, for wave vectors $\kappa = yp_f$ in the range $0 < y < 2$,

$$S(\kappa, \omega) = (M/\hbar^2\kappa)(3/4p_f^3)\left\{p_f^2\left[1 - \left(\frac{x - y^2}{2y}\right)^2\right] - p_f^2\left[1 - \left(\frac{x + y^2}{2y}\right)^2\right]\right\}$$

$$= (3x/8y\epsilon_f); \qquad 0 \leq x \leq (2y - y^2)$$

and

$$S(\kappa,\omega)=(M/\hbar^2\kappa)(3/4p_f^3)p_f^2\left[1-\left(\frac{x-y^2}{2y}\right)^2\right]$$

$$=(3/8y\epsilon_f)\left[1-\left(\frac{x-y^2}{2y}\right)^2\right]; \qquad (2y-y^2)\leqslant x\leqslant(y^2+2y),$$

while for $y>2$

$$S(\kappa,\omega)=(3/8y\epsilon_f)\left[1-\left(\frac{x-y^2}{2y}\right)^2\right]; \qquad (y^2-2y)\leqslant x\leqslant(y^2+2y). \quad (3.119)$$

From these results we conclude that the Pauli exclusion principle confines the response of a degenerate Fermi fluid to a continuum bounded above by $(y^2+2y)$, and below by $x=0$ for $0<y<2$ and $(y^2-2y)$ for $y>2$. For $y>2$, the response spectrum forms a parabola whose maximum occurs at $x=y^2$, which is equivalent to $\hbar\omega=E_R$. The spectrum for $0<y<2$ contains a discontinuity in slope at $x=2y-y^2$. In the limit $y\to 0$, the width of the spectrum shrinks to zero, and the integrated intensity vanishes since, from (3.119) and remembering that $S(\kappa,\omega)$ is zero for $\omega<0$,

$$S(\kappa)=\hbar\int_0^\infty d\omega\, S(\kappa,\omega)=\tfrac{3}{4}y(1-y^2/12); \qquad 0\leqslant y\leqslant 2$$

$$=1; \qquad y\geqslant 2.$$

The deviation of $S(\kappa)$ from unity for $y<2$ is a manifestation of the correlations induced in a degenerate Fermi fluid by exchange forces.

We notice that in (3.97) $\hbar^2\kappa^2/2M$ is the recoil energy of the nucleus and that the whole expression is, in fact, independent of $\hbar$ when written in terms of the momenta $\mathbf{p}=\hbar\mathbf{k}$ and $\mathbf{p}'=\hbar\mathbf{k}'$ and energies $E$ and $E'$ of the neutron. Thus we would expect to obtain the same result from the classical approximation to the cross-section given in § 3.5. To demonstrate that this is in fact the case, we proceed as follows.

Our classical approximation to the scattering law is

$$\bar{S}(\boldsymbol{\kappa},\omega)=\frac{1}{2\pi\hbar}\int_{-\infty}^{\infty} dt\,\exp(-i\omega t)\exp\{i\hbar\gamma(t)\}\langle\exp[-i\boldsymbol{\kappa}\cdot\{\mathbf{r}-\mathbf{r}(t)\}]\rangle,$$

where

$$\gamma(t)=\tfrac{1}{2}\{\boldsymbol{\kappa}\cdot\mathbf{r},\boldsymbol{\kappa}\cdot\mathbf{r}(t)\}. \quad (3.120)$$

Clearly, for free motion,

$$\mathbf{r}(t)=\mathbf{r}+t\mathbf{p}/M,$$

so that

$$\langle\exp[-i\boldsymbol{\kappa}\cdot\{\mathbf{r}-\mathbf{r}(t)\}]\rangle=\langle\exp(it\boldsymbol{\kappa}\cdot\mathbf{p}/M)\rangle=\exp(-t^2\kappa^2/2M\beta) \quad (3.121)$$

by (3.94). Also it is quite easy to show that

$$\gamma(t)=\tfrac{1}{2}\sum_{\alpha,\beta}\kappa_\alpha\kappa_\beta\{r_\alpha,\, r_\beta(t)\}=\tfrac{1}{2}\sum_{\alpha,\beta}\kappa_\alpha\kappa_\beta\,\delta_{\alpha,\beta}(t/M)=\tfrac{1}{2}t\kappa^2/M. \tag{3.122}$$

If we insert (3.121) and (3.122) into (3.120) it is obvious that $\bar{S}(\boldsymbol{\kappa}, \omega)$ is identical to $S(\boldsymbol{\kappa}, \omega)$ given by (3.97), as expected.

We use the present example of the scattering by a single free nucleus subject to Boltzmann statistics to demonstrate the need for the discussion given in § 3.5 of how to obtain the classical approximation to $S(\boldsymbol{\kappa}, \omega)$, which in this particular instance is, as we have just shown, identical to the exact quantum mechanical expression. For if we had simply said that in the exact expression for $S(\boldsymbol{\kappa}, \omega)$ (eqn (3.91)) ignore the fact that $\exp\{i\boldsymbol{\kappa}\cdot\hat{\mathbf{r}}(t)\}$ is an operator and replace it by its classical analogue, then (3.91) would read

$$\frac{1}{2\pi\hbar}\int_{-\infty}^{\infty}\mathrm{d}t\,\exp(-\mathrm{i}\omega t)\langle\exp[-\mathrm{i}\boldsymbol{\kappa}\cdot\{\mathbf{r}-\mathbf{r}(t)\}]\rangle,$$

which by (3.121) is equal to

$$\frac{1}{2\pi\hbar}\int_{-\infty}^{\infty}\mathrm{d}t\,\exp(-\mathrm{i}\omega t)\exp(-t^2\kappa^2/2M\beta)=\left(\frac{\beta M}{2\pi\hbar^2\kappa^2}\right)^{1/2}\exp\left(-\frac{\beta M(\hbar\omega)^2}{2\hbar^2\kappa^2}\right).$$

This is clearly incorrect because the term giving the recoil energy of the nucleus is absent; the latter appears in $\bar{S}(\boldsymbol{\kappa}, \omega)$, eqn (3.120), through the function $\gamma(t)$.

A final point is to evaluate the correlation function $\bar{G}(\mathbf{r}, t)$ defined in § 3.5. For the present case, using (3.94), and $[\hat{\mathbf{r}}_\alpha, \hat{\mathbf{r}}_\beta(t)]=\delta_{\alpha,\beta}(\mathrm{i}\hbar t/M)$,

$$\begin{aligned}\bar{G}(\mathbf{r}, t)&=\left(\frac{1}{2\pi}\right)^3\int \mathrm{d}\boldsymbol{\kappa}\,\exp(-\mathrm{i}\boldsymbol{\kappa}\cdot\mathbf{r})\langle\exp[-\mathrm{i}\boldsymbol{\kappa}\cdot\{\hat{\mathbf{r}}-\hat{\mathbf{r}}(t)\}]\rangle\\&=\left(\frac{1}{2\pi}\right)^3\int \mathrm{d}\boldsymbol{\kappa}\,\exp(-\mathrm{i}\boldsymbol{\kappa}\cdot\mathbf{r})\exp(-t^2\kappa^2/2M\beta)\\&=\left(\frac{\beta M}{2\pi t^2}\right)^{3/2}\exp\left(-\frac{\beta M}{2t^2}r^2\right).\end{aligned}$$

In contrast to $G(\mathbf{r}, t)$, eqn (3.96), this expression for $\bar{G}(\mathbf{r}, t)$ is purely real and independent of $\hbar$.

### 3.6.2. *Harmonic oscillator*

The Hamiltonian of a particle of mass $M$ in an isotropic, harmonic oscillator potential is

$$\mathcal{H}=\frac{1}{2M}\hat{\mathbf{p}}^2+\frac{M\omega_0^2}{2}\hat{\mathbf{r}}^2, \tag{3.123}$$

where $\omega_0$ is the frequency of vibration. Because of the isotropic nature of the Hamiltonian we need only consider the motion of one cartesian component $x$, say.

As is well known the eigenvalues of the Hamiltonian

$$\hat{\mathcal{H}}_1 = \frac{1}{2M}\hat{p}^2 + \frac{M\omega_0^2}{2}\hat{x}^2 \tag{3.124}$$

are

$$E_n = \hbar\omega_0(n+\tfrac{1}{2}), \quad \text{where} \quad n = 0, 1, 2, \ldots .$$

Furthermore, $\hat{\mathcal{H}}_1$ can be expressed with advantage in terms of Bose operators $\hat{a}$ and $\hat{a}^+$, which satisfy

$$[\hat{a}, \hat{a}^+] = 1, \tag{3.125}$$

all other commutators being zero. Indeed, if we set

$$\hat{x} = \left(\frac{\hbar}{2M\omega_0}\right)^{1/2}(\hat{a}+\hat{a}^+) \tag{3.126}$$

and

$$\hat{p} = -\mathrm{i}\left(\frac{M\hbar\omega_0}{2}\right)^{1/2}(\hat{a}-\hat{a}^+), \tag{3.127}$$

then (3.124) reduces to

$$\hat{\mathcal{H}}_1 = \hbar\omega_0(\hat{a}^+\hat{a}+\tfrac{1}{2}), \tag{3.128}$$

as is easily verified with the aid of (3.125). If the eigenkets of (3.124) are denoted by $|n\rangle$ then

$$\hat{a}^+|n\rangle = (n+1)^{1/2}|n+1\rangle \tag{3.129}$$

and

$$\hat{a}\,|n\rangle = (n)^{1/2}|n-1\rangle, \tag{3.130}$$

and hence

$$\hat{a}^+\hat{a}\,|n\rangle = n\,|n\rangle, \tag{3.131}$$

showing that the eigenvalues of (3.128) are as given above.

To calculate the scattering law for this model we need to evaluate the correlation function

$$\langle\exp(-\mathrm{i}\kappa_x\hat{x})\exp\{\mathrm{i}\kappa_x\hat{x}(t)\}\rangle, \tag{3.132}$$

where we have denoted the component of the scattering vector $\boldsymbol{\kappa}$ that is parallel to $x$ by $\kappa_x$.

The equation of motion for $\hat{a}$ is

$$\mathrm{i}\hbar\partial_t\hat{a} = [\hat{a}, \hat{\mathcal{H}}_1] = \hbar\omega_0[\hat{a}, \hat{a}^+\hat{a}] = \hbar\omega_0\hat{a}.$$

Hence

$$\hat{a}(t) = \hat{a}\exp(-i\omega_0 t). \tag{3.133}$$

and

$$\hat{a}^+(t) = \hat{a}^+\exp(i\omega_0 t). \tag{3.134}$$

so that

$$\hat{x}(t) = \left(\frac{\hbar}{2M\omega_0}\right)^{1/2}\{\hat{a}\exp(-i\omega_0 t) + \hat{a}^+\exp(i\omega_0 t)\}. \tag{3.135}$$

We now make use of the identity (3.81) to write

$$\exp(-i\kappa_x\hat{x})\exp\{i\kappa_x\hat{x}(t)\} = \exp\{\tfrac{1}{2}\kappa_x^2[\hat{x}, \hat{x}(t)]\}\exp[-i\kappa_x\{\hat{x} - \hat{x}(t)\}], \tag{3.136}$$

since $[\hat{x}, \hat{x}(t)]$ is a c-number.

To proceed further we invoke the following identity (frequently referred to as Bloch's identity): if $\hat{Q}$ is a linear function of the operators $\hat{a}$ and $\hat{a}^+$, then

$$\langle\exp\hat{Q}\rangle = \exp(\tfrac{1}{2}\langle\hat{Q}^2\rangle). \tag{3.137}$$

A proof is given at the end of this chapter.

Applying this to (3.136) we find

$$\begin{aligned}\langle\exp(-i\kappa_x\hat{x})\exp\{i\kappa_x\hat{x}(t)\}\rangle &= \exp\{\tfrac{1}{2}\kappa_x^2[\hat{x}, \hat{x}(t)]\}\\ &\times\exp[-\tfrac{1}{2}\kappa_x^2\langle\{\hat{x} - \hat{x}(t)\}^2\rangle] = \exp[-\kappa_x^2\{\langle(\hat{x})^2\rangle - \langle\hat{x}\hat{x}(t)\rangle\}].\end{aligned} \tag{3.138}$$

Now from (3.135) we have

$$\begin{aligned}\langle\hat{x}\hat{x}(t)\rangle &= \frac{\hbar}{2M\omega_0}\langle\hat{a}\hat{a}^+\exp(i\omega_0 t) + \hat{a}^+\hat{a}\exp(-i\omega_0 t)\rangle\\ &= \frac{\hbar}{2M\omega_0}\{(1 + \langle\hat{a}^+\hat{a}\rangle)\exp(i\omega_0 t) + \langle\hat{a}^+\hat{a}\rangle\exp(-i\omega_0 t)\}\end{aligned} \tag{3.139}$$

and

$$\begin{aligned}\langle\hat{a}^+\hat{a}\rangle &= \sum_{n=0}^{\infty}\exp(-\beta E_n)\langle n|\,\hat{a}^+\hat{a}\,|n\rangle\Big/\sum_{n=0}^{\infty}\exp(-\beta E_n)\\ &= \sum_{n=0}^{\infty}\exp(-\beta\hbar\omega_0 n)n\Big/\sum_{n=0}^{\infty}\exp(-\beta\hbar\omega_0 n)\\ &= -\left[\frac{\partial}{\partial\alpha}\left\{\ln\sum_{n=0}^{\infty}e^{-\alpha n}\right\}\right]\Bigg|_{\alpha=\hbar\omega_0\beta},\end{aligned}$$

which gives

$$\langle\hat{a}^+\hat{a}\rangle = \{\exp(\hbar\omega_0\beta) - 1\}^{-1} = n(\omega_0). \tag{3.140}$$

If we use (3.139) in (3.138) the latter becomes

$$\exp\left\{-\frac{\hbar\kappa_x^2}{2M\omega_0}(2\langle\hat{a}^+\hat{a}\rangle+1)\right\}$$
$$\times\exp\left[\frac{\hbar\kappa_x^2}{2M\omega_0}\{(\langle\hat{a}^+\hat{a}\rangle+1)\exp(\mathrm{i}\omega_0 t)+\langle\hat{a}^+\hat{a}\rangle\exp(-\mathrm{i}\omega_0 t)\}\right]. \quad (3.141)$$

The actual correlation function required to evaluate $S(\kappa, \omega)$ is

$$\langle\exp(-\mathrm{i}\boldsymbol{\kappa}\cdot\hat{\mathbf{r}})\exp\{\mathrm{i}\boldsymbol{\kappa}\cdot\hat{\mathbf{r}}(t)\}\rangle=\prod_{\alpha=x,y,z}\langle\exp(-\mathrm{i}\kappa_\alpha\hat{\mathbf{r}}_\alpha)\exp\{\mathrm{i}\kappa_\alpha\hat{\mathbf{r}}_\alpha(t)\}\rangle$$
$$=\exp\{-\langle(\boldsymbol{\kappa}\cdot\hat{\mathbf{r}})^2\rangle+\langle(\boldsymbol{\kappa}\cdot\hat{\mathbf{r}})\{\boldsymbol{\kappa}\cdot\hat{\mathbf{r}}(t)\}\rangle\} \quad (3.142)$$

where we have used (3.138) to obtain the final expression. From (3.140) and (3.141) it follows that, for an isotropic harmonic oscillator,

$$\langle(\boldsymbol{\kappa}\cdot\hat{\mathbf{r}})\{\boldsymbol{\kappa}\cdot\hat{\mathbf{r}}(t)\}\rangle=\frac{\hbar\kappa^2\cosh\{\omega_0(\mathrm{i}t+\frac{1}{2}\hbar\beta)\}}{2M\omega_0\sinh(\frac{1}{2}\hbar\omega_0\beta)}, \quad (3.143)$$

and, on setting $t=0$ on the right-hand side, we find

$$\langle(\boldsymbol{\kappa}\cdot\hat{\mathbf{r}})^2\rangle=(\hbar\kappa^2/2M\omega_0)\coth(\tfrac{1}{2}\hbar\omega_0\beta)=2W(\kappa) \quad (3.144)$$

where the second equality in (3.144) defines $W(\kappa)$. Notice from (3.143) that the imaginary part of the correlation function is an odd function of $t$ and independent of the temperature.

The quantity $\exp(-2W)$ is called the Debye–Waller factor (cf. § 4.3). For temperatures large compared with the vibrational energy,

$$2W(\kappa)=(\kappa^2/M\beta\omega_0^2) \quad (3.145)$$

whereas in the limit $T\to 0$, or $\beta\to\infty$,

$$2W(\kappa)=(\hbar\kappa^2/2M\omega_0). \quad (3.146)$$

The finite value of $W(\kappa)$ at absolute zero is a consequence of the quantum mechanical, zero-point motion of the particle.

If the amplitude of the harmonic oscillations is small compared with $\kappa^{-1}$, so that $(\hbar\kappa^2/M\omega_0)\ll 1$, then the response function $S(\kappa,\omega)$ can be developed as an expansion in the time-dependent correlation function in

(3.142). From the latter result, (3.143), and the definition of $W(\kappa)$

$$S(\kappa,\omega)=\frac{1}{2\pi\hbar}\int_{-\infty}^{\infty}\mathrm{d}t\,\exp(-\mathrm{i}\omega t)\langle\exp(-\mathrm{i}\boldsymbol{\kappa}\cdot\hat{\mathbf{r}})\exp\{\mathrm{i}\boldsymbol{\kappa}\cdot\hat{\mathbf{r}}(t)\}\rangle$$

$$=\exp\{-2W(\kappa)\}\frac{1}{2\pi\hbar}\int_{-\infty}^{\infty}\mathrm{d}t\,\exp(-\mathrm{i}\omega t)\exp\langle(\boldsymbol{\kappa}\cdot\hat{\mathbf{r}})\{\boldsymbol{\kappa}\cdot\hat{\mathbf{r}}(t)\}\rangle$$

$$\doteq\exp\{-2W(\kappa)\}\left\{\delta(\hbar\omega)+\left(\frac{\kappa^2}{2M\omega_0}\right)\{1+n(\omega)\}\left[\delta(\omega-\omega_0)-\delta(\omega+\omega_0)\right]\right\},\tag{3.147}$$

where in the last line we have retained the first two terms in the expansion. The first term represents purely elastic scattering, and the second represents the annihilation and creation of one quanta. In the limit of low temperatures, the behaviour of the detailed balance factor in the second term of (3.147) limits the response to a creation process whereas, at high temperatures $\hbar\omega_0\beta \ll 1$, annihilation and creation contribute to the response more or less equally.

The total elastic scattering is readily obtained, beginning with

$$\left(\frac{\mathrm{d}\sigma}{\mathrm{d}\Omega}\right)^{\mathrm{el}}=\overline{|b|^2}\exp\{-2W(\kappa)\}.\tag{3.148}$$

Since $k=k'$ for elastic scattering, $\kappa^2=2k^2(1-\cos\theta)$ where $\theta$ is the angle between $\mathbf{k}$ and $\mathbf{k}'$. Thus the total elastic cross-section is,

$$\sigma^{\mathrm{el}}=\overline{|b|^2}\,2\pi\int_0^{\pi}\sin\theta\,\mathrm{d}\theta\exp\{-\tfrac{1}{2}H(1-\cos\theta)\}$$

$$=(4\pi\overline{|b|^2}/H)\{1-\exp(-H)\}$$

where

$$H=(4mE/M\hbar\omega_0)\coth(\tfrac{1}{2}\hbar\omega_0\beta).\tag{3.150}$$

If $H\ll 1$, because $E\ll\hbar\omega_0$ or $m\ll M$, then the cross-section reduces to that for a bound particle, as expected. For the opposite limit, $H\gg 1$, the elastic scattering is small compared with the bound cross-section.

It is probably no surprise to learn that $S(\kappa,\omega)$ for a harmonic oscillator can be evaluated in closed form for arbitrary values of the various parameters. To this end we make use of the identity

$$\exp\{y\cosh(x)\}=\sum_{n=-\infty}^{\infty}\exp(nx)\,I_n(y),\tag{3.151}$$

where $n$ is an integer and $I_n(y)=I_{-n}(y)$ is a Bessel function of the first kind. This identity is applied to the time-dependent correlation function in (3.142), that is given explicitly in (3.143), using

$$x = \omega_0(\mathrm{i}t + \tfrac{1}{2}\hbar\beta)$$

and

$$y = \hbar\kappa^2/\{2M\omega_0 \sinh(\tfrac{1}{2}\hbar\omega_0\beta)\} \tag{3.152}$$

for then

$$\exp\langle(\boldsymbol{\kappa}\cdot\hat{\mathbf{r}})\{\boldsymbol{\kappa}\cdot\hat{\mathbf{r}}(t)\}\rangle = \sum_{n=-\infty}^{\infty} I_n(y)\exp\{n\omega_0(\mathrm{i}t + \tfrac{1}{2}\hbar\beta)\}, \tag{3.153}$$

and we obtain

$$S(\kappa, \omega) = \exp\{-2W(\kappa) + \tfrac{1}{2}\hbar\omega\beta\} \sum_{n=-\infty}^{\infty} I_n(y)\,\delta\{\hbar\omega - n\hbar\omega_0\}. \tag{3.154}$$

The integer $n$ in (3.154) measures the number of units of energy $\hbar\omega_0$ lost ($n>0$) or gained ($n<0$) by the neutron. The approximate result (3.147) is obtained from (3.154) by taking the three terms with $n=0, \pm 1$, and noting that for $y \to 0$, $I_0(y) \to 1$ and $I_1(y) \to (y/2)$.

We draw attention to the elastic contribution in the exact result (3.154); the amplitude of the delta function $\delta(\hbar\omega)$ is $I_0(y)\exp(-2W)$. This result is to be compared with the result obtained by arguing that the elastic contribution arises from the correlation function evaluated for $t \to \infty$, and that in this limit

$$\begin{aligned}\langle\exp(-\mathrm{i}\boldsymbol{\kappa}\cdot\hat{\mathbf{r}})\exp\{\mathrm{i}\boldsymbol{\kappa}\cdot\hat{\mathbf{r}}(t)\}\rangle &\to \langle\exp(-\mathrm{i}\boldsymbol{\kappa}\cdot\hat{\mathbf{r}})\rangle\langle\exp(\mathrm{i}\boldsymbol{\kappa}\cdot\hat{\mathbf{r}})\rangle \\ &= \exp\{-2W(\kappa)\},\end{aligned}$$

where the equality results from the definition of $W(\kappa)$. The comparison reveals that the exact result contains the additional factor $I_0(y)$. The reason for the discrepancy is that the argument based on the long-time behaviour of the correlation function is valid for bulk samples, e.g. lattice vibrations. In fact the error involved in the limiting procedure is of order $(1/N)$ where $N$ is the number of participating particles, and so the argument is invalid for a single isolated particle. It follows that the argument is also not valid for an Einstein model of lattice vibrations since this model is equivalent to an assembly of independent oscillators.

To verify this explanation consider an assembly of $N$ particles whose motion can be described in terms of (harmonic) normal coordinates. We denumerate the coordinate states by $\nu$; in a crystal this label would index the modes and wave vectors. For such an assembly the result (3.143) is replaced by

$$\langle(\boldsymbol{\kappa}\cdot\hat{\mathbf{r}})\{\boldsymbol{\kappa}\cdot\hat{\mathbf{r}}(t)\}\rangle = \frac{1}{N}\sum_{\nu}^{N} y_\nu \cosh\{\omega_\nu(\mathrm{i}t + \tfrac{1}{2}\hbar\beta)\},$$

where $\omega_\nu$ is the $\nu$th normal coordinate frequency, and $y_\nu$ is defined as in (3.152) with $\omega_0$ replaced by $\omega_\nu$. Using (3.151) it follows that the elastic response is

$$\exp\{-2W(\kappa)\}\prod_{\nu=1}^{N} I_0(y_\nu/N)\,\delta(\hbar\omega),$$

which reduces to the elastic contribution in (3.154) for the special case $N=1$.

A bulk sample corresponds to large $N$. Because $I_0(y)\to\{1+(y/2)^2\}$ for $y\to 0$, the elastic contribution for $N\to\infty$ tends to the result

$$\exp\{-2W(\kappa)\}\left\{1+(1/2N)^2\sum_\nu y_\nu^2\right\}\delta(\hbar\omega).$$

The sum is of order $N/N^2$, and it vanishes for $N\to\infty$. We conclude that, for this model of a bulk sample, the amplitude of the elastic contribution to the response function is indeed the Debye–Waller factor $\exp(-2W)$. The argument used here is a particular case of the central-limit theorem.

The physical origin of the different results for elastic scattering from a single particle and a bulk sample is one of thermal fluctuations. The latter are minimal in a bulk sample in which the particles participate in a collective motion described by normal coordinates.

For $y\ll 1$ the amplitudes of the higher $|n|$ terms in (3.154) are much smaller than those retained in (3.147) since, for $y\to 0$ and $n>0$,

$$I_n(y)\to\frac{1}{n!}(y/2)^n. \tag{3.155}$$

Thus, when $y\ll 1$, very few terms in (3.154) are significant, and only few quanta processes contribute to the response. In contrast, for $y\to\infty$, $I_n(y)$ is independent of $n$ and therefore many terms contribute to $S(\kappa,\omega)$. Clearly the expansion has no virtue in this instance. The appropriate expression for $S(\kappa,\omega)$ is obtained by noting that, for $(\hbar\kappa^2/M\omega_0)\gg 1$, the Fourier transform of the correlation function (3.142) is dominated by its initial behaviour. Keeping terms to order $t^2$ we obtain, from (3.143),

$$\langle(\boldsymbol{\kappa}\cdot\hat{\mathbf{r}})\{\boldsymbol{\kappa}\cdot\hat{\mathbf{r}}(t)\}-(\boldsymbol{\kappa}\cdot\hat{\mathbf{r}})^2\rangle \doteq (\hbar\kappa^2/2M)\left\{it-\frac{t^2\omega_0}{2}\coth(\tfrac{1}{2}\hbar\omega_0\beta)\right\}. \tag{3.156}$$

The approximation (3.156) amounts to including the leading order contributions from the real and imaginary parts of the correlation function. The corresponding $S(\kappa,\omega)$ is obtained from (3.156) using (3.75). We find that, in the limit $(\hbar\kappa^2/M\omega_0)\gg 1$, the response function for a harmonic oscillator tends to the result

$$S(\kappa, \omega) = \{2\pi\hbar^2\Delta^2\}^{-1/2} \exp\left\{-\left(\omega - \frac{\hbar\kappa^2}{2M}\right)^2 \Big/ 2\Delta^2\right\} \tag{3.157}$$

where

$$\Delta^2 = (\hbar\kappa^2\omega_0/2M)\coth(\tfrac{1}{2}\hbar\omega_0\beta) \equiv 2\omega_0^2 W(\kappa).$$

In this result the frequency shift arises from the imaginary part of (3.156), and $\Delta^2$ arises from the real part. From the result (3.157) we deduce that in the limit $(\hbar\kappa^2/M\omega_0) \gg 1$ all the terms in the expansion (3.154) combine to give a broad spectrum response centred about the recoil energy. If $\omega_0\beta \ll 1$, which is achieved at high temperatures,

$$\Delta^2 = (\kappa^2/M\beta)$$

and then (3.157) is identical to the response function for a free Boltzmann particle (eqn (3.97)). Finally, observe that the result (3.157) is identical to the impulse approximation for a harmonic system, (3.76); the correlation function (3.77) is readily evaluated with the aid of (3.127) and the subsequent results.

The argument leading to (3.157) makes light of subtle mathematical features of the asymptotic expansion of the Fourier transform of (3.142) which arise because the correlation function is periodic (Gunn and Warner 1983). Strictly speaking the result (3.157) is correct in the limit of large $\kappa$ and $\omega$, and an energy resolution too poor to resolve the multitude of contributions. In general, we know that the limit $\kappa \to \infty$ corresponds to vanishing wavelength, and an infinitesimal displacement of the scattering particle so that the nature of the chemical binding is essentially irrelevant in determining the particle response.

## 3.7. Total cross-section for a perfect Boltzmann fluid

From § 3.6.1, the partial differential cross-section for a Boltzmann fluid of $N$ non-interacting nuclei, each of mass $M$, is clearly

$$\frac{d^2\sigma}{d\Omega\, dE'} = N\,\overline{|b|^2}\,\frac{k'}{k}\frac{1}{2\pi\hbar}\int_{-\infty}^{\infty} dt\, \exp\left\{\frac{\hbar\kappa^2}{2M}\left(it - \frac{t^2}{\hbar\beta}\right) - i\omega t\right\}. \tag{3.158}$$

To calculate the total cross-section from this equation we first note that

$$\hbar\omega = E - E' = -\frac{\hbar^2}{2m}(\kappa^2 - 2\mathbf{k}\cdot\boldsymbol{\kappa}). \tag{3.159}$$

Hence (3.158) reads

$$\frac{d^2\sigma}{d\Omega\, dE'} = N\,\overline{|b|^2}\,\frac{k'}{k}\frac{1}{2\pi\hbar}\int_{-\infty}^{\infty} dt\, \exp\left\{-\frac{\hbar\kappa^2}{2m}\left(\frac{t^2 E}{\lambda\hbar} - it\mu\right) - \frac{it\hbar}{m}\mathbf{k}\cdot\boldsymbol{\kappa}\right\}, \tag{3.160}$$

where for ease of notation we have defined

$$\mu = 1 + \frac{m}{M} \quad \text{and} \quad \lambda = \frac{EM\beta}{m}. \tag{3.161}$$

It follows from (3.160) that

$$\begin{aligned}\sigma &= N\overline{|b|^2}\frac{\hbar}{2\pi km}\int_{-\infty}^{\infty}\mathrm{d}t\int \mathrm{d}\boldsymbol{\kappa}\exp\left\{-\frac{\hbar\kappa^2}{2m}\left(\frac{t^2E}{\lambda\hbar}-\mathrm{i}t\mu\right)-\frac{\mathrm{i}t\hbar}{m}\mathbf{k}\cdot\boldsymbol{\kappa}\right\}\\ &= N\overline{|b|^2}\left(\frac{\pi\hbar}{E}\right)^{1/2}\int_{-\infty}^{\infty}\mathrm{d}t\left(\frac{t^2E}{\lambda\hbar}-\mathrm{i}t\mu\right)^{-3/2}\exp\left[-\frac{t^2E}{\hbar\{(t^2E/\lambda\hbar)-\mathrm{i}t\mu\}}\right]\end{aligned} \tag{3.162}$$

where in obtaining the second line we have made use of the integral (3.75).

The integral

$$\left(\frac{\pi\hbar}{E}\right)^{1/2}\int_{-\infty}^{\infty}\mathrm{d}t\left(\frac{t^2E}{\lambda\hbar}-\mathrm{i}t\mu\right)^{-3/2}\exp\left[-\frac{t^2E}{\hbar\{(t^2E/\lambda\hbar)-\mathrm{i}t\mu\}}\right] \tag{3.163}$$

can be rearranged into the form

$$\sqrt{\pi}\,\lambda^{3/2}\mathrm{e}^{-\lambda}\int_{-\infty}^{\infty}\frac{\mathrm{d}t}{(\delta-\mathrm{i}t)^{3/2}(\mu\lambda+\mathrm{i}t)^{3/2}}\exp\left(\frac{\lambda^2\mu}{\mu\lambda+\mathrm{i}t}\right),$$

where we have introduced a small positive constant $\delta$ to define the integral at $t=0$, which we shall later let go to zero. Since

$$\begin{aligned}&\int_{-\infty}^{\infty}\frac{\mathrm{d}t}{(\delta-\mathrm{i}t)^{3/2}(\lambda\mu+\mathrm{i}t)^{3/2}}\exp\left(\frac{\lambda^2\mu}{\lambda\mu+\mathrm{i}t}\right)\\ &\quad=\int_{-\infty}^{\infty}\mathrm{d}x\left\{\frac{1}{2\pi}\int_{-\infty}^{\infty}\mathrm{d}t(\lambda\mu+\mathrm{i}t)^{-3/2}\exp\left(\frac{\lambda^2\mu}{\lambda\mu+\mathrm{i}t}\right)\mathrm{e}^{\mathrm{i}xt}\right\}\\ &\quad\quad\times\int_{-\infty}^{\infty}\mathrm{d}t'(\delta-\mathrm{i}t')^{-3/2}\mathrm{e}^{-\mathrm{i}xt'},\end{aligned} \tag{3.164}$$

and

$$\begin{aligned}&\frac{1}{2\pi}\int_{-\infty}^{\infty}\mathrm{d}t(\lambda\mu+\mathrm{i}t)^{-3/2}\exp\left(\frac{\lambda^2\mu}{\lambda\mu+\mathrm{i}t}\right)\mathrm{e}^{\mathrm{i}xt}\\ &\quad=\begin{cases}\dfrac{1}{\lambda(\mu\pi)^{1/2}}\mathrm{e}^{-\mu\lambda x}\sinh\{2\lambda\sqrt{(\mu x)}\} & (x>0),\\ 0 & (x<0),\end{cases}\end{aligned} \tag{3.165}$$

and

$$\int_{-\infty}^{\infty}\mathrm{d}t(\delta-\mathrm{i}t)^{-3/2}\mathrm{e}^{-\mathrm{i}xt}=\begin{cases}4\sqrt{\pi}\,x^{1/2}\mathrm{e}^{-\delta x} & (x>0),\\ 0 & (x<0),\end{cases} \tag{3.166}$$

we finally obtain

$$\sigma = N\,\overline{|b|^2}\,4\sqrt{\left(\frac{\lambda\pi}{\mu}\right)}\mathrm{e}^{-\lambda}\int_0^\infty \mathrm{d}x\,x^{1/2}\mathrm{e}^{-\mu\lambda x}\sinh\{2\lambda\sqrt{(\mu x)}\}. \quad (3.167)$$

Alternatively this expression can be rewritten in the form

$$\sigma = N\,\overline{|b|^2}\,\frac{4\pi}{\mu^2}\left\{\left(1+\frac{1}{2\lambda}\right)\Phi(\sqrt{\lambda})+\frac{\mathrm{e}^{-\lambda}}{\sqrt{(\pi\lambda)}}\right\}, \quad (3.168)$$

where $\Phi(x)$ is the probability function

$$\Phi(x) = \frac{2}{\sqrt{\pi}}\int_0^x \mathrm{d}t\,\mathrm{e}^{-t^2}. \quad (3.169)$$

When $\lambda \ll 1$,

$$\sigma \simeq N\,\overline{|b|^2}\,\frac{8}{\mu^2}(\pi/\lambda)^{1/2}. \quad (3.170)$$

This result is seen to diverge as $\lambda \to 0$. This is simply a consequence of the fact that the limit $\lambda = 0$ corresponds to zero-energy neutrons and these are struck by the free-moving nuclei and thus scattered in spite of the absence of the apparent incident flux.

In the opposite limit, $\lambda \gg 1$, we obtain from (3.168)

$$\sigma \simeq \frac{N4\pi\,\overline{|b|^2}}{\mu^2} \equiv \frac{N4\pi\,\overline{|b|^2}}{(1+m/M)^2}. \quad (3.171)$$

which is simply the scattering from $N$ completely free and independent nuclei.

## 3.8. Proof of Bloch's identity

We give here a proof of the identity

$$\langle \exp \hat{Q}\rangle = \exp\{\tfrac{1}{2}\langle \hat{Q}^2\rangle\}, \quad (3.172)$$

where $\hat{Q}$ is a linear combination of the Bose operators $\hat{a}$ and $\hat{a}^+$.

Let us choose

$$\hat{Q} = A\hat{a} + B\hat{a}^+ \quad (3.173)$$

where $A$ and $B$ are arbitrary constants, and use the identity (3.81) to write

$$\exp \hat{Q} = \exp A\hat{a}\, \exp B\hat{a}^+ \exp(-\tfrac{1}{2}AB). \quad (3.174)$$

Thus we need to calculate, for a harmonic oscillator,

$$\langle \exp A\hat{a}\, \exp B\hat{a}^+\rangle = \frac{1}{Z}\sum_{n=0}^{\infty} \exp(-\beta E_n)\,\langle n|\exp A\hat{a}\,\exp B\hat{a}^+|n\rangle. \quad (3.175)$$

Here

$$Z = \sum_{n=0}^{\infty} \exp(-\beta E_n) = \exp(-\tfrac{1}{2}\hbar\omega_0\beta) \sum_{n=0}^{\infty} \exp(-\hbar\omega_0\beta n)$$
$$= \frac{\exp(-\tfrac{1}{2}\hbar\omega_0\beta)}{1-\exp(-\hbar\omega_0\beta)}. \tag{3.176}$$

With the aid of the relation

$$|n\rangle = \left(\frac{1}{n!}\right)^{1/2} (\hat{a}^+)^n \,|0\rangle \tag{3.177}$$

it is straightforward to show that

$$\exp(B\hat{a}^+)\,|n\rangle = \sum_{m=0}^{\infty} \left\{\frac{(n+m)!}{n!}\right\}^{1/2} \frac{B^m}{m!}\,|n+m\rangle. \tag{3.178}$$

Hence

$$\langle n|\exp A\hat{a} \exp B\hat{a}^+\,|n\rangle = \sum_{m,m'} \{(n+m')!\,(n+m)!\}^{1/2}$$
$$\times \frac{A^{m'}B^m}{n!\,m'!\,m!} \langle n+m'\,|n+m\rangle$$
$$= \sum_{m=0}^{\infty} \frac{(n+m)!}{n!\,(m!)^2} (AB)^m,$$

and so

$$\sum_{n=0}^{\infty} \exp(-\beta E_n)\,\langle n|\exp A\hat{a} \exp B\hat{a}^+\,|n\rangle$$
$$= \exp(-\tfrac{1}{2}\hbar\omega_0\beta) \sum_{n=0}^{\infty}\sum_{m=0}^{\infty} \exp(-\hbar\omega_0\beta n) \frac{(n+m)!}{(n!)(m!)^2} (AB)^m.$$

Now

$$\frac{1}{(1-y)^{m+1}} = \sum_{n=0}^{\infty} \frac{(n+m)!}{n!\,m!}\, y^n,$$

so if we choose $y = \exp(-\hbar\omega_0\beta)$,

$$\sum_{n=0}^{\infty} \exp(-\beta E_n)\,\langle n|\exp A\hat{a} \exp B\hat{a}^+\,|n\rangle$$
$$= \frac{\exp(-\tfrac{1}{2}\hbar\omega_0\beta)}{\{1-\exp(-\hbar\omega_0\beta)\}} \sum_{m=0}^{\infty} \left[\frac{AB}{\{1-\exp(-\hbar\omega_0\beta)\}}\right]^m \frac{1}{m!}$$
$$= Z \exp\left\{\frac{AB}{1-\exp(-\hbar\omega_0\beta)}\right\} = Z \exp\{AB(\langle\hat{a}^+\hat{a}\rangle + 1)\}, \tag{3.179}$$

where in the second line we have used (3.176); we have, finally,

$$\langle \exp \hat{Q} \rangle = \exp\{AB(\langle \hat{a}^{+} \hat{a} \rangle + \tfrac{1}{2})\}, \tag{3.180}$$

which is identical to (3.172).

## REFERENCES

Gunn, J. M. F. and Warner, M. (1983). Rutherford Appleton Laboratory, Report RL-83-119.
Inkson, J. C. (1984). *Many-Body Theory of Solids.* Plenum Press, New York.
Lovesey, S. W. (1980). *Condensed matter physics, dynamic correlations.* Frontiers in Physics, Vol. 49. Benjamin/Cummings, Reading, MA.
Reichl, L. E. (1980). *A Modern Course in Statistical Physics.* Edward Arnold, New York.
Rickayzen, G. (1980). *Green's Functions and Condensed Matter.* Academic Press, London.
Van Hove, L. (1954). *Phys. Rev.* **95,** 249.

# 4
# LATTICE DYNAMICS

Neutron scattering studies pervade most aspects of lattice dynamics since thermal neutrons are almost ideal for the determination of the phonon density of states and lattice dispersion relations. A limitation is the relatively low energy or wave vector resolution of neutron scattering compared with that readily available with light scattering (Cummins and Levanyuk 1983). In consequence neutron studies are not so useful at very long wavelengths. The neutron scattering method applied to lattice dynamics has been reviewed by several groups of authors (Dolling 1974; Smith and Wakabayashi 1977; Currat and Pynn 1979; Bilz and Kress 1979; Dorner 1982).

The studies referred to exploit the one-phonon contribution to the neutron cross-section. For a harmonic lattice, in which the phonons have an infinite lifetime, the one-phonon cross-sections vanish unless the neutron energy transfer matches the lattice dispersion exactly. Anharmonic effects that are present in all materials to some extent modify the cross-sections in a nontrivial manner. Moreover, multiphonon effects are significant at intermediate and large scattering vectors and, for example, produce structure in the one-phonon line shapes even in harmonic lattices. Some anharmonic and multiphonon effects which contribute to the one-phonon cross-sections are discussed following the derivation of the one-phonon cross-sections in § 4.4. Defects in the target sample can also modify the one-phonon cross-sections, and the important case of mass defects is discussed from various viewpoints in § 4.7.

## 4.1. Harmonic lattice vibrations (Brüesch 1982)

We need to calculate the displacements and vibrational frequencies of the atoms in a crystal about their equilibrium configuration. In Chapter 2 we discussed how the latter can be defined in terms of basic vectors $\mathbf{a}_1$, $\mathbf{a}_2$, and $\mathbf{a}_3$. We shall first consider the problem of determining the vibrational states of $N$ atoms with an equilibrium configuration that coincides with a Bravais lattice, i.e. that is given by the lattice vectors

$$\mathbf{l} = l_1\mathbf{a}_1 + l_2\mathbf{a}_2 + l_3\mathbf{a}_3, \tag{4.1}$$

and then consider the general case of more than one atom per unit cell.

The displacement of the atom at the $l$th site is denoted by $\mathbf{u}(l)$. Thus the position vector of this atom is

$$\mathbf{R}_l = \mathbf{l} + \mathbf{u}(l). \tag{4.2}$$

The potential energy of the atoms $U$ is assumed to be lowest when all **u**'s are zero. We can expand $U$ in a Taylor series

$$U-U_0=\tfrac{1}{2}\sum_{\alpha,l}\sum_{\beta,l'} u_\alpha(l)A_{\alpha\beta}(l,l')u_\beta(l')$$
$$+\tfrac{1}{6}\sum_{\alpha,l}\sum_{\beta,l'}\sum_{\gamma,l''} u_\alpha(l)u_\beta(l')u_\gamma(l'')B_{\alpha\beta\gamma}(l,l',l'')+\ldots, \qquad (4.3)$$

where $U_0$ is the potential energy in equilibrium. We are concerned here with the terms quadratic in the displacements, i.e. the first term on the right-hand side of (4.3). The $A_{\alpha\beta}(l,l')$, the second-order derivatives of the potential energy, clearly satisfy

$$A_{\alpha\beta}(l,l')=A_{\beta\alpha}(l',l) \qquad (4.4a)$$

and can depend only on the relative positions of the relevant cells, i.e.

$$A_{\alpha\beta}(l,l')=A_{\alpha\beta}(l-l'). \qquad (4.4b)$$

Also, since every atom is a centre of inversion symmetry,

$$A_{\alpha\beta}(l,l')=A_{\alpha\beta}(l',l)=A_{\beta\alpha}(l,l'). \qquad (4.4c)$$

The Hamiltonian of the classical motion of the atoms in the harmonic approximation is, from (4.3),

$$\mathcal{H}=\sum_{\alpha,l}\frac{1}{2M}p_\alpha^2(l)+\tfrac{1}{2}\sum_{\alpha,l}\sum_{\beta,l'} u_\alpha(l)A_{\alpha\beta}(l,l')u_\beta(l'), \qquad (4.5)$$

where $M$ is the mass of a nucleus and $p_\alpha(l)$ the momentum conjugate to $u_\alpha(l)$.

The equation of motion for the $l$th atom is

$$\dot{p}_\alpha(l)=-\frac{\partial\mathcal{H}}{\partial u_\alpha(l)}=M\ddot{u}_\alpha(l),$$

or

$$M\ddot{u}_\alpha(l)=-\sum_{\beta,l'}A_{\alpha\beta}(l,l')u_\beta(l'), \qquad (4.6)$$

with the aid of (4.4a). Since the right-hand side of (4.6) gives the force acting on atom $l$, and this force must vanish if all atoms are translated by an identical amount, it follows that $\sum_{l'}A_{\alpha\beta}(l,l')=0$. The set of equations (4.6) can be reduced to a set of equations for $3N$ independent harmonic oscillators by introducing a set of normal coordinates, which we shall now calculate.

If we allow the displacements to have the time dependence $\exp(-i\omega_0 t)$, then (4.6) becomes

$$M\omega_0^2 u_\alpha(l)=\sum_{l',\beta}A_{\alpha\beta}(l,l')u_\beta(l'). \qquad (4.7)$$

The condition that the set of $3N$ linear equations (4.7) for the displacements should have a non-trivial solution is that the determinant of the coefficients vanishes,

$$\det |M\omega_0^2 \,\delta_{\alpha,\beta}\, \delta_{l,l'} - A_{\alpha\beta}(l, l')| = 0. \tag{4.8}$$

The roots of this equation are the normal-mode frequencies of the crystal lattice.

Because of the translational symmetry of the lattice, let us make the further substitution

$$u_\alpha(l) = \frac{\sigma_\alpha}{\sqrt{M}} \exp(i\mathbf{q}\cdot\mathbf{l}). \tag{4.9}$$

Then the eigenvalue equation for the normal-mode frequencies can be written

$$\begin{aligned} M\omega_0^2\sigma_\alpha &= \exp(-i\mathbf{q}\cdot\mathbf{l}) \sum_{l',\beta} A_{\alpha\beta}(l, l')\exp(i\mathbf{q}\cdot\mathbf{l}')\sigma_\beta \\ &= \sum_{l,\beta} A_{\alpha\beta}(l)\exp(-i\mathbf{q}\cdot\mathbf{l})\sigma_\beta, \end{aligned} \tag{4.10}$$

where the last line follows because of (4.4b).

For each value of $\mathbf{q}$ there are three solutions for $\omega_0^2$ corresponding to the three values of $\alpha$. We denote this fact by writing the lattice dispersion frequencies as $\omega_j(\mathbf{q})$, the index $j$ having three values. To each $\omega_j(\mathbf{q})$ there corresponds a polarization vector $\sigma_\alpha$, and we shall henceforth write these as $\sigma_\alpha^j(\mathbf{q})$. Thus the eigenvalue equation (4.10) now reads

$$\omega_j^2(\mathbf{q})\sigma_\alpha^j(\mathbf{q}) = \sum_\beta \mathscr{A}_{\alpha\beta}(\mathbf{q})\sigma_\beta^j(\mathbf{q}). \tag{4.11a}$$

Here

$$\mathscr{A}_{\alpha\beta}(\mathbf{q}) = \frac{1}{M} \sum_l A_{\alpha\beta}(l)\exp(-i\mathbf{q}\cdot\mathbf{l}) \tag{4.11b}$$

and is usually called the dynamical matrix.

From (4.11b) we deduce that in general

$$\mathscr{A}^*_{\alpha\beta}(\mathbf{q}) = \mathscr{A}_{\alpha\beta}(-\mathbf{q}). \tag{4.11c}$$

If every atom is a centre of inversion symmetry (as it is for a Bravais crystal) so that (4.4c) is satisfied, then

$$\mathscr{A}^*_{\alpha\beta}(\mathbf{q}) = \mathscr{A}_{\alpha\beta}(\mathbf{q}), \tag{4.11d}$$

i.e. the dynamical matrix is purely real.

From (4.11a) and (4.11c) it follows that $\sigma_\alpha^j(\mathbf{q})$ and $\sigma_\alpha^{*j}(-\mathbf{q})$ satisfy the same equation (4.11a) and are therefore identical except for an arbitrary phase factor. We adopt a convention that makes $\sigma_\alpha^j(\mathbf{q})$ real and even in $\mathbf{q}$

for Bravais crystals, i.e. in the general case we take

$$\sigma_{\alpha}^{*j}(\mathbf{q}) = \sigma_{\alpha}^{j}(-\mathbf{q}). \tag{4.11e}$$

The vectors $\sigma_{\alpha}^{j}(\mathbf{q})$ can be constructed so as to satisfy the following closure and orthonormality conditions,

$$\sum_{j} \sigma_{\alpha}^{*j}(\mathbf{q})\sigma_{\beta}^{j}(\mathbf{q}) = \delta_{\alpha,\beta} \tag{4.12a}$$

and

$$\boldsymbol{\sigma}^{*j}(\mathbf{q}) \cdot \boldsymbol{\sigma}^{j'}(\mathbf{q}) = \delta_{j,j'}. \tag{4.12b}$$

The number of vectors $\mathbf{q}$ is not unlimited. The vectors $\mathbf{q}$ and $\mathbf{q}+\boldsymbol{\tau}$ ($\boldsymbol{\tau}$ we recall is a reciprocal lattice vector, i.e. $\mathbf{l}\cdot\boldsymbol{\tau} = 2\pi \times \text{integer}$) give the same displacement, as is seen by reference to (4.9). The matrix $\mathscr{A}_{\alpha\beta}(\mathbf{q})$, the dynamical matrix, remains unchanged under the transformation $\mathbf{q} \to \mathbf{q}+\boldsymbol{\tau}$, and hence its vectors $\boldsymbol{\sigma}^{j}(\mathbf{q})$ must also be invariant under the same transformation. This means that we can restrict our values of $\mathbf{q}$ to lie within the first Brillouin zone. The first Brillouin zones for body-centred and face-centred cubic lattices are shown in Fig. 4.1. By applying periodic boundary conditions to the crystal we find that the density of vectors $\mathbf{q}$ is, for each value of $j$, $V/(2\pi)^3$, where $V$ is the volume of the crystal. Hence

$$\sum_{\mathbf{q}} \to \frac{V}{(2\pi)^3} \int d\mathbf{q}. \tag{4.13}$$

The frequencies $\omega_j(\mathbf{q})$ are shown in Fig. 4.2 for the particular case of a face-centred cubic lattice. The frequencies are shown for three directions of $\mathbf{q}$, each parallel to a symmetry direction. These are clearly shown in the diagram of the first Brillouin zone for a face-centred lattice given in Fig. 4.1. Because the directions of $\mathbf{q}$ are along symmetry directions, the

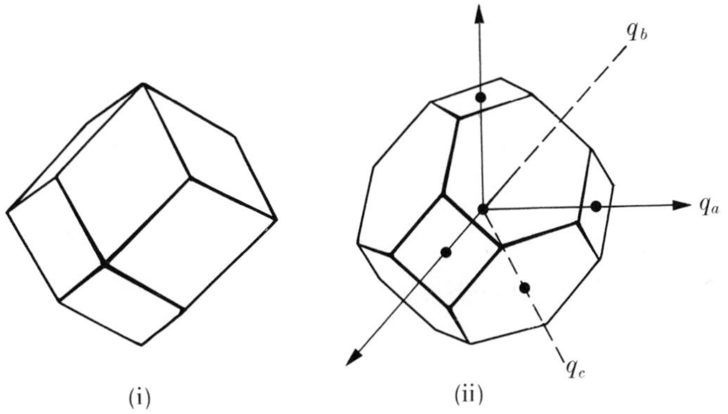


FIG. 4.1. The first Brillouin zone for (i) body-centred cubic and (ii) face-centred cubic lattices.

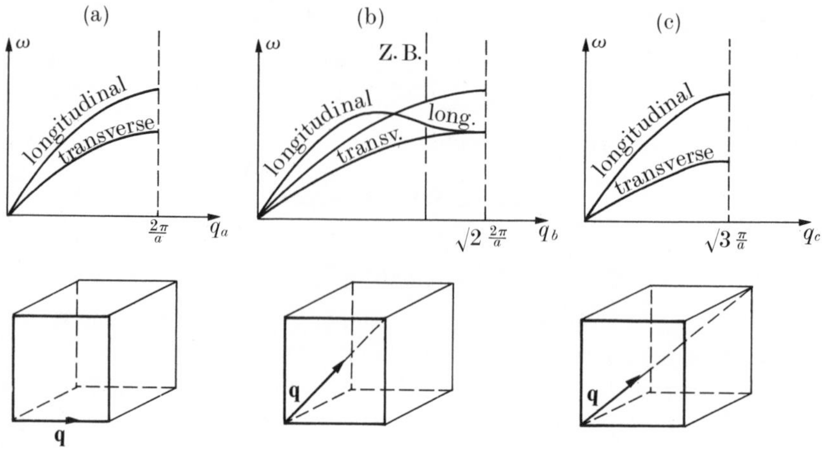


FIG. 4.2. The frequencies $\omega_j(\mathbf{q})$ are shown for $\mathbf{q}$ along special symmetry directions for a face-centred cubic structure. The three wave vectors $q_a$, $q_b$, and $q_c$ are also shown in the first Brillouin zone of the lattice given in Fig. 4.1. Because the vectors $\mathbf{q}$ are taken along symmetry directions, the vectors $\boldsymbol{\sigma}^j(\mathbf{q})$ are either parallel or perpendicular to them, and the corresponding frequencies are denoted as longitudinal or transverse. In the particular example shown in the figure the two transverse modes are degenerate in cases (a) and (c). In (b) the curves have been extended into the next Brillouin zone to a total distance $\sqrt{2}(2\pi/a)$. For more details see, for example, Brüesch (1982).

polarization vectors $\boldsymbol{\sigma}^j(\mathbf{q})$ have a simple relation to $\mathbf{q}$, namely two of them are perpendicular to $\mathbf{q}$ and the other is parallel to $\mathbf{q}$. Hence, the corresponding frequencies have been labelled as transverse and longitudinal, the two transverse modes being degenerate in the examples chosen except for the case of $\mathbf{q}$ parallel to a face diagonal.

We now show that the quantum mechanical Hamiltonian that corresponds to (4.5) can be reduced to the sum of $3N$ independent harmonic oscillators. In Chapter 3 we noted that a component $x$ of the displacement vector for a particle in a harmonic oscillator can be written in terms of Bose operators $\hat{a}$ and $\hat{a}^+$ in the form

$$\hat{x} = \left(\frac{\hbar}{2M\omega_0}\right)^{1/2} (\hat{a} + \hat{a}^+) \tag{4.14a}$$

and, consequently, the momentum operator $\hat{p}$ conjugate to $\hat{x}$ as

$$\hat{p} = -\mathrm{i}\left(\frac{\hbar\omega_0 M}{2}\right)^{1/2} (\hat{a} - \hat{a}^+). \tag{4.14b}$$

In the present case we can introduce Bose operators associated with a given lattice site, namely $\hat{\mathbf{a}}(l)$ and $\hat{\mathbf{a}}^+(l)$. We then define phonon operators $\hat{a}_j(\mathbf{q})$ and $\hat{a}_j^+(\mathbf{q})$ through the transformations

$$\hat{a}_\alpha(l) = \frac{1}{\sqrt{N}} \sum_{j,\mathbf{q}} \sigma_\alpha^j(\mathbf{q}) \exp(\mathrm{i}\mathbf{q}\cdot\mathbf{l})\hat{a}_j(\mathbf{q}) \tag{4.15a}$$

and

$$\hat{a}_\alpha^+(l) = \frac{1}{\sqrt{N}} \sum_{j,\mathbf{q}} \sigma_\alpha^{*j}(\mathbf{q}) \exp(-\mathrm{i}\mathbf{q}\cdot\mathbf{l})\hat{a}_j^+(\mathbf{q}) \tag{4.15b}$$

and thus, by analogy with (4.14),

$$\hat{u}_\alpha(l) = \sum_{j,\mathbf{q}} \left\{\frac{\hbar}{2NM\omega_j(\mathbf{q})}\right\}^{1/2} \{\sigma_\alpha^j(\mathbf{q})\exp(\mathrm{i}\mathbf{q}\cdot\mathbf{l})\hat{a}_j(\mathbf{q}) + \sigma_\alpha^{*j}(\mathbf{q})\exp(-\mathrm{i}\mathbf{q}\cdot\mathbf{l})\hat{a}_j^+(\mathbf{q})\} \tag{4.16a}$$

and

$$\hat{p}_\alpha(l) = -\mathrm{i}\sum_{j,\mathbf{q}} \left\{\frac{\hbar\omega_j(\mathbf{q})M}{2N}\right\}^{1/2} \times\{\sigma_\alpha^j(\mathbf{q})\exp(\mathrm{i}\mathbf{q}\cdot\mathbf{l})\hat{a}_j(\mathbf{q}) - \sigma_\alpha^{*j}(\mathbf{q})\exp(-\mathrm{i}\mathbf{q}\cdot\mathbf{l})\hat{a}_j^+(\mathbf{q})\}. \tag{4.16b}$$

From

$$[\hat{u}_\alpha(l), \hat{p}_\beta(l')] = \mathrm{i}\hbar\, \delta_{\alpha,\beta}\, \delta_{l,l'} \tag{4.17}$$

it follows that

$$[\hat{a}_j(\mathbf{q}), \hat{a}_{j'}^+(\mathbf{q}')] = \delta_{j,j'}\, \delta_{\mathbf{q},\mathbf{q}'}. \tag{4.18}$$

The operators $\hat{a}_j(\mathbf{q})$ and $\hat{a}_j^+(\mathbf{q})$ commute with themselves, of course.

By analogy with the discussion of a simple harmonic oscillator in Chapter 3 it follows that $\hat{a}_j^+(\mathbf{q})\hat{a}_j(\mathbf{q})$ is the number operator for the phonon described by $\mathbf{q}$ and $j$. Hence†

$$\langle \hat{a}_j^+(\mathbf{q})\hat{a}_{j'}(\mathbf{q}')\rangle = \delta_{jj'}\, \delta_{\mathbf{q}\mathbf{q}'}\langle \hat{a}_j^+(\mathbf{q})\hat{a}_j(\mathbf{q})\rangle = \delta_{jj'}\, \delta_{\mathbf{q}\mathbf{q}'} n_j(\mathbf{q}), \tag{4.19}$$

where

$$n_j(\mathbf{q}) = [\exp\{\beta\hbar\omega_j(\mathbf{q})\} - 1]^{-1}.$$

† If $\hat{O}$ is any symmetry operator that commutes with the Hamiltonian and $\hat{P}$ an arbitrary operator then

$$\langle[\hat{O}, \hat{P}]\rangle = 0 \tag{a}$$

by the cyclic properties of the trace of two operators. Let us choose for $\hat{O}$

$$\sum_{j,\mathbf{q}} \mathbf{q}\hat{a}_j^+(\mathbf{q})\hat{a}_j(\mathbf{q}),$$

which clearly commutes with (4.20), and for $\hat{P}$ we take $\hat{a}_{j_1}^+(\mathbf{q}_1)\hat{a}_{j_2}(\mathbf{q}_2)$. Then

$$[\hat{O}, \hat{P}] = \hat{a}_{j_1}^+(\mathbf{q}_1)\hat{a}_{j_2}(\mathbf{q}_2)(\mathbf{q}_1 - \mathbf{q}_2), \tag{b}$$

so that from (a) it follows that we must have $\mathbf{q}_1 = \mathbf{q}_2$ in $\langle \hat{a}_j^+(\mathbf{q}_1)\hat{a}_j(\mathbf{q}_2)\rangle$. We can similarly prove that we must also have $j_1 = j_2$ by choosing $\hat{O} = \sum_{j,\mathbf{q}} j\hat{a}_j^+(\mathbf{q})\hat{a}_j(\mathbf{q})$.

With the transformations (4.16) we shall show that the Hamiltonian

$$\hat{\mathcal{H}} = \sum_{\alpha,l} \frac{1}{2M} \{\hat{p}_\alpha(l)\}^2 + \tfrac{1}{2} \sum_{\alpha,l} \sum_{\beta,l'} \hat{u}_\alpha(l) A_{\alpha\beta}(l, l') \hat{u}_\beta(l') \tag{4.20a}$$

becomes

$$\hat{\mathcal{H}} = \sum_{j,\mathbf{q}} \hbar\omega_j(\mathbf{q}) \{\hat{a}_j^+(\mathbf{q}) \hat{a}_j(\mathbf{q}) + \tfrac{1}{2}\}. \tag{4.20b}$$

That is to say (4.20a), which is the quantum analogue of the classical Hamiltonian (4.5), can be represented as the sum of $3N$ independent harmonic oscillators; the operators $\hat{a}_j(\mathbf{q})$ and $\hat{a}_j^+(\mathbf{q})$ annihilate and create, respectively, phonons defined by the quantum numbers $j$ and $\mathbf{q}$. It follows immediately from (4.20b) that

$$i\hbar\, \partial_t \hat{a}_j(\mathbf{q}, t) = [\hat{a}_j(\mathbf{q}, t), \hat{\mathcal{H}}] = \hbar\omega_j(\mathbf{q}) \hat{a}_j(\mathbf{q}, t)$$

and hence

$$\hat{a}_j(\mathbf{q}, t) = \hat{a}_j(\mathbf{q}) \exp\{-\mathrm{i}\omega_j(\mathbf{q})t\}. \tag{4.21a}$$

Similarly we find

$$\hat{a}_j^+(\mathbf{q}, t) = \hat{a}_j^+(\mathbf{q}) \exp\{\mathrm{i}\omega_j(\mathbf{q})t\}. \tag{4.21b}$$

In verifying that the Hamiltonian (4.20b) is equivalent to that given by (4.20a) we find it convenient to use (4.11e) and rewrite (4.16) in the alternative form

$$\hat{u}_\alpha(l) = \sum_{j,\mathbf{q}} \left\{\frac{\hbar}{2NM\omega_j(\mathbf{q})}\right\}^{1/2} \sigma_\alpha^j(\mathbf{q}) \exp(\mathrm{i}\mathbf{q}\cdot\mathbf{l}) \{\hat{a}_j(\mathbf{q}) + \hat{a}_j^+(-\mathbf{q})\} \tag{4.22a}$$

and

$$\hat{p}_\alpha(l) = -\mathrm{i} \sum_{j,\mathbf{q}} \left\{\frac{\hbar\omega_j(\mathbf{q})M}{2N}\right\}^{1/2} \sigma_\alpha^j(\mathbf{q}) \exp(\mathrm{i}\mathbf{q}\cdot\mathbf{l}) \{\hat{a}_j(\mathbf{q}) - \hat{a}_j^+(-\mathbf{q})\}. \tag{4.22b}$$

The kinetic energy is then

$$\begin{aligned}
\sum_{\alpha,l} \frac{1}{2M} \hat{p}_\alpha(l)\hat{p}_\alpha(l) &= \sum_{\alpha,l} \frac{1}{2M} \hat{p}_\alpha^+(l)\hat{p}_\alpha(l) \\
&= \sum_{j,\mathbf{q}} \sum_{j',\mathbf{q}'} \frac{\hbar}{4N} \{\omega_j(\mathbf{q})\omega_{j'}(\mathbf{q}')\}^{1/2} \sum_{\alpha,l} \sigma_\alpha^{*j}(\mathbf{q}) \sigma_\alpha^{j'}(\mathbf{q}') \exp\{-\mathrm{i}(\mathbf{q}-\mathbf{q}')\cdot\mathbf{l}\} \\
&\quad \times \{\hat{a}_j^+(\mathbf{q}) - \hat{a}_j(-\mathbf{q})\}\{\hat{a}_{j'}(\mathbf{q}') - \hat{a}_{j'}^+(-\mathbf{q}')\} \\
&= \tfrac{1}{4}\hbar \sum_{j,\mathbf{q}} \omega_j(\mathbf{q}) \{2\hat{a}_j^+(\mathbf{q})\hat{a}_j(\mathbf{q}) + 1 - \hat{a}_j(-\mathbf{q})\hat{a}_j(\mathbf{q}) - \hat{a}_j^+(\mathbf{q})\hat{a}_j^+(-\mathbf{q})\},
\end{aligned} \tag{4.23}$$

where we have used (4.12b).

Similarly the potential energy is

$$\tfrac{1}{2}\sum_{\alpha,l}\sum_{\beta,l'} \hat{u}_\alpha^+(l)A_{\alpha\beta}(l,l')\hat{u}_\beta(l') = \frac{\hbar}{4NM}\sum_{j,\mathbf{q}}\sum_{j',\mathbf{q}'}\left\{\frac{1}{\omega_j(\mathbf{q})\omega_{j'}(\mathbf{q}')}\right\}^{1/2}$$
$$\times\sum_{\alpha,l}\sum_{\beta,l'}\sigma_\alpha^{*j}(\mathbf{q})A_{\alpha\beta}(l,l')\exp(-i\mathbf{q}\cdot\mathbf{l}+i\mathbf{q}'\cdot\mathbf{l}')\sigma_\beta^{j'}(\mathbf{q}')$$
$$\times\{\hat{a}_j^+(\mathbf{q})+\hat{a}_j(-\mathbf{q})\}\{\hat{a}_{j'}(\mathbf{q}')+\hat{a}_{j'}^+(-\mathbf{q}')\}$$
$$=\tfrac{1}{4}\hbar\sum_{j,\mathbf{q}}\omega_j(\mathbf{q})\{2\hat{a}_j^+(\mathbf{q})\hat{a}_j(\mathbf{q})+1+\hat{a}_j(-\mathbf{q})\hat{a}_j(\mathbf{q})+\hat{a}_j^+(\mathbf{q})\hat{a}_j^+(-\mathbf{q})\}. \quad (4.24)$$

Here we have used (4.11a) and (4.11b). If we now add together the kinetic and potential energy terms we get the result (4.20b) immediately.

The analysis we have given so far has been for the special case of a Bravais lattice, but it is quite easy to extend it to the general case of a lattice consisting of several interpenetrating Bravais lattices of identical structure each containing nuclei of different mass. The position vector of an atom is now

$$\mathbf{R}_{ld}=\mathbf{l}+\mathbf{d}+\mathbf{u}\begin{pmatrix}l\\d\end{pmatrix} \quad (4.25)$$

Here $\mathbf{d}$ is the position vector to the $d$th atom within the $l$th cell, there being $r$ atoms in each unit cell. The total number of degrees of freedom is now $3rN$ and hence the quantum number $j$ takes $3r$ values.

The displacement operator for the general case is conveniently written in terms of a normal coordinate displacement operator $\hat{Q}(\mathbf{q}j;t)$ (cf. eqn (4.22a)). We denote the polarization vectors by $\boldsymbol{\sigma}_d^j(\mathbf{q})$, with $\{\boldsymbol{\sigma}_d^j(\mathbf{q})\}^*=\boldsymbol{\sigma}_d^j(-\mathbf{q})$, and find the following expression for the displacement operator for a harmonic lattice,

$$\hat{\mathbf{u}}\begin{pmatrix}l\\d\end{pmatrix}, t\Big) = (NM_d)^{-1/2}\sum_{j\mathbf{q}}\boldsymbol{\sigma}_d^j(\mathbf{q})\exp(i\mathbf{q}\cdot\mathbf{l})\hat{Q}(\mathbf{q}j;t) \quad (4.26)$$

where

$$\hat{Q}(\mathbf{q}j;t)=\left(\frac{\hbar}{2\omega_j(\mathbf{q})}\right)^{1/2}\{\hat{a}_j(\mathbf{q})\exp\{-it\omega_j(\mathbf{q})\}+\hat{a}_j^+(-\mathbf{q})\exp\{it\omega_j(\mathbf{q})\}\}=\hat{Q}^+(-\mathbf{q}j;t).$$

Because $\hat{Q}(-\mathbf{q}j;0)=\{\hat{Q}(\mathbf{q}j;0)\}^+$, the displacement operator is real. The orthonormality and closure conditions for the polarization vectors are now (cf. eqn (4.12))

$$\sum_d\{\boldsymbol{\sigma}_d^j(\mathbf{q})\}^*\cdot\boldsymbol{\sigma}_d^{j'}(\mathbf{q})=\delta_{jj'}$$

and

$$\sum_j\{\sigma_{\alpha d}^j(\mathbf{q})\}^*\sigma_{\beta d'}^j(\mathbf{q})=\delta_{\alpha\beta}\,\delta_{dd'}. \quad (4.27)$$

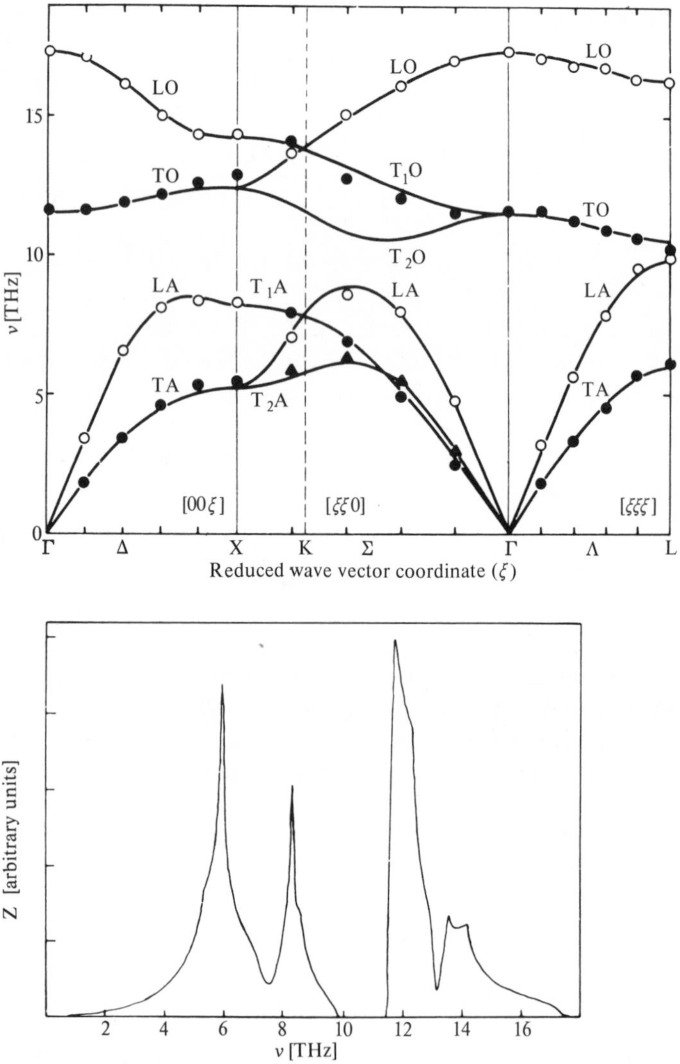


FIG. 4.3. The dispersion curves and density of states for the compound NiO (Bilz and Kress 1979).

From expression (4.26) for the displacement operator we notice that, if we can choose $\boldsymbol{\sigma}_d^i(\mathbf{q}) = \boldsymbol{\sigma}_d^{*j}(\mathbf{q})$, then the phonons are linearly polarized; otherwise the phonons are elliptically polarized. Hence if each lattice point is a centre of inversion symmetry the phonons are linearly polarized; otherwise they are elliptically polarized, in general. Notice that this question of polarization is quite different from that for electromagnetic radiation. For the latter, in free space, the two polarization directions are always degenerate and the polarization is therefore undetermined; the radiation may be linearly polarized or elliptically polarized depending upon the way it is produced. For phonons a degeneracy between the two transverse modes can exist only for $\mathbf{q}$ along special symmetry directions. For all other directions of $\mathbf{q}$ there is no degeneracy and the polarization vectors are completely determined (apart from an unimportant phase factor) by the eigenvalue equation.

An alternative definition of the polarization vectors is

$$\boldsymbol{\sigma}_d^j(\mathbf{q}) = \exp(\mathrm{i}\mathbf{q}\cdot\mathbf{d})\mathbf{e}_d^j(\mathbf{q}). \tag{4.28}$$

The vectors $\mathbf{e}_d^j(\mathbf{q})$ satisfy relations analogous to (4.27). An advantage of the choice (4.28) is that, for crystals such as the alkali halides where each atom is at a centre of inversion symmetry, the $\mathbf{e}_d^j(\mathbf{q})$ and the corresponding dynamic matrix are purely real. For either representation the lattice dispersion frequency $\omega_j(\mathbf{q})$ is periodic in reciprocal space, so that

$$\omega_j(\mathbf{q}) = \omega_j(\mathbf{q}+\boldsymbol{\tau}) \tag{4.29}$$

for any reciprocal lattice vector $\boldsymbol{\tau}$. The normal-mode displacement is also periodic, and this implies that $\boldsymbol{\sigma}_d^j(\mathbf{q}) = \boldsymbol{\sigma}_d^j(\mathbf{q}+\boldsymbol{\tau})$ whereas

$$\mathbf{e}_d^j(\mathbf{q}) = \exp(-\mathrm{i}\boldsymbol{\tau}\cdot\mathbf{d})\mathbf{e}_d^j(\mathbf{q}-\boldsymbol{\tau}). \tag{4.30}$$

The next few sections are devoted to a study of the elastic and inelastic one-phonon neutron cross-sections associated with harmonic lattice vibrations. Several aspects are more easily explained with the aid of an illustrative example of a set of dispersion relations and the associated density of states. The example given in Fig. 4.3 is for NiO, and this compound has the rock salt structure.

## 4.2. Elastic scattering (Bacon 1977; Brown 1979)

We now calculate the cross-sections for the elastic coherent and incoherent scattering of neutrons by nuclei undergoing harmonic vibrations. Initially we assume the equilibrium positions of the nuclei to form a Bravais lattice and then later extend the formulae to general lattice structures.

Consider first the elastic coherent cross-section. From the previous chapter,

$$\left(\frac{\mathrm{d}\sigma}{\mathrm{d}\Omega}\right)^{\mathrm{el}}_{\mathrm{coh}} = \frac{\sigma_{\mathrm{c}}}{4\pi}\left|\int \mathrm{d}\mathbf{r}\exp(\mathrm{i}\boldsymbol{\kappa}\cdot\mathbf{r})\langle\hat{\rho}(\mathbf{r})\rangle\right|^2, \tag{4.31}$$

where in the present case

$$\langle\hat{\rho}(\mathbf{r})\rangle=\sum_{l}\langle\delta\{\mathbf{r}-\hat{\mathbf{R}}_l\}\rangle$$
$$=\left(\frac{1}{2\pi}\right)^3\int \mathrm{d}\mathbf{k}\sum_{l}\exp\{\mathrm{i}\mathbf{k}\cdot(\mathbf{r}-\mathbf{l})\}\langle\exp\{-\mathrm{i}\mathbf{k}\cdot\hat{\mathbf{u}}(l)\}\rangle. \tag{4.32}$$

If we invoke the identity (3.172),

$$\langle\exp\{-\mathrm{i}\mathbf{k}\cdot\hat{\mathbf{u}}(l)\}\rangle=\exp[-\tfrac{1}{2}\langle\{\mathbf{k}\cdot\hat{\mathbf{u}}(l)\}^2\rangle]. \tag{4.33}$$

From the expansion of $\hat{\mathbf{u}}(l)$ in terms of the phonon operators $\hat{a}_j(\mathbf{q})$ and $\hat{a}_j^+(\mathbf{q})$ (eqn (4.16a)) it follows that

$$\langle\{\mathbf{k}\cdot\hat{\mathbf{u}}(l)\}^2\rangle=\frac{\hbar}{2NM}\sum_{j,\mathbf{q}}\sum_{j',\mathbf{q}'}\{\omega_j(\mathbf{q})\omega_{j'}(\mathbf{q}')\}^{-1/2}\sum_{\alpha,\beta}k_\alpha k_\beta\sigma^j_\alpha(\mathbf{q})\sigma^{j'}_\beta(\mathbf{q}')$$
$$\times[\exp\{-\mathrm{i}\mathbf{l}\cdot(\mathbf{q}-\mathbf{q}')\}\langle\hat{a}_j^+(\mathbf{q})\hat{a}_{j'}(\mathbf{q}')\rangle+\exp\{\mathrm{i}\mathbf{l}\cdot(\mathbf{q}-\mathbf{q}')\}\langle\hat{a}_j(\mathbf{q})\hat{a}_{j'}^+(\mathbf{q}')\rangle]$$
$$=\frac{\hbar}{2NM}\sum_{j,\mathbf{q}}\frac{|\mathbf{k}\cdot\boldsymbol{\sigma}^j(\mathbf{q})|^2}{\omega_j(\mathbf{q})}\{2n_j(\mathbf{q})+1\}. \tag{4.34}$$

The second line of (4.34) follows from (4.18a). Thus, on combining (4.33) and (4.34) it follows that

$$\langle\exp\{-\mathrm{i}\mathbf{k}\cdot\hat{\mathbf{u}}(l)\}\rangle=\exp\left[-\frac{\hbar}{4NM}\sum_{j,\mathbf{q}}\frac{|\mathbf{k}\cdot\boldsymbol{\sigma}^j(\mathbf{q})|^2}{\omega_j(\mathbf{q})}\{2n_j(\mathbf{q})+1\}\right]$$
$$=\exp\{-W(\mathbf{k})\}, \tag{4.35}$$

which is the well-known Debye–Waller factor.

It now follows that

$$\int \mathrm{d}\mathbf{r}\exp(\mathrm{i}\boldsymbol{\kappa}\cdot\mathbf{r})\langle\hat{\rho}(\mathbf{r})\rangle$$
$$=\sum_{l}\left(\frac{1}{2\pi}\right)^3\int\mathrm{d}\mathbf{k}\int\mathrm{d}\mathbf{r}\exp\{\mathrm{i}\mathbf{r}\cdot(\boldsymbol{\kappa}+\mathbf{k})-\mathrm{i}\mathbf{k}\cdot\mathbf{l}\}\exp\{-W(\mathbf{k})\}$$
$$=\sum_{l}\exp(\mathrm{i}\boldsymbol{\kappa}\cdot\mathbf{l})\exp\{-W(\boldsymbol{\kappa})\} \tag{4.36}$$

and thus, finally, using (2.9),

$$\left(\frac{\mathrm{d}\sigma}{\mathrm{d}\Omega}\right)^{\mathrm{el}}_{\mathrm{coh}}=\frac{N\sigma_{\mathrm{c}}}{4\pi}\frac{(2\pi)^3}{v_0}\sum_{\boldsymbol{\tau}}\delta(\boldsymbol{\kappa}-\boldsymbol{\tau})\exp\{-2W(\boldsymbol{\kappa})\}. \tag{4.37}$$

This result is the cross-section for coherent elastic scattering by a static lattice multiplied by the factor $\exp\{-2W(\boldsymbol{\kappa})\}$. By definition,

$$2W(\boldsymbol{\kappa})=\langle\{\boldsymbol{\kappa}\cdot\hat{\mathbf{u}}(l)\}^2\rangle, \tag{4.38}$$

so that $2W$ is merely the mean square displacement of a nucleus multiplied by $\kappa^2$. The presence of the factor $\exp\{-2W(\boldsymbol{\kappa})\}$ means that the intensity of the Bragg peaks decreases with increasing $|\boldsymbol{\kappa}|$.

The formula for $W(\boldsymbol{\kappa})$ can be greatly simplified if we introduce the normalized density of states (see Fig. 4.3)

$$Z(\omega)=\frac{1}{3N}\sum_{j,\mathbf{q}}\delta\{\omega-\omega_j(\mathbf{q})\} \tag{4.39}$$

for then, from (4.35),

$$W(\boldsymbol{\kappa})=3N\int_0^{\omega_m}\mathrm{d}\omega\, Z(\omega)\frac{\hbar}{4NM\omega}\{2n(\omega)+1\}\{|\boldsymbol{\kappa}\cdot\boldsymbol{\sigma}|^2\}_{\mathrm{av}}. \tag{4.40}$$

In (4.40), $\omega_m$ denotes the maximum frequency of the phonons and in the last factor the subscript av means an average over a surface with a given $\omega$ must be taken. In cubic symmetry this average is $\frac{1}{3}\kappa^2$, and this result will be a fair approximation even for non-cubic crystals in most cases. Hence

$$W(\boldsymbol{\kappa})=\frac{\hbar\kappa^2}{4M}\int_0^{\omega_m}\mathrm{d}\omega\frac{Z(\omega)}{\omega}\{2n(\omega)+1\}=\frac{\hbar\kappa^2}{4M}\int_0^{\omega_m}\mathrm{d}\omega\frac{Z(\omega)}{\omega}\coth(\tfrac{1}{2}\hbar\omega\beta). \tag{4.41}$$

It is interesting to invert (4.36) to give

$$\langle\hat{\rho}(\mathbf{r})\rangle=\frac{1}{(2\pi)^3}\int\mathrm{d}\mathbf{k}\exp(-\mathrm{i}\mathbf{k}\cdot\mathbf{r})\sum_l\exp(\mathrm{i}\mathbf{k}\cdot\mathbf{l})\exp\{-W(\mathbf{k})\}.$$

If for simplicity we assume a cubic lattice, then, from (4.38),

$$W(\mathbf{k})=\tfrac{1}{2}\langle(\mathbf{k}\cdot\hat{\mathbf{u}})^2\rangle=\tfrac{1}{6}k^2\langle\hat{\mathbf{u}}^2\rangle,$$

and we get

$$\langle\hat{\rho}(\mathbf{r})\rangle=\left\{\frac{3}{2\pi\langle\hat{\mathbf{u}}^2\rangle}\right\}^{3/2}\sum_l\exp\{-3\,|\mathbf{r}-\mathbf{l}|^2/2\langle\hat{\mathbf{u}}^2\rangle\}.$$

This shows that $\langle\hat{\rho}(\mathbf{r})\rangle$ is a set of Gaussian functions centred on the lattice points $\mathbf{l}$. Each Gaussian has a mean square width $\langle\hat{\mathbf{u}}^2\rangle$, and, by analogy with the magnetic scattering of neutrons, we see that the Debye–Waller factor is effectively the 'form factor' of this nuclear density.

Anticipating (4.45), which shows that $\exp\{-2W(\mathbf{k})\}$ is the Fourier transform of the self-correlation function $G_s(\mathbf{r},\infty)$, we conclude that

$$G_s(\mathbf{r},\infty)=\left\{\frac{3}{4\pi\langle\hat{\mathbf{u}}^2\rangle}\right\}^{3/2}\exp\{-3r^2/4\langle\hat{\mathbf{u}}^2\rangle\}.$$

We recall from Chapter 3 that $G_s(\mathbf{r},0)$ is simply $\delta(\mathbf{r})$, so as the time increases to infinity $G_s(\mathbf{r},t)$ approaches the Gaussian distribution described above, whereas for a liquid $G_s(\mathbf{r},t)$ approaches zero, because the motion of the particle is unconfined.

To obtain some idea of the temperature dependence of $W(\boldsymbol{\kappa})$ we can replace $Z(\omega)$ by the Debye spectrum, i.e.

$$Z(\omega)\to\frac{3\omega^2}{\omega_D^3}, \tag{4.42}$$

where $\omega_D$ is the Debye frequency, which is related to the Debye temperature $\theta_D$ through $\hbar\omega_D=k_B\theta_D$. If $T\gg\theta_D$, then in the integrand of (4.41) we can take

$$\coth(\tfrac{1}{2}\hbar\omega\beta)\sim\frac{2}{\hbar\omega\beta},$$

so that, with (4.42),

$$W(\boldsymbol{\kappa})\sim 3\left(\frac{\hbar^2\kappa^2}{2M}\right)\left(\frac{1}{\hbar\omega_D}\right)\left(\frac{T}{\theta_D}\right); \qquad T\gg\theta_D. \tag{4.43}$$

Note that $\hbar^2\kappa^2/2M$ is the energy of recoil of a nucleus of mass $M$, initially at rest.

In the limit $T\to 0$, $W(\boldsymbol{\kappa})$ is not zero because of the zero-point motion of the nuclei. Indeed in this limit we obtain, with the use of (4.42),

$$W(\boldsymbol{\kappa})=\frac{3}{4}\left(\frac{\hbar^2\kappa^2}{2M}\right)\left(\frac{1}{\hbar\omega_D}\right); \qquad T=0. \tag{4.44}$$

The evaluation of the incoherent elastic cross-section is quite straightforward. From Chapter 3,

$$\left(\frac{d\sigma}{d\Omega}\right)^{\text{el}}_{\text{incoh}}=\frac{\sigma_i}{4\pi}\sum_l |\langle\exp\{i\boldsymbol{\kappa}\cdot\hat{\mathbf{R}}_l\}\rangle|^2,$$

which from (4.35) reduces to

$$\left(\frac{d\sigma}{d\Omega}\right)^{\text{el}}_{\text{incoh}}=\frac{N\sigma_i}{4\pi}\exp\{-2W(\boldsymbol{\kappa})\}. \tag{4.45}$$

Again this is the result for the static lattice multiplied by the factor $\exp\{-2W(\boldsymbol{\kappa})\}$. Because of this factor, the incoherent elastic scattering is no longer isotropic but slightly larger in the forward direction than in the backward direction.

If we use (4.38) and assume a cubic crystal, then the total elastic incoherent cross-section is readily derived from (4.45). We have

$$\begin{aligned}\sigma_i^{\text{el}}&=\frac{N\sigma_i}{4\pi}2\pi\int_{-1}^{1}d\mu\,\exp\{-\tfrac{1}{3}\langle\hat{\mathbf{u}}^2\rangle(k^2+k'^2-2kk'\mu)\}\\&=\frac{N\sigma_i}{4\pi}2\pi\exp\{-\tfrac{2}{3}k^2\langle\hat{\mathbf{u}}^2\rangle\}\int_{-1}^{1}d\mu\,\exp\{\tfrac{2}{3}k^2\langle\hat{\mathbf{u}}^2\rangle\mu\}\\&=N\sigma_i\frac{3\hbar^2}{8mE\langle\hat{\mathbf{u}}^2\rangle}\left[1-\exp\left\{-\frac{8}{3}\frac{mE\langle\hat{\mathbf{u}}^2\rangle}{\hbar^2}\right\}\right].\end{aligned} \tag{4.46}$$

This formula replaces the simple result $N\sigma_i$ quoted for a rigid lattice in Chapter 1. Note that when the nuclei are rigidly fixed, i.e. $\langle \hat{\mathbf{u}}^2 \rangle \rightarrow 0$, then the simple result is reproduced.

The total elastic coherent cross-section for the scattering from a powder sample is now, in place of (2.42),

$$\sigma_c^{el} = \frac{N\pi^2\hbar^2}{2mEv_0}\sigma_c \sum_{\tau \neq 0}^{\tau<2k} \exp\{-\tfrac{1}{3}\tau^2\langle \hat{\mathbf{u}}^2 \rangle\}/\tau \tag{4.47}$$

where we have again used (4.38) and assumed a cubic crystal structure In the limit when the incident energy $E$ is large, then, provided $\langle \hat{\mathbf{u}}^2 \rangle$ is not too large, the summation can be replaced by an integral to give, as a reasonable approximation,

$$\begin{aligned}\sigma_c^{el} &= \frac{N\pi^2\hbar^2\sigma_c}{2mEv_0}\frac{v_0}{(2\pi)^3}4\pi\int_0^{(8mE/\hbar^2)^{1/2}} d\tau\, \tau \exp\{-\tfrac{1}{3}\tau^2\langle \hat{\mathbf{u}}^2 \rangle\} \\ &= N\sigma_c \frac{3\hbar^2}{8mE\langle \hat{\mathbf{u}}^2 \rangle}\left[1 - \exp\left\{-\frac{8}{3}\frac{mE\langle \hat{\mathbf{u}}^2 \rangle}{\hbar^2}\right\}\right].\end{aligned} \tag{4.48}$$

We note that both (4.46) and (4.48) become zero as the incident energy $E$ tends to infinity. This is just as we expect: the elastic scattering must approach zero as the energy of the incident neutrons increases, because all the scattering is becoming inelastic. At these high energies there are no interference effects in the elastic coherent cross-section, because the wavelength of the incident neutrons is very much smaller than the internuclear spacing.

Let us now consider how eqns (4.37) and (4.45) for the elastic coherent and incoherent cross-sections become modified when we remove the restriction of a Bravais lattice structure.

Equations (4.31) and (4.32) now give for the elastic coherent cross-section†

$$\begin{aligned}\left(\frac{d\sigma}{d\Omega}\right)_{coh}^{el} &= \left|\sum_{l,d} \bar{b}_d \exp\{i\boldsymbol{\kappa}\cdot(\mathbf{l}+\mathbf{d})\}\left\langle \exp\left\{-i\boldsymbol{\kappa}\cdot\hat{\mathbf{u}}\binom{l}{d}\right\}\right\rangle\right|^2 \\ &= N\frac{(2\pi)^3}{v_0}\sum_{\tau}|F_N(\boldsymbol{\kappa})|^2\,\delta(\boldsymbol{\kappa}-\boldsymbol{\tau}),\end{aligned} \tag{4.49}$$

where

$$F_N(\boldsymbol{\kappa}) = \sum_d \bar{b}_d \exp(i\boldsymbol{\kappa}\cdot\mathbf{d})\exp\{-W_d(\boldsymbol{\kappa})\} \tag{4.50}$$

† We assume that there is no correlation between the scattering length and Debye–Waller factor.

is the nuclear unit-cell structure factor. The exponent in the Debye–Waller factor is given by

$$W_d(\boldsymbol{\kappa}) = \frac{1}{2}\left\langle\left\{\boldsymbol{\kappa}\cdot\hat{\mathbf{u}}\begin{pmatrix} l \\ d \end{pmatrix}\right\}^2\right\rangle = \frac{\hbar}{4NM_d}\sum_{j,\mathbf{q}} \omega_j^{-1}(\mathbf{q})\,|\boldsymbol{\kappa}\cdot\boldsymbol{\sigma}_d^j(\mathbf{q})|^2\,\{2n_j(\mathbf{q})+1\}. \tag{4.51}$$

We notice that $F_N(\boldsymbol{\kappa})$ depends upon the temperature through the Debye–Waller factor and, furthermore, that this does not appear as a simple scaling factor. The relative contributions to $F_N(\boldsymbol{\kappa})$ of different species of atoms can therefore change with temperature.

The elastic incoherent cross-section for a general lattice structure reads

$$\left(\frac{d\sigma}{d\Omega}\right)^{el}_{incoh} = N\sum_d (\overline{b_d^2} - \bar{b}_d^2)\exp\{-2W_d(\boldsymbol{\kappa})\}. \tag{4.52}$$

## 4.3. Debye–Waller factor (Willis and Pryor 1975)

We now turn to a more detailed discussion of the Debye–Waller factor. It is self-evident from (4.41) that $W(\boldsymbol{\kappa})$ is dependent on the phonon frequency spectrum; however, it is not immediately apparent that, for all practical purposes, $W$ has only a single-parameter dependence on the spectrum. To demonstrate this we recall that at low temperatures $W$ is small and unimportant and therefore we examine (4.41) in the high-temperature region. Expanding in powers of $\beta$ we obtain

$$W(\boldsymbol{\kappa}) = \frac{\hbar\kappa^2}{4M}\int_0^{\omega_m} d\omega\,\frac{Z(\omega)}{\omega}\left\{\frac{2}{\hbar\omega\beta} + \frac{\hbar\omega\beta}{6} - \frac{(\hbar\omega\beta)^3}{360} + \cdots\right\}$$

$$= \frac{\hbar^2\kappa^2}{2M}k_BT\left\{\overline{(\hbar\omega)^{-2}} + \frac{1}{12k_B^2T^2} - \frac{\overline{(\hbar\omega)^2}}{720k_B^4T^4} + \cdots\right\}, \tag{4.53}$$

where we define the moments of the frequency distribution by

$$\overline{\omega^n} = \int_0^{\omega_m} d\omega\,\omega^n Z(\omega). \tag{4.54}$$

We notice that the second term of (4.53) is independent of the frequency spectrum and the third term has such a small coefficient that over a wide temperature range it can be neglected. Hence for all practical purposes $W(\boldsymbol{\kappa})$ depends on the single parameter $\overline{\omega^{-2}}$. To get a quantitative estimate of the terms we can use the Debye spectrum (4.42) to give

$$\overline{(\hbar\omega)^{-2}} = 3/(k_B\theta_D)^2, \tag{4.55}$$

$$\overline{(\hbar\omega)^2} = \tfrac{3}{5}(k_B\theta_D)^2. \tag{4.56}$$

Hence

$$W(\boldsymbol{\kappa}) \sim \frac{\hbar^2\kappa^2}{2M}\frac{3k_B T}{(k_B\theta_D)^2}\left\{1+\frac{1}{36}\left(\frac{\theta_D}{T}\right)^2-\frac{1}{3600}\left(\frac{\theta_D}{T}\right)^4\right\}. \tag{4.57}$$

This shows that for all $T \geqslant 0.4\theta_D$ the third term is less than 1 per cent of the leading term.

This result is of the following value. Many thermodynamic quantities depend on $Z(\omega)$, the specific heat for example, but the dependence is not simple and depends upon the detailed shape of $Z(\omega)$. In contrast to this we see that the Debye–Waller factor is quite insensitive to the details of $Z(\omega)$; it depends only upon the single parameter $\overline{\omega^{-2}}$. We deduce that measurements of the Debye–Waller factor cannot be used to give the details of $Z(\omega)$, but they can be used to give $\overline{\omega^{-2}}$.

It is worthwhile giving a brief discussion of how the Debye–Waller factor is affected by anharmonic effects. The most important effect is that the spectrum $Z(\omega)$ and hence the parameter $\overline{\omega^{-2}}$ become themselves temperature dependent.

In addition there are explicit corrections to be considered because (4.33) is not valid in the presence of anharmonic terms. To display these corrections we notice from (4.35) that, in general,

$$e^{-W} = \langle \exp(-i\boldsymbol{\kappa}\cdot\hat{\mathbf{u}})\rangle, \tag{4.58}$$

and we look for $W$ as a power series in the vibration amplitudes, which by dimensional arguments must be identical to a series in powers of the vector $\boldsymbol{\kappa}$. To express this formally we write

$$\hat{A} = -i\boldsymbol{\kappa}\cdot\hat{\mathbf{u}}$$

and

$$e^{-W(x)} = \langle e^{x\hat{A}}\rangle,$$

and look for power series in the parameter $x$, which we later set equal to unity. Hence

$$W(x) = -\ln\langle e^{x\hat{A}}\rangle. \tag{4.59}$$

The first derivative is

$$W'(x) = -\frac{\langle \hat{A}e^{x\hat{A}}\rangle}{\langle e^{x\hat{A}}\rangle}.$$

For a Bravais lattice $\langle\hat{A}\rangle = 0$. A nonvanishing $\langle\hat{A}\rangle$ must correspond to a rigid-body translation of the crystal as a whole, put this would contradict the condition that there is no net force acting on any atom. This result holds for crystals with more than one atom per unit cell, provided that each atom is at a centre of inversion symmetry. Similarly,

$$W''(x) = -\frac{\langle \hat{A}^2 e^{x\hat{A}} \rangle}{\langle e^{x\hat{A}} \rangle} + \frac{\langle \hat{A} e^{x\hat{A}} \rangle^2}{\langle e^{x\hat{A}} \rangle^2},$$

$$W'''(x) = -\frac{\langle \hat{A}^3 e^{x\hat{A}} \rangle}{\langle e^{x\hat{A}} \rangle} + \frac{3\langle \hat{A}^2 e^{x\hat{A}} \rangle \langle \hat{A} e^{x\hat{A}} \rangle}{\langle e^{x\hat{A}} \rangle^2} - \frac{2\langle \hat{A} e^{x\hat{A}} \rangle}{\langle e^{x\hat{A}} \rangle^3}, \tag{4.60}$$

and

$$W^{\mathrm{iv}}(x) = -\frac{\langle \hat{A}^4 e^{x\hat{A}} \rangle}{\langle e^{x\hat{A}} \rangle} + \frac{4\langle \hat{A}^3 e^{x\hat{A}} \rangle \langle \hat{A} e^{x\hat{A}} \rangle}{\langle e^{x\hat{A}} \rangle^2} + \frac{3\langle \hat{A}^2 e^{x\hat{A}} \rangle^2}{\langle e^{x\hat{A}} \rangle^2}$$
$$- \frac{12\langle \hat{A}^2 e^{x\hat{A}} \rangle \langle \hat{A} e^{x\hat{A}} \rangle^2}{\langle e^{x\hat{A}} \rangle^3} + \frac{6\langle \hat{A} e^{x\hat{A}} \rangle^4}{\langle e^{x\hat{A}} \rangle^4}.$$

Evaluating the derivatives at $x = 0$ and assuming $\langle \hat{A} \rangle = 0$ gives

$$W'(0) = 0,$$
$$W''(0) = -\langle \hat{A}^2 \rangle,$$
$$W'''(0) = -\langle \hat{A}^3 \rangle,$$
$$W^{\mathrm{iv}}(0) = -\langle \hat{A}^4 \rangle + 3\langle \hat{A}^2 \rangle^2, \tag{4.61}$$

and

$$W(x) = -x^2 \frac{1}{2!} \langle \hat{A}^2 \rangle - x^3 \frac{1}{3!} \langle \hat{A}^3 \rangle - x^4 \frac{1}{4!} \{\langle \hat{A}^4 \rangle - 3\langle \hat{A}^2 \rangle^2\} + \cdots$$

Hence, finally,

$$W(\boldsymbol{\kappa}) = \tfrac{1}{2}\langle (\boldsymbol{\kappa} \cdot \hat{\mathbf{u}})^2 \rangle - \tfrac{1}{6}\mathrm{i}\langle (\boldsymbol{\kappa} \cdot \hat{\mathbf{u}})^3 \rangle - \tfrac{1}{24}\{\langle (\boldsymbol{\kappa} \cdot \hat{\mathbf{u}})^4 \rangle - 3\langle (\boldsymbol{\kappa} \cdot \hat{\mathbf{u}})^2 \rangle^2\} + \cdots. \tag{4.62}$$

The leading term in this expression is the usual harmonic form for $W$. The second term vanishes if every atom is a centre of inversion symmetry (if every atom is not at a centre of symmetry it can give contributions to the coherent cross-section). The third term vanishes if the vibration amplitudes have the Gaussian distribution due to harmonic forces. But in the presence of anharmonic forces this term does not vanish. The behaviour of the Debye–Waller factor at a structural phase transition is discussed in § 4.6.

### 4.4. Inelastic one-phonon scattering

As in § 4.2 we shall first discuss the inelastic one-phonon cross-sections for Bravais lattices and then later generalize our results.

We recall that the inelastic coherent cross-section can be written (cf. § 3.2)

$$\left(\frac{\mathrm{d}^2\sigma}{\mathrm{d}\Omega\,\mathrm{d}E'}\right)_{\mathrm{coh}}^{\mathrm{inel}} = \frac{N\sigma_{\mathrm{c}}}{4\pi} \frac{k'}{k} \frac{1}{2\pi\hbar} \int_{-\infty}^{\infty} \mathrm{d}t \exp(-\mathrm{i}\omega t) \int \mathrm{d}\mathbf{r} \exp(\mathrm{i}\boldsymbol{\kappa} \cdot \mathbf{r}) G'(\mathbf{r}, t). \tag{4.63}$$

In (4.63),

$$G'(\mathbf{r}, t) = G(\mathbf{r}, t) - G(\mathbf{r}, \infty),$$

where

$$G(\mathbf{r}, t) = \frac{1}{(2\pi)^3 N} \sum_{l,l'} \int \mathrm{d}\mathbf{k} \exp(-\mathrm{i}\mathbf{k} \cdot \mathbf{r}) \langle \exp\{-\mathrm{i}\mathbf{k} \cdot \hat{\mathbf{R}}_l\} \exp\{\mathrm{i}\mathbf{k} \cdot \hat{\mathbf{R}}_{l'}(t)\} \rangle. \tag{4.64}$$

For a Bravais lattice the correlation function in $G(\mathbf{r}, t)$ reduces to

$$\exp\{\mathrm{i}\mathbf{k} \cdot (\mathbf{l}' - \mathbf{l})\} \langle \exp\{-\mathrm{i}\mathbf{k} \cdot \hat{\mathbf{u}}(l)\} \exp\{\mathrm{i}\mathbf{k} \cdot \hat{\mathbf{u}}(l', t)\} \rangle. \tag{4.65}$$

If we set

$$\hat{A} = -\mathrm{i}\mathbf{k} \cdot \hat{\mathbf{u}}(l)$$

and

$$\hat{B} = \mathrm{i}\mathbf{k} \cdot \hat{\mathbf{u}}(l', t),$$

then, since $[\hat{A}, \hat{B}]$ is a c-number,

$$\langle \exp \hat{A} \exp \hat{B} \rangle = \exp\{\tfrac{1}{2}[\hat{A}, \hat{B}]\} \langle \exp(\hat{A} + \hat{B}) \rangle$$

from (3.53), and further, using the identity (3.172), we obtain

$$\begin{aligned} \langle \exp \hat{A} \exp \hat{B} \rangle &= \exp\{\tfrac{1}{2}[\hat{A}, \hat{B}]\} \exp\{\tfrac{1}{2}\langle(\hat{A} + \hat{B})^2\rangle\} \\ &= \exp\{\tfrac{1}{2}\langle \hat{A}^2 + \hat{B}^2 + 2\hat{A}\hat{B} \rangle\}. \end{aligned} \tag{4.66}$$

Thus the correlation function in (4.65) can be written

$$\begin{aligned} &\langle \exp\{-\mathrm{i}\mathbf{k} \cdot \hat{\mathbf{u}}(l)\} \exp\{\mathrm{i}\mathbf{k} \cdot \hat{\mathbf{u}}(l', t)\} \rangle \\ &\quad = \exp[\tfrac{1}{2}\langle -\{\mathbf{k} \cdot \hat{\mathbf{u}}(l)\}^2 - \{\mathbf{k} \cdot \hat{\mathbf{u}}(l')\}^2 + 2\mathbf{k} \cdot \hat{\mathbf{u}}(l) \mathbf{k} \cdot \hat{\mathbf{u}}(l', t) \rangle] \\ &\quad = \exp\{-2W(\mathbf{k})\} \exp\{\langle \mathbf{k} \cdot \hat{\mathbf{u}}(l) \mathbf{k} \cdot \hat{\mathbf{u}}(l', t) \rangle\}. \end{aligned} \tag{4.67}$$

Clearly, for a bulk sample,

$$\begin{aligned} G(\mathbf{r}, \infty) &= \frac{1}{(2\pi)^3 N} \sum_{l,l'} \int \mathrm{d}\mathbf{k} \exp(-\mathrm{i}\mathbf{k} \cdot \mathbf{r}) \langle \exp\{-\mathrm{i}\mathbf{k} \cdot \hat{\mathbf{R}}_l\} \rangle \langle \exp\{\mathrm{i}\mathbf{k} \cdot \hat{\mathbf{R}}_{l'}\} \rangle \\ &= \frac{1}{(2\pi)^3 N} \sum_{l,l'} \int \mathrm{d}\mathbf{k} \exp(-\mathrm{i}\mathbf{k} \cdot \mathbf{r}) \exp\{-\mathrm{i}\mathbf{k} \cdot (\mathbf{l} - \mathbf{l}')\} \exp\{-2W(\mathbf{k})\}, \end{aligned}$$

so that the inelastic coherent cross-section is

$$\left(\frac{\mathrm{d}^2\sigma}{\mathrm{d}\Omega\, \mathrm{d}E'}\right)_{\mathrm{coh}}^{\mathrm{inel}} = \frac{\sigma_{\mathrm{c}}}{4\pi} \frac{k'}{k} \frac{1}{2\pi\hbar} \int_{-\infty}^{\infty} \mathrm{d}t \exp(-\mathrm{i}\omega t) \exp\{-2W(\boldsymbol{\kappa})\} \times \sum_{l,l'} \exp\{-\mathrm{i}\boldsymbol{\kappa} \cdot (\mathbf{l} - \mathbf{l}')\} [\exp\{\langle \boldsymbol{\kappa} \cdot \hat{\mathbf{u}}(l) \boldsymbol{\kappa} \cdot \hat{\mathbf{u}}(l', t) \rangle\} - 1]. \tag{4.68}$$

Furthermore, it is readily seen from § 3.2 that the inelastic incoherent cross-section is

$$\left(\frac{d^2\sigma}{d\Omega\, dE'}\right)^{\text{inel}}_{\text{incoh}} = \frac{\sigma_i}{4\pi}\frac{k'}{k}\frac{1}{2\pi\hbar}\int_{-\infty}^{\infty} dt\, \exp(-i\omega t)\exp\{-2W(\boldsymbol{\kappa})\}$$

$$\times\sum_l [\exp\{\langle\boldsymbol{\kappa}\cdot\hat{\mathbf{u}}(l)\boldsymbol{\kappa}\cdot\hat{\mathbf{u}}(l,t)\rangle\}-1]. \tag{4.69}$$

These two equations for the inelastic coherent and incoherent partial differential cross-sections are exact within the harmonic approximation of the lattice vibrations. If we expand the exponentials in (4.68) and (4.69) containing the displacement correlation functions, the first nonzero terms give the one-phonon cross-sections, the second terms the two-phonon cross-sections, etc. For the present we shall consider in detail the one-phonon cross-sections. With regard to obtaining information on the dynamical motion of the nuclei these are by far the most important.

We need now to evaluate the correlation function

$$\langle\boldsymbol{\kappa}\cdot\hat{\mathbf{u}}(l)\boldsymbol{\kappa}\cdot\hat{\mathbf{u}}(l',t)\rangle.$$

From (4.22a) this is

$$\sum_{\alpha,\beta}\kappa_\alpha\kappa_\beta\sum_{j,\mathbf{q}}\sum_{j',\mathbf{q}'}\left(\frac{\hbar}{2NM}\right)\sigma^j_\alpha(\mathbf{q})\sigma^{j'}_\beta(\mathbf{q}')\{\omega_j(\mathbf{q})\omega_{j'}(\mathbf{q}')\}^{-1/2}$$
$$\times[\exp(i\mathbf{q}\cdot\mathbf{l}-i\mathbf{q}'\cdot\mathbf{l}')\exp\{it\omega_{j'}(\mathbf{q}')\}\langle\hat{a}_j(\mathbf{q})\hat{a}^+_{j'}(\mathbf{q}')\rangle$$
$$+\exp(-i\mathbf{q}\cdot\mathbf{l}+i\mathbf{q}'\cdot\mathbf{l}')\exp\{-it\omega_{j'}(\mathbf{q}')\}\langle\hat{a}^+_j(\mathbf{q})\hat{a}_{j'}(\mathbf{q}')\rangle]$$
$$=\sum_{\alpha,\beta}\kappa_\alpha\kappa_\beta\sum_{j,\mathbf{q}}\left(\frac{\hbar}{2NM}\right)\sigma^j_\alpha(\mathbf{q})\sigma^j_\beta(\mathbf{q})\omega_j^{-1}(\mathbf{q})$$
$$\times[\exp\{i\mathbf{q}\cdot(\mathbf{l}-\mathbf{l}')\}\exp\{it\omega_j(\mathbf{q})\}\{1+n_j(\mathbf{q})\}+\exp\{-i\mathbf{q}\cdot(\mathbf{l}-\mathbf{l}')\}\exp\{-it\omega_j(\mathbf{q})\}n_j(\mathbf{q})]$$

because of (4.18a). Hence

$$\langle\boldsymbol{\kappa}\cdot\hat{\mathbf{u}}(l)\boldsymbol{\kappa}\cdot\hat{\mathbf{u}}(l',t)\rangle$$
$$=\frac{\hbar}{2NM}\sum_{j,\mathbf{q}}\frac{|\boldsymbol{\kappa}\cdot\boldsymbol{\sigma}^j(\mathbf{q})|^2}{\omega_j(\mathbf{q})}[\exp\{i\mathbf{q}\cdot(\mathbf{l}-\mathbf{l}')\}\exp\{it\omega_j(\mathbf{q})\}\{1+n_j(\mathbf{q})\}$$
$$+\exp\{-i\mathbf{q}\cdot(\mathbf{l}-\mathbf{l}')\}\exp\{-it\omega_j(\mathbf{q})\}n_j(\mathbf{q})], \tag{4.70}$$

and using (2.9) it immediately follows that the one-phonon coherent partial differential cross-section

$$\left(\frac{d^2\sigma}{d\Omega\, dE'}\right)^{\text{inel}}_{\text{coh}} = \frac{\sigma_c}{4\pi}\frac{k'}{k}\frac{1}{2\pi\hbar}\int_{-\infty}^{\infty} dt \exp(-i\omega t)\exp\{-2W(\boldsymbol{\kappa})\}$$
$$\times\sum_{l,l'}\exp\{i\boldsymbol{\kappa}\cdot(\mathbf{l}'-\mathbf{l})\}\langle\boldsymbol{\kappa}\cdot\hat{\mathbf{u}}(l)\boldsymbol{\kappa}\cdot\hat{\mathbf{u}}(l',t)\rangle$$
$$=\frac{\sigma_c}{4\pi}\frac{k'}{k}\frac{(2\pi)^3}{v_0}\frac{1}{2M}\sum_{\boldsymbol{\tau}}\exp\{-2W(\boldsymbol{\kappa})\}\sum_{j,\mathbf{q}}\frac{|\boldsymbol{\kappa}\cdot\boldsymbol{\sigma}^j(\mathbf{q})|^2}{\omega_j(\mathbf{q})}$$
$$\times[n_j(\mathbf{q})\,\delta\{\omega+\omega_j(\mathbf{q})\}\,\delta(\boldsymbol{\kappa}+\mathbf{q}-\boldsymbol{\tau})+\{n_j(\mathbf{q})+1\}\,\delta\{\omega-\omega_j(\mathbf{q})\}\,\delta(\boldsymbol{\kappa}-\mathbf{q}-\boldsymbol{\tau})]. \quad (4.71)$$

The cross-section (4.71) is the sum of two terms. The first, which contains the expression $\delta\{\omega+\omega_j(\mathbf{q})\}\delta(\boldsymbol{\kappa}+\mathbf{q}-\boldsymbol{\tau})$, represents a scattering process in which one phonon is annihilated and the second term, containing $\delta\{\omega-\omega_j(\mathbf{q})\}\delta(\boldsymbol{\kappa}-\mathbf{q}-\boldsymbol{\tau})$, represents a process in which one phonon is created. In the limit $T\to 0$ only the second process occurs, since there are clearly no phonons to be annihilated at absolute zero. Let us consider in detail the structure of the one-phonon annihilation cross-section (Dolling 1974; Dorner 1982).

The two $\delta$-functions associated with this scattering process represent conservation of energy and momentum and tell us that the scattering process obeys the conditions

$$E' = E + \hbar\omega_j(\mathbf{q}),$$
$$\mathbf{k}' = \mathbf{k}+\mathbf{q}-\boldsymbol{\tau}. \quad (4.72)$$

These conditions are so restrictive that for given scattering angle only phonons of a particular $\mathbf{q}$ and $\omega_j(\mathbf{q})$ can give scattering. We can make use of this to determine the phonon dispersion $\omega_j(\mathbf{q})$, as a function of $\mathbf{q}$. Suppose we have a monochromatic beam incident on a crystal and we measure the energy of the neutrons scattered through a given angle by using a time of flight apparatus or by using another crystal as analyser. From (4.72) these scattered neutrons will in general have one of three energies corresponding to the three choices of $j$. Choose one of these energies; knowing the scattered energy we have $\mathbf{k}'$ and from the second of (4.72) we have $\mathbf{k}'-\mathbf{k}=\mathbf{q}-\boldsymbol{\tau}$. But $\mathbf{q}$ must lie in the first Brillouin zone, so we get both $\mathbf{q}$ and $\boldsymbol{\tau}$. Hence from this experiment we have both $\mathbf{q}$ and $\omega_j(\mathbf{q})$ for a phonon, i.e. we have a point on the phonon dispersion curve.

The conservation conditions (4.72) are illustrated in Fig. 4.4. Starting from the origin $O$ we draw the vector $\mathbf{k}$ and the vector $-\boldsymbol{\tau}$ to give an end-point at $C$. The dotted line through the origin is the direction of $\mathbf{k}'$ as determined by the position of the neutron counters. Then any vector such as $OA'$, $OA$, or $OA''$ can represent $\mathbf{k}'$ while $CA'$, $CA$, $CA''$ respectively

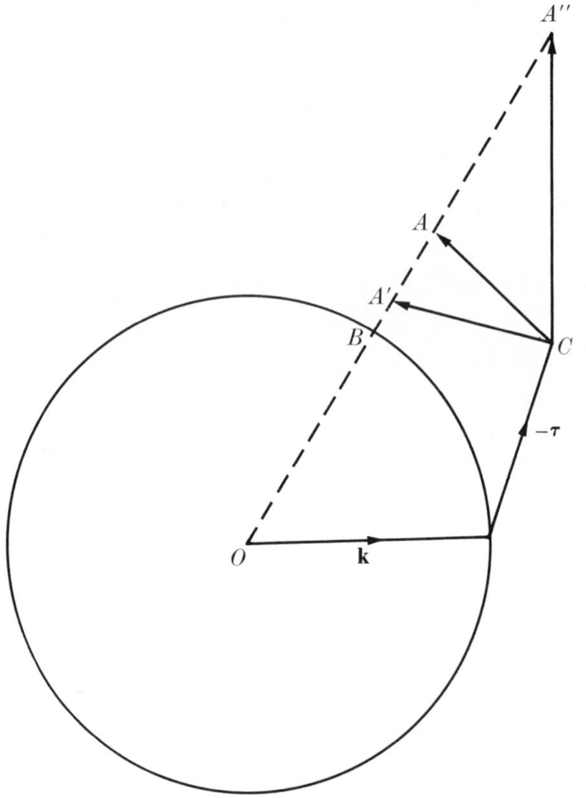


FIG. 4.4. Conservation of momentum and energy in a one-phonon scattering event.

represent $\mathbf{q}$. The momentum conservation law is obeyed automatically, but energy conservation is given only by a special choice of the position $A$. In the figure a circle of radius $|\mathbf{k}|$ has been drawn about the origin; it intersects the direction of $\mathbf{k}'$ at $B$. The energy conservation condition then states that $(\hbar^2/2m)\{(OA)^2-(OB)^2\}$ must equal the energy associated with the vector $\mathbf{q}$, i.e. $\omega_j(\mathbf{q})$. For a position such as $A'$ the neutron energy gain is too small compared with $\omega_j(\mathbf{q})$, and for a position such as $A''$ the neutron energy gain is too large. Hence it is only at some intermediate position, say $A$, that the conservation conditions can both be satisfied. The same construction has to be repeated for each choice of $j$. Therefore, in general, scattering at a fixed angle can correspond to annihilation of any one of the $3r$ phonon modes.

The numerical value of the cross-section can be obtained by first performing the sum over $\mathbf{q}$,

$$\left(\frac{d^2\sigma}{d\Omega\, dE'}\right)^{\text{inel}}_{\text{coh},-1} = \frac{N\sigma_c}{8\pi M}\frac{\hbar k'}{k}\exp\{-2W(\boldsymbol{\kappa})\}\sum_j \delta\{\hbar\omega + \hbar\omega_j(\mathbf{q})\}\frac{n_j(\mathbf{q})}{\omega_j(\mathbf{q})}|\boldsymbol{\kappa}\cdot\boldsymbol{\sigma}^j(\mathbf{q})|^2, \quad (4.73)$$

where it is now understood that, for each $j$, $\mathbf{q}$ and $\boldsymbol{\tau}$ are determined by (4.72).

We have chosen to denote the one-phonon annihilation cross-section by the suffix $-1$; the one-phonon creation cross-section will be denoted by the suffix $+1$.

When we integrate (4.73) over $E'$ we must now remember that $\omega_j(\mathbf{q})$ is a function of $E'$. Since

$$\frac{\partial}{\partial E'}\{\hbar\omega + \hbar\omega_j(\mathbf{q})\} = -1 + \frac{\hbar}{2E'}\mathbf{k}'\cdot\nabla\omega_j(\mathbf{q}),$$

where $\nabla\omega_j(\mathbf{q})$ is the gradient of $\omega_j(\mathbf{q})$ with respect to $\mathbf{q}$, we have (Dolling 1974)

$$\left(\frac{d\sigma}{d\Omega}\right)^{\text{inel}}_{\text{coh},-1} = \frac{N\sigma_c\hbar}{8\pi M}\exp\{-2W(\boldsymbol{\kappa})\}\sum_j \frac{k'}{k}|\boldsymbol{\kappa}\cdot\boldsymbol{\sigma}^j(\mathbf{q})|^2\frac{n_j(\mathbf{q})}{\omega_j(\mathbf{q})}\left\{1 - \frac{\hbar}{2E'}\mathbf{k}'\cdot\nabla\omega_j(\mathbf{q})\right\}^{-1}. \quad (4.74)$$

Using $n_j(\mathbf{q}) \simeq k_B T/\hbar\omega_j(\mathbf{q})$ and putting $T = 300$ K we find this cross-section is typically of the order of 10 millibarns. Notice that for scattering near the forward direction, $\boldsymbol{\tau} = 0$ and the polarization factor $|\boldsymbol{\kappa}\cdot\boldsymbol{\sigma}^j(\mathbf{q})|^2$ becomes $|\mathbf{q}\cdot\boldsymbol{\sigma}^j(\mathbf{q})|^2$. It follows that purely transverse modes give no scattering with $\boldsymbol{\tau} = 0$.

The coherent cross-section corresponding to the creation of a single phonon is evaluated similarly to give

$$\left(\frac{d\sigma}{d\Omega}\right)^{\text{inel}}_{\text{coh},+1} = \frac{N\sigma_c\hbar}{8\pi M}\exp\{-2W(\boldsymbol{\kappa})\}\sum_j \frac{k'}{k}|\boldsymbol{\kappa}\cdot\boldsymbol{\sigma}^j(\mathbf{q})|^2\frac{\{n_j(\mathbf{q})+1\}}{\omega_j(\mathbf{q})}\left\{1 + \frac{\hbar}{2E'}\mathbf{k}'\cdot\nabla\omega_j(\mathbf{q})\right\}^{-1}, \quad (4.75)$$

where it is understood that $\mathbf{q}$ and $E'$ are determined from

$$E' = E - \hbar\omega_j(\mathbf{q}), \qquad \mathbf{k}' = \mathbf{k} - \mathbf{q} - \boldsymbol{\tau}. \quad (4.76)$$

The $\boldsymbol{\tau}=0$ part of the one-phonon creation process has a special property: eliminating $\mathbf{k}'$ from eqns (4.76) gives

$$\frac{\hbar^2}{2m}(2\mathbf{k}\cdot\mathbf{q}-q^2)=\hbar\omega_j(\mathbf{q}). \tag{4.77}$$

Now

$$\tfrac{1}{2}(2\mathbf{k}\cdot\mathbf{q}-q^2)<\mathbf{k}\cdot\mathbf{q}<kq,$$

and for small $q$

$$\omega_j(\mathbf{q})\simeq cq,$$

where $c$ is the velocity of sound. It follows that there is only a solution of (4.77) (other than the trivial solution $q=0$) if

$$\frac{\hbar k}{m}>c,$$

i.e. if the velocity of the oncoming neutron is greater than the velocity of sound in the material. Hence, if the incoming neutrons have such a low energy that their velocity is less than the velocity of sound in the crystal, then the one-phonon coherent creation process has zero cross-section for $\boldsymbol{\tau}=0$, i.e. near the forward direction. It is possible to take advantage of this when looking for magnetic inelastic scattering; for long wavelength neutrons there is very little non-magnetic scattering in the forward direction (because the one-phonon creation process is forbidden and the one-phonon annihilation process can be made small by having a low temperature) and therefore this is a good place to look for magnetic scattering.

The one-phonon incoherent cross-section is given from (4.69) by

$$\left(\frac{\mathrm{d}^2\sigma}{\mathrm{d}\Omega\,\mathrm{d}E'}\right)^{\mathrm{inel}}_{\mathrm{incoh}}=\frac{\sigma_{\mathrm{i}}}{4\pi}\frac{k'}{k}\frac{1}{2\pi\hbar}\int_{-\infty}^{\infty}\mathrm{d}t\,\exp(-\mathrm{i}\omega t)\exp\{-2W(\boldsymbol{\kappa})\}$$
$$\times\sum_l\langle\boldsymbol{\kappa}\cdot\hat{\mathbf{u}}(l)\boldsymbol{\kappa}\cdot\hat{\mathbf{u}}(l,t)\rangle,$$

which with the aid of the result (4.70) with $l=l'$ reduces to

$$\left(\frac{\mathrm{d}^2\sigma}{\mathrm{d}\Omega\,\mathrm{d}E'}\right)^{\mathrm{inel}}_{\mathrm{incoh}}=\frac{\sigma_{\mathrm{i}}}{8\pi M}\exp\{-2W(\boldsymbol{\kappa})\}\frac{k'}{k}\sum_{j,\mathbf{q}}\frac{|\boldsymbol{\kappa}\cdot\boldsymbol{\sigma}^j(\mathbf{q})|^2}{\omega_j(\mathbf{q})}$$
$$\times[n_j(\mathbf{q})\,\delta\{\omega+\omega_j(\mathbf{q})\}+\{n_j(\mathbf{q})+1\}\,\delta\{\omega-\omega_j(\mathbf{q})\}]. \tag{4.78}$$

We notice that this formula contains $\delta$-functions to ensure energy conservation, the first term representing one-phonon annihilation scattering and the second one-phonon creation scattering, but there are no momentum conservation conditions associated with them. Thus, one-phonon incoherent scattering cannot give us the detailed information that coherent scattering can give. Nevertheless it is valuable, because for Bravais lattices it can be used to determine the vibrational density of states $Z(\omega)$ directly. In terms of $Z(\omega)$ the creatioin part of (4.78) can be written

$$\left(\frac{d^2\sigma}{d\Omega\, dE'}\right)^{\text{inel}}_{\text{incoh},+1} = \frac{\sigma_i}{8\pi M}\exp\{-2W(\boldsymbol{\kappa})\}\frac{k'}{k}3N\int_0^{\omega_m} d\omega' \frac{Z(\omega')}{\omega'} \times\{|\boldsymbol{\kappa}\cdot\boldsymbol{\sigma}^j(\mathbf{q})|^2\}_{\text{av}}\{n(\omega')+1\}\,\delta(\omega-\omega'), \tag{4.79}$$

where the subscript av means that an average over modes with given $\omega'$ has been taken. For cubic crystals $|\boldsymbol{\kappa}\cdot\boldsymbol{\sigma}^j(\mathbf{q})|^2_{\text{av}}$ is rigorously $\frac{1}{3}\kappa^2$. Using this in (4.79),

$$\left(\frac{d^2\sigma}{d\Omega\, dE'}\right)^{\text{inel}}_{\text{incoh},+1} = \frac{N\sigma_i}{8\pi M}\frac{k'}{k}\kappa^2\exp\{-2W(\boldsymbol{\kappa})\}\frac{Z(\omega)}{\omega}\{n(\omega)+1\}. \tag{4.80}$$

This formula tells us that incoherent one-phonon scattering takes place into all directions and that for any direction the incoherently scattered neutrons have a continuous energy distribution that is proportional to $Z(\omega)$ (where $\hbar\omega$ is the energy lost by the neutron) and other factors. If the final energy $E'$ is measured, then $k'$ and $k$ are both known, $\omega$ is easily calculated, and the intensity leads directly to $Z(\omega)$ as a function of $\omega$.

The one-phonon annihilation part of (4.78) can be written in a similar way:

$$\left(\frac{d^2\sigma}{d\Omega\, dE'}\right)^{\text{inel}}_{\text{incoh},-1} = \frac{N\sigma_i}{8\pi M}\frac{k'}{k}\kappa^2\exp\{-2W(\boldsymbol{\kappa})\}\frac{Z(-\omega)}{(-\omega)}n(-\omega)$$
$$= \frac{N\sigma_i}{8\pi M}\frac{k'}{k}\kappa^2\exp\{-2W(\boldsymbol{\kappa})\}\frac{Z(\omega)}{\omega}\{n(\omega)+1\}, \tag{4.81}$$

if we define

$$Z(\omega)=Z(-\omega). \tag{4.82}$$

Hence, from (4.80) and (4.81) it follows that the cross-section (4.78) for $\omega>0$ and $\omega<0$ if of exactly the same form, so that we can drop the suffixes $\pm1$ on (4.80) and (4.81).

The generalization of the preceding results of this section to non-Bravais lattices is relatively straightforward. First

$$\langle\exp(-i\boldsymbol{\kappa}\cdot\hat{\mathbf{R}}_{ld})\exp\{i\boldsymbol{\kappa}\cdot\hat{\mathbf{R}}_{l'd'}(t)\}\rangle=\exp\{i\boldsymbol{\kappa}\cdot(\mathbf{l}'+\mathbf{d}'-\mathbf{l}-\mathbf{d})\} \times\left\langle\exp\left\{-i\boldsymbol{\kappa}\cdot\hat{\mathbf{u}}\binom{l}{d}\right\}\exp\left\{i\boldsymbol{\kappa}\cdot\hat{\mathbf{u}}\left(\begin{matrix}l'\\d'\end{matrix},t\right)\right\}\right\rangle, \tag{4.83}$$

and the correlation function occurring on the right-hand side is simply

$$\exp\{-W_d(\boldsymbol{\kappa})\}\exp\{-W_{d'}(\boldsymbol{\kappa})\}\exp\left\{\left\langle \boldsymbol{\kappa}\cdot\hat{\mathbf{u}}\binom{l}{d}\boldsymbol{\kappa}\cdot\hat{\mathbf{u}}\binom{l'}{d'},t\right)\right\rangle\right\}. \tag{4.84}$$

as is clear from (4.66). $W_d(\boldsymbol{\kappa})$ is given by eqn (4.51). Thus it follows from (4.83) and (4.84) that in place of eqn (4.68) for the inelastic coherent cross-section we obtain, for a general lattice structure,

$$\left(\frac{\mathrm{d}^2\sigma}{\mathrm{d}\Omega\,\mathrm{d}E'}\right)^{\mathrm{inel}}_{\mathrm{coh}}=\frac{k'}{k}\frac{1}{2\pi\hbar}\int_{-\infty}^{\infty}\mathrm{d}t\,\exp(-\mathrm{i}\omega t)\sum_{d,d'}\bar{b}^*_d\bar{b}_{d'}\exp\{-W_d(\boldsymbol{\kappa})\}$$
$$\times\exp\{-W_{d'}(\boldsymbol{\kappa})\}\exp\{\mathrm{i}\boldsymbol{\kappa}\cdot(\mathbf{d}'-\mathbf{d})\}\sum_{l,l'}\exp\{\mathrm{i}\boldsymbol{\kappa}\cdot(\mathbf{l}'-\mathbf{l})\}$$
$$\times\left[\exp\left\{\left\langle\boldsymbol{\kappa}\cdot\hat{\mathbf{u}}\binom{l}{d}\boldsymbol{\kappa}\cdot\hat{\mathbf{u}}\left(\begin{matrix}l'\\d'\end{matrix},t\right)\right\rangle\right\}-1\right]. \tag{4.85}$$

For the inelastic incoherent cross-section we find

$$\left(\frac{\mathrm{d}^2\sigma}{\mathrm{d}\Omega\,\mathrm{d}E'}\right)^{\mathrm{inel}}_{\mathrm{incoh}}=\frac{k'}{k}\frac{1}{2\pi\hbar}\int_{-\infty}^{\infty}\mathrm{d}t\,\exp(-\mathrm{i}\omega t)\sum_{d}\{\overline{|b_d|^2}-|\bar{b}_d|^2\}\exp\{-2W_d(\boldsymbol{\kappa})\}$$
$$\times\sum_l\left[\exp\left\{\left\langle\boldsymbol{\kappa}\cdot\hat{\mathbf{u}}\binom{l}{d}\boldsymbol{\kappa}\cdot\hat{\mathbf{u}}\left(\begin{matrix}l\\d\end{matrix},t\right)\right\rangle\right\}-1\right]. \tag{4.86}$$

In order to obtain explicit expressions for the one-phonon cross-sections we use (4.26) to calculate the correlation function

$$\left\langle\boldsymbol{\kappa}\cdot\hat{\mathbf{u}}\binom{l}{d}\boldsymbol{\kappa}\cdot\hat{\mathbf{u}}\left(\begin{matrix}l'\\d'\end{matrix},t\right)\right\rangle=\frac{1}{N}\sum_{j\mathbf{q}}\sum_{j'\mathbf{q}'}\left(\frac{1}{M_dM_{d'}}\right)^{1/2}\{\boldsymbol{\kappa}\cdot\boldsymbol{\sigma}^{*j}_d(\mathbf{q})\boldsymbol{\kappa}\cdot\boldsymbol{\sigma}^{j'}_{d'}(\mathbf{q}')\}$$
$$\times\exp(\mathrm{i}\mathbf{q}'\cdot\mathbf{l}'-\mathrm{i}\mathbf{q}\cdot\mathbf{l})\langle\hat{Q}^+(\mathbf{q}j)\hat{Q}(\mathbf{q}'j';t)\rangle. \tag{4.87}$$

Here $\hat{Q}(\mathbf{q}j)\equiv\hat{Q}(\mathbf{q}j;t=0)$, and we have exploited the fact that $\boldsymbol{\kappa}\cdot\hat{\mathbf{u}}\binom{l}{d}$ is purely real. The correlation function is zero unless $\mathbf{q}=\mathbf{q}'$ and $j=j'$; see eqn (4.18a), for example. Using (2.9) and the definition,

$$H^j_{\mathbf{q}}(\boldsymbol{\kappa})=\sum_d\bar{b}_d\exp(-W_d(\boldsymbol{\kappa})+\mathrm{i}\boldsymbol{\kappa}\cdot\mathbf{d})\{\boldsymbol{\kappa}\cdot\boldsymbol{\sigma}^j_d(\mathbf{q})\}M_d^{-1/2}, \tag{4.88}$$

the one-phonon inelastic coherent cross-section is

$$\left(\frac{\mathrm{d}^2\sigma}{\mathrm{d}\Omega\,\mathrm{d}E'}\right)^{\mathrm{inel}}_{\mathrm{coh}}=\frac{k'}{k}\frac{(2\pi)^3}{v_0}\sum_{\boldsymbol{\tau}}\sum_{j\mathbf{q}}\delta(\boldsymbol{\kappa}+\mathbf{q}-\boldsymbol{\tau})\,|H^j_{\mathbf{q}}(\boldsymbol{\kappa})|^2$$
$$\times\frac{1}{2\pi\hbar}\int_{-\infty}^{\infty}\mathrm{d}t\,\exp(-\mathrm{i}\omega t)\langle\hat{Q}^+(\mathbf{q}j)\hat{Q}(\mathbf{q}j;t)\rangle. \tag{4.89}$$

The correlation function in (4.89) is readily evaluated for harmonic lattice vibrations. From the definition of the normal coordinate displacement operator we obtain

$$\langle \hat{Q}^{+}(\mathbf{q}j)\hat{Q}(\mathbf{q}j;t)\rangle = \frac{\hbar}{2\omega_j(\mathbf{q})}[n_j(\mathbf{q})\exp\{-it\omega_j(\mathbf{q})\} + \{1+n_j(\mathbf{q})\}\exp\{it\omega_j(\mathbf{q})\}] \tag{4.90}$$

and therefore, for a harmonic lattice,

$$\begin{aligned} S_j(\mathbf{q},\omega) &= \frac{1}{2\pi\hbar}\int_{-\infty}^{\infty} \mathrm{d}t\,\exp(-\mathrm{i}\omega t)\langle \hat{Q}^{+}(\mathbf{q}j)\hat{Q}(\mathbf{q}j;t)\rangle \\ &= \frac{1}{2\omega_j(\mathbf{q})}[n_j(\mathbf{q})\,\delta\{\omega+\omega_j(\mathbf{q})\} + \{1+n_j(\mathbf{q})\}\,\delta\{\omega-\omega_j(\mathbf{q})\}]. \end{aligned} \tag{4.91}$$

Note that, since $H^j_{\mathbf{q}}(\boldsymbol{\kappa})$ occurs in the cross-section multiplied by the delta function $\delta(\boldsymbol{\kappa}+\mathbf{q}-\boldsymbol{\tau})$, we may replace $\mathbf{q}$ by $\boldsymbol{\tau}-\boldsymbol{\kappa}$ in the definition (4.87) where the phonon wave vector appears in the polarization vector. Because $\boldsymbol{\sigma}^j_d(\mathbf{q}+\boldsymbol{\tau}) = \boldsymbol{\sigma}^j_d(\mathbf{q}) = \{\boldsymbol{\sigma}^j_d(-\mathbf{q})\}^*$, we conclude that in (4.87)

$$\boldsymbol{\sigma}^j_d(\mathbf{q}) \equiv \boldsymbol{\sigma}^j_d(\boldsymbol{\tau}-\boldsymbol{\kappa}) = \{\boldsymbol{\sigma}^j_d(\boldsymbol{\kappa})\}^*.$$

If we use the alternative definition of the polarization vectors (4.28), then in $H^j_{\mathbf{q}}(\boldsymbol{\kappa})$ we have, using (4.30),

$$\begin{aligned} \exp(\mathrm{i}\boldsymbol{\kappa}\cdot\mathbf{d})\boldsymbol{\sigma}^j_d(\mathbf{q}) &\equiv \exp\{\mathrm{i}\mathbf{d}\cdot(\boldsymbol{\kappa}+\mathbf{q})\}\mathbf{e}^j_d(\mathbf{q}) \\ &\equiv \exp(\mathrm{i}\boldsymbol{\tau}\cdot\mathbf{d})\mathbf{e}^j_d(\boldsymbol{\tau}-\boldsymbol{\kappa}) = \mathbf{e}^j_d(-\boldsymbol{\kappa}) = \{\mathbf{e}^j_d(\boldsymbol{\kappa})\}^*. \end{aligned}$$

We conclude that $H^j_{\mathbf{q}}(\boldsymbol{\kappa})$ does not depend explicitly on $\mathbf{q}$ if $\mathbf{q}$ and $\boldsymbol{\kappa}$ differ by a reciprocal lattice vector as is the case in the one-phonon coherent cross-section.

The one-phonon inelastic incoherent cross-section is simply derived from (4.87) by setting $d=d'$ and $l=l'$. The result is

$$\left(\frac{\mathrm{d}^2\sigma}{\mathrm{d}\Omega\,\mathrm{d}E'}\right)^{\text{inel}}_{\text{incoh}} = \frac{k'}{k}\sum_d \{\overline{|b_d|^2} - |\bar{b}_d|^2\}\exp\{-2W_d(\boldsymbol{\kappa})\}\frac{1}{M_d}\sum_{j\mathbf{q}} |\boldsymbol{\kappa}\cdot\boldsymbol{\sigma}^j_d(\mathbf{q})|^2\, S_j(\mathbf{q},\omega), \tag{4.92}$$

where the correlation function is given in (4.91).

This incoherent scattering is not as useful as that for a simple lattice because, in general, it is not possible to estimate $|\boldsymbol{\kappa}\cdot\boldsymbol{\sigma}^j_d(\mathbf{q})|^2$ for the different $d$ values and therefore the relative contributions made by each species of atom remain unknown. For a hydrogenous target sample the incoherent scattering will be dominated by the contributions from the protons, for not only is the incoherent proton cross-section large but the proton mass is relatively small (cf. § 6.4).

## 4.5. Anharmonic phonon theory

Let us now consider how the one-phonon coherent cross-section derived in the preceding section is affected by the third- and higher-order interactions in the Hamiltonian of the translational motion of the nuclei (eqn (4.3)). Macroscopically, these anharmonic interactions give rise to thermal expansion, a difference between the specific heats at constant pressure and volume, and a finite thermal conductivity. The presence of anharmonicity prevents the diagonalization of the Hamiltonian in terms of the phonon operators introduced in the discussion of the harmonic part of the Hamiltonian. When the third- and higher-order derivatives of the potential energy are considerably smaller than the second, it is still meaningful to use the concept of phonons, but they can no longer be regarded as forming a perfect Bose gas. The terms in the Hamiltonian involving three or more displacements represent interactions of the phonons and we must therefore associate with a given phonon a finite lifetime. Furthermore, in general we can expect the phonon interactions to lead to a shift in the dispersion relation from the value $\omega_j(\mathbf{q})$. Thus, in the presence of weak anharmonic interactions, the frequency of a given phonon can be written

$$\omega_j(\mathbf{q})+\Delta_j(\mathbf{q},\omega)+\mathrm{i}\Gamma_j(\mathbf{q},\omega), \tag{4.93}$$

where $\Delta$ represents the shift from the non-interacting phonon frequency $\omega_j(\mathbf{q})$ and $\Gamma$ the damping or decay constant.

Part of the shift in the phonon frequency arises from the thermal expansion of the crystal. This contribution can be assessed in terms of Grüneisen parameters that are proportional to the cubic interaction coefficients in weakly anharmonic crystals. The macroscopic Grüneisen parameter is the coefficient of proportionality between the volume expansion coefficient and the specific heat at constant volume. In this guise, the Grüneisen parameter $\gamma$ characterizes the thermal expansion in much the same way that the Debye temperature characterizes the specific heat. The Grüneisen relation, in which the volume expansion coefficient has the same temperature dependence as the specific heat, is approximately satisfied by many weakly anharmonic compounds with values of $\gamma$ between 1 and 2. Strongly anharmonic materials, like the inert gas solids, have values of $\gamma$ as large as 3.0–3.5. Moreover, $\gamma$ can display a pronounced temperature dependence, particularly at low temperatures, $T \ll \theta_D$, when it may even become negative.

Pursuing the development of the thermal expansion contribution to the shift, $\Delta_j(\mathbf{q},\omega)$, in terms of a macroscopic Grüneisen parameter we obtain a frequency-independent result

$$\Delta_j^{(0)}(\mathbf{q}) = -3\gamma\epsilon\omega_j(\mathbf{q}), \tag{4.94}$$

where $\epsilon$ is the linear expansivity. This result amounts to renormalizing the phonon dispersion frequency by a factor $(1-3\gamma\epsilon)$. The expansivity is proportional to the temperature for $T > \theta_D$, and a simple analysis leads to a result in which $\epsilon$ is proportional to $\gamma$. We are led to conclude that, for a simple macroscopic model, thermal expansion always leads to a softening of the phonon frequency, independent of the sign of $\gamma$, and that the softening increases linearly with temperature. For a Debye model,

$$\gamma = -\frac{V}{\theta_D}\left(\frac{\partial \theta_D}{\partial V}\right) \tag{4.95}$$

where $V$ is the sample volume.

A microscopic model assigns a Grüneisen parameter to each phonon mode. The parameters depend on the wave vector and the temperature, and the dependence can vary significantly from one mode to another.

For weakly anharmonic materials, the frequency-dependent shift and damping in (4.93) can be calculated by perturbation theory The leading-order effects that arise from cubic and quartic anharmonic contributions are analysed in § B.9.2. The anharmonic contributions to the Hamiltonian are expressed in terms of interaction coefficients $J$ and $K$, and the normal coordinate displacement operators† (eqn (4.26)). We use a shorthand for the mode index and wave vector of a normal coordinate, and write $1 \equiv \mathbf{q}_1 j_1$, etc. With this notation the anharmonic contribution to the Hamiltonian is taken to be

$$\begin{aligned}\hat{\mathcal{H}}_1 = &-\sum_{123} J(1,2,3)\hat{Q}(1)\hat{Q}(2)\hat{Q}(3)\\ &-\sum_{1234} K(1,2,3,4)\hat{Q}(1)\hat{Q}(2)\hat{Q}(3)\hat{Q}(4).\end{aligned} \tag{4.96}$$

Each interaction term in (4.96) conserves momentum, so that $J(1,2,3)$ is zero unless $\mathbf{q}_1+\mathbf{q}_2+\mathbf{q}_3=0$, and $K(1,2,3,4)$ is zero unless $\mathbf{q}_1+\mathbf{q}_2+\mathbf{q}_3+\mathbf{q}_4=0$. The interaction coefficients are completely symmetric in the indices. The calculation given in § B.9b is based on the assumption that $\langle\hat{Q}(\mathbf{q}j)\rangle = 0$ for all $\mathbf{q}$ and $j$.

The quantity calculated is the generalized susceptibility $\chi_{\mathbf{q}j}[\omega]$, from which we obtain $S_j(\mathbf{q}, \omega)$ (eqn (4.91)) from the relation

$$S_j(\mathbf{q}, \omega) = \{1+n(\omega)\}\frac{1}{\pi}\chi''_{\mathbf{q}j}[\omega]. \tag{4.97}$$

† The operators (4.26) differ from the normal coordinate operators used in Appendix B by a factor $M^{1/2}$; the definition used here is useful for the general case of more than one atom per unit cell. Because of this change of notation, the interaction coefficients in (4.96) differ from those used in Appendix B by factors of $M^{1/2}$.

Here, $\{1+n(\omega)\}$ is the detailed balance factor, and the dissipative part of the susceptibility is obtained from the expression,

$$\chi_{\mathbf{q}j}[\omega]=\{\omega^2-\omega_j^2(\mathbf{q})-2\omega_j(\mathbf{q})\Delta_j^{(0)}(\mathbf{q}) \\ -2\omega_j(\mathbf{q})\Delta_j^{(1)}(\mathbf{q})-\Sigma_{\mathbf{q}j}(\omega)\}^{-1}, \quad (4.98)$$

where $\Delta_j^{(0)}(\mathbf{q})$ is the shift in the phonon frequency due to thermal expansion. The shift $\Delta_j^{(1)}(\mathbf{q})$ and the complex self-energy $\Sigma_{\mathbf{q}j}(\omega)$ are calculated in § B.9.2, where it is shown that the leading-order contributions from (4.96) are of order $K$ and $J^2$; the shift $\Delta_j^{(1)}(\mathbf{q})$ is of order $K$, and the self-energy is of order $J^2$. The latter is ($\epsilon\to 0^+$),

$$\Sigma_{\mathbf{q}j}(\omega)=\Sigma'_{\mathbf{q}j}(\omega)+\mathrm{i}\Sigma''_{\mathbf{q}j}(\omega) \\ =\frac{9\hbar}{2}\sum_{12}\frac{|J(-\mathbf{q}j,1,2)|^2}{\omega_1\omega_2}\{(1+n_1+n_2)[(\omega-\Omega_+-\mathrm{i}\epsilon)^{-1}-(\omega+\Omega_+-\mathrm{i}\epsilon)^{-1}] \\ +(n_2-n_1)[(\omega-\Omega_--\mathrm{i}\epsilon)^{-1}-(\omega+\Omega_--\mathrm{i}\epsilon)^{-1}]\} \quad (4.99)$$

where the frequencies

$$\Omega_\pm=\omega_1\pm\omega_2$$

and $n_1$ is the Bose occupation function for the mode $1\equiv\mathbf{q}_1j_1$. The damping in (4.93) is related to the imaginary part of the self-energy by

$$2\omega_j(\mathbf{q})\Gamma_j(\mathbf{q},\omega)=\Sigma''_{\mathbf{q}j}(\omega) \quad (4.100)$$

and we note that $\Gamma_j(\mathbf{q},\omega)$ is an odd function of $\omega$. The remaining components of the frequency shift are

$$2\omega_j(\mathbf{q})\{\Delta_j^{(1)}(\mathbf{q})+\Delta_j^{(2)}(\mathbf{q},\omega)\} \\ =-12\sum_2 K(-\mathbf{q}j,\mathbf{q}j,2,-2)\langle|\hat{Q}(2)|^2\rangle+\Sigma'_{\mathbf{q}j}(\omega). \quad (4.101)$$

In § B.6.2 the shift $\Delta_j^{(1)}(\mathbf{q})$ is included in the susceptibility through a temperature-dependent frequency (cf. eqn (B.181)).

The susceptibility (4.98) is written in terms of a shift $\Delta_j(\mathbf{q},\omega)=\{\Delta_j^{(0)}(\mathbf{q})+\Delta_j^{(1)}(\mathbf{q})+\Delta_j^{(2)}(\mathbf{q},\omega)\}$ and a damping constant

$$\chi_{\mathbf{q}j}[\omega]=[\omega^2-\omega_j^2(\mathbf{q})-2\omega_j(\mathbf{q})\{\Delta_j(\mathbf{q},\omega)+\mathrm{i}\Gamma_j(\mathbf{q},\omega)\}]^{-1} \quad (4.102)$$

and, provided that $\Delta_j(\mathbf{q},\omega)$ and $\Gamma_j(\mathbf{q},\omega)$ are small quantities and benign functions of $\mathbf{q}$ and $\omega$, we can safely write the susceptibility in the approximate form

$$\chi_{\mathbf{q}j}[\omega]=\frac{1}{2\omega_j(\mathbf{q})}\{[\omega-\omega_j(\mathbf{q})-\Delta_j(\mathbf{q},\omega)-\mathrm{i}\Gamma_j(\mathbf{q},\omega)]^{-1} \\ -[\omega+\omega_j(\mathbf{q})+\Delta_j(\mathbf{q},\omega)+\mathrm{i}\Gamma_j(\mathbf{q},\omega)]^{-1}\}. \quad (4.103)$$

We are consistent in our treatment of the shift and damping functions if we evaluate them at the phonon frequency $\omega_j(\mathbf{q})$, for the first contribution in (4.103) is expected to be negligible except at $\omega = \omega_j(\mathbf{q})$ and the second contribution is expected to be negligible except at $\omega = -\omega_j(\mathbf{q})$. We will use the fact that $\Delta_j$ and $\Gamma_j$ are even and odd functions of $\omega$, respectively. Let us define a renormalized and temperature-dependent phonon frequency through the relation

$$\omega_j(\mathbf{q}, T) = \omega_j(\mathbf{q}) + \Delta_j(\mathbf{q}, \omega_j(\mathbf{q})) \tag{4.104}$$

and a decay constant

$$\gamma_j(\mathbf{q}) = \Gamma_j(\mathbf{q}, \omega_j(\mathbf{q})) \equiv \Sigma''_{\mathbf{q}j}(\omega_j(\mathbf{q}))/2\omega_j(\mathbf{q}). \tag{4.105}$$

The approximation to the susceptibility is now

$$\chi_{\mathbf{q}j}[\omega] = \frac{1}{2\omega_j(\mathbf{q})}\{[\omega - \omega_j(\mathbf{q}, T) - \mathrm{i}\gamma_j(\mathbf{q})]^{-1} - [\omega + \omega_j(\mathbf{q}, T) - \mathrm{i}\gamma_j(\mathbf{q})]^{-1}\}$$

and the dissipative component required in (4.97) is

$$\chi''_{\mathbf{q}j}[\omega] = \left(\frac{\gamma_j(\mathbf{q})}{2\omega_j(\mathbf{q})}\right)\{[(\omega - \omega_j(\mathbf{q}, T))^2 + \gamma_j^2(\mathbf{q})]^{-1} - [(\omega + \omega_j(\mathbf{q}, T))^2 + \gamma_j^2(\mathbf{q})]^{-1}\}. \tag{4.106}$$

We emphasize that (4.106) is obtained from the result (4.98) on the assumption that the quantities calculated by perturbation theory are small corrections to the susceptibility for all values of $\mathbf{q}$ and $\omega$.

The calculation presented does not include polarization mixing induced by the anharmonic forces whereby phonon states corresponding to the same wave vector but different polarization branches are coupled together. This mixing is believed to generally be a small effect in the neutron cross-section, and it vanishes at high symmetry points in the Brillouin zone. However, it does mean that eqn (4.89) is not completely general because it assumes that polarization mixing does not occur.

In the limit $\Delta_j$, $\gamma_j \to 0$ the expression (4.106) reduces to

$$\chi''_{\mathbf{q}j}[\omega] = \frac{\pi}{2\omega_j(\mathbf{q})}[\delta\{\omega - \omega_j(\mathbf{q})\} - \delta\{\omega + \omega_j(\mathbf{q})\}] \tag{4.107}$$

and, when this is inserted in (4.97), we recover the harmonic result (4.91) for $S_j(\mathbf{q}, \omega)$. The susceptibility for a damped harmonic oscillator is obtained from (4.102) with the approximation

$$2\omega_j(\mathbf{q})\Gamma_j(\mathbf{q}, \omega) = \gamma\omega,$$

where $\gamma$ is the friction coefficient (see § B.3).

In order to use the preceding results for the self-energy and frequency shift it is necessary to have specific expressions for the interaction coefficients and the harmonic phonon frequencies. Notwithstanding the general lack of such a detailed knowledge of the sample we can reach several useful conclusions. In the limit of absolute zero the Bose occupation factors in the self-energy (4.99) can be set to zero. The self-energy is finite in this limit because of the zero-point motion of the lattice vibrations. The shift $\Delta_j^{(1)}(\mathbf{q})$ is also finite at absolute zero, as can be seen from (4.101) using (4.90) for the correlation function. In the opposite limit of high temperatures the Bose factors can be approximated by

$$n_1 \doteq k_B T/\hbar\omega_{j_1}(\mathbf{q}_1); \qquad T \gg \theta_D$$

and $(n_1+1) \doteq n_1$. From this observation it follows at once that $\Delta_j^{(1)}$, $\Delta_j^{(2)}$, and $\Gamma_j$ are linear functions of the temperature for $T \gg \theta_D$.

There are major contributions to the self-energy (4.99) when the phonon frequencies $\Omega_\pm$ coincide with $\omega$. The actual value of the contribution depends on the joint density of states, as well as the interaction coefficient $J(1, 2, 3)$, and therefore the self-energy depends on the details of the phonon dispersions in a complicated manner. The self-energy will, in most cases, have its maxima at frequencies that are not close to $\omega_j(\mathbf{q})$. Marked structure in the self-energy is reflected in the susceptibility, and therefore it is usually prudent to use the full form (4.98) rather than (4.103). Note that the real and imaginary parts of the self-energy are related by a dispersion relation, viz. § B.2. Hence, $\Delta_j^{(2)}(\mathbf{q}, \omega)$ can be obtained from $\Gamma_j(\mathbf{q}, \omega)$, and vice versa. We should expect low-frequency phonons to be little effected by anharmonic interactions because the phonon density of states is small for frequencies much less than $\omega_D$, and thus the probability of the scattering of phonons is itself also small. The behaviour is borne out for simple models of lattice vibrations because the damping is proportional to $\omega_j^2(\mathbf{q})$. Simple models also have $\Gamma_j$ proportional to the square of the macroscopic Grüneisen parameter.

The approximation (4.106) is frequently used in the analysis of phonon line-shapes in samples that are far from a structural phase transition. The line-shape in this instance is a symmetric Lorentz function centred at the renormalized phonon frequency $\omega_j(\mathbf{q}, T)$. However, it is well established that the phonon line-shape can be highly asymmetric (Glyde 1974; Meyer *et al.* 1976). The physical origin of the asymmetry is an interference between one-phonon and multiphonon processes. For example, a single phonon initially excited in neutron scattering can decay, via the anharmonic interactions, into two phonons. The contribution of this process to the cross-section contains the one-phonon susceptibility and the approximate conservation of wave vector, and it is therefore to be included in the one-phonon neutron cross-section. There is also the possibility of two phonons combining into a single phonon, and equivalent processes involving three or more phonons and decay or creation of a single phonon. In general, the interference terms lead to both an asymmetry and change in the intensity of the one-phonon line-shape. However, the leading-order contribution to the cross-section vanishes for $\boldsymbol{\kappa}$ equal to a reciprocal lattice vector, or a vector midway between two Bragg positions. The interference contribution to the cross-section changes sign as $\boldsymbol{\kappa}$ passes through a reciprocal lattice position. We conclude that, although interference terms are small contributions to the cross-section, they must be accommodated in a high-resolution study of the phonon lineshape as functions of temperature or pressure, for example (Meyer *et al.* 1976; Cowley *et al.* 1976).

## 4.6. Structural phase transitions (Currat and Pynn 1979; Bruce and Cowley 1981; Cummins and Levanyuk 1983)

Anharmonic interactions are believed to be responsible for a large number of observed structural phase changes. The fundamental concept is that of a soft-phonon-induced instability. If a phonon frequency reaches zero at some critical temperature $T_0$, the phonon is regarded as condensing into the lattice to provoke a structural phase transition. In other words below $T_0$ the crystal is distorted by a 'frozen in' amplitude of the incipient soft-phonon mode. The vanishing of a phonon frequency is equivalent to a divergence of the corresponding static isothermal susceptibility, since they are reciprocal functions (cf. § B.9.2). A physically intuitive picture is that the force constants associated with the unstable phonon decrease, or soften, with decreasing temperature and tend eventually to zero at $T_0$. There is a corresponding increase in the size of the lattice displacement arising from the $\{\omega_j(\mathbf{q})\}^{-1/2}$-dependence of the normal coordinate operator (eqn (4.26)). In the low-symmetry phase below $T_0$, the unstable phonon is stabilized by higher-order anharmonic terms.

The development outlined in the preceding paragraph can be made more precise with the aid of some of the results from § 4.5. The generalized susceptibility (4.98) possesses a pole at a frequency determined by the phonon dispersion frequency and the shift due to anharmonic interactions; in § 4.5 we consider three specific contributions to the shift. The renormalized phonon frequency is approximately

$$\omega = \{\omega_j^2(\mathbf{q}) + 2\omega_j(\mathbf{q})\Delta_j(\mathbf{q}, \omega_j(\mathbf{q}))\}^{1/2},$$

where the shift $\Delta_j$ is evaluated at the phonon frequency $\omega_j(\mathbf{q})$. If the phonon labelled by $\mathbf{q}$ and $j$ is unstable then $\omega_j^2(\mathbf{q})$ is negative. We found in § 4.5 that the contributions to the shift are linear functions of the temperature for $T > \theta_D$. Hence, the renormalized phonon frequency can be written

$$\omega = \alpha(T - T_0)^{1/2},$$

where $\alpha$ is a constant.

This picture of a phase transition is satisfactory for a wide variety of ferroelectric crystals which undergo a first-order phase transition at $T_0$. In most of these crystals the soft mode is overdamped even at temperatures well above $T_0$ (cf. § B.3). A ferroelectric must involve condensation of a phonon that is both polar and long-wavelength because the ferroelectric state must have a macroscopic polarization. The connection between the onset of ferroelectricity and the existence of a soft mode is seen in the relation between the zero frequency dielectric function $\epsilon(0)$ and the frequency of transverse and longitudinal optic modes, $\omega_0$ and $\omega_L$, respectively. In its simplest form the relation is

$$\epsilon(0) \propto (\omega_L/\omega_0)^2.$$

Hence the onset of ferroelectricity, characterized by $\epsilon \to \infty$ as $T \to T_0$, can arise from the softening of a long-wavelength transverse optic phonon. An example is the cubic perovskite $BaTiO_3$ in which an optic phonon, which is coupled to an acoustic phonon, becomes soft at the phase transition.

In contrast, the perovskite $SrTiO_3$ undergoes a non-ferroelectric, continuous-phase transition which involves the softening of a short-wavelength optic mode. The soft mode makes a distinct contribution to the response function $S_j(\mathbf{q}, \omega)$ near the phase transition, unlike the situation in $BaTiO_3$ and most other ferroelectric transitions. However, in the immediate vicinity of the transition, at about 102 K, the soft-mode contribution is not a distinct feature of the spectrum; this applies as the transition is approached from above and below $T_0$ where the soft modes are of F, and A and E symmetry, respectively. The spectrum contains significant low-energy, quasi-elastic scattering, and the phonon frequency does not appear to go to zero as $T \to T_0$. These features of the spectrum are not accounted for by the susceptibility of a damped harmonic oscillator, viz.

$$\chi[\omega] = [\omega^2 - \omega_0^2 - i\gamma\omega]^{-1}$$

where $\omega_0$ and $\gamma$ are the soft-mode frequency and friction coefficients, respectively. For this model of a damped harmonic oscillator the phonon contribution is underdamped or overdamped for $\omega_0 > \gamma/2$ or $\omega_0 < \gamma/2$, respectively. Note that, if $\gamma$ is temperature-independent, the phonon mode always becomes overdamped as $T \to T_0$ since $\omega_0 \to 0$.

The quasi-elastic scattering can be interpreted in terms of a frequency-dependent model for the self-energy in (4.98). The frequency dependence is ascribed to a coupling of the soft mode to phonon density fluctuations, or a slowly relaxing variable that is required to describe the defect structure of the sample, say. Let the relaxation time of the phonons that contribute to the structure of the self-energy be $\tau$. If $\omega\tau \ll 1$, we have a collision-dominated structure akin to the damped harmonic oscillator, whereas in the opposite limit of collisionless behaviour, $\omega\tau \gg 1$, a viscoelastic structure occurs. The actual form of the displacement susceptibility is ($s = i\omega + \epsilon$ with $\epsilon \to 0^+$)

$$-\chi[\omega] = \left\{s^2 + \omega_0^2 + 2\omega_0\Delta + s\gamma_0 - \frac{\Omega^2}{1+s\tau}\right\}^{-1}$$

$$\equiv \left\{s^2 + \omega_0^2 + 2\omega_0\Delta - \Omega^2 + s\gamma_0 + \frac{s\tau\Omega^2}{1+s\tau}\right\}^{-1}. \tag{4.108}$$

Here $\Delta$, $\gamma_0$, $\Omega^2$, and $\tau$ are constants associated with a specific soft-phonon mode. If the observed spectrum contains contributions from more than one soft mode, as is the case in the low-temperature phase of $SrTiO_3$, it is assumed that the response can be approximated by the sum of an appropriate number of independent susceptibilities with, in general, different constants.

The static isothermal susceptibility associated with the model susceptibility (4.108) is

$$\chi = -\chi[0] = (\omega_0^2 + 2\omega_0\Delta - \Omega^2)^{-1} \tag{4.109}$$

and it is usual to use the notation $\omega_\infty^2 = \omega_0^2 + 2\omega_0\Delta$. If the quantity $(\omega_\infty^2 - \Omega^2)$ tends to zero as the temperature approaches $T_0$, the susceptibility diverges, as it must for a continuous phase change, and the divergence is characterized by a power-law behaviour

$$\chi = (\omega_\infty^2 - \Omega^2)^{-1} \propto (T - T_0)^{-\gamma}; \qquad T > T_0. \tag{4.110}$$

For $SrTiO_3$ the exponent $\gamma = 2.0 \pm 0.5$, whereas classical mean field, or Landau, theory predicts $\gamma = 1$ (Shirane 1974).

In the collisionless region, $\omega\tau \gg 1$, the susceptibility (4.108) shows a strong response at the frequency $\omega_\infty$, while for $\omega\tau \ll 1$ the response frequency shifts to the lower value $\chi^{-1/2} \doteq (\omega_\infty - \Omega^2/2\omega_0)$. For the quasi-elastic response we observe that

$$\lim_{\omega \to 0} \chi''[\omega] = \omega(\gamma_0 + \tau\Omega^2)\chi^2 \tag{4.111}$$

where $\chi$ is given in (4.109). Hence, when $\chi$ diverges as the transition is approached, a strong central peak develops in $S_j(\mathbf{q}, \omega)$ and the total intensity diverges. The half-width of the central peak vanishes as $(\omega_\infty^2 - \Omega^2)$ tends to zero. Indeed, the quasi-elastic response is a Lorentz function

$$\left\{\frac{\chi''[\omega]}{\omega}\right\} = \frac{(\gamma_0 + \tau\Omega^2)^{-1}}{\{\chi(\gamma_0 + \tau\Omega^2)\}^{-2} + \omega^2} \tag{4.112}$$

in which the width is $\{\chi(\gamma_0 + \tau\Omega^2)\}^{-1}$.

The modification of the Debye–Waller factor near a structural phase transition follows from the relation

$$W_d(\boldsymbol{\kappa}) = \frac{\hbar}{2NM_d} \sum_{j\mathbf{q}} |\boldsymbol{\kappa} \cdot \boldsymbol{\sigma}_d^j(\mathbf{q})|^2 \int_{-\infty}^{\infty} \mathrm{d}\omega\, S_j(\mathbf{q}, \omega)$$

$$= \frac{\hbar}{2\pi NM_d} \sum_{j\mathbf{q}} |\boldsymbol{\kappa} \cdot \boldsymbol{\sigma}_d^j(\mathbf{q})|^2 \int_{-\infty}^{\infty} \mathrm{d}\omega\, \{1 + n(\omega)\}\chi''_{\mathbf{q}j}[\omega]. \tag{4.113}$$

In deriving (4.113) from the definition (4.51) we have used (4.91) and (4.97). The contribution of the soft modes to (4.113) will increase as $T \rightarrow T_0$ because the integrals are proportional to the mode static susceptibilities at high temperatures. However, the sums in (4.113), which arise because $W_d(\boldsymbol{\kappa})$ is a single-site property, will nullify the changes to a large extent.

Structural phase transitions can involve transformations driven by a collective instability of the conduction electrons. A key concept is that of a Kohn anomaly in the phonon frequency dispersion which arises from the modulation of the lattice force constants by the conduction electrons. The anomalies are often a very weak feature in the dispersion relations. Striking exceptions are materials whose electronic properties are very anisotropic, and essentially quasi-one-dimensional. The electronic instability which causes the abnormally large Kohn anomalies in such materials can prompt a displacive transition. Because the wave vector of the Kohn anomaly, and the distortion if it occurs, is determined by the geometry of the Fermi surface the effects need not be commensurate with the crystal periodicity. Clearly a complete interpretation of an incommensurable phase transition requires a detailed knowledge of the electron–phonon interaction and the electron distribution on the Fermi surface.

The physics behind Kohn anomalies in the dispersion curves is as follows. In a real metal the motions of the conduction electrons and ions are clearly not independent of each other; when ions move from their equilibrium positions they scatter the electrons. Because typical phonon frequencies are much less than electronic frequencies of importance, the motion of the ions appears to the electrons as a static perturbation varying in space with a wave vector $\mathbf{q}$. A consequence of this is that the Coulomb potential between the ions is screened by the electrons, the 'bare' potential being essentially multiplied by the susceptibility of the conduction electrons of the static perturbation. For a free electron gas the susceptibility possesses a singularity at $|\mathbf{q}|$ equal to twice the Fermi wave vector (see § 3.6, for example). It follows that the ion–ion potential, and therefore the phonon frequency, reflects this singularity. In real metals the effect on the phonon frequency depends on both the strength of the electron–phonon interaction and the sharpness and shape of the Fermi surface. For example, for pieces of the Fermi surfaces possessing cylindrical form, the singularity in the derivative of the dispersion curve is of the inverse square root kind and, for planar sections of the Fermi surface, the most singular part becomes logarithmic.

Electron–phonon interactions are particularly important in understanding the onset of superconductivity. For example, $Nb_3Sn$ is one of several A-15 compounds that are high-temperature superconductors, and it exhibits a structural phase transition at 45 K which is approximately 26 K above the critical temperature for the onset of superconductivity (in zero field). Neutron-scattering studies of the structural phase transition reveal a number of interesting features which are, at first sight, akin to those observed in $SrTiO_3$. In particular, a central peak is observed which grows in intensity as the transition is approached from above. However, the instability in $Nb_3Sn$ is associated with an acoustic phonon of zero wave vector. Moreover, only acoustic modes are involved in the central peak in $Nb_3Sn$, in contrast to the situation in $SrTiO_3$ where the peak arises from the decay of an optic phonon into acoustic phonons. A microscopic picture for the mechanism driving the transition is currently lacking (Ranninger 1975).

## 4.7. **Mixed harmonic systems** (Elliott *et al.* 1974; Taylor 1975; Nicklow 1979)

The spectra of neutrons scattered by lattice vibrations usually display an intrinsic width due to the finite lifetime of the vibrations. One contribution to the width arises from the anharmonic forces described in § 4.5. Another contribution arises from disturbances created by defects in the target sample. Here we consider the important case of mass defects.

In some instances the dynamics of the mass defect is of central interest. We must then ask how the vibrations of the host lattice, or macromolecule, participate in the dynamic response of the defect under investigation. The opposite viewpoint is to assess the effect of mass contaminants on the propagation of the host vibtations. Because defects scatter the lattice vibrations they contribute to the phonon lifetime.

A very convenient method for calculating the dynamics of mixed harmonic systems is based on the Green function for lattice displacements introduced in Appendix B, and we adopt this method throughout this section.

*4.7.1. Particle hybridized with phonon environment*

We calculate the cross-section for scattering by a single particle that is hybridized with a phonon heat bath. The cross-section possesses many of the features found in mathematically more complicated models of mixed harmonic systems considered in later parts of this section. Our calculation is couched in the same language that we adopt to discuss the more complicated models, namely the Green function introduced in § B.8. Hence the present calculation serves as something of an introduction to subsequent developments. The model might be realized in the scattering of neutrons from a proton, say, which inhabits a macromolecule. Because the cross-section for hydrogen is exceptionally large, the scattering from a hydrogen-bearing macromolecule is dominated by the hydrogen contribution. Whether it is appropriate to model the environment of the hydrogen by a phonon heat bath depends on the physico-chemical aspects of the target sample.

The general form of the one-phonon cross-section is obtained from (4.92). Let the mass of the scattering particle be $M$ and let its environment be isotropic to a good approximation. The one-phonon cross-section is then

$$\left(\frac{d^2\sigma}{d\Omega\, dE'}\right)^{\text{inel}} = \frac{k'}{k}\left(\frac{\sigma}{4\pi M}\right)\exp\{-2W(\kappa)\}\kappa^2 S(\omega). \tag{4.114}$$

Here,

$$S(\omega) = \frac{1}{2\pi\hbar}\int_{-\infty}^{\infty} dt \exp(-i\omega t)\langle \hat{Q}^+\hat{Q}(t)\rangle \tag{4.115}$$

and the normal coordinate operator follows from (4.26)

$$\hat{Q} = (\hbar/2\omega_0)^{1/2}(\hat{a} + \hat{a}^+), \tag{4.116}$$

where $\hat{a}$, $\hat{a}^+$ are the usual Bose operators for harmonic phonons. We will calculate $S(\omega)$ assuming that the particle is coupled to many phonons that make up a heat bath. The temperature does not appear explicitly in the model, but it would largely control the strength of the coupling to the phonon bath.

The model is described by the Hamiltonian

$$\hat{\mathcal{H}} = \hbar\omega_0\hat{a}^+\hat{a} + \sum_\nu \hbar\omega_\nu \hat{b}_\nu^+\hat{b}_\nu + \sum_\nu g_\nu(\hat{a}\hat{b}_\nu^+ + \hat{a}^+\hat{b}_\nu). \tag{4.117}$$

The operators $\hat{b}_\nu$, $\hat{b}_\nu^+$ describe the phonons in the heat bath and $\omega_\nu$ is the frequency of the $\nu$th phonon. The final term in (4.117) describes a hybridization of the scattering particle and the heat bath through a temperature-dependent interaction coefficient $g_\nu$. Because the coupling is linear and the Hamiltonian is quadratic, the response function (4.115) can be calculated exactly. One method of making the calculation is to diagonalize (4.117) by a linear operator transformation. We choose to use a Green function for the calculation because it is a very convenient language for more complicated models that will be discussed.

Referring to § B.8, we define a Green function

$$G(\omega) = \langle\langle \hat{a}; \hat{a}^+ \rangle\rangle \tag{4.118}$$

that satisfies the equation of motion

$$\hbar\omega G(\omega) = \hbar\langle[\hat{a}, \hat{a}^+]\rangle + \langle\langle[\hat{a}, \hat{\mathcal{H}}]; \hat{a}^+\rangle\rangle. \tag{4.119}$$

Given $G(\omega)$, we obtain

$$\frac{1}{2\pi}\int_{-\infty}^{\infty} dt \exp(-i\omega t)\langle \hat{a}\hat{a}^+(t)\rangle = \frac{-1}{\pi}\{1+n(\omega)\}G''(\omega), \tag{4.120}$$

where $G''(\omega)$ is the imaginary part of (4.118) calculated with $\omega \to (\omega + i\epsilon)$ and $\epsilon \to 0^+$. From (4.120) we can readily deduce that the response function (4.115) is

$$S(\omega) = \frac{-1}{2\pi\omega_0}\{1+n(\omega)\}\{G''(\omega) - G''(-\omega)\}. \tag{4.121}$$

Consider the two terms on the right-hand side of the equation-of-motion (4.119). The inhomogeneous term is $\hbar$ because the Bose operators satisfy the commutation relation $[\hat{a}, \hat{a}^+] = 1$ (cf. eqn (4.18)). The commutator in the second term is equal to

$$\hbar\omega_0 G(\omega) + \sum_\nu g_\nu \langle\langle \hat{b}_\nu; \hat{a}^+\rangle\rangle,$$

so that $G(\omega)$ satisfies

$$(\omega - \omega_0)G(\omega) = 1 + (1/\hbar)\sum_\nu g_\nu \langle\langle \hat{b}_\nu; \hat{a}^+\rangle\rangle. \tag{4.122}$$

Our next step is to calculate the Green function on the right-hand side of (4.122) using an equation of motion akin to (4.119). For this second Green function the inhomogeneous term is zero because $\hat{b}_\nu$ and $\hat{a}^+$ commute. We find

$$(\omega - \omega_\nu)\langle\langle \hat{b}_\nu; \hat{a}^+\rangle\rangle = g_\nu G(\omega)/\hbar, \tag{4.123}$$

and hence, from (4.122),

$$\left\{\omega - \omega_0 - \sum_\nu (g_\nu/\hbar)^2(\omega - \omega_\nu)^{-1}\right\}G(\omega) = 1, \tag{4.124}$$

which is the solution to the problem in hand.

Let us consider first the nature of the solution for $g_\nu = 0$. With $\omega \to (\omega + i\epsilon)$ and $\epsilon \to 0^+$ we find that $G''(\omega)$ is a delta function with an argument $(\omega - \omega_0)$, viz.

$$G''(\omega) = -\pi\,\delta(\omega - \omega_0); \qquad g = 0.$$

Using this result in (4.121) we recover an expression for $S(\omega)$ that is analogous to (4.91), i.e. the motion of an undamped harmonic oscillator.

The damping in our model arises from the imaginary part of the sum in (4.124). We calculate this by replacing the interaction coefficient by an average value g such that

$$\begin{aligned}\mathrm{Im}\sum_{\nu} (g_\nu/\hbar)^2(\omega + i\epsilon - \omega_\nu)^{-1} &= -\pi g^2 Z(\omega)\\ &= -\Gamma(\omega),\end{aligned} \tag{4.125}$$

where $Z(\omega)$ is a phonon density of states, defined in eqn (4.39) and illustrated in Fig. 4.3. The real part of the sum is now determined by (4.125) through the standard dispersion relation. We define

$$\Delta(\omega) = \mathrm{P}\int_{-\infty}^{\infty} \frac{du\, Z(u)}{(\omega - u)} = 2\omega \mathrm{P}\int_0^{\infty} \frac{du\, Z(u)}{(\omega^2 - u^2)}, \tag{4.126}$$

where we have used the relation $Z(\omega) = Z(-\omega)$ and the principal part of the integral P is taken. With these expressions for the real and imaginary parts of the sum in the denominator of $G(\omega)$ (eqn (4.124)) our final result for the Green function is

$$G(\omega) = \{\omega - \omega_0 + i\Gamma(\omega) - g^2\Delta(\omega)\}^{-1}$$

and therefore

$$G''(\omega) = \frac{-\Gamma(\omega)}{\{\omega - \omega_0 - g^2\Delta(\omega)\}^2 + \{\Gamma(\omega)\}^2}. \tag{4.127}$$

Because the damping in this model is determined by the density of phonon states $Z(\omega)$, we conclude that the damping vanishes for $\omega \to 0$ and frequencies which exceed the maximum phonon frequency (see Fig. 4.3). Moreover, the lineshape can be expected to show considerable structure as $\omega$ passes through the extrema of $Z(\omega)$.

The main structure in the response is likely to occur at frequencies which satisfy the equation

$$\omega - \omega_0 - g^2\Delta(\omega) = 0. \tag{4.128}$$

For very small $g^2$ the solution is $\omega = \omega_0$ to a good approximation. The response will be peaked at the frequency $\omega_0$, and the linewidth in this limit is $\Gamma(\omega_0)$. The latter is finite if $\omega_0$ lies in the phonon frequency band determined by the density of states, whereas if $\omega_0$ lies outside the band the response is a delta function and therefore very much like that for an undamped harmonic oscillator. The difference between the undamped harmonic oscillator, $g^2 = 0$, and the present case is that the weight, or amplitude, of the delta function response depends on the function $\Delta(\omega)$. Before we discuss this point further we study the solutions of (4.128) in slightly more detail.

From the definition of $\Delta(\omega)$ in eqn (4.126) it follows that for $\omega \to \infty$

$$\Delta(\omega) \to \frac{2}{\omega} \int_0^\infty du\, Z(u) = \frac{2}{\omega}; \qquad \omega \to \infty, \tag{4.129}$$

where we have used the normalization of the density of states. In the opposite frequency limit,

$$\Delta(\omega) \to -2\omega \int_0^\infty \frac{du\, Z(u)}{u^2}; \qquad \omega \to 0. \tag{4.130}$$

The integral is positive, and finite for three-dimensional materials since $Z(\omega) \propto \omega^2$ for $\omega \to 0$. We conclude that $\Delta(\omega)$ is negative for small $\omega$, and positive for large $\omega$ where it decreases as $1/\omega$. Hence $\Delta(\omega)$ passes through zero at least once, and there is at least one minimum and one maximum below and above the zero, respectively. Numerical calculations (Nicklow 1979) show that a main maximum occurs in the vicinity of the maximum phonon frequency $\omega_m$ and we will later show that this feature is reproduced by a Debye model of the phonon heat bath where $\omega_m = \omega_D$. If $\omega_0 > \omega_m$, then there is at least one solution of (4.128) at a frequency which exceeds $\omega_0$. The departure of the solution from $\omega_0$ increases with increasing $g^2$, as should be expected. There might also be a pair of solutions at frequencies less than $\omega_m$ which arise from the intersection of $(\omega - \omega_0)/g^2$ with the low-frequency minimum in $\Delta(\omega)$. However, the response from these solutions will be heavily damped because they occur in the frequency range in which $Z(\omega)$, and therefore $\Gamma(\omega)$, is finite.

To extract the response due to a solution of (4.128) at $\omega > \omega_m$, often called a resonance of the model, we proceed as follows. Let the denominator of $G(\omega)$ be denoted by $F(z)$ where the complex variable $z = (\omega + i\epsilon)$. The function satisfies $F^*(z) = F(z^*)$. We need to evaluate

$$\lim_{\epsilon \to 0^+} \mathrm{Im}\{1/F(z)\},$$

given that $F(z)$ is purely real at the resonance frequency $\omega_l$ which satisfies $F(\omega_l) = 0$. A Taylor expansion about $\omega_l$ gives

$$\begin{aligned} F(z) &= F(\omega_l) + (\omega + i\epsilon - \omega_l)F'(\omega_l) + \cdots \\ &= (\omega + i\epsilon - \omega_l)F'(\omega_l) + \cdots \end{aligned} \tag{4.131}$$

where $F'(\omega_l)$ is the derivative of $F(z)$ evaluated at the resonance frequency. From (4.131) we find

$$\mathrm{Im}\{1/F(z)\} = \frac{1}{2\mathrm{i}}\left\{\frac{1}{F(z)} - \frac{1}{F^*(z)}\right\}$$
$$= \frac{1}{2\mathrm{i}}\left\{\frac{1}{\omega + \mathrm{i}\epsilon - \omega_l} - \frac{1}{\omega - \mathrm{i}\epsilon - \omega_l}\right\}\frac{1}{F'(\omega_l)}$$
$$= \left\{\frac{-\epsilon}{(\omega - \omega_l)^2 + \epsilon^2}\right\}\frac{1}{F'(\omega_l)}$$

and therefore

$$\lim_{\epsilon \to 0^+} \mathrm{Im}\{1/F(z)\} = -\pi\,\delta(\omega - \omega_l)/F'(\omega_l). \tag{4.132}$$

We conclude that the response is a delta function at the resonance frequency with an amplitude which is modified from the undamped value by a factor $1/F'(\omega_l)$. Applied to (4.128)

$$F(\omega) = \omega - \omega_0 - g^2\,\Delta(\omega)$$

and

$$F(\omega_l) = \omega_l - \omega_0 - g^2\,\Delta(\omega_l) = 0 \tag{4.133}$$

with

$$F'(\omega_l) = 1 - g^2\,\Delta'(\omega_l). \tag{4.134}$$

Assembling the results we have, for $\omega_l > \omega_\mathrm{m}$,

$$G''(\omega) = -\pi\,\delta(\omega - \omega_l)/\{1 - g^2\,\Delta'(\omega_l)\}, \tag{4.135}$$

and, from (4.121),

$$S(\omega) = \frac{\{1 + n(\omega)\}}{2\omega_0\{1 - g^2\,\Delta'(\omega_l)\}}\{\delta(\omega - \omega_l) - \delta(\omega + \omega_l)\}$$
$$= (2\omega_0\{1 - g^2\,\Delta'(\omega_l)\})^{-1}\{[1 + n(\omega_l)]\,\delta(\omega - \omega_l) + n(\omega_l)\,\delta(\omega + \omega_l)\}. \tag{4.136}$$

From this last expression for the response function $S(\omega)$ it is evident that $F'(\omega_l)$ (eqn (4.134)) must be positive if the resonance is physically significant. The result (4.129) implies that for $\omega_l \gg \omega_\mathrm{m}$, $F'(\omega_l) = (1 + 2g^2/\omega_l^2)$.

When the natural frequency of the harmonic oscillator is less than $\omega_\mathrm{m}$, there are several solutions of eqn (4.128). The significant solutions will have $F'(\omega_l)$ small and positive if $\omega_l > \omega_\mathrm{m}$ while, for $\omega_l < \omega_\mathrm{m}$, the response will be significant if $Z(\omega_l)$ is small, e.g. if $\omega_l \ll \omega_\mathrm{m}$. We can derive some insight into the features of the response for $\omega_l < \omega_m$ by approximating $Z(\omega)$ by the Debye density of states

$$Z(\omega) = (3\omega^2/\omega_D^3); \qquad \omega < \omega_D$$
$$= 0; \qquad \omega > \omega_D. \tag{4.137}$$

With a dimensionless frequency variable $x = (\omega/\omega_D)$, we find

$$\Delta(\omega) = (3/\omega_D)\left\{x^2 \ln\left|\frac{1+x}{1-x}\right| - 2x\right\}. \tag{4.138}$$

The divergence of $\Delta(\omega)$ at $\omega = \omega_D$ is an artefact of the Debye model; numerical calculations (Nicklow 1979) for a realistic density of states show a main maximum at $\omega_m$. Even so the model based on the Debye spectrum contains all the main features of the response $S(\omega)$ that we have deduced in the general discussions, and thus it is a useful model that is completely specified in terms of elementary functions.

We conclude this subsection with the observation that, in principle, it is straightforward to calculate the full cross-section for scattering from a harmonically bound particle, whereas up to now we have focused on the one-phonon contribution (4.114). From the identity (4.67) we have, for a single harmonically bound particle,

$$\frac{d^2\sigma}{d\Omega\, dE'} = \frac{k'}{k}\frac{\sigma}{4\pi}\frac{1}{2\pi\hbar}\int_{-\infty}^{\infty} dt \exp(-i\omega t)\exp\langle \boldsymbol{\kappa}\cdot\hat{\mathbf{u}}\,\boldsymbol{\kappa}\cdot\hat{\mathbf{u}}(t) - (\boldsymbol{\kappa}\cdot\hat{\mathbf{u}})^2\rangle. \tag{4.139}$$

If the particle inhabits an isotropic environment, the correlation function in the second exponential in (4.139) is

$$\tfrac{1}{3}\kappa^2\langle\hat{\mathbf{u}}\cdot\hat{\mathbf{u}}(t) - \hat{\mathbf{u}}\cdot\hat{\mathbf{u}}\rangle = (\hbar\kappa^2/M)\int_{-\infty}^{\infty} du\{\exp(iut) - 1\}S(u), \tag{4.140}$$

where the second equality follows from the definition (4.115). For the model of present interest the response function in (4.140) is obtained from (4.121) with Green function (4.127). In the limit $g^2 = 0$ we recover from (4.139) and (4.140) the cross-section for scattering from an undamped harmonic oscillator with a natural frequency $\omega_0$, and this is discussed in detail in § 3.6.2. For modest values of the scattering vector, such that $(\hbar\kappa^2/M\omega_0) < 1$, the cross-section consists of a sum of delta functions with arguments $(\omega - n\omega_0)$ where the integer $n = 0, \pm 1, \pm 2$, etc. denumerates the number of quanta participating in the response. In the limit of large $\kappa$ the cross-section approaches the result for scattering from a free particle, subject to Boltzmann statistics, which is a Gaussian function of $\omega$. It is not possible to cast the cross-section in a closed form which is useful for general $\kappa$, and a numerical calculation is required even for the undamped harmonic oscillator.

### 4.7.2. *Mass defect*

The response of a purely harmonic lattice in which one atom has a mass that is different from the rest can be calculated explicitly in terms of the response of the pure lattice. In this subsection we will calculate the total partial differential cross-section for scattering from the mass defect, i.e. we will not invoke the one-phonon approximation. Special attention will subsequently be given to the case of a light mass defect particle. Many of the results derived here are required for a later discussion of the scattering from a finite concentration of mass defects, and in view of this we formulate the calculation in a slightly more general way than is necessary for the immediate discussion.

We restrict our attention to a Bravais lattice. The ions have position vectors

$$\mathbf{R}_l = \mathbf{l} + \mathbf{u}(l),$$

where $\mathbf{l}$ is the lattice position and $u(\mathbf{l})$ the displacement from this position. The unperturbed Hamiltonian is (cf. § 4.1)

$$\hat{\mathscr{H}}_0 = \sum_l \frac{1}{2M} \{\hat{\mathbf{p}}(l)\}^2 + \tfrac{1}{2} \sum_{ll'} \sum_{\alpha\beta} \hat{u}_\alpha(l) A_{\alpha\beta}(l, l') \hat{u}_\beta(l'), \tag{4.141}$$

where $A_{\alpha\beta}(l, l')$ are the second-order atomic force constants of the crystal. If we remove the ions at the sites $l = s$ and replace them by ions of mass $M'$, then the total Hamiltonian is

$$\hat{\mathscr{H}} = \hat{\mathscr{H}}_0 + \hat{\mathscr{H}}_1$$

where

$$\hat{\mathscr{H}}_1 = \tfrac{1}{2}\left\{\frac{1}{M'} - \frac{1}{M}\right\} \sum_s \{\hat{\mathbf{p}}(s)\}^2$$

$$= \frac{\lambda}{2M'} \sum_s \{\hat{\mathbf{p}}(s)\}^2. \tag{4.142}$$

We shall find later on that an important parameter is

$$(1/\lambda) = \frac{1}{1 - (M'/M)} \tag{4.143}$$

and we note that if $M' > M$, $\lambda^{-1}$ is negative and ranges from $-\infty$ to zero; whereas if $M' < M$, $\lambda^{-1}$ is positive and ranges from $+\infty$ to $+1$.

The correlation function required in (4.139) is calculated in terms of a Green function

$$G_{\alpha\beta}(l, l'; \omega) = \langle\langle \hat{u}_\alpha(l); \hat{u}_\beta(l') \rangle\rangle, \tag{4.144}$$

which satisfies the equation of motion

$$\hbar\omega G_{\alpha\beta}(l, l'; \omega) = \hbar\langle[\hat{u}_\alpha(l), \hat{u}_\beta(l')]\rangle + \langle\langle [\hat{u}_\alpha(l), \hat{\mathscr{H}}]; \hat{u}_\beta(l') \rangle\rangle. \tag{4.145}$$

The relation between the Green function and the correlation function is

$$\langle \hat{u}_\beta(l')\hat{u}_\alpha(l, t)\rangle = -\frac{\hbar}{\pi}\int_{-\infty}^{\infty} d\omega \exp(i\omega t)\{1+n(\omega)\}G''_{\alpha\beta}(l, l'; \omega) \tag{4.146}$$

where the imaginary part of the Green function is calculated with $\omega \to (\omega + i\epsilon)$ and $\epsilon \to 0^+$.

First, we calculate the Green function of the unperturbed system; this is denoted by $P_{\alpha\beta}(l, l'; \omega)$. From the equation of motion

$$\hbar\omega\langle\langle \hat{u}_\alpha(l); \hat{u}_\beta(l')\rangle\rangle = \hbar\langle[\hat{u}_\alpha(l), \hat{u}_\beta(l')]\rangle + \langle\langle[\hat{u}_\alpha(l), \hat{\mathcal{H}}_0]; \hat{u}_\beta(l')\rangle\rangle$$

we obtain

$$\omega P_{\alpha\beta}(l, l'; \omega) = \frac{i}{M}\langle\langle \hat{p}_\alpha(l); \hat{u}_\beta(l')\rangle\rangle \tag{4.147}$$

In deriving this equation we have used the commutation relations

$$[\hat{u}_\alpha(l), \hat{u}_\beta(l')] = 0, \qquad [\hat{p}_\alpha(l), \hat{p}_\beta(l')] = 0, \tag{4.148}$$

and

$$[\hat{u}_\alpha(l), \hat{p}_\beta(l')] = i\hbar\, \delta_{\alpha\beta}\, \delta_{ll'}.$$

Since

$$\begin{aligned}[\hat{p}_\gamma(l''), \hat{\mathcal{H}}_0] &= \tfrac{1}{2}\sum_{l,l'}\sum_{\alpha,\beta} A_{\alpha\beta}(l, l')[\hat{p}_\gamma(l''), \hat{u}_\alpha(l)\hat{u}_\beta(l')]\\ &= -\frac{i\hbar}{2}\left\{\sum_{\beta,l'} A_{\gamma\beta}(l'', l')\hat{u}_\beta(l') + \sum_{\alpha,l} A_{\alpha\gamma}(l, l'')\hat{u}_\alpha(l)\right\}\\ &= -i\hbar\sum_{\beta,l'} A_{\gamma\beta}(l'', l')\hat{u}_\beta(l'),\end{aligned}$$

where in the last line we have used the identity

$$A_{\alpha\beta}(l, l') = A_{\beta\alpha}(l', l),$$

the equation of motion for the Green function on the right-hand side of (4.147) is

$$\hbar\omega\langle\langle \hat{p}_\alpha(l); \hat{u}_\beta(l')\rangle\rangle = -i\hbar\, \delta_{\alpha\beta}\, \delta_{ll'} - i\hbar\sum_{\gamma,l''} A_{\alpha\gamma}(l, l'')P_{\gamma\beta}(l'', l'; \omega). \tag{4.149}$$

If the result (4.149) is substituted into eqn (4.147) we have, finally,

$$M\omega^2 P_{\alpha\beta}(l, l'; \omega) = \delta_{\alpha\beta}\, \delta_{l,l'} + \sum_{\gamma,l''} A_{\alpha\gamma}(l, l'')P_{\gamma\beta}(l'', l'; \omega),$$

or

$$\sum_{\gamma,l''}\{M\omega^2\, \delta_{\alpha,\gamma}\, \delta_{l,l''} - A_{\alpha\gamma}(l, l'')\}P_{\gamma\beta}(l'', l'; \omega) = \delta_{\alpha\beta}\, \delta_{ll'}. \tag{4.150}$$

To solve this equation we expand $P_{\alpha\beta}(l, l'; \omega)$ in terms of the eigenvectors of the force-constant matrix $A_{\alpha\beta}(l, l'')$. These are discussed in detail in § 4.1. We recall that the polarization vectors $\boldsymbol{\sigma}^j(\mathbf{q})$ are real and are defined such that (cf. (4.10))

$$\frac{1}{M}\sum_{l'',\gamma} A_{\alpha\gamma}(l, l'')\exp(\mathrm{i}\mathbf{q}\cdot\mathbf{l}'')\sigma^j_\gamma(\mathbf{q}) = \omega_j^2(\mathbf{q})\exp(\mathrm{i}\mathbf{q}\cdot\mathbf{l})\sigma^j_\beta(\mathbf{q}), \qquad (4.151)$$

the the normalization

$$\sum_j \sigma^j_\alpha(\mathbf{q})\sigma^j_\beta(\mathbf{q}) = \delta_{\alpha,\beta} \qquad (4.152)$$

and

$$\sum_\alpha \sigma^j_\alpha(\mathbf{q})\sigma^{j'}_\alpha(\mathbf{q}) = \delta_{j,j'}.$$

In (4.151), $\omega_j(\mathbf{q})$ is the frequency of the $j$th phonon branch.

From (4.151) it follows that

$$\sum_{l'',\gamma} \{M\omega^2\,\delta_{l,l''}\,\delta_{\alpha\gamma} - A_{\alpha\gamma}(l, l'')\}\exp(\mathrm{i}\mathbf{q}\cdot\mathbf{l}'')\sigma^j_\gamma(\mathbf{q})$$
$$= M\{\omega^2 - \omega_j^2(\mathbf{q})\}\exp(\mathrm{i}\mathbf{q}\cdot\mathbf{l})\sigma^j_\alpha(\mathbf{q}), \qquad (4.153)$$

so that, if

$$P_{\alpha\beta}(l, l'; \omega) = \frac{1}{NM}\sum_{j,\mathbf{q}} \sigma^j_\alpha(\mathbf{q})\sigma^j_\beta(\mathbf{q})\exp\{\mathrm{i}\mathbf{q}\cdot(\mathbf{l}-\mathbf{l}')\}\mathscr{P}^j(\mathbf{q}) \qquad (4.154)$$

then, from eqn (4.150), $\mathscr{P}^j(\mathbf{q})$ satisfies

$$\frac{1}{N}\sum_{j,\mathbf{q}} \{\omega^2 - \omega_j^2(\mathbf{q})\}\sigma^j_\alpha(\mathbf{q})\sigma^j_\beta(\mathbf{q})\exp\{-\mathrm{i}\mathbf{q}\cdot(\mathbf{l}-\mathbf{l}')\}\mathscr{P}^j(\mathbf{q}) = \delta_{\alpha,\beta}\,\delta_{l,l'}. \qquad (4.155)$$

From the orthogonality relation (4.152) it then follows that

$$\mathscr{P}^j(\mathbf{q}) = \{\omega^2 - \omega_j^2(\mathbf{q})\}^{-1}. \qquad (4.156)$$

According to the convention adopted in Appendix B we evaluate the imaginary part of the Green function by giving the energy a small, positive imaginary part and then take the limit as this goes to zero. Hence

$$\mathscr{P}^j(\mathbf{q}) = \lim_{\epsilon\to 0^+} \frac{1}{(\omega + \mathrm{i}\epsilon)^2 - \omega_j^2(\mathbf{q})}$$
$$= \frac{1}{2\omega_j(\mathbf{q})}\lim_{\epsilon\to 0^+}\left\{\frac{1}{\omega + \mathrm{i}\epsilon - \omega_j(\mathbf{q})} - \frac{1}{\omega + \mathrm{i}\epsilon + \omega_j(\mathbf{q})}\right\} \qquad (4.157)$$

and

$$\mathrm{Im}\,\mathscr{P}^j(\mathbf{q}) = -\frac{\pi}{2\omega_j(\mathbf{q})}[\delta\{\omega - \omega_j(\mathbf{q})\} - \delta\{\omega + \omega_j(\mathbf{q})\}].$$

It is readily verified that we recover the result (4.70) from (4.146), (4.154), and (4.157).

It is not difficult to show that the equation of motion of the Green function of the perturbed system, $G_{\alpha\beta}(l, l'; \omega)$, is

$$\sum_{\gamma,l''} \{M\omega^2 \delta_{\alpha\gamma} \delta_{l,l''} - A_{\alpha\gamma}(l, l'')\} G_{\gamma\beta}(l'', l'; \omega) = \delta_{\alpha\beta} \delta_{l,l'} + \sum_{\gamma,l''} V_{\alpha\gamma}(l, l'') G_{\gamma\beta}(l'', l'; \omega),$$

where

$$V_{\alpha\beta}(l, l') = M\lambda\omega^2 \sum_s \delta_{\alpha\beta} \delta_{l,s} \delta_{s,l'}. \tag{4.158}$$

If we write the equation of motion for the unperturbed Green function in matrix form, so that (4.150) reads

$$(M\omega^2 \mathbf{\mathcal{I}} - \mathbf{A})\mathbf{P} = \mathbf{\mathcal{I}}$$

or

$$(M\omega^2 \mathbf{\mathcal{I}} - \mathbf{A}) = \mathbf{P}^{-1},$$

then eqn (4.158) can be written

$$\mathbf{P}^{-1}\mathbf{G} = \mathbf{\mathcal{I}} + \mathbf{VG}, \tag{4.159}$$

i.e.

$$\mathbf{G} = \mathbf{P} + \mathbf{PVG}.$$

Eqn (4.159) is usually referred to as the Dyson equation for the Green function **G**. We have immediately

$$\mathbf{G} = \frac{1}{\mathbf{\mathcal{I}} - \mathbf{PV}} \cdot \mathbf{P} \tag{4.160}$$

so that the poles of the Green function are determined by the condition

$$\text{Re} \det |\mathbf{\mathcal{I}} - \mathbf{PV}| = 0. \tag{4.161}$$

We now examine eqns (4.159) and (4.161) in some detail for the particular case of a single impurity ion at $l = 0$, say. Eqn (4.159) reduces to

$$G_{\alpha\beta}(l, l'; \omega) = P_{\alpha\beta}(l, l'; \omega) + \lambda\omega^2 M \sum_\gamma P_{\alpha\gamma}(l, 0; \omega) G_{\gamma\beta}(0, l'; \omega). \tag{4.162}$$

For orthorhombic, tegragonal, and cubic Bravais crystals $P_{\alpha\beta}(l, l; \omega)$ is diagonal in $\alpha$, $\beta$. Furthermore, for cubic crystals $P_{\alpha\alpha}(l, l; \omega)$ is independent of $\alpha$. For these crystals it is convenient to define

$$\mathcal{P}(\omega) = MP_{\alpha\alpha}(l, l; \omega) = \frac{1}{3N} \sum_{j,\mathbf{q}} \frac{1}{\omega^2 - \omega_j^2(\mathbf{q})}. \tag{4.163}$$

The factor $\frac{1}{3}$ arises because, for all $\alpha$, $|\sigma^j_\alpha|^2 = \frac{1}{3}$ by virtue of eqn (4.152).

With the aid of these results we have immediately

$$G_{\alpha\beta}(0, l'; \omega) = \frac{P_{\alpha\beta}(0, l'; \omega)}{1 - \lambda\omega^2\mathcal{P}(\omega)} \tag{4.164}$$

and thus

$$G_{\alpha\beta}(l, l'; \omega) = P_{\alpha\beta}(l, l'; \omega) + \frac{\lambda\omega^2 M}{1 - \lambda\omega^2\mathcal{P}(\omega)} \sum_\gamma P_{\alpha\gamma}(l, 0; \omega)P_{\gamma\beta}(0, l'; \omega). \tag{4.165}$$

If $Z(\omega)$ is the vibrational density of states for the host lattice, normalized to unity,

$$\begin{aligned}\mathcal{P}(\omega) = \mathcal{P}'(\omega) + \mathrm{i}\mathcal{P}''(\omega) &= \lim_{\epsilon\to 0} \int_0^{\omega_\mathrm{m}} \mathrm{d}\omega' \frac{Z(\omega')}{(\omega + \mathrm{i}\epsilon)^2 - \omega'^2} \\ &= \mathrm{P}\int_0^{\omega_\mathrm{m}} \frac{\mathrm{d}\omega' Z(\omega')}{\omega^2 - \omega'^2} - \frac{\mathrm{i}\pi}{2\omega} Z(\omega),\end{aligned} \tag{4.166}$$

where P $\int$ denotes the principal part integral and $\omega_\mathrm{m}$ is the maximum frequency of the band.

If we compare (4.126) and $\mathcal{P}'(\omega)$ obtained from (4.166) we deduce that

$$\Delta(\omega) = 2\omega\mathcal{P}'(\omega). \tag{4.167}$$

We can therefore take advantage of some of the results of the previous subsection in our analysis of the response of the mixed crystal. The resonance frequencies in the latter are determined by the equation

$$1 - \lambda\omega_l^2\mathcal{P}'(\omega_l) = 1 - \frac{\lambda}{2}\omega_l\,\Delta(\omega_l) = 0 \tag{4.168}$$

or

$$(1/\lambda) = \tfrac{1}{2}\omega_l\,\Delta(\omega_l).$$

Because

$$(1/\lambda) = \frac{1}{1 - (M'/M)}$$

can only have values in the range $(\infty, 1)$, and $(0, -\infty)$, there is a range $0 \leqslant \lambda^{-1} \leqslant 1$ which there are no solutions of (4.168). For both $M'/M > 1$ and $M'/M < 1$, defect modes can exist within the frequency spectrum of the host, the so-called virtual resonance modes. These modes decay into the band because they interact with the vibrations of the host, the strength of the interaction being proportional to the density of states of the unperturbed host $Z(\omega)$. It therefore follows that the virtual modes are well defined only when they occur near the bottom of the frequency spectrum, i.e. for $M'/M \gg 1$. On the other hand, the so-called localized resonance modes, which occur at frequencies greater than the maximum frequency of the host spectrum, exist for only light mass defect ions, $M'/M < 1$. The wave functions associated with these states cannot propagate in the host, so that the disturbance caused by the impurity ion is well localized.

We will now focus attention on the scattering from the mass defect. The correlation function required in the cross-section (4.139) is $(l = l' = 0)$

$$\begin{aligned}
&\langle \boldsymbol{\kappa}\cdot\hat{\mathbf{u}}\boldsymbol{\kappa}\cdot\hat{\mathbf{u}}(t) - (\boldsymbol{\kappa}\cdot\hat{\mathbf{u}})^2\rangle \\
&= -\frac{\hbar}{\pi}\int_{-\infty}^{\infty} \mathrm{d}\omega\{\exp(\mathrm{i}\omega t) - 1\}\{1 + n(\omega)\}\sum_{\alpha\beta} \kappa_\alpha \kappa_\beta G''_{\alpha\beta}(0, 0; \omega) \\
&= -\frac{\hbar\kappa^2}{\pi M}\int_{-\infty}^{\infty} \mathrm{d}\omega\{\exp(\mathrm{i}\omega t) - 1\}\{1 + n(\omega)\}\,\mathrm{Im}\{\mathcal{P}(\omega)/[1 - \lambda\omega^2\mathcal{P}(\omega)]\},
\end{aligned} \tag{4.169}$$

where we have used (4.146) and (4.164).

A case of particular interest is $M' < M$ since this is appropriate for a proton inhabiting a crystal or macromolecule. We have just seen that a light mass defect gives rise to a localized mode. In consequence the integrand in (4.169) separates into two contributions, one from the lattice and a second high-frequency, undamped contribution that arises from the mass defect. The high-frequency contribution is calculated with the aid of the analysis presented in the preceding section. If $\omega_l$ is a solution of (4.168) and $\omega_l > \omega_m$, where $\omega_m$ is the maximum phonon frequency of the host crystal, then

$$\begin{aligned}
&\langle \boldsymbol{\kappa}\cdot\hat{\mathbf{u}}\boldsymbol{\kappa}\cdot\hat{\mathbf{u}}(t) - (\boldsymbol{\kappa}\cdot\hat{\mathbf{u}})^2\rangle \\
&= \frac{\hbar\kappa^2}{2M'\omega_l} h(\omega_l) Q(\omega_l, t) + \frac{\hbar\kappa^2}{2M}\int_0^{\omega_m} \frac{\mathrm{d}\omega\, Z(\omega) Q(\omega, t)/\omega}{|1 - \lambda\omega^2\mathcal{P}(\omega)|^2}
\end{aligned} \tag{4.170}$$

In this expression we have introduced the function

$$\begin{aligned}
Q(\omega, t) &= \mathrm{i}\sin(\omega t) + (\cos(\omega t) - 1)\coth(\tfrac{1}{2}\hbar\omega\beta) \\
&= -\coth(\tfrac{1}{2}\hbar\omega\beta) + \cosh\{\omega(\mathrm{i}t + \tfrac{1}{2}\hbar\beta)\}/\sinh\{\tfrac{1}{2}\hbar\omega\beta\},
\end{aligned} \tag{4.171}$$

and $h(\omega_l)$ is the amplitude of the resonance mode. If $M' = M$, so that $\lambda = 0$, (4.170) reduces to the result for the correlation function used in the one-phonon incoherent cross-section (4.79). For the amplitude $h(\omega_l)$ we find, following the steps that lead to (4.135),

$$h(\omega_l) = \left(\frac{M'}{M}\right)\{\tfrac{1}{2}\,\Delta(\omega_l)\}^2\left\{\int_0^{\omega_m} \frac{\mathrm{d}\omega\,\omega^2 Z(\omega)}{(\omega_l^2 - \omega^2)^2}\right\}^{-1}. \tag{4.172}$$

In the limit $(M'/M) \to 0$ the amplitude tends to unity, and the correlation function (4.170) reduces to the result for a single undamped harmonic oscillator of natural frequency $\omega_l$ because the lattice contribution is minimal. The response function for a harmonic oscillator is discussed in detail in § 3.6.2. An essential step in the discussion is to recognize that the generating function for Bessel functions $I_n(y)$ of order $n$, where $n$ is an integer, is

$$\exp\{y\cosh(x)\} = \sum_{n=-\infty}^{\infty} \exp(nx) I_n(y). \tag{4.173}$$

Applied to the first term of (4.170) and using the second form of $Q(\omega, t)$ given in (4.171) we find

$$\exp\{AQ(\omega_l, t)\} = \exp\{-A\coth(\tfrac{1}{2}\hbar\omega_l\beta)\} \times \sum_{n=-\infty}^{\infty} I_n(A/\sinh(\tfrac{1}{2}\hbar\omega_l\beta))\exp\{n\omega_l(\mathrm{i}t + \tfrac{1}{2}\hbar\beta)\}, \tag{4.174}$$

where $A$ stands for the coefficient of $Q(\omega_l, t)$ in (4.170). When we take the Fourier transform of (4.174) to obtain the response function we obtain delta functions, $\delta(\omega - n\omega_l)$.

The response function for the mass defect

$$\begin{aligned} S(\kappa, \omega) &= \frac{1}{2\pi\hbar}\int_{-\infty}^{\infty} \mathrm{d}t \exp(-\mathrm{i}\omega t)\exp\langle \boldsymbol{\kappa}\cdot\hat{\mathbf{u}}\boldsymbol{\kappa}\cdot\hat{\mathbf{u}}(t) - (\boldsymbol{\kappa}\cdot\hat{\mathbf{u}})^2\rangle \\ &= \frac{1}{2\pi\hbar}\int_{-\infty}^{\infty} \mathrm{d}t \exp(-\mathrm{i}\omega t) I_r(\kappa, t) I_l(\kappa, t) \\ &= \hbar\int_{-\infty}^{\infty} \mathrm{d}\omega' S_r(\kappa, \omega') S_l(\kappa, \omega - \omega'), \end{aligned} \tag{4.175}$$

where subscripts denote the resonance and lattice contributions, and the final expression results from the standard convolution theorem for Fourier transforms. We know from § 3.6.2 and eqn (4.174) that for modest values of $(\hbar\kappa^2/M'\omega_l)$ the response function $S_r(\kappa, \omega')$ consists of a sum of delta functions with arguments $(\omega' - n\omega_l)$ where the integer $n = 0$, $\pm 1$, $\pm 2$, etc. From (4.175) we conclude that the full response function is the sum of terms of the form $S_l(\kappa, \omega - n\omega_l)$, i.e. the shape of the response is derived from the lattice and it is located at frequencies that are multiples of $\omega_l$. The lattice response will not be well approximated by a simple Gaussian function for modest values of $\kappa^2$; indeed it is likely to mirror the lattice density of states which is far from being a smooth function of the frequency (cf. Fig. 4.3). An illustrative example of the inelastic response of a light particle is shown in Fig. 4.5 as a function of frequency for two values of the scattering vector. The structure in the

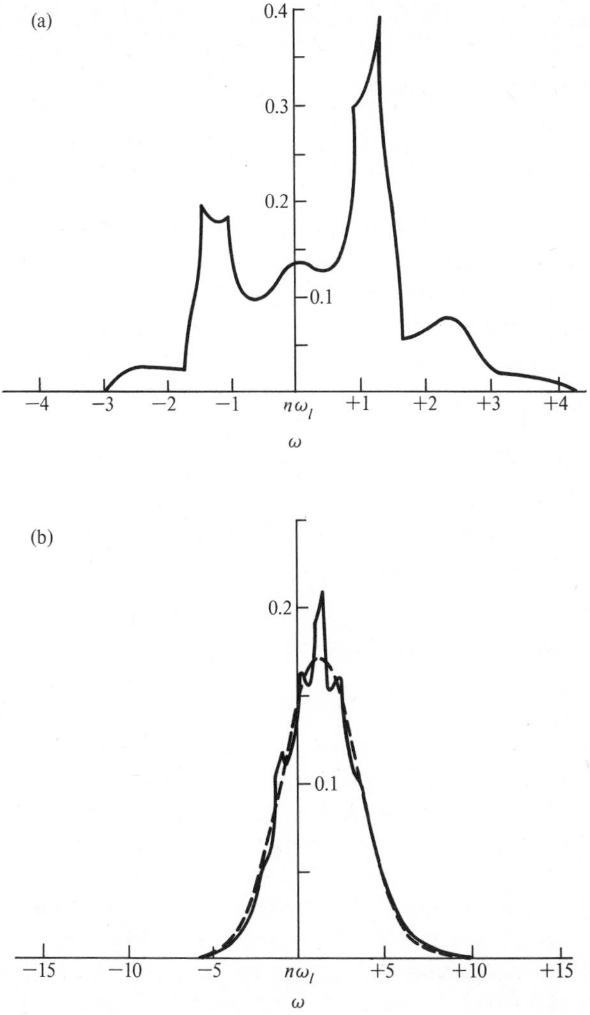


FIG. 4.5. Exact numerical results for the inelastic harmonic response of a light particle embedded in a lattice; reduced units are used, and the scattering vector in (b) is twice the value in (a). The dashed curve in (b) is obtained from the impulse approximation (4.176) (Warner *et al.* 1983).

spectra is essentially an image of the lattice density of states employed in the calculation apart from contributions in (a) at the reduced frequencies ±2.5 that arise from two-phonon processes.

The representation of $S_r(\kappa, \omega')$ as the sum of delta functions is not useful for large values of $\kappa^2$ because each delta function has essentially the same weight. The limit $\kappa^2 \to \infty$ coincides with the impulse approximation discussed in § 3.4. In this instance the response function tends to

$$S(\kappa, \omega) = \frac{1}{2\pi\hbar}\int_{-\infty}^{\infty} dt \exp\left\{-i\omega t + \frac{it\hbar\kappa^2}{2M'} - \frac{t^2}{2}\langle(\boldsymbol{\kappa}\cdot\hat{\mathbf{v}})^2\rangle\right\}$$

$$= \{2\pi\hbar^2\langle(\boldsymbol{\kappa}\cdot\hat{\mathbf{v}})^2\rangle\}^{-1/2} \exp\left\{-\left(\omega - \frac{\hbar\kappa^2}{2M'}\right)^2 \Big/ 2\langle(\boldsymbol{\kappa}\cdot\hat{\mathbf{v}})^2\rangle\right\}, \quad (4.176)$$

where $\hat{\mathbf{v}}$ is the velocity of the particle. Note that the recoil energy in (4.176) involves the mass of the defect particle and that the result, valid for $\kappa^2 \to \infty$, is not restricted to the case of a light mass defect.

A general result is the sum rule,

$$\int_{-\infty}^{\infty} d\omega\, \omega S(\kappa, \omega) = (\kappa^2/2M'). \quad (4.177)$$

Applied to the response of a light mass defect using (4.170) we deduce that

$$\int_0^{\omega_m} \frac{d\omega\, Z(\omega)}{|1 - \lambda\omega^2\mathscr{P}(\omega)|^2} = \left(\frac{M}{M'}\right)\{1 - h(\omega)\}. \quad (4.178)$$

This expression can be interpreted as the normalization condition for an effective lattice density of states

$$Z(\omega)/|1 - \lambda\omega^2\mathscr{P}(\omega)|^2. \quad (4.179)$$

The total number of degrees of freedom is shared between the lattice and the resonance mode, and (4.178) gives the relative distribution.

It is instructive to compare our exact results with an approximate treatment that appeals to a separation of time scales for the lattice and defect particle responses. When the frequency of the light defect is much larger than the maximum lattice frequency $\omega_m$ we should expect the correlation function to factor into contributions from the lattice and the defect. For a single mass defect this factorization is confirmed by the exact result (4.170) that leads to the representation (4.175). The approximation would therefore stem from using expressions

$$\mathrm{I}_r(\kappa, t) \doteq \exp\left\{\frac{\hbar\kappa^2}{2M'\omega_l} Q(\omega_l, t)\right\} \quad (4.180)$$

and

$$\mathrm{I}_l(\kappa, t) \doteq \exp\left\{\frac{\hbar\kappa^2}{2M}\int_0^{\omega_m} d\omega \frac{Z(\omega)}{\omega} Q(\omega, t)\right\}. \quad (4.181)$$

The error amounts to a neglect of the amplitude $h(\omega_l)$ in the exponent of (4.180), and the use of the lattice density of states in (4.181) instead of the effective density of states (4.179).

The approximation (4.180)–(4.181) overestimates the width of the peaks in $S(\kappa, \omega)$ for modest values of $(\hbar\kappa^2/M'\omega_l)$ because there is too much weight in the integral in the exponent of (4.181). A measure of the width of $S_l(\kappa, \omega)$ can be obtained from the coefficient of $t^2$ in the exponent of (4.181), as can be seen in (4.176), namely

$$\Gamma_l' \doteq \left\{\frac{\hbar\kappa^2}{2M}\int_0^{\omega_m} d\omega\, \omega Z(\omega)\coth(\tfrac{1}{2}\hbar\omega\beta)\right\}^{1/2} = \left\{\frac{\kappa^2}{M\beta}\int_0^{\omega_m} d\omega\, Z(\omega)\right\}^{1/2}, \tag{4.182}$$

where the final expression is appropriate at high temperatures, $\hbar\omega_m\beta < 1$. The integral in (4.182) is the normalization of the density of states which is unity. If we follow the development that leads to (4.182) using the exact lattice response, then $Z(\omega)$ is replaced by the effective density of states (4.179), and the integral which is unity in (4.182) is then the right-hand side of (4.178), i.e.

$$\Gamma_l = \left\{\frac{\kappa^2}{M\beta}\left(\frac{M}{M'}\right)[1 - h(\omega_l)]\right\}^{1/2} \tag{4.183}$$

which is smaller than or equal to (4.182).

We will pursue our discussion of the approximate and exact response of a light mass defect by modelling the lattice density of states by the Debye spectrum (eqn (4.137)). The amplitude $h(\omega)$ defined in (4.172) is found to be $(x = (\omega/\omega_D) > 1)$

$$h(\omega) = \frac{\omega}{2}\Delta(\omega)\left\{\frac{\omega}{2}\Delta(\omega) - 1\right\}\Big/ g(x) \tag{4.184}$$

where $\Delta(\omega)$ is given in (4.138) and

$$g(x) = \tfrac{3}{2}x^2\left\{3 + (x^2 - 1)^{-1} - \frac{3x}{2}\ln\left(\frac{x+1}{x-1}\right)\right\}. \tag{4.185}$$

The function (4.184) vanishes for $\omega \to \omega_D$, and approaches unity for large $\omega$. To study the quantity on the right-hand side of (4.178) we express the ratio of the masses in terms of $\Delta(\omega)$ by using the resonance condition (4.168). For $\omega_l \to \omega_D$, achieved with $M' \to M$, the quantity tends to unity, while for $\omega_l \to \infty (M'/M \to 0)$ it saturates to the value $4/21 = 0.19$. We conclude that, for a Debye spectrum, the width produced with an approximate treatment of the lattice response based on (4.181) is too large by a factor $\sim 2.3$ for $(M'/M) \to 0$. Observe that we might expect the approximation (4.180)–(4.181) to be most accurate for $\omega_l \gg \omega_m$ but it appears that the widths are much too large even in this apparently most favourable case.

### 4.7.3. *Small concentration of mass defects*

We need to know the influence of a small concentration of defects on the one-phonon cross-sections for essentially one of two reasons, either to assess the effects of contamination of the target sample on the phonon cross-sections or because of an explicit interest in the dynamics of defects.

Our discussion of the one-phonon coherent and incoherent cross-sections is based on the single-defect problem discussed in the preceding subsection. If the disturbance caused by the mass defects is very small, because the concentration is very small for example, then a good approximation to the cross-sections is the exact single-defect cross-section multiplied by the defect concentration. This calculation is relatively straightforward given a knowledge of the displacement Green function for all sites in the lattice (eqn (4.165)). The result contains the rich frequency structure which stems from localized and virtual resonance modes. However, there is no allowance for the finite lifetime of the lattice phonons which results from scattering by the mass defects. Lifetime effects can be included for low concentrations by using a $t$-matrix theory, which is derived from the exact solution of the single-defect problem. Our entire discussion is based on a mass defect model with no allowance for changes in the force constants.

If we define

$$\mathscr{b}_l = \bar{b}_l \exp[-\tfrac{1}{2}\langle\{\boldsymbol{\kappa}\cdot\hat{\mathbf{u}}(l)\}^2\rangle], \tag{4.186}$$

then the one-phonon cross-sections are

$$\left(\frac{\mathrm{d}^2\sigma}{\mathrm{d}\Omega\,\mathrm{d}E'}\right)^{(1)}_{\mathrm{coh}} = \frac{k'}{k}\frac{1}{2\pi\hbar}\int_{-\infty}^{\infty} \mathrm{d}t\,\exp(-\mathrm{i}\omega t)\sum_{l,l'} \mathscr{b}_l \mathscr{b}_{l'} \exp\{-\mathrm{i}\boldsymbol{\kappa}\cdot(\mathbf{l}-\mathbf{l}')\}$$

$$\times \sum_{\alpha,\beta} \kappa_\alpha \kappa_\beta \langle \hat{u}_\alpha(l)\hat{u}_\beta(l',t)\rangle \tag{4.187}$$

and

$$\left(\frac{\mathrm{d}^2\sigma}{\mathrm{d}\Omega\,\mathrm{d}E'}\right)^{(1)}_{\mathrm{incoh}} = \frac{k'}{k}\frac{1}{2\pi\hbar}\int_{-\infty}^{\infty} \mathrm{d}t\,\exp(-\mathrm{i}\omega t)\sum_{l} \alpha_l^2 \sum_{\alpha,\beta} \kappa_\alpha \kappa_\beta \langle \hat{u}_\alpha(l)\hat{u}_\beta(l,t)\rangle, \tag{4.188}$$

where

$$\alpha_l^2 = (\overline{b_l^2} - \bar{b}_l^2)\exp[-\langle\{\boldsymbol{\kappa}\cdot\hat{\mathbf{u}}(l)\}^2\rangle]. \tag{4.189}$$

and the coherent cross-section in terms of the Green function (4.144) is,

$$\left(\frac{\mathrm{d}^2\sigma}{\mathrm{d}\Omega\,\mathrm{d}E'}\right)^{(1)}_{\mathrm{coh}}$$

$$= -\frac{1}{\pi}\frac{k'}{k}\sum_{l,l'}\sum_{\alpha,\beta} \kappa_\alpha\kappa_\beta \mathscr{b}_l \mathscr{b}_{l'} \exp\{-\mathrm{i}\boldsymbol{\kappa}\cdot(\mathbf{l}-\mathbf{l}')\}\{1+n(\omega)\}G''_{\alpha\beta}(l,l';\omega). \tag{4.190}$$

Similarly for the incoherent cross-section.

If the impurity ions are allowed to have a scattering length that differs from that of the ions, then we need to calculate not only

$$\sum_{l,l'} \exp\{-i\boldsymbol{\kappa}\cdot(\mathbf{l}-\mathbf{l}')\}G_{\alpha\beta}(l,l';\omega)$$

but also the same expression with either one lattice vector restricted to defect sites or both lattice vectors restricted to just defect sites. To see this, let

$$b_l = b_h + p_l(b_d - b_h), \tag{4.191}$$

where $p_l$ is unity or zero according to whether $l$ is a defect or host site. Then

$$\begin{aligned}\sum_{l,l'} b_l b_{l'} \exp\{-i\boldsymbol{\kappa}\cdot(\mathbf{l}-\mathbf{l}')\}G_{\alpha\beta}(l,l';\omega)\\ = b_h^2\sum_{l,l'} \exp\{-i\boldsymbol{\kappa}\cdot(\mathbf{l}-\mathbf{l}')\}G_{\alpha\beta}(l,l';\omega)\\ +2b_h(b_d-b_h)\sum_{l,l'} p_l \exp\{-i\boldsymbol{\kappa}\cdot(\mathbf{l}-\mathbf{l}')\}G_{\alpha\beta}(l,l';\omega)\\ +(b_d-b_h)^2\sum_{l,l'} p_l p_{l'} \exp\{-i\boldsymbol{\kappa}\cdot(\mathbf{l}-\mathbf{l}')\}G_{\alpha\beta}(l,l';\omega).\end{aligned} \tag{4.192}$$

For a single impurity the function $p_l$ becomes a Kronecker delta function $\delta_{l,0}$ so that to calculate the cross-section we need

$G_{\alpha\beta}(0,0;\omega)$ and $\sum_l \exp(i\boldsymbol{\kappa}\cdot\mathbf{l})G_{\alpha\beta}(0,l;\omega)$ in addition to

$$\sum_{l,l'} \exp\{-\boldsymbol{\kappa}\cdot(\mathbf{l}-\mathbf{l}')\}G_{\alpha\beta}(l,l';\omega).$$

However, the first two functions are readily calculated from eqn (4.164) and the latter from eqn (4.165). We find ($\sigma_\alpha^j(\mathbf{q})$ is real for a cubic Bravais lattice)

$$\begin{aligned}\sum_{l,l'} \exp\{-i\boldsymbol{\kappa}\cdot(\mathbf{l}-\mathbf{l}')\}\{b_l b_{l'} G_{\alpha\beta}(l,l';\omega) - b_h^2 P_{\alpha\beta}(l,l';\omega)\}\\ = \frac{1}{M}\sum_j \sigma_\alpha^j(\boldsymbol{\kappa})\sigma_\beta^j(\boldsymbol{\kappa})[b_h^2\lambda\omega^2\{\mathcal{P}^j(\boldsymbol{\kappa})\}^2\\ +2b_h(b_d-b_h)\mathcal{P}^j(\boldsymbol{\kappa})+(b_d-b_h)^2\mathcal{P}(\omega)]\{1-\lambda\omega^2\mathcal{P}(\omega)\}^{-1}.\end{aligned} \tag{4.193}$$

The frequency of the localized mode $\omega_l(>\omega_m)$ is given by

$$1-\lambda\omega_l^2\mathcal{P}(\omega_l)=0,$$

and to evaluate the cross-section we need

$$\operatorname{Im}\frac{1}{1-\lambda\omega_l^2\mathcal{P}(\omega_l)} = \frac{-\pi}{2\lambda\omega_l B(\omega_l)}\{\delta(\omega-\omega_l)-\delta(\omega+\omega_l)\}. \tag{4.194}$$

and

$$\frac{d}{d\omega}\{1-\lambda\omega^2\mathscr{P}(\omega)\}=2\lambda\omega B(\omega),$$

where

$$B(\omega)=\int_0^{\omega_m}\frac{d\omega'\omega'^2Z(\omega')}{(\omega^2-\omega'^2)^2}\quad(\omega>\omega_m). \tag{4.195}$$

The imaginary part of eqn (4.193) for $\omega>\omega_m$ is, therefore,

$$\frac{-\pi}{2M\lambda\omega_l B(\omega_l)}\{\delta(\omega-\omega_l)-\delta(\omega+\omega_l)\}\lambda\omega_l^2\sum_j\sigma_\alpha^j(\boldsymbol{\kappa})\sigma_\beta^j(\boldsymbol{\kappa})\left[b_h\mathscr{P}^j(\boldsymbol{\kappa})+\frac{b_d-b_h}{\lambda\omega_l^2}\right]^2, \tag{4.196}$$

where we have made use of eqn (4.168) to eliminate $\mathscr{P}(\omega_l)$.

Thus the one-phonon coherent cross-section for the single-impurity case when the neutron energy change is greater than $\hbar\omega_m$ is, from (4.190) and (4.196),

$$\left(\frac{d\sigma^2}{d\Omega\,dE'}\right)^{(1)}_{coh}=\frac{k'}{2kM}[\{1+n(\omega_l)\}\,\delta(\omega-\omega_l)+n(\omega_l)\,\delta(\omega+\omega_l)]$$
$$\times\frac{\omega_l}{B(\omega_l)}\sum_j\{\boldsymbol{\kappa}\cdot\boldsymbol{\sigma}^j(\boldsymbol{\kappa})\}^2\left\{\frac{b_h}{\omega_l^2-\omega_j^2(\boldsymbol{\kappa})}+\frac{b_d-b_h}{\lambda\omega_l^2}\right\}^2. \tag{4.197}$$

If this expression is multiplied by the number of impurities $Nc$, we have an approximate expression for a small finite concentration of impurities.

The cross-section given by eqn (4.197) has a simple form if the scattering vector $\boldsymbol{\kappa}$ lies along one of the symmetry directions of the cubic lattice, i.e. along either the $(1,0,0)$, $(1,1,0)$, or $(1,1,1)$ directions. For lattice waves propagating along these directions are purely longitudinal and purely transverse, so that $\boldsymbol{\kappa}\cdot\boldsymbol{\sigma}^j(\boldsymbol{\kappa})=0$ unless $j$ refers to the longitudinal branch.

The calculation of the one-phonon incoherent cross-section for the single-impurity case when $\omega>\omega_m$, follows along similar lines to that given above for the coherent cross-section. If

$$\alpha_l^2=\alpha_h^2+(\alpha_d^2-\alpha_h^2)\,\delta_{l,0},$$

then

$$\left(\frac{d\sigma^2}{d\Omega\,dE'}\right)^{(1)}_{incoh}=-\frac{1}{\pi}\frac{k'}{k}\sum_{\alpha,\beta}\kappa_\alpha\kappa_\beta\{1+n(\omega)\}$$
$$\times\left\{\alpha_h^2\sum_l \text{Im}\,G_{\alpha\beta}(l,l;\omega)+(\alpha_h^2-\alpha_d^2)\text{Im}\,G_{\alpha\beta}(0,0;\omega)\right\}. \tag{4.198}$$

The term in curly brackets is

$$\delta_{\alpha\beta}\left\{\alpha_{\rm h}^2\,{\rm Im}\left[\frac{\lambda\omega^2}{1-\lambda\omega^2\mathcal{P}(\omega)}\frac{1}{M}\frac{1}{3N}\sum_{j,\mathbf{q}}\{\mathcal{P}^j(\mathbf{q})\}^2\right]+(\alpha_{\rm d}^2-\alpha_{\rm h}^2){\rm Im}\left[\frac{\mathcal{P}(\omega)/M}{1-\lambda\omega^2\mathcal{P}(\omega)}\right]\right\}.$$

But

$$\frac{1}{3N}\sum_{j,\mathbf{q}}\{\mathcal{P}^j(\mathbf{q})\}^2=-\frac{1}{2\omega}\frac{\mathrm{d}\mathcal{P}(\omega)}{\mathrm{d}\omega}$$

so that (4.198) is

$$\left(\frac{\mathrm{d}\sigma^2}{\mathrm{d}\Omega\,\mathrm{d}E'}\right)^{(1)}_{\rm incoh}=\frac{k'}{2kM}\frac{\kappa^2}{B(\omega_l)}[\{1+n(\omega_l)\}\,\delta(\omega-\omega_l)+n(\omega_l)\,\delta(\omega+\omega_l)]$$
$$\times\left\{\frac{\alpha_{\rm d}^2-\alpha_{\rm h}^2}{\lambda^2\omega_l^3}-\frac{\alpha_{\rm h}^2}{2}\left(\frac{\mathrm{d}\mathcal{P}(\omega)}{\mathrm{d}\omega}\right)\bigg|_{\omega=\omega_l}\right\}. \qquad (4.199)$$

The preceding results do not incorporate damping of the lattice phonons due to the defects. We begin our discussion of the phonon lifetime, and the concomitant shift in the phonon frequency, by calculating the configurationally averaged Green function. The result, eqn (4.202), is exact to leading order in the concentration of the mass defects $c$.

Let us assume that the disturbance on the host lattice about each defect is strongly localized so that the cluster formed by a defect and its associated disturbance can be regarded as an isolated scatterer of lattice waves. The probability of a lattice wave scattering from one cluster is proportional to $c$, the probability that it suffers scattering by a second cluster is proportional to $c^2$, and so on and so forth. These scattering events are additive in the displacement Green function, at least for very small $c$, as can be seen from the exact solution for a single scatterer (4.165). If we expand $G_{\alpha\beta}(l, l'; \omega)$ in terms of the polarization vectors as in (4.154), then we find from (4.165) that the spatial Fourier transform of the exact Green function is

$$G^j(\mathbf{q})=\mathcal{P}^j(\mathbf{q})\left\{1+\frac{1}{N}T(\omega)\mathcal{P}^j(\mathbf{q})\right\} \qquad (4.200)$$

where the $t$-matrix is

$$T(\omega)=\lambda\omega^2/\{1-\lambda\omega^2\mathcal{P}(\omega)\}. \qquad (4.201)$$

Because $c=1/N$ for a single defect, (4.200) shows that a single scattering event adds a term to the Green function which is proportional to $c\{\mathcal{P}^j(\mathbf{q})\}^2$. The Green function for a finite concentration of defects has full translational invariance after it has been averaged over the configuration of defects, and it can therefore be spatially Fourier transformed in terms of a function $\bar{G}^j(\mathbf{q})$, where the bar denotes the configurational average. The foregoing discussion and the form of the Green function for a single defect (4.200) lead to the result

$$\begin{aligned}\bar{G}^i(\mathbf{q}) &= \mathscr{P}^i(\mathbf{q})\{1+cT(\omega)\mathscr{P}^i(\mathbf{q})+[cT(\omega)\mathscr{P}^i(\omega)]^2+\cdots\}\\ &= \mathscr{P}^i(q)\{1-cT(\omega)\mathscr{P}^i(\mathbf{q})\}^{-1}\\ &= \{\omega^2-\omega_j^2(\mathbf{q})-cT(\omega)\}^{-1}.\end{aligned} \tag{4.202}$$

The dispersion of the collective modes of the mixed system are now determined by solutions of the equation

$$\omega_0^2=\omega_j^2(\mathbf{q})+c\ \mathrm{Re}\ T(\omega_0). \tag{4.203}$$

If the width of the mode $\Gamma(\omega_0)$ is small compared to $\omega_j(\mathbf{q})$, then to a good approximation

$$\Gamma(\omega_0)=\frac{\pi c\lambda^2\omega_0^2 Z(\omega_0)/4}{\{1-\omega_0^2\mathscr{P}'(\omega_0)\}^2+\{\lambda\omega_0^2\mathscr{P}''(\omega_0)\}^2}. \tag{4.204}$$

Here we have expressed $\mathscr{P}''(\omega)$ in the numerator in terms of the density of states using (4.166). Notice that the width induced by the impurities (4.204) is proportional to the unperturbed phonon density of states and the impurity concentration, and that it is explicitly of second order in the perturbation parameter $\lambda$. The width, or decay constant, depends on the phonon wave vector through $\omega_0$.

One shortcoming of the previous calculation is that it allows more than one defect at a given site. When this is corrected, the $t$-matrix depends on the concentration. The following discussion leads to the desired result, and generates some results that are required in the calculation of the neutron cross-section.

First we simplify our notation. For the purpose of the following discussion the Cartesian component labels $\alpha$, $\beta$ and the frequency argument of the Green function are redundant. Also, we denote the site positions simply by 1, 2, 3, etc.

It is to be expected that the behaviour of the Green function $G(1,2)$ is mainly determined by whether or not either, or both, its site labels 1 and 2 refer to defect sites. The first site label of the Green function on the right-hand side of eqn (4.159) always refers to a defect site. We define two additional Green functions according to whether their first site label refers to defect or host sites: namely

$$\begin{aligned}G(1,2)&=G^{\mathrm{d}}(1,2) \qquad \text{(defect at site 1)},\\ G(1,2)&=G^{\mathrm{h}}(1,2) \qquad \text{(host atom at site 1)}.\end{aligned} \tag{4.205}$$

Thus eqn (4.159) becomes

$$\mathbf{G}=\mathbf{P}+M\lambda\omega^2\mathbf{P}\mathbf{G}^{\mathrm{d}}. \tag{4.206}$$

The equations for $G^{\mathrm{d}}$ and $G^{\mathrm{h}}$ are

$$G^{\mathrm{d}}(1,2)=P(1,2)+M\lambda\omega^2\sum_3 P(1,3)G^{\mathrm{d}}(3,2) \tag{4.207}$$

and

$$G^{\mathrm{h}}(1,2)=P(1,2)+M\lambda\omega^2\sum_3 P(1,3)G^{\mathrm{d}}(3,2)-M\lambda\omega^2P(1,1)G^{\mathrm{d}}(1,2). \tag{4.208}$$

As in the previous subsection, we are concerned with cubic Bravais lattices; thus $\mathscr{P}=MP(1,1)$.

If we assume that $G$ is completely determined by the type of atom at site 1 we can equate the two summations in (4.207) and (4.208); this neglects the influence of the configuration of all the remaining defects in the system. This approximation gives immediately the following relationship between $G^{\mathrm{d}}$ and $G^{\mathrm{h}}$:

$$\{1-\lambda\omega^2\mathscr{P}(\omega)\}G^{\mathrm{d}}(1,2)=G^{\mathrm{h}}(1,2). \tag{4.209}$$

Further

$$\bar{G}(1,2)=cG^{\mathrm{d}}(1,2)+(1-c)G^{\mathrm{h}}(1,2) \tag{4.210}$$

is the Green function averaged over all configurations of the defect atoms. (4.209) and (4.210) give

$$G^{\mathrm{d}}(1,2)=\{1-(1-c)\lambda\omega^2\mathscr{P}(\omega)\}^{-1}\bar{G}(1,2), \tag{4.211}$$

which together with (4.206) leads to

$$\bar{G}(1,2)=P(1,2)+\frac{M\lambda\omega^2}{1-(1-c)\lambda\omega^2\mathscr{P}(\omega)}c\sum_3 P(1,3)\bar{G}(3,2), \tag{4.212}$$

where the summation over defect sites has been replaced by $c$ times the summation over all sites. If

$$T'(\omega)=\frac{\lambda\omega^2}{1-(1-c)\lambda\omega^2\mathscr{P}(\omega)}, \tag{4.213}$$

then eqn (4.212) in full is

$$\bar{G}_{\alpha\beta}(l,l')=P_{\alpha\beta}(l,l')+cT'M\sum_{\gamma,l''}P_{\alpha\gamma}(l,l'')\bar{G}_{\gamma\beta}(l'',l'). \tag{4.214}$$

The configurational averaging has restored the translational aymmetry, so that eqn (4.214) can be solved for $\bar{G}_{\alpha\beta}(l,l')$ by expressing the latter in terms of the polarization vectors of the force-constant matrix. Thus we define

$$\bar{G}_{\alpha\beta}(l, l') = \frac{1}{N}\sum_{\mathbf{q}} \exp(-\mathrm{i}\mathbf{q}\cdot\mathbf{l})\bar{G}_{\alpha\beta}(\mathbf{q})\exp(\mathrm{i}\mathbf{q}\cdot\mathbf{l}') \tag{4.215}$$

and

$$\bar{G}_{\alpha\beta}(\mathbf{q}) = \frac{1}{M}\sum_{j} \sigma^{j}_{\alpha}(\mathbf{q})\sigma^{j}_{\beta}(\mathbf{q})\bar{G}^{j}(\mathbf{q}). \tag{4.216}$$

From eqn (4.214) we deduce that

$$\bar{G}^{j}(\mathbf{q}) = \mathscr{P}^{j}(\mathbf{q}) + cT'(\omega)\bar{G}^{j}(\mathbf{q})$$

or

$$\bar{G}^{j}(\mathbf{q}) = \{\omega^2 - \omega_j^2(\mathbf{q}) - cT'(\omega)\}^{-1}. \tag{4.217}$$

This result is an improvement on (4.202). Using the expression (4.213) for the improved $t$-matrix we find that the width is proportional to $c(1-c)$ and therefore vanishes for a pure lattice of particles of either mass, i.e. $\Gamma(\omega_0)$ vanishes for $c=0$ and $c=1$, as it should.

From eqn (4.206) we can obtain $G^{\mathrm{d}}$. On performing the configuration average and Fourier transforming,

$$G^{j,\mathrm{d}}(\mathbf{q}) = \frac{cT'}{\lambda\omega^2}\bar{G}^{j}(\mathbf{q}). \tag{4.218}$$

We also need $G^{\mathrm{dd}}$ to calculate the neutron cross-section. From eqn (4.159)

$$\begin{aligned} G_{\alpha\beta}(l, l') = P_{\alpha\beta}(l, l') &+ M\lambda\omega^2 \sum_{\lambda,s} P_{\alpha\gamma}(l, s)P_{\gamma\beta}(s, l') \\ &+ (M\lambda\omega^2)^2 \sum_{\gamma_1,\gamma_2}\sum_{s_1,s_2} P_{\alpha\gamma_1}(l, s_1)G_{\gamma_1\gamma_2}(s_1, s_2)P_{\gamma_2\beta}(l_2, l'). \end{aligned} \tag{4.219}$$

If in the second term on the right-hand side we replace the restricted summation by $c$ times a summation over all sites and then Fourier transform, we obtain

$$\bar{G}^{j}(\mathbf{q}) = \mathscr{P}^{j}(\mathbf{q}) + c\lambda\omega^2\{\mathscr{P}^{j}(\mathbf{q})\}^2 + (\lambda\omega^2)^2\{\mathscr{P}^{j}(\mathbf{q})\}^2 G^{j,\mathrm{dd}}(\mathbf{q}).$$

Thus

$$\begin{aligned} G^{j,\mathrm{dd}}(\mathbf{q}) &= \frac{\bar{G}^{j}(\mathbf{q}) - \mathscr{P}^{j}(\mathbf{q}) - c\lambda\omega^2\{\mathscr{P}^{j}(\mathbf{q})\}^2}{(\lambda\omega^2)^2\{\mathscr{P}^{j}(\mathbf{q})\}^2} \\ &= c(1-c)\frac{T'}{\lambda\omega^2}\mathscr{P}(\omega) + \left(\frac{cT'}{\lambda\omega^2}\right)^2 \bar{G}^{j}(\mathbf{q}). \end{aligned} \tag{4.220}$$

Now let us examine the analytic behaviour of $\bar{G}^{j}(\mathbf{q})$ as a function of $\omega$. If

$$cT'(\omega) = \omega^2\{\Delta(\omega) + \mathrm{i}\Gamma(\omega)\}, \tag{4.221}$$

then

$$\operatorname{Im} \bar{G}^j(\mathbf{q}) = \frac{\omega^2\Gamma(\omega)}{\{\omega^2 - \omega_j^2(\mathbf{q}) - \omega^2\,\Delta(\omega)\}^2 + \{\omega^2\Gamma(\omega)\}^2}. \tag{4.222}$$

Since $\Gamma(\omega)$ is proportional to the density of states $Z(\omega)$ in the unperturbed crystal, there are additional contributions to (4.222) when, for $\omega_l > \omega_m$,

$$\omega_l^2 - \omega_j^2(\mathbf{q}) - \omega^2\,\Delta(\omega_l) = 0. \tag{4.223}$$

Eqn (4.223) is the expression analogous to the condition for the appearance of localized modes for the single-impurity problem, eqn (4.168). Since

$$\frac{\mathrm{d}}{\mathrm{d}\omega}\{\omega^2 - \omega_j^2(\mathbf{q}) - \omega^2\,\Delta(\omega)\}|_{\omega=\omega l} = 2\omega_l\left[1 - \Delta(\omega_l)\left\{1 - \frac{\omega_l^2}{c}(1-c)\Delta(\omega_l)B(\omega_l)\right\}\right],$$

where $B(\omega)$ is defined by eqn (4.195),

$$\operatorname{Im} \bar{G}^j(\mathbf{q}) = \frac{-\pi}{2\omega_l[1-\Delta(\omega_l)\{1-(\omega_l^2/c)(1-c)\Delta(\omega_l)B(\omega_l)\}]} \times\{\delta(\omega-\omega_l) - \delta(\omega+\omega_l)\}, \tag{4.224}$$

when $\omega > \omega_m$. For very small $c$,

$$\begin{aligned}\operatorname{Im} \bar{G}^j(\mathbf{q}) &= \frac{-c\omega_l\pi}{2\{\omega_l^2 - \omega_j^2(\mathbf{q})\}^2 B(\omega_l)}\{\delta(\omega-\omega_l) - \delta(\omega+\omega_l)\} \\ &= \frac{c\pi}{4B(\omega_l)}\left(\frac{\mathrm{d}}{\mathrm{d}\omega}\mathcal{P}^j(\mathbf{q})\right)\bigg|_{\omega=\omega_l}\{\delta(\omega-\omega_l) - \delta(\omega+\omega_l)\}.\end{aligned} \tag{4.225}$$

Where in the single-impurity problem the ratio $M'/M$ is such as to give a localized mode, the present theory predicts an impurity band with a well-defined frequency for each $j$, $\mathbf{q}$, which spans the frequency of the isolated mode.

With the aid of eqns (4.218) and (4.220),

$$\begin{aligned}&\frac{1}{N}\sum_{l,l'} b_l b_{l'} \exp\{-\mathrm{i}\boldsymbol{\kappa}\cdot(\mathbf{l}-\mathbf{l}')\}\sum_{\alpha\beta}\kappa_\alpha\kappa_\beta G_{\alpha\beta}(l,l') \\ &= \frac{1}{N^2}\sum_{l,l'}\sum_{\alpha,\beta}\sum_{j,\mathbf{q}} \exp\{-\mathrm{i}\boldsymbol{\kappa}\cdot(\mathbf{l}-\mathbf{l}') + \mathrm{i}\mathbf{q}\cdot(\mathbf{l}-\mathbf{l}')\}\frac{\kappa_\alpha\kappa_\beta}{M}\sigma_\alpha^j(\mathbf{q})\sigma_\beta^j(\mathbf{q}) \\ &\quad\times\{b_\mathrm{h}^2\bar{G}^j(\mathbf{q}) + 2b_\mathrm{h}(b_\mathrm{d}-b_\mathrm{h})G^{j,\mathrm{d}}(\mathbf{q}) + (b_\mathrm{d}-b_\mathrm{h})^2 G^{j,\mathrm{dd}}(\mathbf{q})\} \\ &= \frac{1}{M}\sum_{\alpha,\beta}\sum_j \kappa_\alpha\kappa_\beta\sigma_\alpha^j(\boldsymbol{\kappa})\sigma_\beta^j(\boldsymbol{\kappa})\left[c(1-c)\mathcal{P}(\omega)\frac{T'}{\lambda\omega^2}(b_\mathrm{d}-b_\mathrm{h})^2\right. \\ &\quad\left. + \bar{G}^j(\boldsymbol{\kappa})\left\{b_\mathrm{h} + (b_\mathrm{d}-b_\mathrm{h})\left(\frac{cT'}{\lambda\omega^2}\right)\right\}^2\right].\end{aligned} \tag{4.226}$$

Hence, the one-phonon coherent cross-section is

$$\left(\frac{d\sigma^2}{d\Omega\, dE'}\right)^{(1)}_{\text{coh}} = -\frac{N}{\pi M}\{1+n(\omega)\}\frac{k'}{k}\,\text{Im}\Bigg[\kappa^2 c(1-c)\mathscr{P}(\omega)\left(\frac{T'}{\lambda\omega^2}\right)(b_d - b_h)^2$$
$$+\left\{b_h + (b_d - b_h)\left(\frac{cT'}{\lambda\omega^2}\right)\right\}^2 \sum_j \{\boldsymbol{\kappa}\cdot\boldsymbol{\sigma}^j(\boldsymbol{\kappa})\}^2 \bar{G}^j(\boldsymbol{\kappa})\Bigg]. \quad (4.227)$$

The imaginary part of the first term in the square brackets in eqn (4.227) can be reduced to

$$\frac{\kappa^2(b_d - b_h)^2\Gamma(\omega)}{(\lambda\omega)^2}. \quad (4.228)$$

Similarly, the one-phonon incoherent cross-section is

$$\left(\frac{d\sigma^2}{d\Omega\, dE'}\right)^{(1)}_{\text{incoh}} = -\frac{N}{\pi M}\{1+n(\omega)\}\frac{k'}{k}\kappa^2$$
$$\times\,\text{Im}\left[\alpha_h^2\bar{G}(\omega) + (\alpha_d^2 - \alpha_h^2)\left\{c(1-c)\mathscr{P}(\omega)\left(\frac{T'}{\lambda\omega^2}\right) + \left(\frac{cT'}{\lambda\omega^2}\right)^2\bar{G}(\omega)\right\}\right], \quad (4.229)$$

where

$$\bar{G}(\omega) = \frac{1}{3N}\sum_{j,\mathbf{q}} \bar{G}^j(\mathbf{q}).$$

An examination of (4.226) shows that the first term is a diffuse contribution arising solely from the impurities. The second term is a *modified coherent* contribution, and we shall refer to this part of the cross-section (4.226) as such. The first term could be more properly grouped with the terms coming from the incoherent cross-section (eqn (4.229)).

The modified coherent cross-section has a peak shifted from that corresponding to the pure host by

$$\frac{\Delta\omega}{\omega} \simeq \tfrac{1}{2}\Delta\{\omega_j(\boldsymbol{\kappa})\} \quad (4.230)$$

and a relative width

$$\frac{\gamma}{\omega} \simeq \tfrac{1}{2}\Gamma\{\omega_j(\boldsymbol{\kappa})\}.$$

When the neutron energy change is greater than $\hbar\omega_m$ the additional contributions to $\text{Im}\,\bar{G}^j(\boldsymbol{\kappa})$ given by eqn (4.224) must be taken into account. For small $c$ we can use (4.225) and obtain

$$\left(\frac{d\sigma^2}{d\Omega\, dE'}\right)^{(1)}_{\text{coh}} = cN\frac{k'}{2kM}[\{1+n(\omega_l)\}\,\delta(\omega-\omega_l) + n(\omega_l)\,\delta(\omega+\omega_l)]$$
$$\times\frac{\omega_l}{B(\omega_l)}\sum_j \{\boldsymbol{\kappa}\cdot\boldsymbol{\sigma}^j(\boldsymbol{\kappa})\}^2\left\{\frac{b_h}{\omega_l^2 - \omega_j^2(\boldsymbol{\kappa})} + \frac{b_d - b_h}{\lambda\omega_l^2}\right\}^2. \quad (4.231)$$

This has an identical form to the corresponding single-defect cross-section (eqn (4.197) but here the $\omega_l$ is determined from eqn (4.223).

Let us now examine the incoherent cross-section (eqn (4.229) for the same circumstances. In view of the second expression for Im $\bar{G}^i$ given in eqn (4.225),

$$\operatorname{Im} \bar{G}(\omega) = \frac{c\pi}{4B(\omega_l)}\{\delta(\omega-\omega_l)-\delta(\omega+\omega_l)\}\left(\frac{\mathrm{d}\mathscr{P}(\omega)}{\mathrm{d}\omega}\right)\bigg|_{\omega=\omega_l} \tag{4.232}$$

for $\omega > \omega_{\mathrm{m}}$.

The other contribution to (4.229) for $\omega > \omega_{\mathrm{m}}$ is

$$\begin{aligned}
&\operatorname{Im}\left(\frac{cT'}{\lambda\omega^2}\right)^2 \bar{G}(\omega)\\
&= -\frac{c\omega_l\pi}{2B(\omega_l)}\{\delta(\omega-\omega_l)-\delta(\omega+\omega_l)\}\frac{1}{3N}\sum_{j,\mathbf{q}}\left(\frac{\omega_l^2-\omega_j^2(\mathbf{q})}{\lambda\omega_l^2}\right)^2\left(\frac{1}{\omega_l^2-\omega_j^2(\mathbf{q})}\right)^2\\
&= -\frac{c\pi}{2B(\omega_l)\lambda^2\omega_l^3}\{\delta(\omega-\omega_l)-\delta(\omega+\omega_l)\}.
\end{aligned} \tag{4.233}$$

On combining eqns (4.232) and (4.233) in the expression for the incoherent cross-section, the latter becomes

$$\begin{aligned}
\left(\frac{\mathrm{d}^2\sigma}{\mathrm{d}\Omega\,\mathrm{d}E'}\right)^{(1)}_{\mathrm{incoh}} &= Nc\frac{k'\kappa^2}{2kMB(\omega_l)}[\{1+n(\omega_l)\}\,\delta(\omega-\omega_l)+n(\omega_l)\,\delta(\omega+\omega_l)]\\
&\quad\times\left\{\frac{\alpha_{\mathrm{d}}^2-\alpha_{\mathrm{h}}^2}{\lambda^2\omega_l^3}-\frac{\alpha_{\mathrm{h}}^2}{2}\left(\frac{\mathrm{d}\mathscr{P}(\omega)}{\mathrm{d}\omega}\right)\bigg|_{\omega=\omega_l}\right\}.
\end{aligned} \tag{4.234}$$

This expression has exactly the same form as the corresponding cross-section for the single-defect problem (eqn (4.199) as expected: as with the coherent cross-section, $\omega_l$ here satisfies eqn (4.223).

The absence of damping in expressions (4.231) and (4.234) is an indication that the theory is limited to the description of very low-concentration mixed systems. For high concentrations the environment about each mass defect is highly variable being composed of various types of clusters. Hence, for finite defect concentrations the localized modes should form a band whose width varies with the concentration. An appropriate theory can be constructed with the coherent potential approximation or one of its many variants (Elliott *et al.* 1974; Taylor 1975).

Figure 4.6 shows a comparison between neutron scattering measurements and theoretical calculations based on the coherent potential approximation for rubidium–potassium alloys. The concentration of potassium varies between 0.06 and 0.29. The two peaks observed in the scattered spectrum are ascribed to a phonon and a localized mode, and the latter dominates the spectrum at the highest concentration. We conclude from the comparison that the theory is at best qualitative. Perhaps the inadequacy of the theory is due to a neglect of changes in the force constants, or the exclusion of defect clusters formed from pairs, triplets, etc. of potassium atoms.

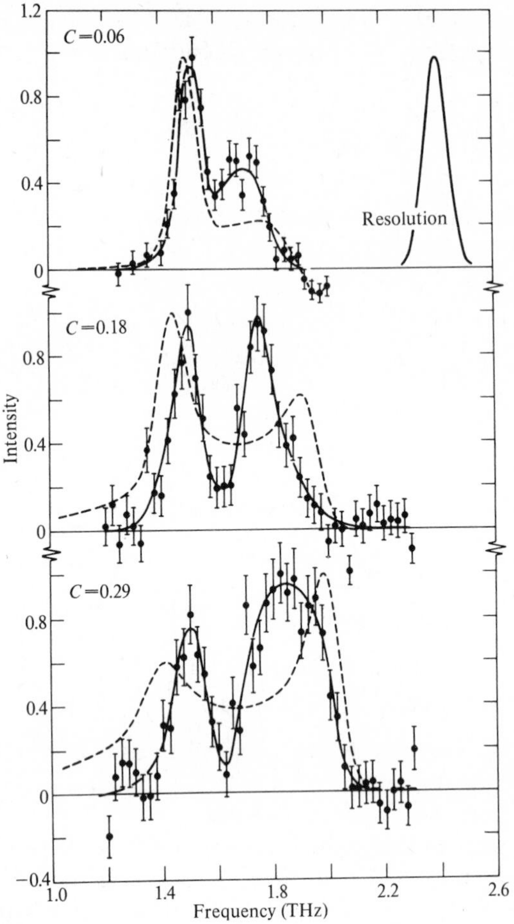


FIG. 4.6. Experimental (solid line) and theoretical (broken line) spectra for neutrons scattered by rubidium–potassium alloys with three different concentrations of potassium. The spectra are for a fixed scattering vector (Kamitakahara and Copley 1978).

## 4.8. Multi-phonon effects

The one-phonon cross-sections derived in § 4.4 are good approximations to the complete partial differential cross-sections for coherent and incoherent scattering provided the scattering vector is not too large. The higher-order terms in the expansion of (4.68) and (4.69), which represent multi-phonon contributions, are significant for sufficiently large scattering vectors and, with $\kappa \to \infty$, the complete cross-sections no longer resemble the one-phonon terms. Even if the relative wave-vector resolution is preserved with increasing $\kappa$ it will eventually become insufficient to discriminate between the wave vectors in the one-phonon coherent cross-section. In the extreme limit, $\kappa \to \infty$, the coherent and incoherent cross-sections are the same, and the phonon expansion is no longer useful.

For a harmonic lattice the identity (4.67), or more generally (4.83) and (4.84), enables us to express the neutron cross-section in terms of the correlation function of the displacements. From (4.83), (4.84), and (4.87) we find the exact expressions

$$\langle \exp(-i\boldsymbol{\kappa}\cdot\hat{\mathbf{R}}_{ld})\exp\{i\boldsymbol{\kappa}\cdot\hat{\mathbf{R}}_{l'd'}(t)\}\rangle = \exp\{i\boldsymbol{\kappa}\cdot(\mathbf{l}'+\mathbf{d}'-\mathbf{l}-\mathbf{d})\}$$
$$\times\exp\left\{\left\langle \boldsymbol{\kappa}\cdot\hat{\mathbf{u}}\begin{pmatrix} l \\ d \end{pmatrix}\boldsymbol{\kappa}\cdot\hat{\mathbf{u}}\begin{pmatrix} l' \\ d' \end{pmatrix}, t\right\rangle - W_d(\boldsymbol{\kappa}) - W_{d'}(\boldsymbol{\kappa})\right\}, \quad (4.235)$$

with $W_d(\boldsymbol{\kappa})$ defined in eqn (4.51) and

$$\left\langle \boldsymbol{\kappa}\cdot\hat{\mathbf{u}}\begin{pmatrix} l \\ d \end{pmatrix}\boldsymbol{\kappa}\cdot\hat{\mathbf{u}}\begin{pmatrix} l' \\ d' \end{pmatrix}, t\right\rangle = \frac{(M_d M_{d'})^{-1/2}}{N}\sum_{j\mathbf{q}}\{\boldsymbol{\kappa}\cdot\boldsymbol{\sigma}_d^{j*}(\mathbf{q})\boldsymbol{\kappa}\cdot\boldsymbol{\sigma}_{d'}^{j}(\mathbf{q})\}$$
$$\times\exp\{i\mathbf{q}\cdot(\mathbf{l}'-\mathbf{l})\}\langle \hat{Q}^{+}(\mathbf{q}j)\hat{Q}(\mathbf{q}j;t)\rangle, \quad (4.236)$$

where the normal coordinate displacement correlation is given explicitly in (4.90). Given the polarization vectors $\boldsymbol{\sigma}_d^j(\mathbf{q})$, and the dispersion relations, $\omega_j(\mathbf{q})$, the correlation function (4.235) can be evaluated, and by Fourier transformation we obtain the partial differential cross-section. These various steps may involve considerable numerical computation but the numerical analysis is not demanding. Hence, if the lattice dynamics are specified, the neutron cross-section for a harmonic solid can be evaluated completely for arbitrary values of the scattering vector and energy transfer.

We can develop an understanding of multi-phonon effects in the neutron cross-section by examining the Einstein model of lattice vibrations. In this model the vibrations have a single frequency, and not a spectrum of values. Hence, an Einstein solid is equivalent to many independent harmonic oscillators with a common frequency $\omega_0$, say. From the analysis of the scattering from an isotropic harmonic oscillator given in § 3.6.2 we know that for modest values of $(\hbar\kappa^2/M\omega_0)$ the response consists of delta function contributions located at frequencies that are multiples of $\omega_0$. In consequence the response can be attributed to one, two, . . . quanta scattering processes. For large $\kappa^2$ this description is no longer useful because very many quanta participate in the response with more or less equal weight. In the limit $(\hbar\kappa^2/M\omega_0) \to \infty$, the observed response approaches a Gaussian function centred at the recoil energy $(\hbar^2\kappa^2/2M)$. The many quanta engaged in the response at large scattering vectors contrive to give a non-oscillatory correlation function that decays with a short relaxation time which is proportional to $\{\langle(\boldsymbol{\kappa}\cdot\hat{\mathbf{v}})^2\rangle\}^{-1/2}$ where $\hat{\mathbf{v}}$ is the particle velocity (Gunn and Warner 1983).

The behaviour described in the preceding paragraph is illustrated in Fig. 4.5 which shows the inelastic response for two values of $\kappa$ of a particle embedded in a harmonic lattice; the only parameter changed in the calculations is the scattering vector, and the value used in (b) is twice that used in (a). Increasing $\kappa$ engages multi-phonon processes in the response that virtually wipe out the detailed structure apparent at the smallest scattering vector. Most of the structure in the response (Fig. 4.5) can be attributed to an image of the lattice density of states, although it can be shown that the response generated at the reduced frequencies $\pm 2.5$ is due mainly to two-phonon processes. The increase in the width of the response spectra observed in Fig. 4.5(a), (b) arises because the first frequency moment of the spectra is $(\hbar\kappa^2/2M)$. We emphasize, however, that the marked difference in the two spectra is largely a multi-phonon effect.

In the remainder of this section we consider semi-analytic approximations to the complete cross-section for harmonic solids. The analysis is really of limited value since we cannot expect to reproduce the rich structure illustrated in Fig. 4.5 by simple analytic expressions. Hence, almost from the outset we see the need for numerical work, in which case it is reasonable to argue that the entire calculation should be made numerically rather than adopt a hybrid formulation that is only approximate.

Notwithstanding these reservations we analyse the multi-phonon cross-section at large $\kappa^2$ using two different methods. Both methods exploit the so-called incoherent approximation in which the difference between the coherent and incoherent cross-sections is neglected. The approximation is valid in the limit of large scattering vectors where interference effects vanish. An immediate consequence of this approximation is that the total cross-section is readily expressed in terms of the phonon density of states (cf. (4.80)).

To see how this is achieved consider the correlation function (4.236). For $l = l'$ and $d = d'$ the displacement correlation function reduces to

$$\begin{aligned}
\left\langle \boldsymbol{\kappa}\cdot\hat{\mathbf{u}}\binom{l}{d}\boldsymbol{\kappa}\cdot\hat{\mathbf{u}}\left(\begin{matrix}l\\d\end{matrix}, t\right)\right\rangle &= \frac{1}{NM_d}\sum_{j\mathbf{q}} |\boldsymbol{\kappa}\cdot\boldsymbol{\sigma}_d^j(\mathbf{q})|^2 \langle \hat{Q}^+(\mathbf{q}j)\hat{Q}(\mathbf{q}j;t)\rangle \\
&\doteq \left(\frac{1}{3rNM_d}\right)\kappa^2\sum_{j\mathbf{q}} \langle \hat{Q}^+(\mathbf{q}j)\hat{Q}(\mathbf{q}j;t)\rangle \\
&= \left(\frac{\hbar\kappa^2}{2M_d}\right)\int_0^\infty \mathrm{d}\omega \frac{Z(\omega)}{\omega}\{\mathrm{i}\sin(\omega t) + \cos(\omega t)\coth(\tfrac{1}{2}\hbar\omega\beta)\} \\
&= \left(\frac{\hbar\kappa^2}{2M_d}\right)\int_{-\infty}^\infty \mathrm{d}\omega \frac{Z(\omega)}{\omega} n(\omega)\exp(-\mathrm{i}\omega t). \qquad (4.237)
\end{aligned}$$

The approximation of the polarization vectors by their average value is exact for a single Bravais lattice in which each atom is at a centre of inversion symmetry. We will henceforth restrict ourselves to Bravais lattices, and introduce the notation

$$\langle \boldsymbol{\kappa} \cdot \hat{\mathbf{u}}(l) \boldsymbol{\kappa} \cdot \hat{\mathbf{u}}(l, t) \rangle = \left( \frac{\hbar \kappa^2}{2M} \right) \gamma(t) \tag{4.238}$$

where $\gamma(t)$ is defined through (4.237). The total partial differential cross-section is now

$$\begin{aligned} \frac{\mathrm{d}^2\sigma}{\mathrm{d}\Omega\, \mathrm{d}E'} &= \frac{k'}{k} \frac{\sigma}{4\pi} \frac{N}{2\pi\hbar} \int_{-\infty}^{\infty} \mathrm{d}t \exp(-\mathrm{i}\omega t) \exp\left\{ \frac{\hbar\kappa^2}{2M} [\gamma(t) - \gamma(0)] \right\} \\ &= \frac{k'}{k} \frac{\sigma}{4\pi} \frac{N}{2\pi\hbar} \exp\{-2W(\kappa)\} \\ &\quad \times \sum_{p=0}^{\infty} \frac{1}{p!} \int_{-\infty}^{\infty} \mathrm{d}t \exp(-\mathrm{i}\omega t) \left\{ \frac{\hbar\kappa^2}{2M} \gamma(t) \right\}^p . \end{aligned} \tag{4.239}$$

The terms in the last expression with $p = 0$ and $p = 1$ are the elastic and one-phonon contributions, respectively, and the remaining terms give the multi-phonon contribution. The approximation of the coherent elastic and one-phonon contributions by a suitably weighted incoherent contribution is valid at large wave vectors where pronounced interference effects in the coherent response are negligible. For sufficiently large $\kappa^2$, we can invoke the impulse approximation in which the correlation function $\gamma(t)$ is replaced by its value near $t = 0$.

For short times we can expand

$$\gamma(t) = \gamma(0) + \mathrm{i}t - \frac{2t^2}{3\hbar} \mathscr{E} + \cdots, \tag{4.240}$$

where

$$\mathscr{E} = \tfrac{3}{4} \int_0^{\infty} \mathrm{d}\omega\, Z(\omega) \hbar\omega \coth(\tfrac{1}{2}\hbar\omega\beta) \tag{4.241}$$

is the mean kinetic energy per atom and approaches $3k_{\mathrm{B}}T/2$ at high temperatures.

If we use (4.240) as an approximation for all $p$ in (4.239) we can, of course, sum the series over $p$ once again and recover the formula

$$\frac{\mathrm{d}^2\sigma}{\mathrm{d}\Omega\, \mathrm{d}E'} = \frac{N\sigma}{4\pi} \frac{k'}{k} \frac{1}{2\pi\hbar} \int_{-\infty}^{\infty} \mathrm{d}t \exp(-\mathrm{i}\omega t) \exp\left[ \frac{\hbar\kappa^2}{2M} \left\{ \mathrm{i}t - \frac{2t^2 \mathscr{E}}{3\hbar} \right\} \right], \tag{4.242}$$

which was first given in Chapter 3. However, this gives a poor representation of the elastic and one-phonon terms particularly. We therefore treat these two exactly and use the approximation (4.240) for $p \geqslant 2$ only. We therefore need to consider

$$\sum_{p=2}^{\infty} \frac{1}{p!} \left\{\frac{\hbar\kappa^2}{2M}\right\}^p \left\{\gamma(0) + it - \frac{2t^2\mathscr{E}}{3\hbar} + \cdots\right\}^p$$

$$= \sum_{p=2}^{\infty} \frac{1}{p!} \left\{\frac{\hbar\kappa^2\gamma(0)}{2M}\right\}^p \left\{1 + \frac{it}{\gamma(0)} - \frac{2t^2\mathscr{E}}{3\hbar\gamma(0)} + \cdots\right\}^p. \tag{4.243}$$

We now write

$$e^x = 1 + \frac{it}{\gamma(0)} - \frac{2t^2\mathscr{E}}{3\hbar\gamma(0)} + \cdots$$

and solve for $x$ to leading order in $t$. Then (4.243) becomes

$$\sum_{p=2}^{\infty} \frac{1}{p!} \left\{\frac{\hbar\kappa^2\gamma(0)}{2M}\right\}^p e^{xp}$$

$$= \sum_{p=2}^{\infty} \frac{1}{p!} \left\{\frac{\hbar\kappa^2\gamma(0)}{2M}\right\}^p \exp\left[p\left\{\frac{it}{\gamma(0)} - \tfrac{1}{2}\Delta^2 t^2 + \cdots\right\}\right], \tag{4.244}$$

where

$$\Delta^2 = \frac{4\mathscr{E}}{3\hbar\gamma(0)} - \frac{1}{\gamma^2(0)}. \tag{4.245}$$

Using (4.244) in (4.239) and performing the integral over $t$ we get as the multi-phonon cross-section

$$\frac{\sigma}{4\pi} \frac{k'}{k} \frac{N}{\hbar} \exp(-2W) \sum_{p=2}^{\infty} \frac{1}{\Delta} \frac{1}{p!} \frac{(2W)^p}{(2\pi p)^{1/2}} \exp\left(-\frac{\omega^2}{2\Delta^2 p} - \frac{p}{2\Delta^2\gamma^2(0)} + \frac{\omega}{\gamma(0)\Delta^2}\right). \tag{4.246}$$

Adding this contribution to the elastic ($p = 0$) and one-phonon ($p = 1$) terms of (4.239) gives the final result

$$\frac{d^2\sigma}{d\Omega\, dE'} = \frac{N\sigma}{4\pi} \frac{k'}{k} \exp\{-2W(\kappa)\}$$

$$\times \left[\delta(\hbar\omega) + \frac{\hbar^2\kappa^2}{2M} \frac{Z(\omega)}{\omega} \{1 + n(\omega)\} + \frac{1}{\hbar\Delta} \exp\{\omega/\Delta^2\gamma(0)\} F(x, y)\right],$$

where

$$x = \omega/\Delta,$$

$$y = 2W \exp\left\{\frac{-1}{2\Delta^2\gamma^2(0)}\right\},$$

and

$$F(x, y) = \sum_{p=2}^{\infty} \frac{1}{p!(2\pi p)^{1/2}} y^p \exp\left(-\frac{x^2}{2p}\right). \tag{4.247}$$

In this approximation each multi-phonon term in (4.246) is represented as a Gaussian function of $\omega$ with mean values and widths depending upon $p$. The final result (4.247) has the following merits.

(a) It has treated the elastic and one-phonon terms exactly and therefore gives good results for low incident energy.

(b) Although the multi-phonon terms have been approximated, the method has treated accurately the correlation terms linear and quadratic in $t$. Hence automatically the formula gives the correct asymptotic limit at high energies.

(c) Because the linear and quadratic terms in $t$ are treated exactly, the mean energy transfer and mean square energy transfer are correctly given by the formula at all incident energies.

It is worth noting the sensitivity of the three terms in (4.247) to the spectrum $Z(\omega)$. The second term, representing the one-phonon cross-section, is directly proportional to $Z(\omega)$ and therefore is dependent upon the precise shape of $Z(\omega)$. The first term, however, is just elastic scattering and, as we saw in § 4.3, depends only on the single parameter $\overline{\omega^{-2}}$. The third term, describing the multi-phonon cross-section, depends upon two parameters: $\gamma(0)$, which in turn involves only $\overline{\omega^{-2}}$, and $\mathscr{E}$, the mean kinetic energy of the target nuclei, which becomes independent of $Z(\omega)$ at high enough temperatures.

We now consider a procedure for obtaining an approximation to the total cross-section containing all multi-phonon terms. As in our discussion of the phonon expansion above, we shall concern ourselves only with crystals with a Bravais lattice structure and employ the incoherent approximation.

Consider the inelastic contribution to (4.239). Denoting $\hbar^2\kappa^2/2M$ by $E_{\mathrm{R}}$, this contains

$$\frac{1}{2\pi\hbar}\int_{-\infty}^{\infty} \mathrm{d}t \exp(-\mathrm{i}\omega t)[\exp\{E_{\mathrm{R}}(\gamma(t)-\gamma(0))/\hbar\}-\exp\{-E_{\mathrm{R}}\gamma(0)/\hbar\}]$$

$$=\frac{1}{2\pi\hbar}\int_{-\infty}^{\infty} \mathrm{d}t \exp(-\mathrm{i}\omega t)\exp\{-E_{\mathrm{R}}\gamma(0)/\hbar\}\sum_{l=1}^{\infty}\frac{1}{l!}\left(\frac{E_{\mathrm{R}}}{\hbar}\right)^{l}\gamma^{l}(t). \quad (4.248)$$

Now in this expression we write

$$\exp\{-E_{\mathrm{R}}\gamma(0)/\hbar\}=\sum_{n=l}^{\infty}\frac{1}{(n-l)!}\left(\frac{E_{\mathrm{R}}}{\hbar}\right)^{n-l}\{-\gamma(0)\}^{n-l}$$

and rearrange the resultant expression into an expansion in $E_{\mathrm{R}}/\hbar$. It is simple to show that (4.248) can then be written

$$\sum_{n=1}^{\infty}(E_{\mathrm{R}}/\hbar)^{n}\sum_{l=1}^{n}\frac{\{-\gamma(0)\}^{n-l}}{l!(n-l)!}\frac{1}{2\pi\hbar}\int_{-\infty}^{\infty}\exp(-\mathrm{i}\omega t)\gamma^{l}(t).$$

This is an expansion $1/M$ and it is a very rapidly convergent sum.

Now, from (4.237),

$$\frac{1}{2\pi}\int_{-\infty}^{\infty} \mathrm{d}t \exp(-\mathrm{i}\omega t)\gamma^{l}(t) = \prod_{i=1}^{l}\int_{-\infty}^{\infty} \mathrm{d}\omega_i \frac{Z(\omega_i)}{\omega_i} n(\omega_i)\, \delta\Big(\omega + \sum_{i=1}^{l}\omega_i\Big).$$

Hence

$$\left(\frac{\mathrm{d}^2\sigma}{\mathrm{d}\Omega\,\mathrm{d}E'}\right)^{\mathrm{inel}} = \frac{N\sigma}{4\pi}\frac{k'}{\hbar k}\sum_{n=1}^{\infty}\left(\frac{E_{\mathrm{R}}}{\hbar}\right)^{n}\sum_{l=1}^{n}\frac{\{-\gamma(0)\}^{n-l}}{l!(n-l)!}$$
$$\times\prod_{i=1}^{l}\int_{-\infty}^{\infty}\mathrm{d}\omega_i\frac{Z(\omega_i)}{\omega_i}n(\omega_i)\,\delta\Big(\omega+\sum_{i=1}^{l}\omega_i\Big). \quad (4.249)$$

Integrating over $E'$ gives

$$\left(\frac{\mathrm{d}\sigma}{\mathrm{d}\Omega}\right)^{\mathrm{inel}} = \frac{N\sigma}{4\pi}\frac{k'}{k}\sum_{n=1}^{\infty}\left(\frac{E_{\mathrm{R}}}{\hbar}\right)^{n}\sum_{l=1}^{n}\frac{\{-\gamma(0)\}^{n-l}}{l!(n-l)!}\left\{\prod_{i=1}^{l}\int_{-\infty}^{\infty}\mathrm{d}\omega_i\frac{Z(\omega_i)}{\omega_i}n(\omega_i)\right\},$$

where it is now understood that

$$\sum_{i=1}^{l}\omega_i > -\frac{\hbar^2 k^2}{2m}, \qquad k' = \left\{k^2 + \frac{2m}{\hbar}\sum_{i=1}^{l}\omega_i\right\}^{1/2},$$

and $\kappa^2$, which appears in $E_{\mathrm{R}}$, is

$$\kappa^2 = k^2 + k'^2 - 2kk'\cos\theta,$$

where $\theta$ is the scattering angle.

Integrating again,

$$\sigma^{\mathrm{inel}} = \frac{N\sigma}{4k^2}\sum_{n=1}^{\infty}\left(\frac{\hbar}{2M}\right)^{n}\frac{1}{(n+1)}\sum_{l=1}^{n}\frac{\{-\gamma(0)\}^{n-l}}{l!(n-l)!}$$
$$\times\prod_{i=1}^{l}\int_{-\infty}^{\infty}\mathrm{d}\omega_i\frac{Z(\omega_i)}{\omega_i}n(\omega_i)\{(k+k')^{2n+2}-(k-k')^{2n+2}\}. \quad (4.250)$$

Eqn (4.250) is easily computed, using, for example, the Debye model to give $Z(\omega)$, (4.42). With

$$\theta = \frac{T}{\theta_{\mathrm{D}}}, \qquad x = \sqrt{\left(\frac{\hbar^2 k^2}{2mk_{\mathrm{B}}\theta_{\mathrm{D}}}\right)}, \qquad \xi_i = \frac{\hbar\omega_i}{k_{\mathrm{B}}\theta_{\mathrm{D}}},$$

and, finally,

$$\phi(\theta) = 6\left\{\tfrac{1}{4} + \int_0^1 \frac{\mathrm{d}u\,u}{\mathrm{e}^{u/\theta}-1}\right\} = 3\int_{-1}^{1}\frac{\mathrm{d}u\,u}{\mathrm{e}^{u/\theta}-1},$$

we obtain

$$\sigma^{\text{inel}} = \frac{N\sigma}{4x^2} \sum_{n=1}^{\infty} \left(\frac{m}{M}\right)^n \frac{1}{(n+1)} \sum_{l=1}^{n} \frac{3^l\{-\phi(\theta)\}^{n-l}}{l!(n-l)!}$$
$$\times \prod_{i=1}^{l} \int_{-1}^{1} \frac{\mathrm{d}\xi_i\, \xi_i}{\mathrm{e}^{\xi_i/\theta}-1} \{(x+x')^{2n+2}-(x-x')^{2n+2}\}, \tag{4.251}$$

where

$$x' = \left\{x^2 + \sum_{i=1}^{l} \xi_i\right\}^{1/2} \quad \text{for} \quad \sum_{i=1}^{l} \xi_i > -x^2$$
$$= 0 \quad \text{for} \quad \sum_{i=1}^{l} \xi_i < -x^2.$$

In this notation the elastic cross-section is

$$\sigma^{\text{el}} = N\sigma \sum_{n=0}^{\infty} \left(\frac{m}{M}\right)^n \frac{1}{(n+1)!} \{-4x^2\phi(\theta)\}^n. \tag{4.252}$$

Thus if we add this to the total inelastic cross-section given by (4.251), we obtain an expression for the total cross-section.

We have shown that the total elastic coherent cross-section for a powder sample (eqn (4.47)) has exactly the same form as the total elastic incoherent cross-section in the limit of high energies. Thus, in this limit we obtain the total coherent cross-section merely by replacing $\sigma_{\mathrm{i}}$ by $\sigma_{\mathrm{c}}$. However, this approximation is clearly a poor one at low energies where the total elastic coherent cross-section has much structure. In general it is given by

$$\sigma_{\mathrm{c}}^{\text{el}} = \frac{N\sigma_{\mathrm{c}}\pi^2\hbar^2}{2mEv_0} \sum_{\tau \neq 0}^{\tau<2k} \{z(\tau)/\tau\} \exp\left\{-\frac{\hbar\gamma(0)\tau^2}{2M}\right\}, \tag{4.253}$$

where $z(\tau)$ is the multiplicity of each $\tau$. In terms of a Debye spectrum (4.253) becomes

$$\sigma_{\mathrm{c}}^{\text{el}} = \frac{N\sigma_{\mathrm{c}}\pi^2\hbar^2}{2mEv_0} \sum_{\tau \neq 0}^{\tau<2k} \{z(\tau)/\tau\} \exp\left\{-\frac{\hbar^2\beta\phi(\theta)\tau^2}{2M}\right\},$$

Total cross-sections for vanadium at $T = 293$ K are shown in Fig. 4.7 as functions of the incident energy $E$ (Mayers 1983). Vanadium is an incoherent scatterer, with $\sigma_{\mathrm{i}} = 4.98$ bn, and the free-atom value $\sigma_{\mathrm{i}}/(1+m/M)^2 = 4.79$ bn is approached at the highest energy $E = 1$ eV. The results shown are obtained from (4.247), using the measured density of states for the one-phonon contribution and a Debye spectrum with $\omega_{\mathrm{D}} = 33.6$ meV to calculate $W(\kappa)$ and the mean kinetic energy per atom. For $T = 293$ K the maximum $p = 20$, whereas for $T = 77$ K comparable

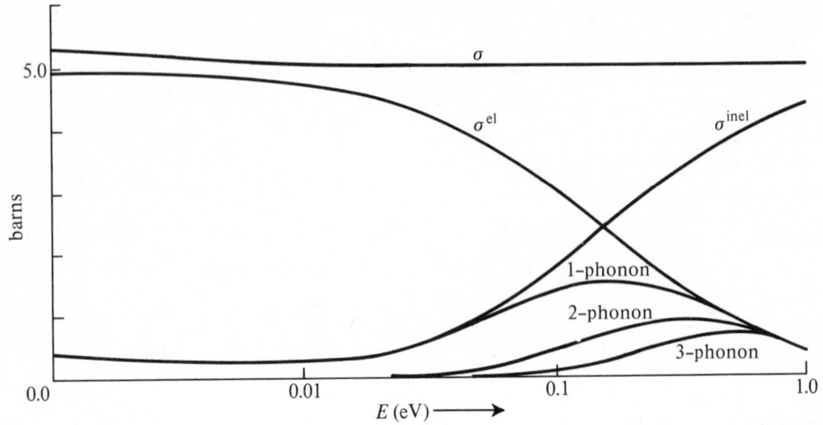


FIG. 4.7. The total cross-sections for vanadium at $T = 293$ K calculated with (4.247) are shown as a function of the incident energy $E$. The one-phonon, and $p = 2$ and 3 contributions are also included (Mayers, 1984).

numerical accuracy is obtained with a maximum $p = 10$, although the difference in the total cross-sections is negligible except at low energies. The first term of the mass expansion gives values that differ by a fraction of a per cent.

## REFERENCES

Bacon, G. E. (1977). *Neutron scattering in chemistry*. Butterworths, London.

Bilz, H. and Kress, W. (1979). *Phonon dispersion relations in insulators*. Solid-state sciences, Vol. 10. Springer-Verlag, Berlin.

Brown, P. J. (1979). In G. Kostorz (ed.) *Treatise on materials science and technology*, Vol. 15. Academic Press, New York.

Bruce, A. D. and Cowley, R. A. (1981). *Structural phase transitions*. Taylor and Francis, London.

Brüesch, P. (1982). *Phonons: theory and experiments I*, Solid-state sciences Vol. 34. Springer-Verlag, Heidelberg.

Cowley, E. R., Jacucci, G., Klein, M. L., and McDonald, I. R. (1976). *Phys. Rev.* **B14,** 1758.

Cummins, H. Z. and Levanyuk, A. P. (1983). *Light Scattering Near Phase Transitions*. North-Holland, Amsterdam.

Currat, R. and Pynn, R. (1979). In G. Kostorz (ed.) *Treatise on materials science and technology*, Vol. 15. Academic Press, New York.

Dolling, G. (1974). In G. K. Horton and A. A. Maradudin (ed.) *Dynamical properties of solids*, Vol. 1. North-Holland, Amsterdam.

Dorner, B. (1982). *Coherent inelastic neutron scattering in lattice dynamics*. Tracts in Modern Physics, Vol. 93. Springer-Verlag, Heidelberg.

Elliott, R. J., Krumhansl, J. A., and Leath, P. L. *Rev. mod. Phys.* **46,** 465.

Glyde, H. R. (1974). *Can. J. Phys.* **52,** 2281.

Gunn, J. M. F. and Warner, M. (1983). *Rutherford Appleton Laboratory Report RL-83-119.*
Kamitakahara, W. A. and Copley, J. R. D. (1978). *Phys. Rev.* **B18,** 3772.
Mayers, J. (1984). *Nucl. Inst. Meth. Phys.* **221,** 609.
Meyer, J., Dolling, G., Scherm, R., and Glyde, H. R. (1976). *J. Phys.* **F6,** 943.
Nicklow, R. M. (1979). In G. Kostorz (ed.) *Treatise on materials science and technology*, Vol. 15. Academic Press, New York.
Ranninger, J. (1975). *J. Phys.* **F5,** 1083.
Shirane, G. (1974). *Rev. mod. Phys.* **46,** 437.
Smith, H. G. and Wakabayashi, N. (1977). In S. W. Lovesey and T. Springer (ed.) *Dynamics of solids and liquids by neutron scattering*. Topics in current physics, Vol. 3. Springer–Verlag, Berlin.
Taylor, D. W. (1975). In G. K. Horton and A. A. Maradudin (ed.) *Dynamical properties of solids,* Vol. 2. North-Holland, Amsterdam.
Warner, M., Lovesey, S. W., and Smith, J. (1983). *Z. Phys.* **B, 51,** 109.
Willis, B. T. M. and Pryor, A. W. (1975). *Thermal vibrations in crystallography*. Cambridge University Press.

# 5
# DENSE FLUIDS

A combination of neutron scattering and computer simulation studies of dense fluids in the past decade has led to major advancements in our understanding of the atomic correlations (Hansen and McDonald 1976; March and Tosi 1977; Boon and Yip 1980; Hoover 1983). Neutron scattering and computer simulation studies are complementary techniques to a large extent, although it can rightly be argued that computer simulations have set a standard for the experimental work. Scattering experiments are expensive in terms of manpower and neutrons because much data is required to map out the relatively structureless response spectrum in detail (Copley and Lovesey 1975). For classical fluids, the spectrum as a function of energy for fixed scattering vector $\kappa$ has a width of the order of $(\hbar^2 k_B T \kappa^2/M)^{1/2}$ where $M$ is the atomic mass. The scale of relevant values of $\kappa$ is set essentially by the position of the main peak in the static structure factor, and this is typically in the range 1.5–2.5 $\text{Å}^{-1}$. Hence, for a dense fluid it is not unusual to require accurate experimental data for a fairly wide spread of energy and wave vector transfers.

We describe the properties of classical and quantal fluids relevant to the interpretation of neutron-scattering experiments. There is some emphasis on the study of atomic motions, using inelastic neutron scattering. The discussion is restricted throughout to the scattering of unpolarized neutrons and nuclear scattering lengths are taken to be energy independent. Theoretical developments are based on models in which the atoms interact through velocity-independent potentials between pairs of atoms.

Theoretical studies in the literature are, by and large, either for models of classical fluids or degenerate quantum fluids. Certainly most computer simulation studies are based on particle motions described by classical equations (Powles and Rickayzen 1979). Moreover, for most fluids it is an excellent approximation to use classical, Boltzmann statistics to describe the atomic correlations. The exceptions are few, and amount really to dense fluids of the two isotopes $^3$He and $^4$He at temperatures of a few degrees above absolute zero. The question of the relation between the measured response and models of classical motion is discussed in Chapter 3.

Throughout this chapter we adopt a prescription based on the representation of the response function in terms of the Fourier transform of the relaxation function (eqn (5.24)). The two functions are related through the detailed balance factor. The relaxation function is an even function of the energy transfer $\hbar\omega$ and it is easy to demonstrate that it reduces to a simple classical correlation function in the limit $\hbar \to 0$. Hence, in all theoretical developments for classical fluids we obtain our approximation to the measured response by using a classical form for the relaxation function; the approximation then automatically satisfies the condition of detailed balance. We propose that this prescription is adopted in the comparison of the measured response with computer simulation data.

One reason for emphasizing our adopted prescription at the beginning of the chapter concerns a thorny problem of notation. Since we reserve $S(\kappa, \omega)$ for the measured response function, most of the theoretical developments for classical fluids are couched in terms of $R(\kappa, \omega)$ which is the Fourier transform of the relaxation function. Readers therefore face the possibility of finding $R(\kappa, \omega)$ where they might expect to find the widely used $S(\kappa, \omega)$, for the latter is often used indiscriminately in the literature. From the point of view of presenting material there is much to be said for using $R(\kappa, \omega)$, because we do not have to keep inserting caveats in the developments about the use of classical approximations, for this step is made in a natural and precise manner at the outset. Moreover, we are able to discuss classical and quantal fluids within the same framework, since for the latter case the relaxation function is the, widely used, generalized susceptibility. The final point we make in this comment about our notation for classical fluids is that, as befits the discussion, we will use the term response (function) when referring to $R(\kappa, \omega)$.

## 5.1. Cross-section for unpolarized neutrons

Following the developments given in Chapter 3, the most general form of the partial differential cross-section in terms of correlation functions is found to be

$$\frac{\mathrm{d}^3\sigma}{\mathrm{d}\Omega\,\mathrm{d}E'} = \frac{k'}{k}\sum_{jj'}\frac{1}{2\pi\hbar}\int_{-\infty}^{\infty}\mathrm{d}t\,\exp(-\mathrm{i}\omega t)\,\overline{\langle\exp(-\mathrm{i}\boldsymbol{\kappa}\cdot\hat{\mathbf{R}}_j)\hat{b}_j^{+}\hat{b}_{j'}(t)\exp\{\mathrm{i}\boldsymbol{\kappa}\cdot\hat{\mathbf{R}}_{j'}(t)\}\rangle}. \tag{5.1}$$

Here $\hat{b}_j$ and $\hat{\mathbf{R}}_j$ are, respectively, the scattering amplitude operator and position operator associated with the $j$th nucleus, and the horizontal bar denotes all relevant averages over and above the thermal, or ensemble, average denoted by the angular brackets.

Expression (5.1) allows for a dynamic coupling between nucleon scattering properties and positions. In the absence of any such coupling the scattering amplitude operator $\hat{b}_{j'}(t)$ is replaced by $\hat{b}_{j'}(0) \equiv \hat{b}_{j'}$. We will not introduce this assumption until after we have averaged the cross-section over the spin states of the incident neutron beam.

We restrict all discussions in this chapter to the scattering of unpolarized neutrons. The state of polarization enters the cross-section (5.1) through the scattering amplitude operator which is a linear function of the neutron spin $\hat{\boldsymbol{\sigma}}/2$. From § 1.6 we have

$$\hat{b}_j = A_j + \tfrac{1}{2}B_j\hat{\boldsymbol{\sigma}} \cdot \hat{\mathbf{i}}_j, \tag{5.2}$$

where $\hat{\mathbf{i}}_j$ is the nuclear spin operator, and $A_j$ and $B_j$ are functions of the scattering lengths. For unpolarized neutrons the average of the product $\hat{b}_j^+\hat{b}_{j'}$ over the neutron spin states is

$$\overline{\hat{b}_j^+\hat{b}_{j'}} = A_j^*A_{j'} + \tfrac{1}{4}B_j^*B_{j'}\hat{\mathbf{i}}_j \cdot \hat{\mathbf{i}}_{j'}. \tag{5.3}$$

Using this result in conjunction with the definition

$$Y_{jj'}(\kappa, t) = \langle \exp(-i\boldsymbol{\kappa} \cdot \hat{\mathbf{R}}_j)\exp\{i\boldsymbol{\kappa} \cdot \hat{\mathbf{R}}_{j'}(t)\}\rangle, \tag{5.4}$$

the cross-section (5.1) reduces to

$$\frac{d^2\sigma}{d\Omega\, dE'} = \frac{k'}{k}\sum_{jj'}\frac{1}{2\pi\hbar}\int_{-\infty}^{\infty} dt\, \exp(-i\omega t)$$
$$\times \overline{\{A_j^*A_{j'}Y_{jj'}(\kappa, t) + \tfrac{1}{4}B_j^*B_{j'}\langle \exp(-i\boldsymbol{\kappa} \cdot \hat{\mathbf{R}}_j)\hat{\mathbf{i}}_j \cdot \hat{\mathbf{i}}_{j'}(t)\exp\{i\boldsymbol{\kappa} \cdot \hat{\mathbf{R}}_{j'}(t)\}\rangle\}}. \tag{5.5}$$

The nuclear spin operators are included in the second correlation function because, in general, there is correlation between the allowed spin states of the nuclei and the states of the target sample (cf. § 1.7). The correlation is induced by quantum mechanical exchange forces, and it can be significant for identical light mass nuclei at low temperatures.

However, for many situations of interest the sample temperature is relatively high and quantum mechanical exchange effects can be safely neglected. In this instance, the nuclei have random spin orientations, and averaging over the nuclear spin states gives

$$\begin{aligned}\overline{\hat{\mathbf{i}}_j \cdot \hat{\mathbf{i}}_{j'}} &= 0, \qquad j \neq j' \\ &= i_j(i_j + 1), \qquad j = j'.\end{aligned} \tag{5.6}$$

This first result is valid when there is no coupling between the nuclear spins, i.e. the spins behave as an ideal paramagnet.

We conclude that the cross-section for the scattering of unpolarized neutrons from a classical fluid is

$$\frac{d^2\sigma}{d\Omega\, dE'} = \frac{k'}{k}\sum_{jj'}\frac{1}{2\pi\hbar}\int_{-\infty}^{\infty} dt\, \exp(-i\omega t)\overline{\{A_j^*A_{j'} + \tfrac{1}{4}i_j(i_j+1)\,|B_j|^2\,\delta_{jj'}\}Y_{jj'}(\kappa, t)}. \tag{5.7}$$

The remaining average in (5.7) refers to the distribution of atomic species and isotopes.

In this chapter we will for the most part be concerned with monatomic fluids. The only exception is the discussion of classical multi-component fluids in § 5.5.

For a monatomic fluid, the average in (5.7) refers to the distribution of isotopes, and the averaging procedure is discussed in § 1.3. First, we will assume that the dynamic properties of the sample are independent of the isotope distribution. This means in (5.7) that the average over the isotope distribution does not include the correlation function $Y_{jj'}(\kappa, t)$. Moreover, values of $A_j$ for different isotopes are not correlated. Thus, the averaging amounts to the evaluation of

$$\begin{aligned}\overline{A_j^* A_{j'} + \tfrac{1}{4} i_j (i_j + 1)\, |B_j|^2\, \delta_{jj'}} \\ = \overline{A_j^* A_{j'}} + \delta_{jj'} \tfrac{1}{4} \overline{i(i+1)\, |B|^2} \\ = |\bar{A}|^2 + \delta_{jj'} \{\overline{|A|^2} - |\bar{A}|^2 + \tfrac{1}{4}\overline{i(i+1)\, |B|^2}\} \\ = \frac{1}{4\pi} \{\sigma_c + \delta_{jj'} \sigma_i\}\end{aligned} \tag{5.8}$$

where the last equality defines the single atom coherent and incoherent bound scattering cross-sections in terms of the averages of $A$, $|A|^2$, and $|B|^2$. Note that the averaged quantities are independent of the label $j$.

For completeness we record here the specific forms of the averaged quantities in (5.8). The isotopes are labelled by $\xi$, and the fractional concentrations are $c_\xi$. The quantities $A$ and $B$ for a single isotope are given in § 1.6. Beginning with the averaged scattering length

$$\bar{b} = \sum_\xi c_\xi A_\xi \equiv \bar{A} \tag{5.9}$$

and

$$\begin{aligned}\overline{|b|^2} = \sum_\xi c_\xi \{|A_\xi|^2 + \tfrac{1}{4} i_\xi (i_\xi + 1)\, |B_\xi|^2\} \\ \equiv \overline{|A|^2} + \tfrac{1}{4}\overline{i(i+1)\, |B|^2},\end{aligned} \tag{5.10}$$

we have

$$\sigma_c = 4\pi\, |\bar{b}|^2 \tag{5.11}$$

and

$$\sigma_i = 4\pi \{\overline{|b|^2} - |\bar{b}|^2\}. \tag{5.12}$$

For the scattering lengths the horizontal bar denotes an average over both the random nuclear spins and the isotope distribution, whereas for $A$ and $B$ the horizontal bar signifies an average over the isotope distribution.

Returning to the cross-section (5.7) and using (5.8), we find for a (classical) monatomic fluid of $N$ particles,

$$\frac{\mathrm{d}^2\sigma}{\mathrm{d}\Omega\,\mathrm{d}E'}=\frac{k'}{k}\frac{N}{4\pi}\{\sigma_c S(\kappa,\omega)+\sigma_i S_i(\kappa,\omega)\}. \tag{5.13}$$

Here $S(\kappa, \omega)$ and $S_i(\kappa, \omega)$ are, respectively, the coherent and incoherent, or single-particle, response functions introduced in Chapter 3. Since the scattering from a fluid is isotropic, the response functions depend on the magnitude of the scattering vector $\kappa = |\boldsymbol{\kappa}|$ and not its direction; this is implied in the definition (5.4). In terms of the correlation function $Y_{jj'}(\kappa, t)$ the response functions are

$$S(\kappa,\omega)=\frac{1}{2\pi\hbar N}\int_{-\infty}^{\infty}\mathrm{d}t\,\exp(-\mathrm{i}\omega t)\sum_{jj'}Y_{jj'}(\kappa,t), \tag{5.14}$$

and

$$\begin{aligned}S_i(\kappa,\omega)&=\frac{1}{2\pi\hbar N}\int_{-\infty}^{\infty}\mathrm{d}t\,\exp(-\mathrm{i}\omega t)\sum_{j}Y_{jj}(\kappa,t)\\&=\frac{1}{2\pi\hbar}\int_{-\infty}^{\infty}\mathrm{d}t\,\exp(-\mathrm{i}\omega t)Y_{jj}(\kappa,t).\end{aligned} \tag{5.15}$$

It should be noted that $S_i(\kappa, \omega)$ is contained in $S(\kappa, \omega)$ and, for sufficiently large $\kappa$, the difference is minimal because coherence effects vanish at short wavelengths. For perfect fluids there are no coherence effects and the two response functions are identical (cf. § 3.6).

For many discussions there is a real advantage in expressing the response function in the form ($\beta = 1/k_B T$)

$$S(\kappa,\omega)=\frac{\omega\beta}{\{1-\exp(-\hbar\omega\beta)\}}R(\kappa,\omega), \tag{5.16}$$

with a corresponding expression for $S_i(\kappa, \omega)$. We will demonstrate that the function $R(\kappa, \omega)$ is the time Fourier transform of the relaxation function discussed in Appendix B.† For the moment we note that in the classical limit, obtained with $\hbar \to 0$,

$$S(\kappa,\omega)\to\frac{1}{\hbar}R(\kappa,\omega)$$

which implies that $R(\kappa, \omega)$ is an even function of $\omega$. Given that this is a general property of $R(\kappa, \omega)$ the condition of detailed balance, which expresses the relation between $S(\kappa, \omega)$ and $S(\kappa, -\omega)$, follows directly

† Here we have introduced a factor $\beta$ in the definition of the relaxation function for ease of notation in subsequent discussions of classical fluids. In view of this modification, the dimension of $R(\kappa, \omega)$ is (frequency)$^{-1}$.

from the relation (5.16). In view of this, the function

$$\{1-\exp(-\hbar\omega\beta)\}^{-1}=\{1+n(\omega)\} \tag{5.17}$$

is often referred to as the detailed balance factor. Another feature of the expression (5.16) is that the odd-frequency moments of $S(\kappa, \omega)$ are related directly to even-frequency moments of $R(\kappa, \omega)$, i.e. the detailed balance factor reduces to unity in the integrand. For example,

$$k_B T\int_{-\infty}^{\infty} d\omega\, \omega S(\kappa, \omega)=\int_0^{\infty} d\omega\, \omega^2\{1+n(\omega)\}R(\kappa, \omega)$$
$$+\int_{-\infty}^{0} d\omega\, \omega^2\{1+n(\omega)\}R(\kappa, \omega),$$

and the second term is reduced by changing the sign of the integration variable to

$$\int_0^{\infty} d\omega\, \omega^2\{1+n(-\omega)\}R(\kappa, -\omega)=\int_0^{\infty} d\omega\, \omega^2\{-n(\omega)\}R(\kappa, \omega)$$

since

$$1+n(\omega)+n(-\omega)=0,$$

and thus,

$$\int_{-\infty}^{\infty} d\omega\, \omega S(\kappa, \omega)=(\beta/2)\int_{-\infty}^{\infty} d\omega\, \omega^2 R(\kappa, \omega). \tag{5.18}$$

A similar reduction can be made for any odd-frequency moment of $S(\kappa, \omega)$.

We will now review the demonstration that $R(\kappa, \omega)$ in (5.16) is the time Fourier transform of the relaxation function $R(\kappa, t)$ defined by

$$i\hbar\beta\partial_t R(\kappa, t)=\langle[\hat{B}_{\boldsymbol{\kappa}}^{+}(t), \hat{B}_{\boldsymbol{\kappa}}]\rangle \tag{5.19}$$

where,

$$\hat{B}_{\boldsymbol{\kappa}}(t)=\frac{1}{\sqrt{N}}\sum_j \exp\{-i\boldsymbol{\kappa}\cdot\hat{\mathbf{R}}_j(t)\};\ \text{coherent response}$$
$$=\frac{1}{\sqrt{N}}\hat{\rho}_{\boldsymbol{\kappa}}(t)=\frac{1}{\sqrt{N}}\hat{\rho}^{+}_{-\boldsymbol{\kappa}}(t) \tag{5.20}$$

or

$$\hat{B}_{\boldsymbol{\kappa}}(t)=\exp\{-i\boldsymbol{\kappa}\cdot\hat{\mathbf{R}}_j(t)\};\quad \text{incoherent response.} \tag{5.21}$$

If we define the Fourier transform

$$R(\kappa, \omega)=\frac{1}{2\pi}\int_{-\infty}^{\infty} dt\, \exp(-i\omega t)R(\kappa, t) \tag{5.22}$$

and use the identity (Appendix B)

$$\langle \hat{B}^+_{\boldsymbol{\kappa}}(t)\hat{B}_{\boldsymbol{\kappa}}\rangle = \langle \hat{B}_{\boldsymbol{\kappa}}\hat{B}^+_{\boldsymbol{\kappa}}(t+\mathrm{i}\hbar\beta)\rangle, \tag{5.23}$$

we readily obtain the result

$$\{1+n(\omega)\}\omega\beta R(\kappa,\omega) = \frac{1}{2\pi\hbar}\int_{-\infty}^{\infty} \mathrm{d}t\, \exp(-\mathrm{i}\omega t)\langle \hat{B}_{\boldsymbol{\kappa}}\hat{B}^+_{\boldsymbol{\kappa}}(t)\rangle \tag{5.24}$$

and the right-hand side is the response function $S(\kappa, \omega)$ or $S_{\mathrm{i}}(\kappa, \omega)$ depending on the choice of operator, (5.20) or (5.21).

We now establish some important properties of $R(\kappa, t)$ and $R(\kappa, \omega)$. First, from (5.19) we can readily prove that $R(\kappa, t)$ is an even function of time; to do so observe that $\hat{B}^+_{\boldsymbol{\kappa}} \equiv \hat{B}_{-\boldsymbol{\kappa}}$ and the correlation function $\langle \hat{B}_{\boldsymbol{\kappa}}\hat{B}_{-\boldsymbol{\kappa}}(t)\rangle$, for example, is an even function of $\kappa = |\boldsymbol{\kappa}|$, i.e. it is unchanged by the transformation $\boldsymbol{\kappa} \to -\boldsymbol{\kappa}$. In consequence, $R(\kappa, \omega)$ is an even function of the frequency, as implied during the introductory development. Also, $R(\kappa, t)$, and therefore $R(\kappa, \omega)$, is purely real as can be verified directly from (5.19).

A useful alternative expression for $R(\kappa, t)$ is

$$\beta R(\kappa, t) = \int_0^{\infty} \mathrm{d}\mu \langle \hat{B}_{\boldsymbol{\kappa}}\hat{B}^+_{\boldsymbol{\kappa}}(t+\mathrm{i}\hbar\mu)\rangle. \tag{5.25}$$

The equivalence of (5.19) and (5.25) is established with the aid of the equation of motion for Heisenberg operators,

$$\mathrm{i}\hbar\partial_t\hat{B}^+_{\boldsymbol{\kappa}}(t) = [\hat{B}^+_{\boldsymbol{\kappa}}(t), \hat{\mathcal{H}}] \tag{5.26}$$

where $\hat{\mathcal{H}}$ is the Hamiltonian for the target sample.

From (5.25) it is evident that in the classical limit, $\hbar \to 0$, the relaxation function reduces to the correlation function of the dynamical variables $\hat{B}_{\boldsymbol{\kappa}}$, $\hat{B}^+_{\boldsymbol{\kappa}}$. We have

$$\lim_{\hbar\to 0} R(\kappa, t) = \langle \hat{B}_{\boldsymbol{\kappa}}\hat{B}^+_{\boldsymbol{\kappa}}(t)\rangle|_{\hbar=0}$$

$$= \langle B_{\boldsymbol{\kappa}}B^*_{\boldsymbol{\kappa}}(t)\rangle, \tag{5.27}$$

where the last equality contains a purely classical correlation function. In view of this result, we will obtain the cross-section for scattering from a target sample whose various properties are described by classical statistical mechanics, to a good approximation, by using (5.16), or more generally (5.24), with $R(\kappa, \omega)$ derived from the appropriate classical correlation function. The resulting approximation for the cross-section will satisfy the condition of detailed balance. The prescription described is useful in the comparison of neutron scattering data with results obtained in the computer simulation of classical systems.

The quantity $R(\kappa, t=0)$ is, apart from a factor $\beta$, identical to the isothermal response $\chi_\kappa$. This relation is established in Appendix B. To be precise the relation holds when the operators in (5.25) are $\{\hat{B}_{\boldsymbol{\kappa}} - \langle \hat{B}_{\boldsymbol{\kappa}} \rangle\}$ which describe fluctuations. However, for a fluid $\langle \hat{B}_{\boldsymbol{\kappa}} \rangle$ is zero except for $\boldsymbol{\kappa} = 0$, and since this corresponds to scattering in the forward direction it is not of interest. Thus, the interpretation of neutron scattering data from fluids is unaffected by the use of the fluctuation variables $\{\hat{B}_{\boldsymbol{\kappa}} - \langle \hat{B}_{\boldsymbol{\kappa}} \rangle\}$ in the theoretical development. The main reason for using these variables is not one of nicety but the fact that the relations derived from theory are in a form suitable for numerical evaluation, e.g. the relation between the static structure factor and pair distribution function, (5.33).

The following three sections review the properties of classical monatomic fluids observed in neutron spectroscopy. The review will be couched in terms of the relaxation function $R(\kappa, \omega)$ and $R_s(\kappa, \omega)$ appropriate for coherent and single particle properties of classical fluids, respectively. These functions are the time Fourier transforms of

$$R(\kappa, t) = \frac{1}{N} \langle \rho_{\boldsymbol{\kappa}} \rho^*_{\boldsymbol{\kappa}}(t) \rangle; \quad \text{coherent} \tag{5.28}$$

and

$$R_s(\kappa, t) = \langle \exp\{i\boldsymbol{\kappa} \cdot [\mathbf{R}(t) - \mathbf{R}]\} \rangle; \quad \text{single particle} \tag{5.29}$$

## 5.2. Classical monatomic fluid: sum rules

The frequency moments of the response functions, often referred to as sum rules, provide valuable information on the general features of $R(\kappa, \omega)$ and $R_s(\kappa, \omega)$ as functions of $\omega$ for fixed $\kappa$. Some of the results given here are contained in Chapter 3.

From (5.29) it is clear that, since $\mathbf{R} \equiv \mathbf{R}(0)$,

$$R_s(\kappa, t=0) = 1 \tag{5.30}$$

and therefore

$$\int_{-\infty}^{\infty} d\omega \, R_s(\kappa, \omega) = R_s(\kappa, t=0) = 1. \tag{5.31}$$

For coherent scattering the corresponding result is

$$\int_{-\infty}^{\infty} d\omega \, R(\kappa, \omega) = \frac{1}{N} \langle |\rho_{\boldsymbol{\kappa}}|^2 \rangle$$
$$= S(\kappa) \tag{5.32}$$

where the last equality defines the static structure factor $S(\kappa)$. The latter is related to the spatial Fourier transform of the pair distribution function

$g(r)$, introduced in (3.28). Denoting the particle density† by $\rho_0 = N/V$,

$$S(\kappa) = 1 + \rho_0 \int d\mathbf{r} \exp(i\boldsymbol{\kappa} \cdot \mathbf{r}) g(r)$$
$$= (2\pi)^3 \rho_0 \delta(\boldsymbol{\kappa}) + 1 + \rho_0 \int d\mathbf{r} \exp(i\boldsymbol{\kappa} \cdot \mathbf{r}) \{g(r) - 1\}.$$

In the second expression we have extracted the asymptotic behaviour of $g(r)$, namely $g(\infty) = 1$, so that the remaining Fourier integral is well behaved. Because the delta function does not contribute to $S(\kappa)$ except in the forward direction it is of no interest, and we will take

$$S(\kappa) = 1 + \rho_0 \int d\mathbf{r} \exp(i\boldsymbol{\kappa} \cdot \mathbf{r}) \{g(r) - 1\}. \tag{5.33}$$

This choice of definition is consistent with the use of fluctuation variables discussed at the end of the previous section.

Values of $S(\kappa)$ and $g(r)$ for liquid sodium are shown in Fig. 5.1 by way of an illustrative example. The main features of $S(\kappa)$ for classical fluids are a small, nonzero value of $S(0)$ and a main peak at $\kappa \simeq 1.5$–$2.5$ Å$^{-1}$, followed by rapidly decaying oscillations about the asymptotic value $S(\infty) = 1$. The value $S(0)$ is related to the isothermal compressibility $\chi_T$, through (cf. (5.117))

$$S(0) = \rho_0 \chi_T / \beta = \langle (\Delta N)^2 \rangle / N, \tag{5.34}$$

which also shows that $S(0)$ is proportional to the mean-square fluctuation in the particle density. These fluctuations are small in a dense fluid except in the vicinity of the liquid–gas phase transition where they take macroscopic values. Note that $(\beta/\rho_0)$ is the isothermal compressibility of a perfect fluid. The main peak in $S(\kappa)$ occurs at a wave vector $\kappa \simeq 2\pi/$(mean nearest-neighbour distance). It is known that the shape of this peak and much of the structure on the high-$\kappa$ side is determined largely by the repulsive part of the pair potential (Hansen and McDonald 1976). This is demonstrated by comparing data with computer simulation results for hard spheres. Many measured structure factors are tabulated by Waseda (1980).

A useful model for $S(\kappa)$ can be derived from an approximate solution for the hard-sphere fluid. Defining a reduced density by $6\eta/\pi$, and $\alpha = (1+2\eta)^2/(1-\eta)^4$, $\alpha_1 = -6\eta(1+\frac{1}{2}\eta)^2/(1-\eta)^4$, and $\alpha_2 = \eta\alpha/2$, the result is most conveniently expressed in the form of a linear equation for $S(\kappa)$

$$S(\kappa) - 1 = -24 S(\kappa) \eta \int_0^1 du\, u^2 \left\{ \frac{\sin(\kappa\, du)}{\kappa\, du} \right\} (\alpha + \alpha_1 u + \alpha_2 u^3), \tag{5.35}$$

† In other parts of the book we use $\rho$ with no subscript for the particle density.

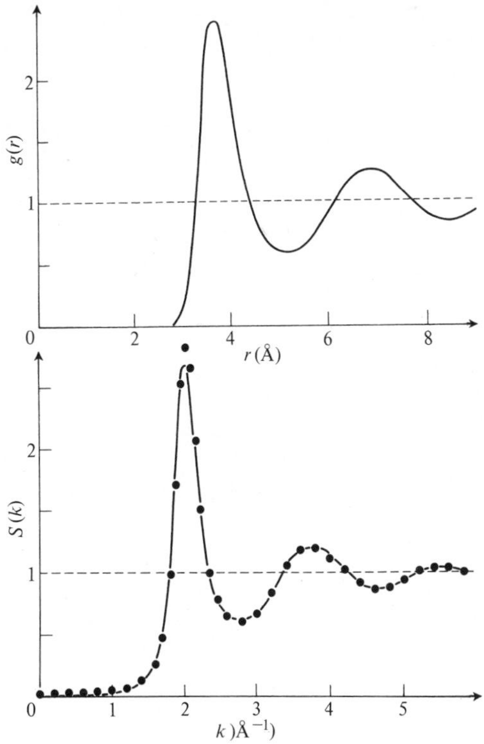


FIG. 5.1. Static structure factor and pair distribution function for liquid sodium (Hansen and McDonald 1976).

where $d$ is the diameter of the spheres. For this model $S(0)=1/\alpha$. Excellent agreement between (5.35) and data is usually obtained by choosing $6\eta/\pi$ to give the height of the main peak in $S(\kappa)$ correctly, and the diameter $d$ such that the zeros of $S(\kappa)-1$ away from the main peak are correct.

The physical significance of the pair distribution function is made apparent by noting that $r^2g(r)$ is the probability distribution for the particle density about the origin. Hence, the number of particles within a sphere of radius $R$ prescribed about a given particle is

$$N(R)=4\pi\rho_0\int_0^R \mathrm{d}r\, r^2 g(r). \tag{5.36}$$

The second moment of $R(\kappa,\omega)$ relates to the mean-square energy transfer. From the relation (5.18) and the result (3.63), we obtain for $R_s(\kappa,\omega)$ the result

$$\int_{-\infty}^{\infty} \mathrm{d}\omega\, \omega^2 R_s(\kappa,\omega)=(\kappa^2/M\beta), \tag{5.37}$$

In view of (5.31), the result (5.37) coincides with the normalized second moment of the single particle response, whereas for coherent scattering the normalized second moment is

$$\omega_0^2 = \frac{1}{S(\kappa)} \int_{-\infty}^{\infty} d\omega\, \omega^2 R(\kappa, \omega) = (\kappa^2/M\beta S(\kappa)). \tag{5.38}$$

From this result we see that the energy spread in the coherent scattering is narrower at those values of $\kappa$ for which the structure factor is large. Hence, the energy spread for coherent scattering depends on the interaction between the atoms, in contrast to the result for single particle scattering.

The fourth moments of the response functions depend explicitly on the atomic potential. Assuming that this potential $u(r)$ involves pairs of atoms only, the normalized fourth moment of the coherent response is

$$\frac{1}{S(\kappa)} \int_{-\infty}^{\infty} d\omega\, \omega^4 R(\kappa, \omega) = \omega_0^2 \omega_l^2 \tag{5.39}$$

where

$$\omega_l^2 = (3\kappa^2/M\beta) + \Omega^2(0) - \Omega^2(\kappa), \tag{5.40}$$

and the squared frequency

$$\Omega^2(\kappa) = (\rho_0/M) \int d\mathbf{r}\, \{\partial_z^2 u(r)\} g(r) \cos(\kappa z). \tag{5.41}$$

Both $\omega_0^2$ and $\omega_l^2$ have a strong $\kappa$-dependence. They rise to a maximum at a wave vector that is half-way to the main peak in $S(\kappa)$, at which they have a deep minimum and thereafter oscillate out of phase with the maxima and minima in $S(\kappa)$. For the single-particle response, the corresponding result is

$$\int_{-\infty}^{\infty} d\omega\, \omega^4 R_s(\kappa, \omega) = \left(\frac{\kappa^2}{M\beta}\right)\left\{\frac{3\kappa^2}{M\beta} + \Omega^2(0)\right\}. \tag{5.42}$$

It is interesting to note that at low temperatures, $\beta \to \infty$, and for a pair distribution

$$g(r) = \frac{1}{\rho_0} \sum_l \delta(\mathbf{r} - \mathbf{R}_l),$$

the relation (5.40) for $\omega_l^2$ reduces to the eigenfrequency relation for a harmonic solid. This result suggests that the coherent response of a cold fluid is governed to a large extent by the behaviour of the fourth frequency moment.

The evaluation of $\Omega^2(\kappa)$ for a given pair potential and $g(r)$ proceeds from the expression

$$\Omega^2(\kappa)=\left(\frac{4\pi\rho_0}{M}\right)\int_0^\infty r^2\,\mathrm{d}r\,g(r)\left[u''\left\{\frac{\sin(\kappa r)}{\kappa r}+\frac{2\cos(\kappa r)}{(\kappa r)^2}-\frac{2\sin(\kappa r)}{(\kappa r)^3}\right\}\right.$$
$$\left.-\frac{2u'}{r}\left\{\frac{\cos(\kappa r)}{(\kappa r)^2}-\frac{\sin(\kappa r)}{(\kappa r)^3}\right\}\right], \quad (5.43)$$

where $u'$ and $u''$ are shorthand for $\partial u/\partial r$ and $\partial^2 u/\partial r^2$, respectively. In many cases the integral in (5.43) is dominated by the term $r^2 g(r)u''$ that describes the spatial range of correlation of forces. This term is very small except in a narrow region centred about a distance $R_0$, which is approximately the position of the main minimum in the potential. In view of this behaviour, the integral is approximated by replacing $(4\pi\rho_0/3M)r^2 g(r)u''$ by a delta function at $R_0$ with a weight

$$\omega_e^2=\left(\frac{4\pi\rho_0}{3M}\right)\int_0^\infty r^2\,\mathrm{d}r\,g(r)u''(r) \quad (5.44)$$

and neglecting terms that involve $u'$. The result is

$$\omega_l^2 \doteq \left(\frac{3\kappa^2}{M\beta}\right)+\omega_e^2\left\{1-\frac{3\sin(\kappa R_0)}{\kappa R_0}-\frac{6\cos(\kappa R_0)}{(\kappa R_0)^2}+\frac{6\sin(\kappa R_0)}{(\kappa R_0)^3}\right\}. \quad (5.45)$$

For a dense fluid $\omega_e$ is of the order of the maximum phonon frequency in the solid. Note that the approximation (5.45) leads to $\Omega(0)=\omega_e$.

Several higher-order moments have been calculated (Boon and Yip 1980). They are much more complicated than the low-order moments considered here. In particular they involve the three-particle distribution function, which is generally not available unless a computer simulation is performed.

## 5.3. Classical monatomic fluid; single-particle response

The response function $R_s(\kappa, \omega)$ is rather structureless when plotted as a function of $\omega$ for a fixed $\kappa$. For large $\kappa$ the fourth moment (5.42) tends to $3(\kappa^2/M\beta)^2$ and $(\kappa^2/M\beta)$ is the second moment (eqn (5.37)). This means that in this limit the fourth moment assumes the same behaviour as in a Gaussian approximation for the frequency dependence. An equivalent reasoning is that for sufficiently large $\kappa$, $R_s(\kappa, \omega)$ must tend to the result for a single particle subject to Boltzmann statistics.

From § 3.6, or directly from (5.29) with $\mathbf{R}(t)-\mathbf{R}(0)\doteq t\mathbf{v}$ where $\mathbf{v}$ is the particle velocity,

$$R_s(\kappa, \omega)=\left(\frac{M\beta}{2\pi\kappa^2}\right)^{1/2}\exp\left(-\frac{M\beta\omega^2}{2\kappa^2}\right); \quad \text{Boltzmann particle} \quad (5.46)$$

and the frequency dependence is indeed Gaussian.

In the opposite limit of small $\kappa$ the fourth moment has a value too large to be compatible with a Gaussian. Hence it is more appropriate for small $\kappa$ to approximate $R_s(\kappa, \omega)$ by a Lorentz curve as obtained from diffusion theory. To see how this comes about, recall that the self-diffusion constant $D_s$ is defined to be

$$D_s = \frac{1}{6} \lim_{t\to\infty} \frac{1}{t} \langle\{\mathbf{R}(t) - \mathbf{R}\}^2\rangle. \tag{5.47}$$

Turning to $R_s(\kappa, t)$ (eqn (5.29)) and considering the limit $\kappa \to 0$, the expansion of the exponential leads to

$$R_s(\kappa, t) = 1 + \langle \mathrm{i}\boldsymbol{\kappa} \cdot \{\mathbf{R}(t) - \mathbf{R}\} - \tfrac{1}{2}[\boldsymbol{\kappa} \cdot \{\mathbf{R}(t) - \mathbf{R}\}]^2 + \cdots \rangle.$$

The second term is zero for an isotropic system, and the third term reduces to

$$-\tfrac{1}{6}\kappa^2\langle\{\mathbf{R}(t) - \mathbf{R}\}^2\rangle$$

which, in the light of (5.47), we approximate for long times by

$$-\kappa^2 t D_s.$$

Hence, for $\kappa \to 0$ and long times and $\kappa^2 D_s t \ll 1$,

$$R_s(\kappa, t) \doteq 1 - \kappa^2 t D_s + \cdots \doteq \exp(-\kappa^2 D_s |t|), \tag{5.48}$$

where we have made $R_s(\kappa, t)$ a function of $|t|$ because the relaxation function is an even function of time. The corresponding approximation for $R_s(\kappa, \omega)$ is

$$R_s(\kappa, \omega) = \frac{1}{\pi} \frac{\kappa^2 D_s}{\omega^2 + (\kappa^2 D_s)^2}; \quad \text{diffusion.} \tag{5.49}$$

The half-width at half-height of the Lorentz curve is $\kappa^2 D_s$.

In view of the structureless behaviour of $R_s(\kappa, \omega)$ for small and large $\kappa$, the main features are summarized in its width $\Delta$ and peak height. For $\kappa \to 0$, we have $\Delta = \kappa^2 D_s$ and $R_s(\kappa, 0) = 1/\pi\Delta$, and in the opposite limit, $\kappa \to \infty$, we find from (5.46) the values $\Delta = (2 \ln 2/\beta M)^{1/2}\kappa$ and $R_s(\kappa, 0) = (\beta M/2\pi)^{1/2}/\kappa$. The ratio $\Delta/\kappa^2 D_s$ as a function of $\kappa$ decreases monotonically from unity to zero. However, it is found in computer simulations of dense fluids that $\Delta/\kappa^2 D_s$ is an oscillatory function of $\kappa$ (Hansen and McDonald 1976). This behaviour is attributed to the backflow of atoms surrounding the observed particle (Wahnström and Sjögren 1982).

The remainder of this section is devoted to a more detailed discussion of the single-particle response function and its relation to other fundamental quantities, such as the velocity auto-correlation function.

### 5.3.1. *Gaussian approximation*

The exact calculation of $R_s(\kappa, t)$ for simple models of interest reveals that it is a Gaussian function of $\kappa$. This observation is the basis for an approximation that assumes the relation

$$R_s(\kappa, t) = \exp\{-\tfrac{1}{2}\kappa^2\mu(t)\}, \tag{5.50}$$

where $\mu(t)$, of dimension $(\text{length})^2$, is to be determined. Because $R_s(\kappa, 0) = 1$, it follows that $\mu(0) = 0$ and, for an unconfined particle, $\mu(t) \to \infty$ for $t \to \infty$. Also, from the known properties of $R_s(\kappa, t)$, $\mu(t)$ must be purely real and an even function of time.

The known exact results are as follows. For a free particle described by Boltzmann statistics,

$$\mu(t) = (t^2/M\beta); \quad \text{free particle} \tag{5.51}$$

while, for a freely diffusing particle,

$$\mu(t) = 2D_s\{|t| - D_s M\beta\}; \quad \text{diffusion.} \tag{5.52}$$

Result (5.51) follows from § 1.6, or directly from (5.29) (cf. (5.46)). Result (5.52) is consistent with our introductory treatment of self-diffusion (eqn (5.48)) and the constant term will be confirmed later. The third, and final, result for a harmonically bound particle is obtained from (3.142) and (3.143) by taking the limit $\hbar \to 0$

$$\mu(t) = \frac{2}{M\beta}\int_0^\infty \mathrm{d}\omega \frac{Z(\omega)}{\omega^2}\{1 - \cos(\omega t)\}; \quad \text{harmonic oscillation} \tag{5.53}$$

where $Z(\omega)$ is the appropriate, normalized vibrational density of states. For a single Einstein oscillator of frequency $\omega_e$,

$$Z(\omega) = \delta(\omega - \omega_e) \tag{5.54}$$

and, for a Debye model of a solid,

$$\begin{aligned} Z(\omega) &= (3\omega^2/\omega_D^3); \qquad && \omega < \omega_D \\ &= 0, && \omega > \omega_D \end{aligned} \tag{5.55}$$

The corresponding $\mu(t)$ is readily shown to be

$$\mu(t) = \mu_\infty\left\{1 - \frac{\sin(t\omega_D)}{t\omega_D}\right\}; \quad \text{Debye solid} \tag{5.56}$$

with

$$\begin{aligned} \mu_\infty = \tfrac{2}{3}\langle \mathbf{R}^2\rangle &= \frac{2}{M\beta}\int_0^\infty \mathrm{d}\omega \frac{Z(\omega)}{\omega^2} \\ &= (6/M\beta\omega_D^2); \qquad \text{Debye solid.} \end{aligned} \tag{5.57}$$

Hence, unlike the case of a free or diffusing particle, $\mu(t)$ for a confined particle tends to a finite limit at long times. The difference in the long-time behaviour of $\mu(t)$ for a single isolated oscillator and a harmonic solid is discussed in § 3.6.1. Note that result (5.51) must be achieved in the limit of $t \to 0$ for any type of motion.

A precise expression for $\mu(t)$ is readily derived. To this end, we recall that $R_s(\kappa, t)$ is the spatial Fourier transform of the probability density,

$$\langle \delta\{\mathbf{r}+\mathbf{R}-\mathbf{R}(t)\}\rangle,$$

that is to say,

$$R_s(\kappa, t) = \int \mathrm{d}\mathbf{r} \exp(\mathrm{i}\boldsymbol{\kappa} \cdot \mathbf{r})\langle \delta\{\mathbf{r}+\mathbf{R}-\mathbf{R}(t)\}\rangle. \tag{5.58}$$

If we differentiate both sides of (5.58) twice with respect to $\kappa$ and then set $\kappa = 0$, it follows that when $R_s(\kappa, t)$ is given by (5.50) the function $\mu(t)$ satisfies the relation

$$\mu(t) = \tfrac{1}{3}\int \mathrm{d}\mathbf{r}\, r^2 \langle \delta\{\mathbf{r}+\mathbf{R}-\mathbf{R}(t)\}\rangle$$
$$= \tfrac{1}{3}\langle\{\mathbf{R}(t)-\mathbf{R}\}^2\rangle. \tag{5.59}$$

From this we conclude that $\mu(t)$ in (5.50) is the mean-square displacement. From (5.59) we can derive an equally important relation between $\mu(t)$ and the velocity auto-correlation function $\langle \mathbf{v} \cdot \mathbf{v}(t)\rangle$. The result is

$$\mu(t) = \tfrac{2}{3}\int_0^t \mathrm{d}\bar{t}\,(t-\bar{t})\langle \mathbf{v} \cdot \mathbf{v}(\bar{t})\rangle. \tag{5.60}$$

Verification of (5.60) from (5.59) follows on observing that, for a stationary system,

$$\langle \mathbf{v} \cdot \mathbf{v}(t)\rangle = \tfrac{1}{2}\partial_t^2 \langle\{\mathbf{R}(t)-\mathbf{R}\}^2\rangle. \tag{5.61}$$

From the definition of the self-diffusion constant (eqn (5.47)) and use of (5.59) and (5.60) it follows that

$$D_s = \tfrac{1}{3}\int_0^\infty \mathrm{d}t \langle \mathbf{v} \cdot \mathbf{v}(t)\rangle. \tag{5.62}$$

We emphasize that (5.62) is an exact relation. The result (5.60) is also exact given that $\mu(t)$ is defined by (5.59). However, the relationship between $\mu(t)$ and $R_s(\kappa, t)$ (eqn (5.50)) is generally an approximation.

Consider now the calculation of the velocity auto-correlation function for an atom diffusing freely in a fluid. The central concept is that of a Brownian motion described by the phenomenological equation of motion (van Kampen 1981)

$$M\dot{v}(t) = -\gamma M v(t) + f(t). \tag{5.63}$$

Here, $\dot{v}(t)$ is the acceleration of the atom, $\gamma$ is the strength of the frictional force exerted by the fluid, and $f(t)$ represents the random force due to the random collisions with other atoms. The period over which $f(t)$ varies is very short compared with the time scale of the observed motion. Thus, the separation of the force on the right-hand side of (5.63) into two components assumes a distinct separation of time scales for the frictional and random forces. Given the separation of time scales there can be no correlation between the velocity and random force at different times. Hence, if in (5.63) we multiply through by $v(0) \equiv v$ and average, we obtain

$$\langle v\dot{v}(t)\rangle = -\gamma\langle vv(t)\rangle$$

from which it follows that

$$\langle \mathbf{v} \cdot \mathbf{v}(t)\rangle = \langle \mathbf{v}^2\rangle \exp(-\gamma\,|t|). \tag{5.64}$$

Note that this result cannot hold for small $t$ because $\langle \mathbf{v} \cdot \mathbf{v}(t)\rangle$ in reality cannot have a discontinuous gradient at $t = 0$ as (5.64) implies.

Inserting (5.64) into (5.62) leads to

$$D_s = \langle \mathbf{v}^2\rangle/3\gamma \tag{5.65}$$

and, assuming the law of equipartition of energy, we recover the Einstein relation

$$\gamma D_s = 1/M\beta. \tag{5.66}$$

Furthermore, the approximation (5.64) inserted into (5.60) gives a more precise form for $\mu(t)$ at long times, namely eqn (5.52).

We will consider the exact long-time behaviour of $\langle \mathbf{v} \cdot \mathbf{v}(t)\rangle$ in the next section. Before doing so we note that the short-time behaviour is

$$\langle \mathbf{v} \cdot \mathbf{v}(t)\rangle = \langle \mathbf{v}^2\rangle \{1 - \tfrac{1}{2}t^2\Omega^2(0) + \cdots\} \tag{5.67}$$

where $\Omega^2(0)$ is obtained from (5.41) with $\kappa = 0$. The result (5.67) can be deduced from (5.42) by noting that from the definition (5.29)

$$\lim_{\kappa \to 0} \frac{1}{\kappa^2} \ddot{R}_s(\kappa, t) = -\tfrac{1}{3}\langle \mathbf{v} \cdot \mathbf{v}(t)\rangle. \tag{5.68}$$

An equivalent relation in terms of $R_s(\kappa, \omega)$ is

$$\frac{1}{6\pi}\int_{-\infty}^{\infty} \mathrm{d}t \exp(-\mathrm{i}\omega t)\langle \mathbf{v} \cdot \mathbf{v}(t)\rangle = \lim_{\kappa \to 0} (\omega^2/\kappa^2) R_s(\kappa, \omega). \tag{5.69}$$

This last result tells us that the time Fourier transform of the velocity auto-correlation function can be deduced from the appropriate limiting form of the response function, $R_s(\kappa, \omega)$, that appears in the cross-section.

### *5.3.2. Hydrodynamic limit*

We finished the previous subsection with an exact relation between $R_s(\kappa, \omega)$ and the velocity auto-correlation function that holds in the limit

$\kappa \to 0$. Velocity correlations have been studied extensively using computer simulations, and the inadequacy of the simple exponential decay (eqn (5.64)) is well established. The most glaring flaw is a failure to reproduce the region of negative correlation observed in nearly all simulations of dense fluids. The negative component of the velocity auto-correlation function is striking evidence of the very highly correlated motion of atoms in dense fluids since no such effect is seen in simulations of gas phase systems.

The general features of the velocity auto-correlation function are described by the approximation (Copley and Lovesey 1975)

$$\langle \mathbf{v} \cdot \mathbf{v}(t) \rangle = \langle \mathbf{v}^2 \rangle \exp(-t/2\tau) \left[ \cos(\epsilon t) + \frac{1}{2\epsilon\tau} \sin(\epsilon t) \right] \tag{5.70}$$

where

$$\epsilon^2 = \Omega^2(0) - (1/2\tau)^2$$

and

$$(1/\tau) = M\beta D_s \Omega^2(0).$$

This approximation reproduces the short-time behaviour (5.67) and it satisfies (5.62). A region of negative correlation occurs provided that $\epsilon$ is real, and this requires

$$M\beta D_s \Omega(0) < 2. \tag{5.71}$$

Simulation data for a Lennard–Jones model, which is an appropriate model for liquid argon, is consistent with the scaling behaviour $\Omega^2(0) \propto T^{1/2}\rho_0^2$ and $D_s \propto T/\rho_0^2$. These results imply that the left-hand side of (5.71) is proportional to $T^{1/4}/\rho_0$, which indicates that the region of negative correlation will vanish for sufficiently low particle density or high enough temperature. For argon near its triple point, $\beta M D_s \Omega(0) \simeq 0.9$ and the inequality is easily satisfied.

In the limit of long times it has been shown that the velocity correlations decrease according to an inverse power law (Résibois and De Leener 1977). The actual behaviour, for $t \to \infty$, is

$$\langle \mathbf{v} \cdot \mathbf{v}(t) \rangle \to \langle \mathbf{v}^2 \rangle A/t^{3/2}, \tag{5.72}$$

where the coefficient

$$A = (2/3\rho_0)\{4\pi(D_s + \eta/M\rho_0)\}^{-3/2}$$

and $\eta$ is the shear viscosity. The corresponding result for the mean-square displacement is

$$\lim_{t\to\infty} \tfrac{1}{6}\langle \{\boldsymbol{R}(t) - \mathbf{R}\}^2 \rangle = tD_s - \frac{4At^{1/2}}{M\beta} + \text{constant}. \tag{5.73}$$

The Fourier transform of the velocity auto-correlation function, which occurs on the left-hand side of (5.69), contains a term in $\omega^{1/2}$. The latter term makes $R_s(\kappa, \omega)$ a nonanalytic function of $\omega$ in the hydrodynamic limit $\kappa, \omega \to 0$.

To investigate the hydrodynamic limit of $R_s(\kappa, \omega)$ in more detail we first establish a relation between this response function and the velocity auto-correlation function. For this, and the ensuing development, it is convenient to work in terms of the Laplace transform of $R_s(\kappa, t)$. We use the definition

$$\tilde{R}_s(\kappa, s) = \int_0^\infty \mathrm{d}t \exp(-st) R_s(\kappa, t). \tag{5.74}$$

Because $R_s(\kappa, t)$ is purely real and an even function of $t$, the Fourier transform $R_s(\kappa, \omega)$ is obtained from (5.74) using the simple relation

$$\pi R_s(\kappa, \omega) = \operatorname{Re} \tilde{R}_s(\kappa, \mathrm{i}\omega). \tag{5.75}$$

If the Laplace transform of the normalized velocity auto-correlation function is denoted by $\tilde{\Phi}(s)$ the relation between $\tilde{R}_s(\kappa, s)$ and $\tilde{\Phi}(s)$ for $\kappa \to 0$ is

$$\tilde{R}_s(\kappa, s) = \{s + \tfrac{1}{3}\kappa^2 \langle \mathbf{v}^2 \rangle \tilde{\Phi}(s)\}^{-1}; \qquad \kappa \to 0. \tag{5.76}$$

This result can be derived by a straightforward application of the generalized Langevin equation discussed in Appendix B. We do not pause to construct the argument, but note that (5.76) is consistent with (5.69).

The form proposed for $\tilde{\Phi}(s)$ is

$$\begin{aligned} \tilde{\Phi}(s) &= \int_0^\infty \mathrm{d}t \exp(-st)\{\langle \mathbf{v} \cdot \mathbf{v}(t) \rangle / \langle \mathbf{v}^2 \rangle\} \\ &= \{s + \tilde{M}(s)\}^{-1} \end{aligned} \tag{5.77}$$

with

$$\tilde{M}(s) = \left\{ \frac{\Omega^2(0)}{s + \left( \dfrac{\tau \delta^2}{1 + s\tau + 2\gamma(s\tau)^{1/2}} \right)} \right\} \tag{5.78}$$

where $\tau$, $\delta^2$, and $\gamma$ are determined from the known properties of the velocity auto-correlation function.

First, the short-time expansion

$$\langle \mathbf{v} \cdot \mathbf{v}(t) \rangle = \langle \mathbf{v}^2 \rangle \left\{ 1 - \frac{1}{2!} t^2 \Omega^2(0) + \frac{1}{4!} t^4 \Omega^2(0)[\delta^2 + \Omega^2(0)] - \cdots \right\} \tag{5.79}$$

is equivalent to an expansion of $\tilde{\Phi}(s)$ in $(1/s)$

$$s\tilde{\Phi}(s) = 1 - \frac{\Omega^2(0)}{s^2} + \frac{\Omega^2(0)[\delta^2 + \Omega^2(0)]}{s^4} - \cdots \tag{5.80}$$

and this relation is satisfied by (5.77) and (5.78) independent of the quantities $\tau$ and $\gamma$.

The relaxation time $\tau$ is determined through (5.62), with the result

$$\tau = \left\{\frac{3D_s\Omega^2(0)}{\langle \mathbf{v}^2\rangle\, \delta^2}\right\}. \tag{5.81}$$

Finally, the dimensionless quantity $\gamma$ is determined by the long-time behaviour (5.72). This result means that for, $s \to 0$ (Boon and Yip 1980),

$$\tilde{\Phi}(s) = \tilde{\Phi}(0) - 2A\sqrt{(\pi s)} + \cdots. \tag{5.82}$$

In obtaining (5.82) we have used the result that, if for $t \to \infty$ the function $g(t) \sim t^{-\mu}$ where the power $\mu > 1$, then the Laplace transform $\tilde{g}(s) \sim \tilde{g}(0) + \pi s^{\mu-1}/\Gamma(\mu)\sin(\pi\mu)$ in the limit of small $s$, and $\Gamma(\mu)$ is the gamma function. Comparing (5.82) with (5.78) expanded for small $(s\tau)$, we find

$$\gamma = \frac{1}{36\pi\rho_0} \frac{\langle \mathbf{v}^2\rangle}{D_s\tau^{1/2}} (D_s + \eta/M\rho_0)^{-3/2}. \tag{5.83}$$

We conclude that the form proposed for $\tilde{\Phi}(s)$ has the correct short- and long-time behaviour. The corresponding result for $R_s(\kappa, \omega)$, obtained from eqns (5.75)–(5.78), is exact in the hydrodynamic limit where $\kappa, \omega \to 0$.

The departure of $R_s(\kappa, \omega)$ from a Lorentz function (eqn (5.49)) is controlled essentially by the parameter $\gamma$ whose magnitude is a measure of the hydrodynamic effects. Moreover, the departure from the simple diffusion form is seen at very small $\omega$, as might be expected. The most noticeable feature in the hydrodynamic result is that the maximum of $R_s(\kappa, \omega)$ is no longer at $\omega = 0$.

Before leaving the subject of hydrodynamic effects in the single-particle response function, we add a few comments about the expression used for $\tilde{M}(s)$ (eqn (5.78)). In the problem of calculating the frictional force on a sphere moving with a constant velocity in a viscous fluid, $\tilde{M}(s)$ appears as a complex friction coefficient, i.e. it relates the frictional force to the velocity (Hansen and McDonald 1976). Hydrodynamic calculations of the frictional force, with various boundary conditions, lead to results that are entirely consistent with the frequency dependence chosen in (5.78). The hydrodynamic results can be used to provide expressions for the quantities in (5.78) in terms of the radius of the sphere $a$ and shear viscosity of the fluid. For the 'stick' boundary condition the fluid velocity at the surface is everywhere taken equal to the sphere velocity, and the result for the diffusion constant is the familiar Stoke's law $D_s = k_B T/6\pi a\eta$. The other quantities in $\tilde{M}(s)$ are,

$$\gamma = \tfrac{1}{2}(a^2 M\rho_0/\tau\eta)^{1/2} \tag{5.84}$$

and

$$\delta = 3D_s\Omega^2(0)/\langle \mathbf{v}^2\rangle, \tag{5.85}$$

provided that $a^3\rho_0 \ll (\langle \mathbf{v}^2\rangle/3\Omega(0)D_s)^2$. The relaxation time $\tau$ is given by (5.81). The result (5.85) can be interpreted as providing a hydrodynamic expression for the sixth frequency moment of $R_s(\kappa, \omega)$ since

$$\lim_{\kappa\to 0}\int_{-\infty}^{\infty} d\omega\, \omega^6 R_s(\kappa, \omega) = \frac{\kappa^2}{M\beta}\Omega^2(0)\{\delta^2 + \Omega^2(0)\}.$$

### 5.3.3. *Intermediate* $\kappa$

For sufficiently large $\kappa$, the response function $R_s(\kappa, \omega)$ approaches the free-particle result (eqn (5.46)). In the hydrodynamic limit, where $\kappa$, $\omega \to 0$, we have the result of the previous section which is very similar to the Lorentz curve (5.49) except for very small $\omega$. Various approximate expressions have been proposed for $R_s(\kappa, \omega)$ at intermediate values of the scattering vector and they are reviewed in several references (Boon and Yip 1980; Wahnström and Sjögren 1982; Copley and Lovesey 1975).

A simple expression that fulfils a number of the exact conditions for $R_s(\kappa, \omega)$ can be derived from (5.76) with a wave-vector dependent quantity replacing $\tilde{\Phi}(s)$. Turning to (5.78) we replace $\tilde{M}(s)$, the complex friction coefficient, by the simpler form

$$\tilde{M}(s) = \delta_2(\kappa)/\{s + 1/\tau(\kappa)\} \tag{5.86}$$

where

$$\delta_2(\kappa) = \frac{2\kappa^2}{M\beta} + \Omega^2(0) \tag{5.87}$$

and the wave-vector dependent relaxation time is to be determined. Independent of the choice of $\tau(\kappa)$, the result for $R_s(\kappa, \omega)$ that corresponds to the choice (5.86) is normalized to unity and satisfies the second and fourth frequency moment relations (eqns (5.37) and (5.42)). The latter is most easily verified by noting that the moment relations give the following result for the short-time expansion of $R_s(\kappa, t)$,

$$R_s(\kappa, t) = 1 - \frac{1}{2!}t^2\left(\frac{\kappa^2}{M\beta}\right) + \frac{1}{4!}t^4\left(\frac{\kappa^2}{M\beta}\right)\left\{\frac{3\kappa^2}{M\beta} + \Omega^2(0)\right\} - \cdots. \tag{5.88}$$

From this we deduce that the Laplace transform $\tilde{R}_s(\kappa, s)$ (eqn (5.74)) satisfies

$$s\tilde{R}_s(\kappa, s) = 1 - \frac{1}{s^2}\left(\frac{\kappa^2}{M\beta}\right) + \frac{1}{s^4}\left(\frac{\kappa^2}{M\beta}\right)\left\{\frac{3\kappa^2}{M\beta} + \Omega^2(0)\right\} - \cdots. \quad (5.89)$$

It is straightforward to show that the expression derived from (5.86)

$$\tilde{R}_s(\kappa, s) = \left\{ s + \frac{(\kappa^2/M\beta)}{\left(s + \dfrac{\delta_2(\kappa)}{s + 1/\tau(\kappa)}\right)} \right\}^{-1} \quad (5.90)$$

agrees with (5.89) if $\delta_2(\kappa)$ is given by (5.87). Because the approximation (5.90) satisfies the moment relations we anticipate that it is good for large frequencies.

There remains the choice of $\tau(\kappa)$. By using a line of reasoning that is given in § 5.4.3 we deduce that

$$1/\tau(\kappa) = \xi\{\delta_2(\kappa)\}^{1/2} \quad (5.91)$$

where $\xi$ will be determined from $R_s(\kappa, \omega = 0)$.

From (5.90) and using (5.75), we obtain

$$\pi R_s(\kappa, \omega) = \frac{\tau\delta_1\delta_2}{[\omega\tau(\omega^2 - \delta_1 - \delta_2)]^2 + (\omega^2 - \delta_1)^2} \quad (5.92)$$

where

$$\delta_1 = (\kappa^2/M\beta). \quad (5.93)$$

The wave-vector dependence of the various quantities appearing on the right-hand side is suppressed for ease of notation.

The quantity $\xi$ in the expression for the relaxation time is estimated to be of order unity. Forming $R_s(\kappa, \omega = 0)$ from (5.93) and comparing the result with the Lorentz curve (5.49) we find the limit $\kappa \to 0$,

$$\xi = M\beta D_s\Omega(0). \quad (5.94)$$

In § 5.3.2 we noted that for argon near its triple point the right-hand side is approximately 0.9. For large $\kappa$ we ask that $R_s(\kappa, \omega = 0)$ formed from (5.92) should agree with the corresponding result for a free particle (eqn (5.46)). This procedure leads to the result

$$\xi = 2/\sqrt{\pi} = 1.13. \quad (5.95)$$

We conclude from this discussion that the approximation (5.92) for $R_s(\kappa, \omega)$ has at small $\omega$ the correct dependence on $\kappa$ at both large and small $\kappa$. Moreover, the estimates of $\xi$ from these two limiting cases are in fair agreement, and $\xi \simeq 1$.

Expression (5.92) is consistent with the approximation (5.70) for the velocity auto-correlation function; this is readily verified using (5.69). A comparison of (5.92) with available data shows that it provides a very satisfactory description (Copley and Lovesey 1975).

## 5.4. Classical monatomic fluid: coherent response

The coherent partial differential cross-section for the scattering of unpolarized neutrons by a monatomic fluid is shown in § 5.1 to be

$$\left(\frac{d^2\sigma}{d\Omega\, dE'}\right)_{\text{coh}} = \frac{k'}{k}\frac{\sigma_c}{4\pi} S(\kappa, \omega). \tag{5.96}$$

The response function $S(\kappa, \omega)$ is expressed in terms of the Fourier transform of the relaxation function for density fluctuations. For a target sample that is described by classical statistical mechanics, to a good approximation, the relaxation function is

$$R(\kappa, \omega) = \frac{1}{2\pi N}\int_{-\infty}^{\infty} dt \exp(-i\omega t)\langle \rho_{\boldsymbol{\kappa}}\rho^*_{\boldsymbol{\kappa}}(t)\rangle, \tag{5.97}$$

where $N$ is the total number of scattering particles. The relation between $S(\kappa, \omega)$ and $R(\kappa, \omega)$ is

$$S(\kappa, \omega) = \{1 + n(\omega)\}\omega\beta R(\kappa, \omega). \tag{5.98}$$

Here, $\beta$ is the inverse temperature and the first term is the detailed balance factor (5.17).

It is usually sensible to subtract from (5.97) the purely elastic contribution. The reason for doing so is that the remaining terms, which describe density fluctuations, lead naturally to expressions that are numerically well behaved, e.g. the static structure factor $S(\kappa)$, expressed as the Fourier transform of the pair distribution function (eqn (5.33)). Moreover, the subtraction of the elastic contribution does not materially alter the cross-section since, for a fluid, the elastic contribution is accompanied by the condition $\boldsymbol{\kappa} = 0$, which corresponds to no scattering. In view of these considerations, in (5.97) we use the variable

$$\rho_{\boldsymbol{\kappa}}(t) = \sum_j [\exp\{-i\boldsymbol{\kappa}\cdot\mathbf{R}_j(t)\} - \delta_{\boldsymbol{\kappa},0}], \tag{5.99}$$

where $\mathbf{R}_j(t)$ is the position of the $j$th particle in the target sample at time $t$. One final point concerning the chosen definition of $\rho_{\boldsymbol{\kappa}}(t)$ is that, since the constant term does not enter the equation-of-motion for $\rho_{\boldsymbol{\kappa}}(t)$, the dynamics of the fluctuations are not changed in any way by our decision to subtract the purely elastic contribution from $R(\kappa, \omega)$.

The fluctuation variable (5.99) is the spatial Fourier transform of the microscopic particle density fluctuation

$$\delta\rho(\mathbf{r}, t) = \sum_j \delta\{\mathbf{r} - \mathbf{R}_j(t)\} - \rho_0 \tag{5.100}$$

where $\rho_0 = N/V$ is the particle density. The relation

$$\rho_{\boldsymbol{\kappa}}(t) = \int d\mathbf{r} \exp(-i\boldsymbol{\kappa} \cdot \mathbf{r})\, \delta\rho(\mathbf{r}, t) \tag{5.101}$$

is consistent with (5.99) since the Dirac and Kronecker delta functions are related by $(2\pi)^3\,\delta(\boldsymbol{\kappa}) = V\delta_{\boldsymbol{\kappa},0}$. The correlation of static fluctuations in the microscopic particle density defines the pair distribution function. Writing $\delta\rho(\mathbf{r}) = \delta\rho(\mathbf{r}, t=0)$, we have

$$\langle \delta\rho(\mathbf{r}_1)\, \delta\rho(\mathbf{r}_2)\rangle = \rho_0\, \delta(\mathbf{r}_1 - \mathbf{r}_2) + \rho_0^2\{g(\mathbf{r}_1 - \mathbf{r}_2) - 1\}. \tag{5.102}$$

For large separations, $|\mathbf{r}_1 - \mathbf{r}_2| \to \infty$, the pair distribution function $g(\mathbf{r}_1 - \mathbf{r}_2)$ tends to unity, and the correlation vanishes.

The behaviour of $R(\kappa, \omega)$ as a function of $\omega$ depends strongly on the value of $\kappa$. We have some indication of this from the frequency moments evaluated in § 5.2. The normalized second moment

$$\omega_0^2 = (\kappa^2/M\beta S(\kappa)) \tag{5.103}$$

shows that the energy spread narrows as $\kappa$ passes through a maximum in the structure factor. The narrowing is pronounced in the vicinity of the main peak of $S(\kappa)$, which can be as large as 2.5–3.0 for dense fluids (cf. Fig. 5.1). For large $\kappa$, defined as being beyond the value at which $\{S(\kappa) - 1\}$ is minimal and coherent effects have vanished, $R(\kappa, \omega)$ approaches the result (5.46) for a single free particle. In the opposite limit, $\kappa \to 0$, the density fluctuations can be identified with acoustic waves and $R(\kappa, \omega)$ features distinct peaks at frequencies that correspond to the propagation of sound; the peaks are usually referred to as the Brillouin doublet. However, this is not the full story of the dynamics of density fluctuations in the hydrodynamic limit, for it is essential to consider the role of entropy fluctuations in addition to the pressure fluctuations that oscillate and give rise to sound propagation.

The thermodynamic state of a system made up of identical particles is determined by the values of any two quantities such as pressure, volume, temperature, entropy, etc. We choose the *statistically independent* variables of pressure $p$ and entropy $S$; the pressure can be regarded as a 'mechanical' quantity and the entropy a 'thermal' quantity. In terms of these variables the fluctuation in the particle density is

$$\delta\rho = \left(\frac{\partial\rho}{\partial p}\right)_S \delta p + \left(\frac{\partial\rho}{\partial S}\right)_p \delta S. \tag{5.104}$$

If we Fourier analyse the density fluctuation $\delta\rho$ each Fourier component consists of two parts, the first part being mechanical and the second thermal. The mechanical part corresponds to the collective oscillations considered above. The thermal part is non-oscillatory and results in a peak in the cross-section centred about $\omega = 0$.

Let us now evaluate the intensity of both the thermal and mechanical parts of the density fluctuations and relate these to the neutron scattering cross-section in the forward direction, i.e. in the limit as $\boldsymbol{\kappa} \to 0$. We can anticipate that, as $\boldsymbol{\kappa} \to 0$, the time dependence of the fluctuations becomes slow and hence the scattering may be treated as quasi-elastic. Hence, we need to consider

$$\lim_{\kappa \to 0} R(\kappa, t=0) = \frac{V^2}{N}\langle(\delta\rho)^2\rangle. \tag{5.105}$$

But, from (5.104),

$$\langle(\delta\rho)^2\rangle = \left(\frac{\partial\rho}{\partial p}\right)_S^2 \langle(\delta p)^2\rangle + \left(\frac{\partial\rho}{\partial S}\right)_p^2 \langle(\delta S)^2\rangle + 2\left(\frac{\partial\rho}{\partial p}\right)_S \left(\frac{\partial\rho}{\partial S}\right)_p \langle\delta p\delta S\rangle, \tag{5.106}$$

where $\langle(\delta p)^2\rangle$ is the mean square thermodynamic fluctuation in the pressure, $\langle(\delta S)^2\rangle$ is the mean square thermodynamic fluctuation in the entropy, and $\langle\delta p\delta S\rangle$ the correlation between them. These have the values (Hansen and McDonald 1976)

$$\langle(\delta p)^2\rangle = -k_B T \Big/ \left(\frac{\partial V}{\partial p}\right)_S, \qquad \langle(\delta S)^2\rangle = k_B C_p, \qquad \langle\delta p\delta S\rangle = 0, \tag{5.107}$$

where the specific heat at constant pressure at

$$C_p = T\left(\frac{\partial S}{\partial T}\right)_p \tag{5.108}$$

and

$$\left(\frac{\partial V}{\partial p}\right)_S = -V\chi_s = -V\frac{C_v}{C_p}\chi_T, \tag{5.109}$$

where $\chi_s$ is the adiabatic compressibility and $\chi_T$ is the isothermal compressibility.

Now

$$\left(\frac{\partial\rho}{\partial p}\right)_S = -\frac{N}{V^2}\left(\frac{\partial V}{\partial p}\right)_S = \rho_0\chi_s \tag{5.110}$$

and

$$\left(\frac{\partial\rho}{\partial S}\right)_p = -\frac{N}{V^2}\left(\frac{\partial V}{\partial S}\right)_p = -\frac{N}{V^2}\left(\frac{\partial V}{\partial T}\right)_p\left(\frac{\partial T}{\partial S}\right)_p = \frac{\rho_0\alpha T}{C_p}, \tag{5.111}$$

where the thermal expansion coefficient is

$$\alpha = -\frac{1}{V}\left(\frac{\partial V}{\partial T}\right)_p. \tag{5.112}$$

Putting all these results together,

$$\langle(\delta\rho)^2\rangle = \frac{\rho_0^2 k_B T}{V}\left(\chi_s + \frac{TV\alpha^2}{C_p}\right). \tag{5.113}$$

We now use the thermodynamic identity

$$C_p - C_v = \frac{TV\alpha^2}{\chi_T} \tag{5.114}$$

to obtain the result

$$\langle(\delta\rho)^2\rangle = \frac{\rho_0^2 k_B T}{V}\left\{\chi_s + \chi_T \frac{C_p - C_v}{C_p}\right\}. \tag{5.115}$$

From this we can immediately deduce that the ratio of intensities between thermal and mechanical fluctuations is

$$\frac{\chi_T}{\chi_s}\frac{C_p - C_v}{C_p} = \frac{C_p - C_v}{C_v}$$

and furthermore the total fluctuation is given by

$$\langle(\delta\rho)^2\rangle = \frac{\rho_0^2 k_B T}{V}\chi_T. \tag{5.116}$$

For a solid $(C_p - C_v)/C_v$ is small and therefore it is usually reasonable to neglect the thermal fluctuations in the density. But for a liquid $(C_p - C_v)/C_v$ is larger and this can no longer be done. Furthermore, at the liquid-gas critical point, $C_p$ diverges while $C_v$ remains finite, and hence in the critical region the thermal fluctuations dominate the mechanical fluctuations. Since $\chi_T$ diverges at the critical point, the neutron cross-section will be large in the critical region.

From (5.116), (5.105), and (5.33), we conclude that

$$\lim_{\kappa\to 0} R(\kappa, t=0) = S(0) = 1 + \rho_0 \int \mathrm{d}\mathbf{r}\{g(r) - 1\}$$
$$= \rho_0 \chi_T/\beta. \tag{5.117}$$

This result is given above in § 5.2.

We continue our discussion of the coherent response by evaluating the static approximation. This approximation is useful when the incident neutron energy exceeds the maximum energy in the response of the target sample, which is of the order of the Debye energy for the solid, and no energy discrimination is applied to the scattered beam (cf. § 1.8). For the present case this amounts to an evaluation of $R(\kappa, t=0)=S(\kappa)$. The discussion in the next subsection emphasizes the development of fluctuations in the vicinity of the liquid–gas phase transition. However, the approach adopted, which is based on an expansion of the free energy, is of general interest in the interpretation of quasi-elastic scattering.

### 5.4.1. *Static approximation*

Although the static approximation

$$R(\kappa, \omega) \doteq R(\kappa, t=0)\, \delta(\omega) = S(\kappa)\, \delta(\omega) \tag{5.118}$$

amounts to the calculation of the structure factor $S(\kappa)$, we are not embarking on a development of theories for $S(\kappa)$. This is a topic in its own right which lies well outside the scope of this book (Hansen and McDonald 1976). The purpose of this subsection is, rather, to obtain $S(\kappa)$ from an expansion of the free energy. The approach is due to Landau, and it is used in the study of a wide range of phenomena. The Landau theory is often referred to as a mean-field theory (Stanley 1971).

Perhaps the first point to make is that $S(\kappa)$ is a wave-vector-dependent susceptibility. In terms of the isothermal susceptibility $\chi_\kappa$ introduced in Appendix B, $\chi_\kappa = \beta S(\kappa)$ for a classical system.

If we perturb a system in equilibrium by applying the interaction described by the Hamiltonian

$$\mathscr{H}_1 = -\int \mathrm{d}\mathbf{r}'\, A(\mathbf{r}')B(\mathbf{r}'), \tag{5.119}$$

the change in the thermal average of $A(\mathbf{r})$ to first order in $B$ is (cf. Appendix B)

$$\begin{aligned}\delta\langle A(\mathbf{r})\rangle &= \beta \int \mathrm{d}\mathbf{r}'\, B(\mathbf{r}')\{\langle A(\mathbf{r})A(\mathbf{r}')\rangle - \langle A(\mathbf{r})\rangle\langle A(\mathbf{r}')\rangle\}\\ &\equiv \beta \int \mathrm{d}\mathbf{r}'\, B(\mathbf{r}')\langle\{A(\mathbf{r}) - \langle A(\mathbf{r})\rangle\}\{A(\mathbf{r}') - \langle A(\mathbf{r}')\rangle\}\rangle.\end{aligned} \tag{5.120}$$

The thermal averages appearing on the right-hand side are taken with respect to the unperturbed system. If for $A(\mathbf{r})$ we choose the particle density $\rho(\mathbf{r}, 0)$ then the thermodynamic conjugate variable $B(\mathbf{r})$ is the change in the equilibrium value of the chemical potential $\delta\mu(\mathbf{r})$. The variables $T$ and $V$ are to be kept constant. Taking $\delta\mu(\mathbf{r}) = \delta\mu$, yields

$$\delta\rho|_{T,V} = \beta\, \delta\mu \int \mathrm{d}\mathbf{r}' \langle \delta\rho(0,0)\, \delta\rho(\mathbf{r}', 0)\rangle$$

from which it follows that

$$\left(\frac{\partial\rho}{\partial\mu}\right)_{T,V} = \beta \int \mathrm{d}\mathbf{r}' \langle \delta\rho(0,0)\, \delta\rho(\mathbf{r}', 0)\rangle. \tag{5.121}$$

Now

$$\left(\frac{\partial \rho}{\partial \mu}\right)_{T,V} = \rho_0\left(\frac{\partial \rho}{\partial p}\right)_T = \rho_0^2 \chi_T \tag{5.122}$$

so making use of (5.102) we obtain the identity (5.117), as required.

Equation (5.121) is important because it relates the thermodynamic derivative $(\partial\rho/\partial\mu)_{T,V}$ to the integral over all space of the density fluctuation correlation. As the critical point is approached, the correlation range must increase in such a way that the integral finally diverges.

The mean-square thermodynamic fluctuation in $N$ is (Hansen and McDonald 1976)

$$\langle(\delta N)^2\rangle = \frac{1}{\beta}\left(\frac{\partial N}{\partial \mu}\right)_{T,V}. \tag{5.123}$$

Hence from (5.117) and (5.122) it follows that

$$\rho_0\chi_T = \frac{\beta}{N}\langle(\delta N)^2\rangle = \beta + \beta\rho_0\int \mathrm{d}\mathbf{r}\{g(\mathbf{r})-1\}. \tag{5.124}$$

The first equality in (5.124) elucidates the physical reason for the observed increase in the isothermal compressibility $\chi_T$ on approaching the liquid-gas transition, for in changing from a liquid (high density) to a gas (low density) the number of particles in a given volume of the fluid undergoes large fluctuations about its mean value and therefore $\langle(\delta N)^2\rangle$ increases markedly over its value far from the critical point. We therefore anticipate that the large fluctuations in the particle density of the target fluid that occur as the critical point is approached will result in a marked increase in the scattering cross-section. Further, it is easy for us to see from this line of argument the general condition for the static approximation to be justified, namely, that the density fluctuations should appear static to the incident neutrons. Of course, as already noted, one way of satisfying this criterion is to use sufficiently high-energy incident neutrons. We find, however, in § 5.4.2 that the density fluctuations slow down at the critical point; this phenomenon is usually called the *thermodynamic slowing down* of fluctuation. In consequence the static approximation becomes increasingly good in the vicinity of the liquid-gas phase transition.

We turn now to evaluating the static cross-section when $\boldsymbol{\kappa}$ is small but not zero. For this we must derive an equation for the fluctuation in $\rho(\mathbf{r}, 0)$ due to the perturbation (5.119). The application of the perturbation described by the Hamiltonian (5.119), with $A(\mathbf{r}) = \rho(\mathbf{r}, 0)$ and $B(\mathbf{r}) = \delta\mu(\mathbf{r})$, will cause the free energy $F$ to deviate from its value in the initial equilibrium state. Regarding the free energy as a function of $\rho(\mathbf{r}, 0)$ the deviation in $F$ (at constant temperature and volume) is

$$\delta F=\int \mathrm{d}\mathbf{r}[a_0\,\delta\rho(\mathbf{r},0)+\tfrac{1}{2}a_1\{\delta\rho(\mathbf{r},0)\}^2+\cdots-\delta\mu(\mathbf{r})\,\delta\rho(\mathbf{r},0)].$$

Since $\int \mathrm{d}\mathbf{r}\rho(\mathbf{r},0)=\int \mathrm{d}\mathbf{r}\rho$, the first term in the Taylor expansion in $\delta\rho(\mathbf{r},0)$ is zero. The coefficient $a_1$ of the second term in the expansion is positive and vanishes at the critical point. For, by definition,

$$a_1=\frac{1}{V}\left(\frac{\partial^2 F}{\partial\rho^2}\right)_{\mathrm{T,V}}$$

and, since

$$\left(\frac{\partial F}{\partial\rho}\right)_{\mathrm{T,V}}=V\mu,$$

it follows from (5.122) that

$$a_1=\left(\frac{\partial\mu}{\partial\rho}\right)_{\mathrm{T,V}}=(\rho_0^2\chi_T)^{-1}. \tag{5.125}$$

The terms included in the Taylor expansion of the free energy per unit volume in $\delta\rho(\mathbf{r},0)$ are not the only ones that contribute. As a consequence of the applied perturbation the fluid is not homogeneous and accordingly we must take into account the spatial derivatives of $\rho(\mathbf{r},0)$. Because the fluid is isotropic the lowest order term of this kind that can enter in the expansion of $F$ is proportional to $\{\boldsymbol{\nabla}\rho(\mathbf{r},0)\}^2$. The coefficient of this expansion is taken to be a constant $\frac{1}{2}r_1^2$. We thus obtain

$$\delta F=\int \mathrm{d}\mathbf{r}[\tfrac{1}{2}a_1\{\delta\rho(\mathbf{r},0)\}^2+\tfrac{1}{2}r_1^2\{\boldsymbol{\nabla}\,\delta\rho(\mathbf{r},0)\}^2-\delta\mu(\mathbf{r})\delta\rho(\mathbf{r},0)]. \tag{5.126}$$

In order that this should be a minimum with respect to $\delta\rho(\mathbf{r},0)$ the latter must satisfy

$$a_1\,\delta\rho(\mathbf{r},0)-r_1^2\nabla^2\,\delta\rho(\mathbf{r},0)-\delta\mu(\mathbf{r})=0. \tag{5.127}$$

This is the desired equation for the fluctuation $\delta\rho$ in the particle density due to the change in the chemical potential from $\mu$ to $\mu+\delta\mu(\mathbf{r})$.

Now (5.120) tells us that

$$\langle\rho(\mathbf{r},0)\rangle-\rho_0=\langle\delta\rho(\mathbf{r},0)\rangle=\beta\int \mathrm{d}\mathbf{r}'\,\delta\mu(\mathbf{r}')\langle\delta\rho(\mathbf{r},0)\,\delta\rho(\mathbf{r}',0)\rangle. \tag{5.128}$$

Operating on (5.128) with $(a_1-r_1^2\nabla_r^2)$ and then using (5.127) yields the equation

$$\beta\int \mathrm{d}\mathbf{r}'\,\delta\mu(\mathbf{r}')(a_1-r_1^2\nabla_1^2)\langle\delta\rho(\mathbf{r},0)\delta\rho(\mathbf{r}',0)\rangle=\delta\mu(\mathbf{r}),$$

which can be rewritten

$$\int d\mathbf{r}'\ \delta\mu(\mathbf{r}')\{\beta(a_1 - r_1^2\nabla_r^2)\langle\delta\rho(\mathbf{r},0)\delta\rho(\mathbf{r}',0)\rangle - \delta(\mathbf{r}-\mathbf{r}')\} = 0. \quad (5.129)$$

Because $\delta\mu(\mathbf{r})$ is arbitrary,

$$\left(\nabla^2 - \frac{a_1}{r_1^2}\right)\langle\delta\rho(\mathbf{r},0)\delta\rho(0,0)\rangle = -\frac{1}{r_1^2\beta}\,\delta(\mathbf{r}). \quad (5.130)$$

By making use of the identity

$$\nabla^2\left(\frac{1}{r}\right) = -4\pi\delta(\mathbf{r}) \quad (5.131)$$

it is readily verified that the solution of (5.130) that is zero for $r\to\infty$ is

$$\langle\delta\rho(\mathbf{r},0)\delta\rho(0,0)\rangle = \frac{1}{4\pi r_1^2\beta}\frac{\exp(-\kappa_1 r)}{r}, \quad (5.132)$$

where

$$\kappa_1^2 = a_1/r_1^2 = 1/(\rho_0^2 r_1^2 \chi_T). \quad (5.133)$$

Near the critical point $\kappa_1$ become very small, so

$$\langle\delta\rho(\mathbf{r},0)\delta\rho(0,0)\rangle \simeq \frac{1}{4\pi r_1^2\beta}\frac{1}{r}.$$

The correlation amongst pairs of atoms is seen to decrease slowly with their separation; that is to say the correlation is much stronger than when the fluid is far from its critical point.

We must inquire after the conditions under which the solution (5.132) is valid. Because we retained in the Taylor expansion of the free energy (per unit volume) only the first non-zero term in $\delta\rho(\mathbf{r},0)$ and the lowest-order contribution involving the derivative of $\delta\rho(\mathbf{r},0)$, it is evident that the theory is limited to small values of $\delta\rho(\mathbf{r},0)$ which vary slowly with $\mathbf{r}$. In other words only those Fourier components of the disturbance that possess a long wavelength are taken into account. From (5.132) it is evident that these conditions are fulfilled for $r \gg \kappa_1^{-1}$.

It is straightforward to evaluate the $S(\kappa)$ corresponding to the solution (5.132). $\langle\rho_{\boldsymbol{\kappa}}\rho_{-\boldsymbol{\kappa}}\rangle$ is given in terms of $\langle\delta\rho(\mathbf{r},0)\delta\rho(0,0)\rangle$ through

$$\langle\rho_{\boldsymbol{\kappa}}\rho_{-\boldsymbol{\kappa}}\rangle = V\int d\mathbf{r}\,\exp(i\boldsymbol{\kappa}\cdot\mathbf{r})\langle\delta\rho(\mathbf{r},0)\delta\rho(0,0)\rangle, \quad (5.134)$$

and, by making use of the result

$$\int d\mathbf{r}\,\frac{\exp(-\kappa_1 r + i\boldsymbol{\kappa}\cdot\mathbf{r})}{r} = \frac{4\pi}{\kappa_1^2+\kappa^2}, \quad (5.135)$$

we find

$$S(\kappa) = \{\rho_0 r_1^2 \beta(\kappa^2 + \kappa_1^2)\}^{-1}. \tag{5.136}$$

Note that in the limit $\kappa \to 0$ we retrieve (5.117).

### 5.4.2. *Hydrodynamic theory*

Here we calculate $R(\kappa, t)$ in the limit of long wavelengths and long times. This is achieved by appealing to the hydrodynamic, or macroscopic, equations of motion for density fluctuations. The linear form of these equations permits us to calculate the variation in time and space of small fluctuations in the particle number density about the thermal equilibrium state of the target sample. An essential feature of the calculation is the assumption that the sample is in local equilibrium, by which we mean that an extension of (5.104) to fluctuations in space and time is valid, for sufficiently small amplitude fluctuations.

For example, we will need to consider the fluctuation in the energy density that accompanies a fluctuation in the particle density. The expression (5.104) shows that density and entropy fluctuations are correlated, and there is a fundamental relation between the pressure $p$, volume $V$, internal energy $U$, and entropy $S$

$$T\,\mathrm{d}S = \mathrm{d}U + p\,\mathrm{d}V. \tag{5.137}$$

Now, $U = eV$, where $e$ is the energy density, and $\mathrm{d}V \equiv -V\,\mathrm{d}\rho/\rho$ for a fixed number of particles. From (5.137) we find,

$$\frac{T}{V}\,\mathrm{d}S = \mathrm{d}e - h\,\mathrm{d}\rho \tag{5.138}$$

where

$$h = (e + p)/\rho \tag{5.139}$$

is the enthalpy, or heat content, per particle. We identify $T\,\mathrm{d}S/V$ as the density of heat energy $Q$ and, from (5.138), the relation between $Q$ and the energy and particle density fluctuations is

$$Q(\mathbf{r}, t) = e(\mathbf{r}, t) - h\rho(\mathbf{r}, t). \tag{5.140}$$

The macroscopic equation for the conservation of energy, expressed in terms of $Q(\mathbf{r}, t)$, is

$$\dot{Q}(\mathbf{r}, t) - \lambda \nabla^2 T(\mathbf{r}, t) = 0, \tag{5.141}$$

where $T(\mathbf{r}, t)$ is the local temperature and $\lambda$ is the thermal conductivity.

The equation of motion for the density $\rho(\mathbf{r}, t)$ is the Navier–Stokes equation

$$\ddot{\rho}(\mathbf{r}, t) - \frac{1}{M}\nabla^2 p(\mathbf{r}, t) + \nu\rho_0 \nabla^2\{\boldsymbol{\nabla} \cdot \mathbf{v}(\mathbf{r}, t)\} = 0. \tag{5.142}$$

Here, $p(\mathbf{r}, t)$ and $\mathbf{v}(\mathbf{r}, t)$ are the local pressure and velocity, respectively, and $\nu$ is the longitudinal kinematic viscosity. The first step is to linearize this equation and then eliminate the pressure in favour of the density and temperature.

The linearization amounts to writing $\rho(\mathbf{r}, t) = \rho_0 + \delta\rho(\mathbf{r}, t)$, and similar relations for $p$, $T$, and $\mathbf{v}$ (the mean fluid velocity is assumed to be zero, so $\mathbf{v} \equiv \delta\mathbf{v}$) and keeping the leading terms in the variations of the fluctuations. The conservation of particle number leads to the relation

$$\boldsymbol{\nabla} \cdot \mathbf{v}(\mathbf{r}, t) \doteq -\frac{1}{\rho_0}\, \delta\dot{\rho}(\mathbf{r}, t) \tag{5.143}$$

and so (5.142) reduces to

$$(\partial_t^2 - \nu\nabla^2\partial_t)\delta\rho(\mathbf{r}, t) = \frac{1}{M}\nabla^2\, \delta p(\mathbf{r}, t). \tag{5.144}$$

The reason we choose to eliminate the pressure in (5.144) in favour of $\rho$ and $T$ is that we aim to find $\delta\rho$, and $\rho$ and $T$ are uncorrelated variables (just as the pressure and entropy are also uncorrelated (eqn (5.107))) so the combination of $\rho$ and $T$ is particularly convenient to work with. From the relation

$$\delta p = \left(\frac{\partial p}{\partial T}\right)_\rho \delta T + \left(\frac{\partial p}{\partial \rho}\right)_T \delta\rho, \tag{5.145}$$

and the identity

$$\left(\frac{\partial p}{\partial \rho}\right)_T = -\frac{V}{\rho_0}\left(\frac{\partial p}{\partial V}\right)_T = Mv_T^2 \tag{5.146}$$

where $v_T$ is the isothermal velocity of sound, we deduce that

$$\frac{1}{M}\nabla^2\, \delta p(\mathbf{r}, t) = \frac{1}{M}\left(\frac{\partial p}{\partial T}\right)_\rho \nabla^2\, \delta T(\mathbf{r}, t) + v_T^2\nabla^2\, \delta\rho(\mathbf{r}, t). \tag{5.147}$$

Using this last result in the equation of motion (5.144) we obtain $\delta\rho$ in terms of $\delta T$

$$(\partial_t^2 - \nu\nabla^2\partial_t - v_T^2\nabla^2)\delta\rho(\mathbf{r}, t) = \frac{1}{M}\left(\frac{\partial p}{\partial T}\right)_\rho \nabla^2\, \delta T(\mathbf{r}, t). \tag{5.148}$$

Next, we use the conservation of energy equation (5.141), to establish a second equation of motion for $\delta T(\mathbf{r}, t)$ in terms of $\delta\rho(\mathbf{r}, t)$. From the identification of $Q$ with $T\,\mathrm{d}S/V$,

$$\begin{aligned}\delta Q &= \frac{T}{V}\left(\frac{\partial S}{\partial \rho}\right)_T \delta\rho + \frac{T}{V}\left(\frac{\partial S}{\partial T}\right)_\rho \delta T \\ &= \frac{T}{V}\left(\frac{\partial S}{\partial \rho}\right)_T \delta\rho + \rho_0 c_v\, \delta T,\end{aligned} \tag{5.149}$$

where $c_v$ is the specific heat per particle at constant volume. Inserting (5.149) in (5.141) gives the desired second equation,

$$\left(\partial_t - \frac{\lambda}{\rho_0 c_v}\nabla^2\right)\delta T(\mathbf{r}, t) = -\frac{T}{Nc_v}\left(\frac{\partial S}{\partial \rho}\right)_T \delta\dot{\rho}(\mathbf{r}, t). \tag{5.150}$$

It is easy to solve the pair of linear equations (5.148) and (5.150) for the fluctuation in the particle density. To obtain $R(\kappa, \omega)$ we actually need to solve for $\rho_\kappa$ (eqn (5.101)). Moreover, $R(\kappa, \omega)$ is readily obtained from the Laplace transform of $\rho_\kappa$. For if

$$\tilde{\rho}_\kappa(s) = \int_0^\infty dt \exp(-st)\rho_\kappa(t), \tag{5.151}$$

then

$$\tilde{R}(\boldsymbol{\kappa}, s) = \int_0^\infty dt \exp(-st)R(\kappa, t) = \frac{1}{N}\langle \rho_\kappa \tilde{\rho}^*_\kappa(s)\rangle \tag{5.152}$$

and

$$\pi R(\kappa, \omega) = \mathrm{Re}\, \tilde{R}(\kappa, \mathrm{i}\omega), \tag{5.153}$$

which is the analogue of (5.75) for single-particle motion. The relation (5.153) between the Fourier and Laplace transform functions arises because $R(\kappa, t)$ is purely real and an even function of the time.

The solution of (5.148) and (5.150) for $\tilde{\rho}_\kappa(s)$ proceeds by a straightforward elimination of $\tilde{T}_\kappa(s)$. We need to use the result for the Laplace transform of a derivative

$$\tilde{\dot{\rho}}_\kappa(s) = -\rho_\kappa + s\tilde{\rho}_\kappa(s), \tag{5.154}$$

where $\rho_\kappa = \rho_\kappa(t=0)$. In addition, the following identity, proved as in eqns (5.108)–(5.114), is used to simplify the result for $\tilde{\rho}_\kappa(s)$,

$$\frac{-1}{M}\left(\frac{\partial p}{\partial T}\right)_\rho \frac{T}{Nc_v}\left(\frac{\partial S}{\partial \rho}\right)_T = (\gamma - 1)v_T^2, \tag{5.155}$$

where $\gamma$ is the ratio $c_p/c_v$. In forming the correlation function $\langle \rho_\kappa \tilde{\rho}^*_\kappa\rangle$ from the equation for $\tilde{\rho}^*_\kappa$, observe that $\rho_\kappa$ and $T_\kappa$ are uncorrelated variables, and that $\langle \rho_\kappa \dot{\rho}^*_\kappa\rangle = 0$.

The result for $\tilde{R}(\kappa, s)$ defined in eqn (5.152) is easily put in the form

$$\tilde{R}(\kappa, s) = S(\kappa)\left\{s + \frac{\kappa^2 v_T^2}{s + \nu\kappa^2 + \dfrac{(\gamma-1)\kappa^2 v_T^2}{\left[s + \left(\dfrac{\kappa^2\lambda}{\rho_0 c_v}\right)\right]}}\right\}^{-1}. \tag{5.156}$$

Following the development given in the introduction to this section, we anticipate two contributions to the response. One contribution arises from non-propagating entropy fluctuations, and the second from propagating pressure fluctuations, or sound waves. These contributions are revealed in the poles of (5.156). We let $s = \mathrm{i}\omega$, and then set the denominator on the right-hand side equal to zero. The resulting cubic equation can be solved in closed form. Even so it is very complicated, so we expand the solutions in $\kappa$ and retain contributions to order $\kappa^2$. This level of approximation is consistent with the validity of the equation, since we have kept first-order terms and gradients on the grounds that the fluctuations have small amplitude and vary slowly in space. The modes that contribute to $\tilde{R}(\kappa, s)$ in the hydrodynamic limit have frequencies,

$$\omega = \mathrm{i}\kappa^2 D_T \tag{5.157}$$

and

$$\omega = \pm\kappa v_s + \mathrm{i}\kappa^2\Gamma \tag{5.158}$$

Here, $v_s = (\gamma v_T^2)^{1/2}$ is the adiabatic sound velocity; the thermal diffusivity

$$D_T = (\lambda/\rho_0 c_p) \tag{5.159}$$

and the sound attenuation coefficient

$$\Gamma = \tfrac{1}{2}[D_T(\gamma - 1) + \nu]. \tag{5.160}$$

The thermal diffusivity is very small in the vicinity of the liquid–gas phase transition because $c_p$ diverges more rapidly than $\lambda$. Note that the latter contributes to $\Gamma$, so that the damping of sound is attributed to both heat exchange between different parts of the fluid and internal frictional effects. If the fluid is supercooled, the viscosity, and therefore $\nu$, becomes extremely large. Indeed, the onset of the glassy-supercooled state is sometimes defined, in an empirical manner, by the attainment of a viscosity that exceeds the viscosity of water by 17 orders of magnitude.

Some algebra is required to write $R(\kappa, \omega)$ in the following appealing form, which is correct to leading-order in $\kappa^2$ and suggested by the development given in the introduction to this section,

$$\begin{aligned}\pi R(\kappa, \omega) &= \mathrm{Re}\,\tilde{R}(\kappa, \mathrm{i}\omega) \\ &= \tfrac{1}{2}S(\kappa)\left\{\left(1 - \frac{1}{\gamma}\right)\frac{2\kappa^2 D_T}{\omega^2 + (\kappa^2 D_T)^2} \right. \\ &\quad \left. + \left(\frac{\kappa^2\Gamma}{\gamma}\right)\left[\frac{1}{(\omega - \kappa v_s)^2 + (\kappa^2\Gamma)^2} + \frac{1}{(\omega + \kappa v_s)^2 + (\kappa^2\Gamma)^2}\right]\right\}.\end{aligned} \tag{5.161}$$

This result is normalized to $S(\kappa)$, as required. If the specific heats are essentially the same $(\gamma - 1) \doteq 0$, and the propagating sound waves dominate the response with an attenuation $\Gamma \doteq \tfrac{1}{2}\nu$. On the other hand, when $\gamma$ is large, the first contribution, called the Rayleigh line, dominates. This occurs when the liquid–gas phase transition is approached. For a given wave vector, density fluctuations decay with a relaxation time that diverges as the critical point is reached. This phenomenon is referred to as critical slowing down.

In the critical region,

$$R(\kappa, \omega) = \frac{1}{\pi} S(\kappa) \frac{\kappa^2 D_T}{\omega^2 + (\kappa^2 D_T)^2} \tag{5.162}$$

with $S(\kappa)$ given by (5.136). There are several points to be noted about the above theory, which belongs to a wide class of theories of critical phenomena collectively referred to as mean-field theories. First, it would appear that the theory is not consistent. For in making the expansion of the free energy (5.126) it was assumed that the fluctuations in the density were small and slowly varying in space, yet we conclude, as a result of the ensuing calculation, that near the critical point the fluctuations are the dominant manifestation of criticality. The error in this argument lies in attempting to apply the theory outside its domain of validity. Clearly the latter is defined such that the theory is consistent; in other words the mean-field theory is not valid close to the critical point. Experiment indicates that, in general, it gives the correct behaviour on approaching the critical point. Putting this argument on a more quantitative basis it has been suggested that the domain of validity is defined by the condition

$$\langle \delta\rho(\mathbf{r})\, \delta\rho \rangle|_{r\kappa_1 \sim 1} \ll \rho^2,$$

which must be examined for each particular system of interest (Kawasaki and Gunton 1978).

As far as the results we have derived for the cross-section are concerned, they will not hold unless $\kappa \ll \kappa_1$. Clearly this condition becomes increasingly more difficult to fulfill the closer the fluid is to its critical point.

Neutron scattering studies can be used to investigate the region of validity of linear hydrodynamic theory since a range of $\kappa$, $\omega$ values can be accessed. By contrast, light scattering experiments are restricted (by neutron scattering standards) to very small $\kappa \sim 10^{-3}$ Å$^{-1}$, where linear hydrodynamic theory is usually quite adequate (Boon and Yip 1980; Cummins and Levanyuk 1983).

At sufficiently high frequencies, a fluid will respond elastically, just as if it were a solid body. The velocity of the collective oscillations of the atoms at high frequencies will be determined by the instantaneous rigidity modulus, $G$, as is appropriate for a solid. The cross-over from viscous to elastic response is determined largely by the Maxwell relaxation time $\tau = \eta/G$, where $\eta$ is the shear viscosity. For many simple liquids, $\tau \sim 10^{-13}$ s. If the applied frequency $\omega$ is such that $\omega\tau \ll 1$, there is viscous

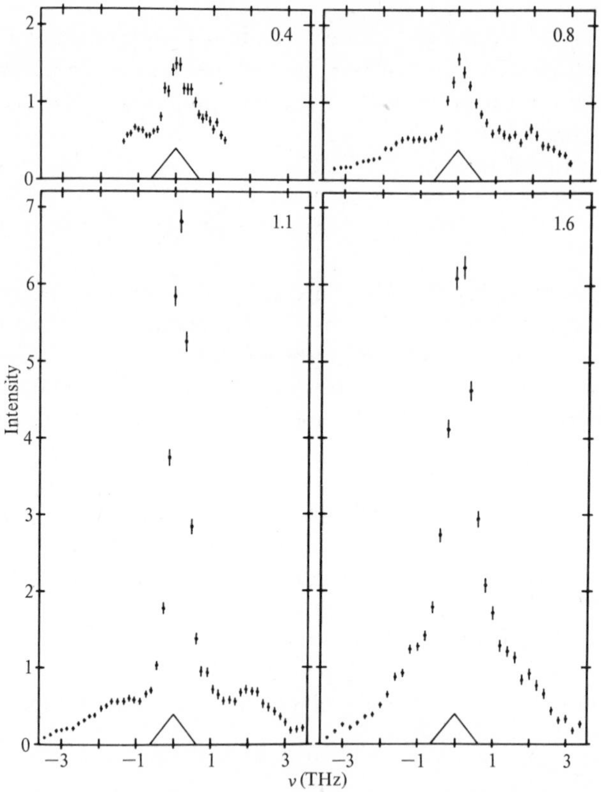


FIG. 5.2. Coherent response of lead at four values of the scattering vector (Söderstrom *et al.* 1980).

flow, whereas the limit $\omega\tau \gg 1$ corresponds to elastic vibrations. Theories that describe the response of a fluid through the cross-over region are usually referred to as viscoelastic theories. The Maxwell relaxation time varies rapidly with temperature for a fluid that forms a supercooled, or glassy, state since $\eta$ may change by many orders of magnitude while $G$ is relatively insensitive to temperature.

An example of the coherent response of a liquid at an intermediate wave vector is shown in Fig. 5.2 (Söderstrom *et al.* 1980). The data spans a range of wave vectors up to a maximum value that corresponds to approximately three-quarters of the wave vector at which the static structure factor displays its main maximum. For wave vectors as large as one-half of this wave vector, the coherent response shows inelastic peaks. The existence of these peaks cannot be realistically attributed to hydrodynamic sound. A satisfactory interpretation can be constructed on the basis of a form of generalized hydrodynamics that is appropriate for the relatively short wavelengths and high frequencies involved. The generalized Langevin equation provides a convenient formalism, and the development is reviewed in the next section.

### 5.4.3. *Viscoelastic theory*

Our objective in this section is to develop a theory appropriate for the interpretation of the coherent response of a monatomic fluid at intermediate wave vectors and frequencies. The behaviour of the response in the hydrodynamic limit, where $\kappa$ and $\omega$ tend to zero, is determined completely from linearized hydrodynamic equations. These equations incorporate the fact that fluctuations in the particle density, created by incident neutrons, decay through variations in both the local pressure and entropy. In generalizing the equations of motion to describe intermediate wave vectors and frequencies, this coupling means that it is prudent to consider fluctuations in the density and energy on an equal footing. Hence, the development of the particle density response, $R(\kappa, \omega)$, proceeds with a consideration of the equation of motion for a dynamic variable that includes the particle density and the energy density.

The approach adopted is based on the notion that the general theory must revert to the result obtained with linear hydrodynamics in the appropriate limit. Given such a general theory, we have a firm basis from which to derive results appropriate for intermediate wave vectors and frequencies—a viscoelastic theory—in the guise of an extended or generalized hydrodynamics. The general theory can also be analysed in the limit of very high frequencies, where the velocity of sound is given in terms of the bulk and rigidity moduli as in an elastic solid.

In the previous section we introduced variables, of space and time, for fluctuations in the local temperature, pressure, and density of heat energy in addition to the particle density, which is of central concern. We also found that it is convenient to work with variables that are uncorrelated. Bearing these considerations in mind, we introduce local thermodynamic variables at the outset of our discussion of generalized hydrodynamics. The variables are constructed so as to satisfy the results of thermodynamic fluctuation theory in the limit of long wavelengths (Schofield 1968). For example, the wave-vector dependent variable

$$T_{\boldsymbol{\kappa}} = \frac{1}{C_v(\kappa)} \left\{ e_{\boldsymbol{\kappa}} - \frac{\langle e_{\boldsymbol{\kappa}} \rho_{\boldsymbol{\kappa}}^* \rangle}{\langle |\rho_{\boldsymbol{\kappa}}|^2 \rangle} \rho_{\boldsymbol{\kappa}} \right\} \tag{5.163}$$

where $e_{\boldsymbol{\kappa}}$ is the energy density, describes local temperature fluctuations. Observe that $T_{\boldsymbol{\kappa}}$ is uncorrelated with $\rho_{\boldsymbol{\kappa}}$ since, by construction, $\langle T_{\boldsymbol{\kappa}} \rho_{\boldsymbol{\kappa}}^* \rangle = 0$; we might equally well describe $T_{\boldsymbol{\kappa}}$ and $\rho_{\boldsymbol{\kappa}}$ as orthogonal variables. Also,

$$\langle |T_{\boldsymbol{\kappa}}|^2 \rangle = k_B T^2 / C_v(\kappa) \tag{5.164}$$

where the wave-vector-dependent specific heat

$$k_B T^2 C_v(\kappa) = \langle |e_\kappa|^2 \rangle - |\langle e_\kappa \rho_\kappa^* \rangle|^2 / \langle |\rho_\kappa|^2 \rangle. \tag{5.165}$$

The concept of a wave-vector-dependent specific heat can be viewed in the same light as the introduction of a wave-vector-dependent susceptibility $\beta S(\kappa)$ for density fluctuations. Both $S(\kappa)$ and $C_v(\kappa)$ are static, wave-vector-dependent response fluctuations; the former describes the spectrum of particle density fluctuations and the latter temperature fluctuations. With regard to (5.164), if $T_\kappa$ is identified with a fluctuation in the temperature then in the long wavelength limit, $\kappa \to 0$, eqn (5.164) is identical with the standard thermodynamic expression (Hansen and McDonald 1976). Note also

$$\langle T_\kappa e_\kappa^* \rangle = k_B T^2, \tag{5.166}$$

which is in accord with thermodynamics. When it comes to constructing an equation of motion for $\rho_\kappa$ we will include energy fluctuations by treating the variables $\rho_\kappa$ and $T_\kappa$ on an equal footing.

Before looking at the dynamics of density fluctuations we complete the introduction of local, wave-vector dependent thermodynamic variables. First, we write the conservation equations for particle density, momentum $M\mathbf{J}_\kappa$, and energy. From

$$\rho_\kappa = \sum_j \exp(-i\boldsymbol{\kappa} \cdot \mathbf{R}_j), \tag{5.167}$$

we have

$$\partial_t \rho_\kappa \equiv \dot{\rho}_\kappa = -i\boldsymbol{\kappa} \cdot \mathbf{J}_\kappa \tag{5.168}$$

with

$$\mathbf{J}_\kappa = \sum_j \dot{\mathbf{R}}_j \exp(-i\boldsymbol{\kappa} \cdot \mathbf{R}_j). \tag{5.169}$$

The variable $\mathbf{J}_\kappa$ is both a flux (eqn (5.168)) and a conserved variable. The conservation of momentum introduces the stress tensor $\boldsymbol{\sigma}$

$$M\dot{\mathbf{J}}_\kappa + i\boldsymbol{\kappa} \cdot \boldsymbol{\sigma} = 0 \tag{5.170}$$

and the conservation of energy is written

$$\dot{e}_\kappa + i\boldsymbol{\kappa} \cdot \mathbf{J}_\kappa^e = 0. \tag{5.171}$$

Expressions for the fluxes $\boldsymbol{\sigma}$ and $\mathbf{J}_\kappa^e$, for particles interacting through a pair potential, are given in Hansen and McDonald (1976), Boon and Yip (1980), and Schofield (1968). The transverse components of the current density are independent of the other variables and they can therefore be treated separately. They do not concern us because we are interested in density fluctuations and these couple to the longitudinal component of the momentum flux (eqn (5.168)). In view of this, we will simplify our notation and write $\mathbf{J}_\kappa$ as a scalar $J_\kappa$, which is the longitudinal component of $\mathbf{J}_\kappa$; hence, the conservation of particle density becomes

$$\dot{\rho}_{\kappa} + \mathrm{i}\kappa J_{\kappa} = 0. \tag{5.172}$$

For the same reasons we need only the diagonal component of the stress tensor, and we denote this by $\sigma_\kappa$, i.e.

$$M\dot{J}_{\kappa} + \mathrm{i}\kappa\sigma_{\kappa} = 0. \tag{5.173}$$

The virial theorem leads to the result

$$\langle \sigma_{\kappa=0} \rangle = pV. \tag{5.174}$$

The correlation functions of the fluxes determine the transport coefficients for internal friction and heat transport (Hansen and McDonald 1976; Boon and Yip 1980). The results of particular interest to us relate the kinematic viscosity $\nu$ to the longitudinal component of the stress tensor, and the thermal conductivity $\lambda$ to the longitudinal component of the heat flux $J^e_\kappa$; they are derived using the hydrodynamic analysis of § 5.4.2 together with the equations that define the fluxes. The relations are usually called Green–Kubo formulae. For the kinematic viscosity we find

$$\nu = (\beta/MN) \int_0^\infty \mathrm{d}t \lim_{\kappa \to 0} \langle \{\sigma_\kappa(t) - \langle\sigma_\kappa\rangle\}\{\sigma_\kappa - \langle\sigma_\kappa\rangle\}^* \rangle \tag{5.175}$$

and for the thermal conductivity

$$\lambda = (Vk_{\mathrm{B}}T^2)^{-1} \int_0^\infty \mathrm{d}t \lim_{\kappa \to 0} \langle J^e_\kappa(t) J^{e*}_\kappa \rangle. \tag{5.176}$$

We emphasize that the results make sense only when the limit $\boldsymbol{\kappa} \to 0$ is taken before the integral is evaluated, since for $\boldsymbol{\kappa} \neq 0$ the integral vanishes.

We are now in a position to complete our set of local thermodynamic variables. The pressure $p_\kappa$ and entropy $S_\kappa$ are taken to be

$$p_{\kappa} = \frac{\langle \sigma_\kappa \rho^*_\kappa \rangle}{\langle |\rho_\kappa|^2 \rangle} \rho_\kappa + \frac{\langle \sigma_\kappa T^*_\kappa \rangle}{\langle |T_\kappa|^2 \rangle} T_\kappa \tag{5.177}$$

and

$$TS_{\kappa} = e_{\kappa} - \frac{\langle e_\kappa \sigma^*_\kappa \rangle}{\langle \rho_\kappa \sigma^*_\kappa \rangle} \rho_\kappa. \tag{5.178}$$

With these definitions we find $\langle \rho_\kappa S^*_\kappa \rangle = 0$, as required, and

$$\langle S_\kappa T^*_\kappa \rangle = k_{\mathrm{B}} T, \tag{5.179}$$

which agrees with the thermodynamic result for the correlation between fluctuations in entropy and temperature. The density of heat energy is (cf. (5.140))

$$Q_\kappa = e_\kappa - h(\kappa)\rho_\kappa \tag{5.180}$$

where the enthalpy per particle,

$$h(\kappa) = \langle e_\kappa p_\kappa^* \rangle / \langle \rho_\kappa \sigma_\kappa^* \rangle. \tag{5.181}$$

From the result

$$\langle |J_\kappa|^2 \rangle = N/M\beta \tag{5.182}$$

which follows directly from (5.169), we obtain

$$\langle \rho_\kappa \sigma_\kappa^* \rangle = N/\beta, \tag{5.183}$$

and

$$\begin{aligned} \omega_0^2 &= \frac{|\langle \dot{\rho}_\kappa J_\kappa^* \rangle|^2}{\langle |J_\kappa|^2 \rangle \langle |\rho_\kappa|^2 \rangle} = \frac{\kappa^2 \langle |J_\kappa|^2 \rangle}{\langle |\rho_\kappa|^2 \rangle} . \\ &= (\kappa^2 / M\beta S(\kappa)). \end{aligned} \tag{5.184}$$

Having completed these preliminary considerations we return to the calculation of the coherent response function. This is accomplished in terms of a generalized Langevin equation for a variable $\mathrm{A}_\kappa$ that is the column vector formed from the mutually orthogonal variables $\rho_\kappa$, $J_\kappa$, and $T_\kappa$. From the theory developed in Appendix B, the Laplace transform

$$\begin{aligned} \tilde{F}_\kappa(s) &= \int_0^\infty \mathrm{d}t \exp(-st) \langle A_\kappa(t) A_\kappa^* \rangle \cdot \langle A_\kappa A_\kappa^* \rangle^{-1} \\ &= \{ s\mathscr{I} - \mathrm{i}\mathbf{\Omega} + \tilde{\mathbf{M}}_\kappa(s) \}^{-1}, \end{aligned} \tag{5.185}$$

where $\mathscr{I}$ is the unit matrix. The remaining $3 \times 3$ matrices in (5.185) are defined as

$$\mathrm{i}\mathbf{\Omega} = \langle \dot{A}_\kappa A_\kappa^* \rangle \cdot \langle A_\kappa A_\kappa^* \rangle^{-1} = \begin{pmatrix} 0 & \dfrac{\langle \dot{\rho}_\kappa J_\kappa^* \rangle}{\langle |J_\kappa|^2 \rangle} & 0 \\ -\dfrac{\langle J_\kappa \dot{\rho}_\kappa^* \rangle}{\langle |\rho_\kappa|^2 \rangle} & 0 & \dfrac{\langle \dot{J}_\kappa T_\kappa^* \rangle}{\langle |T_\kappa|^2 \rangle} \\ 0 & -\dfrac{\langle T_\kappa \dot{J}_\kappa^* \rangle}{\langle |J_\kappa|^2 \rangle} & 0 \end{pmatrix} \tag{5.186}$$

and the memory function matrix

$$\tilde{\mathbf{M}}_\kappa(s) = \int_0^\infty \mathrm{d}t \exp(-st) \langle f_\kappa(t) f_\kappa^* \rangle \cdot \langle A_\kappa A_\kappa^* \rangle^{-1}, \tag{5.187}$$

where the random force

$$f_\kappa = \dot{A}_\kappa - \mathrm{i}\boldsymbol{\Omega} \cdot A_\kappa. \tag{5.188}$$

Using the result for the frequency matrix (5.186), the components of the random force column matrix are

$$f_\kappa^{(1)} = 0,$$

$$f_\kappa^{(2)} = -\frac{\mathrm{i}\kappa}{M}(\sigma_\kappa - p_\kappa), \tag{5.189}$$

and

$$f_\kappa^{(3)} = \dot{T}_\kappa + \frac{\langle T_\kappa \dot{J}_\kappa^* \rangle}{\langle |J_\kappa|^2 \rangle} J_\kappa = -\frac{\mathrm{i}\kappa}{C_v(\kappa)} q_\kappa \tag{5.190}$$

where the last equality defines the flux $q_\kappa$. The latter is identified as the flux of heat current since

$$\dot{Q}_\kappa + \mathrm{i}\kappa q_\kappa = 0, \tag{5.191}$$

with $Q_\kappa$ given by (5.180).

The correlation functions of the random forces (5.189)–(5.190) which occur in (5.187) are related to the Green–Kubo formulae for the transport coefficients. To see this we first note that, in the limit $\kappa \to 0$ the time development of the random forces in (5.187) does not differ from that of any other dynamic variable. For $\kappa \neq 0$ this is not the case, as discussed in Appendix B, because the time development operator associated with the random forces contains a projection operator that complicates the time dependence. An essential feature of the forces in (5.189)–(5.190) is that they are derived from conserved variables and therefore vanish in the long-wavelength limit. It is this feature of the random forces which leads to the result that there are no additional, complicating aspects to their time development in the limit of long wavelengths (Mountain 1976).

If we form $\langle f_\kappa^{(2)}(t) f_\kappa^{(2)*} \rangle / \kappa^2$, for example, and take the limit $\kappa \to 0$, the integral over all time is proportional to the kinematic viscosity $\nu$ given in (5.175). The corresponding result for the thermal conductivity is not so easy to follow through because $f_\kappa^{(3)} \propto q_\kappa$, whereas the Green–Kubo formula (5.176) contains the longitudinal component of the energy flux. However, the time auto-correlation functions of $q_\kappa$ and $J_\kappa^e$ are identical in the present case. The proof of this statement requires us to examine the explicit form of the fluxes in terms of the atomic potential, and for this reason we will not pursue it; the interested reader can consult Boon and Yip (1980), § 2.5, for example.

While the integrals of the $\kappa \to 0$ limit of the auto-correlation functions of $\sigma_\kappa - p_\kappa$ and $q_\kappa$ are finite, their cross-correlation functions vanish.

Hence, the dynamic coupling between the stress and heat current does not contribute in the long-wavelength limit. Moreover, it is straightforward to demonstrate that $\langle(\sigma_{\boldsymbol{\kappa}} - p_{\boldsymbol{\kappa}})q^*_{\boldsymbol{\kappa}}\rangle = 0$, i.e. the instantaneous correlation of the fluxes is zero. Since the cross-correlations of the stress and heat current vanish in the hydrodynamic and instantaneous limits they are dropped in the subsequent development. However, it should be said that our scant knowledge of these cross-correlations does make it difficult to know how to set about including them into the development.

To specify the memory function matrix $\tilde{\mathbf{M}}_\kappa(s)$, we find it useful to define the functions

$$\tilde{\theta}_\kappa(s) = \frac{1}{\langle|J_\kappa|^2\rangle}\int_0^\infty \mathrm{d}t\, \exp(-st)\langle f^{(2)}_{\boldsymbol{\kappa}}(t) f^{(2)*}_{\boldsymbol{\kappa}}\rangle \tag{5.192}$$

and

$$\tilde{\pi}_\kappa(s) = \frac{1}{\langle|T_{\boldsymbol{\kappa}}|^2\rangle}\int_0^\infty \mathrm{d}t\, \exp(-st)\langle f^{(3)}_{\boldsymbol{\kappa}}(t) f^{(3)*}_{\boldsymbol{\kappa}}\rangle. \tag{5.193}$$

All other components of $\tilde{\mathbf{M}}_\kappa(s)$ are zero, either because $f^{(1)}_{\boldsymbol{\kappa}}$ is zero, by definition, or because we have dropped the cross-correlation between the stress and heat current.

Using the simplified memory matrix in (5.185) together with (5.186), we obtain

$$\tilde{R}(\boldsymbol{\kappa}, s) = \frac{1}{N}\int_0^\infty \mathrm{d}t\, \exp(-st)\langle \rho_{\boldsymbol{\kappa}} \rho^*_{\boldsymbol{\kappa}}(t)\rangle \tag{5.194}$$

as the $(1, 1)$ matrix element of $\tilde{F}_\kappa(s)$, apart from the normalization $S(\kappa) = \langle|\rho_{\boldsymbol{\kappa}}|^2\rangle/N$. The result is

$$\tilde{R}(\kappa, s) = S(\kappa)\left\{s + \frac{\omega_0^2}{s + \tilde{K}_\kappa(s)}\right\}^{-1} \tag{5.195}$$

where $\omega_0^2$ is given in (5.184) and

$$\tilde{K}_\kappa(s) = \tilde{\theta}_\kappa(s) + \frac{|\langle \dot{J}_{\boldsymbol{\kappa}} T^*_{\boldsymbol{\kappa}}\rangle|^2}{\langle|J_{\boldsymbol{\kappa}}|^2\rangle\langle|T_{\boldsymbol{\kappa}}|^2\rangle}\{s + \tilde{\pi}_\kappa(s)\}^{-1}. \tag{5.196}$$

We will now investigate some properties of (5.195), and confirm that it is an appropriate basis for a viscoelastic theory.

Result (5.195) and the corresponding hydrodynamic result (5.156) are written so as to facilitate comparison. We observe that

$$\lim_{\kappa \to 0} \omega_0^2 = \kappa^2 v_T^2 \tag{5.197}$$

where $v_T$ is the isothermal sound velocity, and the results

$$\lim_{s \to 0} \lim_{\kappa \to 0} \tilde{\theta}_\kappa(s) = \nu\kappa^2 \tag{5.198}$$

$$\lim_{s \to 0} \lim_{\kappa \to 0} \tilde{\pi}_\kappa(s) = (\lambda\kappa^2/\rho_0 c_v), \tag{5.199}$$

follow from the Green–Kubo formulae for the transport coefficients $\nu$ and $\lambda$. Lastly, using the definitions of $J_{\kappa}$ and $T_{\kappa}$,

$$\frac{|\langle \dot{J}_{\kappa} T^{*}_{\kappa} \rangle|^2}{\langle |J_{\kappa}|^2 \rangle \langle |T_{\kappa}|^2 \rangle} = \frac{N\kappa^2}{MTC_v(\kappa)} \left\{ h(\kappa) - \frac{\langle e_{\kappa} \rho^{*}_{\kappa} \rangle}{\langle |\rho_{\kappa}|^2 \rangle} \right\}^2 \tag{5.200}$$

and in the limit $\kappa \to 0$ the right-hand side reduces to (cf. (5.155))

$$\kappa^2 \left( \frac{1}{Tc_v} \right) \left( \frac{1}{M\rho_0^2} \right) \left\{ T \left( \frac{\partial p}{\partial T} \right)_v \right\}^2 = \kappa^2 v_T^2 (\gamma - 1). \tag{5.201}$$

The results (5.197)–(5.201) establish the equivalence of (5.195)–(5.196) with the hydrodynamic expression (5.156) in the limit of long wavelengths, $\kappa \to 0$.

Let us turn now to the behaviour of the coherent response at intermediate frequencies where effects of thermal conduction are small. We begin by considering the extreme case $\omega \to \infty$ that corresponds to the instantaneous response. In this limit, a fluid behaves elastically and the velocity of the collective oscillation of the atoms will be $\{(B + \frac{4}{3}G)/M\rho_0\}^{1/2}$ where $B$ and $G$ are the bulk and rigidity moduli, respectively. This velocity is greater than, or equal to, the low-frequency velocity of sound $(\chi_s M\rho_0)^{-1/2}$, where $\chi_s$ is the adiabatic compressibility.

The appropriate wave-vector-dependent moduli can be shown to be (Hansen and McDonald 1976; Boon and Yip 1980)

$$C_{11}(\kappa) = (\beta/V)\langle |\sigma_{\kappa}|^2 \rangle \tag{5.202}$$

and

$$C_{44}(\kappa) = (\beta/V)\langle |\sigma^{t}_{\kappa}|^2 \rangle, \tag{5.203}$$

where $\sigma_{\kappa}$ and $\sigma^{t}_{\kappa}$ are the components of the stress tensor that are parallel (longitudinal) and perpendicular (transverse) to the wave vector. The key results for the present discussion are first the identity

$$\kappa^2 C_{11}(\kappa) = M\rho_0 \omega_l^2 \tag{5.204}$$

where $\omega_l^2$ is related to the fourth frequency moment of the coherent response (eqn (5.39)) and

$$\lim_{\kappa \to 0} C_{11}(\kappa) = B + \tfrac{4}{3}G \tag{5.205}$$

$$\lim_{\kappa \to 0} C_{44}(\kappa) = G. \tag{5.206}$$

From (5.204) we note that $C_{11}(\kappa)$ can be expressed in terms of the pair distribution function and the atomic potential as in (5.40), and a similar expression exists for $C_{44}(\kappa)$. Using the previous results, we conclude that, with $f_{\boldsymbol{\kappa}}^{(2)}$ given by (5.189),

$$\frac{\langle|f_{\boldsymbol{\kappa}}^{(2)}|^2\rangle}{\langle|J_{\boldsymbol{\kappa}}|^2\rangle}=\omega_l^2-\omega_0^2-\frac{|\langle\dot{J}_{\boldsymbol{\kappa}}T_{\boldsymbol{\kappa}}^*\rangle|^2}{\langle|J_{\boldsymbol{\kappa}}|^2\rangle\langle|T_{\boldsymbol{\kappa}}|^2\rangle} \tag{5.207}$$

and, in the limit $\kappa\to 0$, the right-hand side reduces to

$$\kappa^2\left\{\frac{1}{M\rho_0}(B+\tfrac{4}{3}G)-v_T^2-v_T^2(\gamma-1)\right\}=\frac{\kappa^2}{M\rho_0}\left(B+\tfrac{4}{3}G-\frac{1}{\chi_s}\right). \tag{5.208}$$

The high-frequency limit of $\tilde{\theta}_\kappa(s)$ (eqn (5.192)) is

$$\lim_{s\to\infty}\tilde{\theta}_\kappa(s)=\langle|f_{\boldsymbol{\kappa}}^{(2)}|^2\rangle/s\langle|J_{\boldsymbol{\kappa}}|^2\rangle, \tag{5.209}$$

with the static correlation function given in (5.207), or in (5.208) in the limit of long wavelengths. Comparing (5.198) with (5.209) we see that the stress tensor correlation is independent of the frequency in the hydrodynamic limit, and it decreases as the inverse of the frequency in the high frequency limit. The change in $\tilde{\theta}_\kappa(s)$ as a function of frequency characterizes the change in the response of the fluid from that of viscous flow to one of elastic vibration, and the frequency scale for observing the two extreme types of response is set effectively by the ratio of the rigidity modulus and shear viscosity.

The concepts can be made a little more precise by introducing a simple form for the frequency dependence of $\tilde{\theta}_\kappa(s)$ which involves a single relaxation time $\tau$. If, for small $\kappa$, we assume the form

$$\tilde{\theta}_\kappa(s)\doteq\frac{\kappa^2(B+\tfrac{4}{3}G-1/\chi_s)\tau}{M\rho_0(1+s\tau)}, \tag{5.210}$$

then we recover the high-frequency result (5.209) when $s\tau\gg 1$ irrespective of our choice for $\tau$. The latter is taken to be

$$\tau=\nu M\rho_0/(B+\tfrac{4}{3}G-1/\chi_s), \tag{5.211}$$

in order to recover (5.198) in the limit $s\tau\to 0$. For many simple fluids the relaxation time $\tau$ is of the order of $10^{-13}$ s. However, for fluids that supercool and form a glassy state, $\tau$ can be many orders of magnitude larger because the shear viscosity (contained in the longitudinal kinematic viscosity) becomes very large on approaching the glass transition.

The form for $\tilde{\theta}_\kappa(s)$ adopted in (5.210) is the starting point of many approximate theories for the response of fluids in the viscoelastic region of intermediate wave vectors and frequencies. One obvious improvement, which extends the development to larger wave vectors, is to make the approximation satisfy high-order frequency moments and to let the relaxation time vary with the wave vector. The form

$$\tilde{\theta}_\kappa(s) = \frac{(\omega_l^2 - \omega_0^2)\tau}{(1 + s\tau)} \qquad (5.212)$$

leads to an expression for $\tilde{R}(\kappa, s)$ that possesses the correct frequency moments up to the fourth; it is easy to verify that

$$s\tilde{R}(\kappa, s) = S(\kappa)\left\{1 - \frac{\omega_0^2}{s^2} + \frac{\omega_0^2\omega_l^2}{s^4} - \cdots\right\}. \qquad (5.213)$$

Substituting (5.212) in (5.195)and continuing to neglect thermal conduction, it is straightforward to show that the corresponding coherent response

$$R(\kappa, \omega) = \frac{1}{\pi}\tilde{R}(\kappa, i\omega), \qquad (5.214)$$

possesses a maximum at nonzero $\omega$ when

$$3\omega_0^2 > \omega_l^2, \qquad (5.215)$$

independent of the value of $\tau$. In other words, the viscoelastic theory based on (5.212) contains a collective density oscillation when the condition (5.215) is satisfied, and it provides a basis for the interpretation of the neutron data in Fig. 5.2. For small wave vectors, (5.215) reduces to the condition

$$3 > \chi_T(B + \tfrac{4}{3}G). \qquad (5.216)$$

In the vicinity of the liquid–gas phase transition, the right-hand side of (5.216) diverges so the viscoelastic mode ceases to exist near the transition. This observation is a telling reminder that in constructing our simple viscoelastic theory we have divorced ourselves from hydrodynamic theory for our theory does not contain ordinary sound propagation.

The short-wavelength viscoelastic mode observed in Fig. 5.2 has been observed in a few other simple fluids near their triple points (Hansen and McDonald 1976; Boon and Yip 1980; Copley and Lovesey 1975) and the findings are described very well by our theory. The ability of a dense fluid to support a short-wavelength density oscillation appears to be intimately connected with the shape of the main minimum in the atomic potential. Extensive computer simulation studies of a model liquid in which the particles interact via a generalized Lennard–Jones potential show that the existence of the short-wavelength collective mode is determined largely by the magnitude of an effective Grüneisen parameter (Lewis and Lovesey 1977, 1978). This parameter can be written as the ratio of the third and second derivatives of the pair potential evaluated at the potential minimum $r_0$, namely

$$\gamma_G \doteq -\frac{r_0}{6}(u'''/u'')_{r_0}. \tag{5.217}$$

It is found that systems with $\gamma_G \leqslant 2$ support a collective mode, whereas systems with $\gamma_G \sim 3$, a value obtained for a standard $(12-6)$ Lennard–Jones potential that is appropriate for liquid argon, do not support a short-wavelength mode. The computer simulation data also provide a demanding test of the ability of our viscoelastic theory to represent the coherent response function at wave vectors in the intermediate range given by $\kappa > r_0^2\rho_0$, and $\kappa$ less than the position of the main peak in the structure factor.

In order to evaluate the expression

$$\pi R(\kappa, \omega) = \frac{S(\kappa)\omega_0^2(\omega_l^2 - \omega_0^2)\tau}{[\omega\tau(\omega^2 - \omega_l^2)]^2 + (\omega^2 - \omega_0^2)^2} \tag{5.218}$$

that follows from (5.194), (5.195), (5.212), and (5.214), we must find a prescription for the relaxation time $\tau$ which is expected to depend on $\kappa$, as do $\omega_0^2$ and $\omega_l^2$. A way forward is based on the fact that the approximation (5.218) is equivalent to a simple termination of the continued fraction expansion of the memory function. This expansion for $\tilde{M}_\kappa(s)$ is discussed in Appendix B; it takes the form

$$\tilde{R}(\kappa, s) = S(\kappa)\{s + \tilde{M}_\kappa(s)\}^{-1} \tag{5.219}$$

with the memory function

$$\tilde{M}_\kappa(s) = \omega_0^2/\{s + \tilde{K}_\kappa^{(1)}(s)\}, \tag{5.220}$$

and

$$\tilde{K}_\kappa^{(n)}(s) = \delta_n/\{s + \tilde{K}_\kappa^{(n+1)}(s)\} \tag{5.221}$$

where the integer $n \geqslant 1$,

$$\delta_n = K_\kappa^{(n)}(t=0), \tag{5.222}$$

and $\delta_1 = \omega_l^2 - \omega_0^2$. All the static correlation functions $\delta_n$ can be expressed in terms of frequency moments of $R(\kappa, \omega)$. The continued fraction relation (5.221) implies that

$$K_\kappa^{(n)}(t) = \delta_n\left\{1 - \frac{t^2}{2}\,\delta_{n+1} + \cdots\right\}. \tag{5.223}$$

The approximation (5.218) is obtained when the continued fraction expansion is terminated at the level $n = 1$ by invoking a procedure that is valid if $K_\kappa^{(2)}(t)$ decays rapidly on the time scale of interest. In this instance,

$$\dot{K}_\kappa^{(1)}(t) = -\int_0^t d\bar{t}\, K_\kappa^{(1)}(t-\bar{t}) K_\kappa^{(2)}(\bar{t})$$
$$\doteq -\frac{1}{\tau} K_\kappa^{(1)}(t) \tag{5.224}$$

with

$$(1/\tau) = \int_0^\infty dt\, K_\kappa^{(2)}(t) \tag{5.225}$$

To estimate the integral in (5.225) we first recognize that the continued fraction expansion enables us to write (5.225) in terms of $\tilde{M}_\kappa(0)$,

$$(1/\tau) = \frac{\delta_1}{\omega_0^2} \tilde{M}_\kappa(0)$$
$$\equiv \frac{\delta_1}{\omega_0^2} \int_0^\infty dt\, M_\kappa(t). \tag{5.226}$$

The integral of the memory function $M_\kappa(t)$ is probably insensitive to the details of the shape of the function, and consequently a reasonable estimate can be obtained from a model function that satisfies the short-time behaviour

$$M_\kappa(t) = \omega_0^2\left\{1 - \frac{t^2}{2}\delta_1 + \cdots\right\}$$

and the requirement

$$|M_\kappa(t)| \leq M_\kappa(0).$$

Taking $M_\kappa(t)$ to be a Gaussian function of time we deduce that

$$(1/\tau) \propto \frac{\delta_1}{\omega_0^2} \cdot \frac{\omega_0^2}{(\delta_1)^{1/2}}.$$

We can choose the proportionality constant such that $R(\kappa, 0)$ coincides with the result for a perfect gas for large $\kappa$, in which case we arrive at the estimate

$$(1/\tau) = 2\{(\omega_l^2 - \omega_0^2)/\pi\}^{1/2}. \tag{5.227}$$

We have now specified our viscoelastic approximation (5.218) completely. The expression can be evaluated given the static structure factor and the frequency moment $\omega_l^2$, and it is found to be in accord with a variety of results for short-wavelength excitations in simple dense liquids (Copley and Lovesey 1975; Lewis and Lovesey 1977, 1978).

The contribution of the collective mode to $R(\kappa, \omega)$ is usually relatively weak when compared with the quasi-elastic contribution as is seen in Fig. 5.2. It is a simple matter to verify that the result (5.218) has a maximum at $\omega = 0$, for $\tau$ given by (5.227), at all values of the scattering vector. We can attribute this quasi-elastic scattering to the existence of a non-propagating mode in the viscoelastic theory; recall that thermal conduction is not included in the theory, so the origin of the quasi-elastic peak is fundamentally different from the Rayleigh line observed in the hydrodynamic limit. The width of the quasi-elastic contribution to $R(\kappa, \omega)$ increases linearly with $\kappa$ for small wave vectors. This result follows from an expression for the half-width at half-height $\Delta$ derived from (5.128) by keeping the leading terms in $\Delta$, namely

$$\Delta^2 \doteq \omega_0^2/\{\tau^2\omega_l^2(\omega_l^2/\omega_0^2) - 2\}. \tag{5.228}$$

Using the result (5.227) for $\tau$ and the fact that both $\omega_0^2$ and $\omega_l^2$ are proportional to $\kappa^2$ for $\kappa \to 0$, we see immediately from (5.228) that $\Delta \propto \kappa$ for small $\kappa$. A similar variation of the lineshape is found in the limit of very large wave vectors when $R(\kappa, \omega)$ approaches the result for a single free particle. We can conclude that the width of the quasi-elastic contribution to the coherent response function varies as $\kappa^2$ in the hydrodynamic region, where it originates from non-propagating entropy fluctuations, while viscoelastic theory and the free-particle behaviour give a line width that increases linearly with the wave vector.

The build up of strong quasi-elastic scattering with increasing elastic response of the target sample is very striking in the variation with temperature of the coherent response of a fluid that forms a supercooled–glassy state. An appropriate theory, valid for long-wavelength fluctuations, can be constructed on the lines adopted in § 5.3.2 in our analysis of the velocity auto-correlation function. For the present case of coherent scattering, the stress correlation function (5.191) is approximated by (Lovesey 1979)

$$\tilde{\theta}_\kappa(s) \doteq \frac{\kappa^2(B + \frac{4}{3}G - 1/\chi_s)\tau}{M\rho_0(1 + s\tau + 2\gamma(s\tau)^{1/2})}. \tag{5.229}$$

For $\gamma = 0$, the approximation possesses the same frequency structure as (5.210). The additional frequency dependence in (5.229) for $\gamma > 0$ is required in order to account for the long time tail in the stress tensor auto-correlation function. Analytic expressions for the long time tail can be used to determine $\gamma$ in terms of transport coefficients. The resulting approximation for the coherent response satisfies the high-frequency expansion (5.213) in addition to incorporating the hydrodynamic structure, and it is therefore well suited for the description of the change in the response of the target sample from viscous flow to elastic vibration as the sample is brought into a glassy state by reducing the temperature. In terms of the theory, the temperature-induced cross-over in the response characteristics is effected by the dramatic change in the frequency $(G/\eta)$, which is the inverse of the Maxwell relaxation time. The coherent response changes, as the sample temperature is decreased, not only in the build up of strong quasi-elastic scattering but also by a significant decrease in the phonon line width and an increase in the phonon frequency.

There is a dearth of information available on the behaviour of the response function in the vicinity of the stability limit of the supercooled state. This situation reflects a poor understanding of the nature of the stability limit itself, e.g. is it a first- or second-order transition and what type of solid state is achieved. In one approach, the stability limit is associated with a soft relaxational mode in the system, and this implies that the structure factor must diverge at the transition for a finite wave vector. Neutron scattering measurements are not in accord with this, but they are probably cognizant of the enormous change in the self-diffusion near the stability limit (Sjölander and Turski 1978; Mountain 1982; Mountain and Basu 1983).

## 5.5. Classical multicomponent fluid

Neutron diffraction has been used with great success to study the structural properties of a variety of multicomponent fluids. Some of the most extensive studies have aimed at understanding ionic liquids and electrolyte solutions (Howe 1978; Enderby and Neilson 1980, 1981; Biggin *et al.* 1984; McGreevy *et al.* 1984). The objective of such studies is a set of pair distribution functions for the components of the fluid, e.g. three pair distribution functions are required for a complete description of a binary fluid. The measured static structure factor is a weighted sum of the spatial Fourier transforms of the pair distribution functions, and the weighting factors are combinations of the scattering lengths for the various isotopes. The requisite experimental data can be obtained by using isotopically enriched target samples. Given sufficient samples, of known concentration, and a knowledge of the isotope scattering lengths, a set of measured structure factors can be resolved into the desired partial structure factors and pair distribution functions. The difficulties inherent to the lengthy procedure are the main reason for the paucity of inelastic neutron scattering studies of the dynamic properties of multicomponent systems.

We begin by discussing the properties of the cross-section for scattering from a binary target sample, which is assumed to obey classical statistical mechanics. The cross-section can be written in terms of the correlation functions†

$$Y_{jj'}(\kappa, t) = \langle \exp\{\mathrm{i}\boldsymbol{\kappa} \cdot [\mathbf{R}_{j'}(t) - \mathbf{R}_j]\}\rangle \tag{5.230}$$

where $\mathbf{R}_{j'}(t)$ is the position of the $j'$th atom at time $t$, the angular brackets denote a thermal average over the classical motions of the $N$ atoms in the sample, and $\mathbf{R}_j \equiv \mathbf{R}_j(t=0)$. Let us define the weighted sum of the correlation functions to be

$$R(\kappa, t) = \frac{1}{N} \sum_{jj'} \bar{b}_j^* \bar{b}_{j'} Y_{jj'}(\kappa, t) \tag{5.231}$$

and

$$R(\kappa, \omega) = \frac{1}{2\pi} \int_{-\infty}^{\infty} \mathrm{d}t \exp(-\mathrm{i}\omega t) R(\kappa, t). \tag{5.232}$$

The coherent cross-section is proportional to $R(\kappa, \omega)$ averaged over the configurations of the atoms and we denote the average by a horizontal bar. The partial differential cross-section is then

$$\frac{\mathrm{d}^2\sigma}{\mathrm{d}\Omega\, \mathrm{d}E'} = N \frac{k'}{k} \{1 + n(\omega)\} \omega\beta \bar{R}(\kappa, \omega), \tag{5.233}$$

where the detailed balance factor is given in (5.17) and $\beta$ is the inverse temperature.

The coherent scattering lengths $\bar{b}_j$ in (5.231) are the same for each species. We label the species by 1 and 2 and write $\bar{b}_j$ in terms of a labelling function $p_j$ that is zero if $j$ labels a type-2 atom and unity for a type-1 atom. Hence,

$$\bar{b}_j = \bar{b}_2 + p_j(\bar{b}_1 - \bar{b}_2) \tag{5.234}$$

and, if the scattering lengths are real,

$$\bar{b}_j^* \bar{b}_{j'} = \bar{b}_2^2 + p_j(\bar{b}_1 - \bar{b}_2)\bar{b}_2 + p_{j'}(\bar{b}_1 - \bar{b}_2)\bar{b}_2 + p_j p_{j'}(\bar{b}_1 - \bar{b}_2)^2. \tag{5.235}$$

Substituting (5.235) in (5.231) leads us to consider three partial correlation functions,

$$cR_{11}(\kappa, t) = \frac{1}{N} \sum_{jj'} p_j p_{j'} Y_{jj'}(\kappa, t), \tag{5.236}$$

$$(1-c)R_{22}(\kappa, t) = \frac{1}{N} \sum_{jj'} (1-p_j)(1-p_{j'}) Y_{jj'}(\kappa, t), \tag{5.237}$$

† We draw attention to our choice of notation for the correlation function, for we usually reserve $Y_{jj'}$ for the general, quantum mechanical correlation function (eqn (5.4)).

and

$$\{c(1-c)\}^{1/2}R_{12}(\kappa, t) = \{c(1-c)\}^{1/2}R_{21}(\kappa, t)$$
$$= \frac{1}{N}\sum_{jj'} p_j(1-p_{j'})Y_{jj'}(\kappa, t). \quad (5.238)$$

In terms of these partial correlation functions,

$$R(\kappa, t) = c(\bar{b}_1)^2 R_{11}(\kappa, t) + 2\{c(1-c)\}^{1/2}\bar{b}_1\bar{b}_2 R_{12}(\kappa, t)$$
$$+ (1-c)(\bar{b}_2)^2 R_{22}(\kappa, t). \quad (5.239)$$

The structure of this function is clearly consistent with the physical properties of the partial correlation functions, for if $c = 1$, say, $R(\kappa, t) = (\bar{b}_1)^2 R_{11}(\kappa, t)$ and $R_{11}(\kappa, t)$ is, by definition, related only to the properties of the type-1 atoms.

The separation of $R(\kappa, t)$ given in (5.239) is just one of many we could choose, and the representation adopted is a matter of taste to some extent. An alternative separation, with considerable physical appeal, is based on the use of density and concentration variables. Let

$$\rho(\mathbf{r}, t) = \rho_1(\mathbf{r}, t) + \rho_2(\mathbf{r}, t) \quad (5.240)$$

and

$$c(\mathbf{r}, t) = (1-c)\rho_1(\mathbf{r}, t) - c\rho_2(\mathbf{r}, t) \quad (5.241)$$

where, for example, we have introduced

$$\rho_1(\mathbf{r}, t) = \sum_j p_j \delta\{\mathbf{r} - \mathbf{R}_j(t)\} \quad (5.242)$$

and a similar expression for $\rho_2(\mathbf{r}, t)$ with $p_j$ replaced by $(1-p_j)$. The partial correlation function $R_{11}(\kappa, t)$ (eqn (5.236)) in terms of $\rho_1(\mathbf{r}, t)$ is

$$cR_{11}(\kappa, t) = \frac{1}{N}\int d\mathbf{r}\, \exp(i\boldsymbol{\kappa}\cdot\mathbf{r}) \int d\mathbf{r}' \langle \rho_1(\mathbf{r}')\rho_1(\mathbf{r}+\mathbf{r}', t)\rangle. \quad (5.243)$$

Using this and the corresponding expressions for $R_{12}$ and $R_{22}$ we can readily construct the correlation functions for the density and concentration variables. We find, with an obvious notation for the three partial correlation functions,

$$R_{\rho\rho}(\kappa, t) = cR_{11}(\kappa, t) + 2\{c(1-c)\}^{1/2}R_{12}(\kappa, t) + (1-c)R_{22}(\kappa, t), \quad (5.244)$$

$$R_{\rho c}(\kappa, t) = c(1-c)\{R_{11}(\kappa, t) - R_{22}(\kappa, t)\} + (1-2c)\{c(1-c)\}^{1/2}R_{12}(\kappa, t),$$

and, finally,

$$R_{cc}(\kappa, t) = c(1-c)[(1-c)R_{11}(\kappa, t) - 2\{c(1-c)\}^{1/2}R_{12}(\kappa, t) + cR_{22}(\kappa, t)].$$

In terms of these partial correlation functions we obtain, in place of eqn (5.239),

$$R(\kappa, t) = \{c\bar{b}_1 + (1-c)\bar{b}_2\}^2 R_{\rho\rho}(\kappa, t) + 2(\bar{b}_1 - \bar{b}_2)\{c\bar{b}_1 + (1-c)\bar{b}_2\} R_{\rho c}(\kappa, t) + (\bar{b}_1 - \bar{b}_2)^2 R_{cc}(\kappa, t). \quad (5.245)$$

From this representation we can deduce the conditions on the scattering lengths that are necessary in order to isolate a particular partial correlation function. Clearly, $R_{\rho\rho}$ and $R_{cc}$ are obtained directly if $\bar{b}_1 = \bar{b}_2$ and $\{c\bar{b}_1 + (1-c)\bar{b}_2\} = 0$, respectively.

The representation (5.245) is particularly useful in the study of ionic binary systems, where there is some additional benefit from introducing the charge denisty $(Z_1 - Z_2)c(\mathbf{r}, t)$ for ions with valence $Z_1$ and $Z_2$. More generally, the partial correlation functions $R_{11}$, $R_{12}$, and $R_{22}$ are most suitable to a treatment of the scattering at large $\kappa$ whereas $R_{\rho\rho}$, $R_{\rho c}$, and $R_{cc}$ are the appropriate choice for small values of $\kappa$ (March and Tosi 1977).

Averaging $R(\kappa, \omega)$ over the configurations of the atoms we arrive at the quantity which appears in the cross-section (eqn (5.233)). The expressions (5.236)–(5.238) show that in order to perform the configurational average it is necessary to evaluate quantities like

$$\overline{p_j p_{j'} Y_{jj'}(\kappa, t)}. \quad (5.246)$$

Unless we introduce some simplifying assumptions about the correlation between the dynamics and configurations of the atoms, or invoke a specific solvable model, no further reduction can be made. An extreme case is that in which there is no correlation between the dynamics and configurations. For this case the quantity (5.246) reduces to

$$\overline{p_j p_{j'}}\, Y_{jj'}(\kappa, t). \quad (5.247)$$

If, in addition, the atomic arrangement is random, the correlation of the labelling functions is easily worked out. From the definition of $p_j$ it follows that

$$\overline{p_j} = c$$

and so

$$\begin{aligned} \overline{p_j p_{j'}} &= c, \quad \text{if} \quad j = j' \\ &= c^2, \quad \text{if} \quad j \neq j' \end{aligned}$$

or

$$\overline{p_j p_{j'}} = c^2 + c(1-c)\,\delta_{jj'}. \quad (5.248)$$

Using this result in (5.247), we obtain

$$c\bar{R}_{11}(\kappa, t) = \frac{1}{N}\sum_{jj'} \overline{p_j p_j} Y_{jj'}(\kappa, t)$$
$$= c^2 R(\kappa, t) + c(1-c)R_s(\kappa, t), \tag{5.249}$$

where $R(\kappa, t)$ and $R_s(\kappa, t)$ are the coherent and single-particle response functions for the entire binary sample of $N$ atoms. A similar calculation for $R_{22}$ and $R_{12}$ leads to the results

$$\bar{R}_{22}(\kappa, t) = (1-c)R(\kappa, t) + cR_s(\kappa, t) \tag{5.250}$$

and

$$\bar{R}_{12}(\kappa, t) = \{c(1-c)\}^{1/2}\{R(\kappa, t) - R_s(\kappa, t)\}. \tag{5.251}$$

The results (5.249)–(5.251) give some insight into the physical significance of the partial correlation functions. Of greater interest, perhaps, are the results for $\bar{R}_{\rho\rho}$, $\bar{R}_{\rho c}$, and $\bar{R}_{cc}$. From (5.244) and the previous results, we find

$$\bar{R}_{\rho\rho}(\kappa, t) = R(\kappa, t), \qquad \bar{R}_{\rho c}(\kappa, t) = 0,$$

and

$$\bar{R}_{cc}(\kappa, t) = c(1-c)R_s(\kappa, t). \tag{5.252}$$

Hence, for the extreme situation where there is no correlation between the dynamics and the random configurations of the atoms, $\bar{R}_{\rho\rho}$ is the total coherent response, and $\bar{R}_{cc}$ is a diffuse contribution that adds to the single-particle, or incoherent, contribution to the cross-section. The results (5.252) are useful for estimating the cross-section for a slightly contaminated target sample.

We conclude the immediate discussion with the total cross-section, appropriate for the uncorrelated, random configuration model, comprised of the sum of the incoherent and coherent contributions. The single-atom incoherent cross-sections for the species are $\sigma_i^{(1)}$ and $\sigma_i^{(2)}$. Substituting (5.252) in (5.245) and adding the incoherent terms we obtain, from (5.233), the result

$$\frac{d^2\sigma}{d\Omega\, dE'} = N\frac{k'}{k}\{1+n(\omega)\}\omega\beta\Big\{[c\bar{b}_1 + (1-c)\bar{b}_2]^2 R(\kappa, \omega)$$
$$+ [c(1-c)(\bar{b}_1 - \bar{b}_2)^2 + \frac{1}{4\pi}(c\sigma_i^{(1)} + (1-c)\sigma_i^{(2)})]R_s(\kappa, \omega)\Big\}. \tag{5.253}$$

There is evidence, particularly from computer simulation studies (Hansen and McDonald 1976; McGreevy *et al.* 1984), that $\bar{R}_{cc}(\kappa, \omega)$ for a binary ionic system contains a well-defined excitation peak which can be viewed as an optic-type collective mode. The peak in $\bar{R}_{cc}(\kappa, \omega)$ persists to higher values of $\kappa$ than the viscoelastic mode in $\bar{R}_{\rho\rho}(\kappa, \omega)$. Optic vibrations, in which the two species tend to move in opposite directions, are not confined to ionic systems. They are likely to exist in metallic alloys, particularly in those alloys within which atoms of type-1 are very strongly attracted to atoms of type-2 and vice versa.

We will now summarize the formulae for diffraction from a multi-component fluid. In discussing binary systems we introduced two representations of the coherent scattering and commented on the merits of each representation. Whether one or the other is a more natural representation depends partly on the objectives of the study, and partly on personal taste. We will not attempt a comparative study of the many representations used in the interpretation of diffraction studies of multi-component fluids, but rather record the salient features of the cross-section. The cross-section of interest can be written in the form

$$\frac{\mathrm{d}\sigma}{\mathrm{d}\Omega} = \sum_{jj'} \overline{b_j b_{j'}} \langle \exp\{\mathrm{i}\boldsymbol{\kappa} \cdot (\mathbf{R}_{j'} - \mathbf{R}_j)\} \rangle \tag{5.254}$$

where the scattering lengths are assumed real.

The cross-section (5.254) obtains to the static limit in which a negligible neutron energy change occurs. For this limit we are justified in integrating the partial differential cross-section over all values of $\omega$ at constant $\boldsymbol{\kappa}$ to obtain (5.254). Unfortunately, the static limit cannot be realized in practice with low energy neutrons, and corrections must be introduced in the data analysis. The corrections, usually referred to as Placzek corrections, are often small but should ideally be assessed for each study (Howe 1978) particularly if large $\kappa$ data is involved.

Returning to (5.254), we first outline the reduction of the cross-section for a monatomic fluid. If there is no correlation between the values of the scattering lengths for different isotopes, the double sum becomes

$$\sum_{j} \overline{b^2} + \sum_{j \neq j'} (\bar{b})^2 \langle \exp\{\mathrm{i}\boldsymbol{\kappa} \cdot (\mathbf{R}_{j'} - \mathbf{R}_j)\} \rangle, \tag{5.255}$$

where terms with $j = j'$ are excluded in the second term. The static correlation function in (5.255) can be written in terms of the pair distribution function (cf. eqn (5.36))

$$\mathrm{g}(r) = \frac{V}{N^2} \sum_{j \neq j'} \langle \delta\{\mathbf{r} + \mathbf{R}_j - \mathbf{R}_{j'}\} \rangle. \tag{5.256}$$

Notice that in the limit $r \to \infty$ the correlation function in (5.256) tends to $\langle \delta(\mathbf{r}) \rangle = 1/V$, and therefore $\mathrm{g}(\mathbf{r}) \to 1$. From (5.256) we can readily deduce the correlation function in (5.255)

$$\sum_{j\neq j'} \langle \exp\{i\boldsymbol{\kappa}\cdot(\mathbf{R}_{j'}-\mathbf{R}_j)\}\rangle = N\rho_0 \int d\mathbf{r}\, \exp(i\boldsymbol{\kappa}\cdot\mathbf{r})g(r)$$
$$= N\rho_0(2\pi)^3\delta(\boldsymbol{\kappa}) + N\rho_0 \int d\mathbf{r}\, \exp(i\boldsymbol{\kappa}\cdot\mathbf{r})\{g(r)-1\}. \quad (5.257)$$

We have rearranged the integral so that it converges for small $\kappa$. The subtracted term, proportional to $\delta(\boldsymbol{\kappa})$, corresponds to no scattering, and is dropped henceforth. The result for the cross-section is

$$\frac{d\sigma}{d\Omega} = N\overline{b^2} + (\bar{b})^2 N\rho_0 \int d\mathbf{r}\, \exp(i\boldsymbol{\kappa}\cdot\mathbf{r})\{g(r)-1\}$$
$$\equiv N\{\overline{b^2}-(\bar{b})^2\} + N(\bar{b})^2\left[1+\rho_0\int d\mathbf{r}\, \exp(i\boldsymbol{\kappa}\cdot\mathbf{r})\{g(r)-1\}\right] \quad (5.258)$$

and

$$S(\kappa) = 1 + \rho_0 \int d\mathbf{r}\, \exp(i\boldsymbol{\kappa}\cdot\mathbf{r})\{g(r)-1\} \quad (5.259)$$

is the static structure factor. The first term in (5.258) is the incoherent cross-section. The result (5.259) agrees with the discussion in § 5.2.

Consider now a multicomponent system of $N$ atoms. The components are labelled by $\nu$ and $\mu$, and the fractional concentration of the $\nu$th type of atom is $c_\nu = N_\nu/N$. We introduce a notation in the sums involved in the cross-section in which $j(\nu)$ are atoms of type-$\nu$. For example, from the definition of $c_\nu$ it follows that

$$\sum_j \Rightarrow \sum_\nu \sum_{j(\nu)} = \sum_\nu Nc_\nu = N.$$

Using this notation in (5.255), we find

$$N\sum_\nu c_\nu \overline{b_\nu^2} + \sum_{\nu\mu} \bar{b}_\nu \bar{b}_\mu \sum_{j(\nu)} \sum_{j'(\mu)}{}' \langle \exp\{i\boldsymbol{\kappa}\cdot(\mathbf{R}_{j'}-\mathbf{R}_j)\}\rangle \quad (5.260)$$

and the prime in the second term means that we omit terms for which $j=j'$ if $\nu=\mu$. The pair distribution function $g_{\nu\mu}(r)$ is defined as

$$N\rho_0 c_\nu c_\mu g_{\nu\mu}(r) = \sum_{j(\nu)} \sum_{j'(\mu)}{}' \langle \delta\{\mathbf{r}+\mathbf{R}_j-\mathbf{R}_{j'}\}\rangle, \quad (5.261)$$

and all $g_{\nu\mu}(r)$ approach unity for $r\to\infty$. We define (Waseda 1980)

$$P_{\nu\mu}(\kappa) = \rho_0 \int d\mathbf{r}\, \exp(i\boldsymbol{\kappa}\cdot\mathbf{r})\{g_{\nu\mu}(r)-1\}, \quad (5.262)$$

in terms of which (5.260) becomes, apart from a term in $\delta(\boldsymbol{\kappa})$,

$$N\sum_\nu c_\nu \overline{b_\nu^2} + N\sum_{\nu\mu} \bar{b}_\nu \bar{b}_\mu c_\nu c_\mu P_{\nu\mu}(\kappa). \quad (5.263)$$

Finally, we define partial static structure factors

$$S_{\nu\mu}(\kappa) = c_\nu \delta_{\nu\mu} + c_\nu c_\mu P_{\nu\mu}(\kappa). \quad (5.264)$$

Substituting this definition in (5.263), the cross-section becomes

$$\frac{d\sigma}{d\Omega}=N\sum_{\nu} c_\nu(\overline{b_\nu^2}-(\bar{b}_\nu)^2)+N\sum_{\nu\mu} \bar{b}_\nu\bar{b}_\mu S_{\nu\mu}(\kappa), \tag{5.265}$$

which should be compared with (5.258) for a single-component fluid. Note that, since $g_{\nu\mu}(r)$ vanishes for $r\to 0$,

$$\int d\boldsymbol{\kappa} P_{\nu\mu}(\kappa)=-(2\pi)^3\rho_0. \tag{5.266}$$

The results (5.263) and (5.265) are the basis for the interpretation of diffraction data for multicomponent classical fluids.

## 5.6. Quantum fluids

One measure of the quantum nature of a fluid is obtained by comparing the thermal wavelength

$$\lambda_T=2\pi\hbar(3Mk_BT)^{-1/2} \tag{5.267}$$

with the mean particle separation denoted by $d$. For fluids that satisfy $\lambda_T>d$, quantum interference effects are significant, and they are manifest in zero-point motions of the atoms. Consider three fluids, $^4$He, Ne, and Ar for which $d$ is expected to be very much the same, namely $d\sim 3$ Å. The values of the thermal wavelength are $\lambda_T(^4\text{He}, 2.1\text{ K})=4.6$ Å, $\lambda_T(\text{Ne}, 27\text{ K})=0.6$ Å, and finally $\lambda_T(\text{Ar}, 85\text{ K})=0.2$ Å. The gradual decrease of $\lambda_T$ through this series of examples results from both increasing mass and temperature. The figures for $\lambda_T$ show that quantum interference effects are negligible for Ar and significant for $^4$He, while Ne at 27 K is a marginal case.

When $\lambda_T>d$ it is necessary to take account of exchange forces that require identical particles of half-integer and integer spin to obey different statistics. The differences in the low-temperature phase diagrams of the isotopes $^3$He and $^4$He are due largely to their different quantum statistics rather than the difference in their masses. Fermi statistics are appropriate for $^3$He with spin $i=\frac{1}{2}$, and Bose statistics describe $^4$He with $i=0$. At sufficiently high temperatures Boltzmann statistics are valid even for these very light isotopes. To be more precise the use of Boltzmann statistics is valid when the chemical potential $\mu$ satisfies $\exp(\mu\beta)\ll 1$, and this is guaranteed if the particle density $\rho_0$ satisfies (cf (3.113))

$$\rho_0\ll (M/2\pi\hbar^2\beta)^{3/2}. \tag{5.268}$$

Let us consider the case of $^3$He; at a density $\rho_0=0.0164$ Å$^{-3}$,

$$2\pi\rho_0^{2/3}(\hbar^2\beta/M)=(6.54/T),$$

with $T$ in degrees K. Taking $T = 1.0$ K, we see that the condition for the validity of Boltzmann statistics (eqn (5.268)) is not satisfied, and we therefore anticipate that the effects of Fermi statistics will be significant in the properties of dense $^3$He at low temperatures. Since the equilibrium density of $^4$He is $\rho_0 = 0.0218$ Å$^{-3}$, we can also expect quantum statistics to be an important feature of $^4$He at low temperatures.

The normal boiling point of pure $^3$He is 3.2 K, and it does not form a solid at $T = 0$ except under a pressure ~30 atm. The other helium isotope $^4$He behaves similarly and remains a liquid at absolute zero unless a pressure ~25 atm is applied. However, it displays the remarkable properties of a superfluid below the $\lambda$-transition, which occurs at 2.17 K. The phase transition is marked by an anomaly in the specific heat $c_p$. Above and below the transition, $^4$He is usually referred to as He I and He II, respectively. The existence of the superfluid state is believed to result from Bose–Einstein condensation, which leaves a macroscopic number of particles in the state of zero momentum. The number of single-particle Bose–Einstein states with zero momentum $= n_0(T)N$, where the fraction

$$n_0(T) = \{1 - (T/T_0)^{3/2}\}; \qquad T < T_0$$

and the Bose–Einstein condensation temperature

$$T_0 = (2\pi\hbar^2/Mk_B)(\rho_0/2.612)^{2/3}.$$

Using $\rho_0 = 0.0218$ Å$^{-3}$ and $M = 4m$, $T_0 = 3.1$ K and the similarity of this figure with the $\lambda$-transition temperature encourages the belief that Bose–Einstein condensation and superfluidity are connected. It must be emphasized that in an interacting fluid the superfluid density is not given by the condensate fraction. For in an interacting fluid the ground state, $T = 0$, has $n_0 < 1$, whereas the superfluid density matches the particle density $\rho_0$ at $T = 0$. Numerous calculations give $n_0(0) \sim 0.1$ and neutron experiments are consistent with $n_0(1.0\ \text{K}) = 0.146 \pm 0.035$ (Sears *et al.* 1982; Mook 1983).

Neutron scattering from $^4$He is purely coherent, whereas for $^3$He the single-atom incoherent cross-section is approximately one-quarter of the total single-atom cross-section. The appropriate form of the cross-section for scattering unpolarized neutrons from a quantum fluid consisting of a single isotope is, from eqn (5.5),

$$\frac{d^2\sigma}{d\Omega\, dE'} = \frac{k'}{k}\frac{1}{4\pi}\frac{1}{2\pi\hbar}\int_{-\infty}^{\infty} dt\, \exp(-i\omega t) \sum_{jj'} \Big\{\sigma_c Y_{jj'}(\kappa, t) + \frac{\sigma_i}{i(i+1)} \langle \exp(-i\boldsymbol{\kappa} \cdot \hat{\mathbf{R}}_j)\hat{\mathbf{i}}_j \cdot \hat{\mathbf{i}}_{j'}(t) \exp\{i\boldsymbol{\kappa} \cdot \hat{\mathbf{R}}_{j'}(t)\}\rangle \Big\}. \quad (5.269)$$

Here, $\sigma_c$ and $\sigma_i$ are, respectively, the single-atom coherent and incoherent cross-sections, and $Y_{jj'}(\kappa, t)$ is defined in eqn (5.4). In a perfect fluid, no spin-dependent forces are present and therefore $\hat{\mathbf{i}}(t) = \hat{\mathbf{i}}(0) \equiv \hat{\mathbf{i}}$ in the second contribution to the cross-section. The spin factor $\hat{\mathbf{i}}_j \cdot \hat{\mathbf{i}}_{j'}$ can be taken outside the correlation, function and averaged separately and, for randomly oriented nuclear spins, the result is $\delta_{jj'}\{i(i+1)\}$. Moreover, for a perfect fluid, $Y_{jj'}(\kappa, t)$ is zero for $j \neq j'$, so the cross-section reduces to

$$\frac{d^2\sigma}{d\Omega\, dE'} = N\frac{k'}{k}\frac{\sigma}{4\pi}\frac{1}{2\pi\hbar}\int_{-\infty}^{\infty} dt \exp(-i\omega t)\, Y_{jj}(\kappa, t), \tag{5.270}$$

where $\sigma = (\sigma_c + \sigma_i)$ is the total single-atom cross-section. The correlation function is evaluated in § 3.6.1 for perfect Bose and Fermi fluids.

Spin-dependent forces are present in an interacting quantum fluid, and in this instance the second contribution to the cross-section does not reduce to $Y_{jj'}(\kappa, t)$. If the nuclear spins are in a paramagnetic state, then we can make the replacement

$$\hat{\mathbf{i}}_j \cdot \hat{\mathbf{i}}_{j'} = 3\hat{i}^z_j \hat{i}^z_{j'}$$

where $z$ denotes any one spatial component.

We now review the representations of the cross-section in terms of response functions. To be concrete in the review we consider the particle-density contribution, but the general framework applies for any auto-correlation function including the spin-density auto-correlation

$$\langle \exp(-i\boldsymbol{\kappa} \cdot \hat{\mathbf{R}}_j)\hat{i}^z_j \hat{i}^z_{j'}(t) \exp\{i\boldsymbol{\kappa} \cdot \hat{\mathbf{R}}_{j'}(t)\}\rangle. \tag{5.271}$$

Thus, we focus for the moment on the response function

$$\begin{aligned} S(\kappa, \omega) &= \frac{1}{2\pi\hbar N}\sum_{jj'}\int_{-\infty}^{\infty} dt \exp(-i\omega t)\, Y_{jj'}(\kappa, t) \\ &\equiv \frac{1}{2\pi\hbar N}\int_{-\infty}^{\infty} dt \exp(-i\omega t)\langle \hat{\rho}_{\boldsymbol{\kappa}} \hat{\rho}^+_{\boldsymbol{\kappa}}(t)\rangle, \end{aligned} \tag{5.272}$$

where the Fourier transform of the particle density is

$$\begin{aligned} \hat{\rho}_{\boldsymbol{\kappa}} &= \int d\mathbf{r} \exp(-i\boldsymbol{\kappa} \cdot \mathbf{r}) \sum_j \delta(\mathbf{r} - \hat{\mathbf{R}}_j) \\ &= \sum_j \exp(-i\boldsymbol{\kappa} \cdot \hat{\mathbf{R}}_j). \end{aligned} \tag{5.273}$$

First, we recall the representation of $S(\kappa, \omega)$ in terms of the relaxation function introduced in § 5.1, $R(\kappa, \omega)$. This function is purely real and an even function of $\omega$, and

$$S(\kappa, \omega) = \{1 + n(\omega)\}\omega\beta R(\kappa, \omega), \tag{5.274}$$

where the detailed balance factor, $\{1 + n(\omega)\}$, is defined in eqn (5.17).

Most theoretical studies of quantum fluids are couched in terms of a generalized susceptibility, $\chi_\kappa[\omega]$, whose imaginary part is proportional to $R(\kappa, \omega)$. We define the susceptibility as

$$\chi_\kappa[\omega] = -\lim_{\eta\to 0^+}\int_0^\infty \mathrm{d}t\,\exp(-\mathrm{i}\omega t-\eta t)\phi_\kappa(t) \tag{5.275}$$

and

$$\hbar\phi_\kappa(t) = \mathrm{i}\langle[\hat{B}^+_{\boldsymbol{\kappa}}(t), \hat{B}_{\boldsymbol{\kappa}}]\rangle$$

and for density fluctuations $\hat{B}_{\boldsymbol{\kappa}} = \hat{\rho}_{\boldsymbol{\kappa}}/\sqrt{N}$. The relation to the relaxation function is established through

$$\phi_\kappa(t) = -\beta\partial_t R(\kappa, t)$$

so that

$$\begin{aligned}\chi_\kappa[\omega] &= \beta\lim_{\eta\to 0^+}\int_0^\infty \mathrm{d}t\,\exp(-\mathrm{i}\omega t-\eta t)\partial_t R(\kappa, t)\\ &= -\beta R(\kappa, 0)+\mathrm{i}\omega\beta\int_0^\infty \mathrm{d}t\,\exp(-\mathrm{i}\omega t)R(\kappa, t).\end{aligned} \tag{5.276}$$

The integral in the second expression is equivalent to the Laplace transform of $R(\kappa, t)$ evaluated for $s=\mathrm{i}\omega$ (cf. eqn (5.74)). Because $R(\kappa, t)$ is real and an even function of $t$, the Fourier transform $R(\kappa, \omega)$ in eqn (5.274) is obtained from the real part of the Laplace transform (cf. eqn (5.75))

$$\begin{aligned}R(\kappa, \omega) &= \frac{1}{2\pi}\int_{-\infty}^\infty \mathrm{d}t\,\exp(-\mathrm{i}\omega t)R(\kappa, t)\\ &= \frac{1}{\pi}\operatorname{Re}\int_0^\infty \mathrm{d}t\,\exp(-\mathrm{i}\omega t)R(\kappa, t).\end{aligned} \tag{5.277}$$

Using this last expression in (5.276) we arrive at the desired relation between $R(\kappa, \omega)$ and the imaginary part of the generalized susceptibility, which is often called the dissipative part and denoted here by $\chi''_\kappa[\omega]$, namely,

$$\omega\beta R(\kappa, \omega) = \frac{1}{\pi}\chi''_\kappa[\omega]. \tag{5.278}$$

The dimension of $\chi''_\kappa[\omega]$, given dimensionless variables $\hat{B}_\kappa$ in (5.275), is $(\text{energy})^{-1}$. On combining (5.274) and (5.278), we have the relation between the observed response function $S(\kappa, \omega)$ and $\chi''_\kappa[\omega]$

$$S(\kappa, \omega) = \{1+n(\omega)\}\frac{1}{\pi}\chi''_\kappa[\omega], \tag{5.279}$$

and we note that

$$S(\kappa,\omega)-S(\kappa,-\omega)=\frac{1}{\pi}\{[1+n(\omega)]\chi''_\kappa[\omega]-[1+n(-\omega)]\chi''[-\omega]\}$$
$$=\frac{1}{\pi}\chi''_\kappa[\omega], \quad (5.280)$$

where we have used the relation $\{1+n(-\omega)\}=-n(\omega)$ and the fact that $\chi''_\kappa[\omega]$ is an odd function of $\omega$. The relation (5.280) expresses the detailed balance condition which is usually given in the form $S(\kappa,\omega)=\exp(\hbar\omega\beta)S(\kappa,-\omega)$.

As the sample temperature is reduced toward absolute zero, the detailed balance factor approximates to a step function and, in the limit $T\to 0$ $(\beta\to\infty)$,

$$S(\kappa,\omega)=\frac{1}{\pi}\chi''_\kappa[\omega]; \qquad \omega>0$$
$$=0; \qquad \omega<0. \quad (5.281)$$

Because the response vanishes at sufficiently large values of $\omega$, it follows that $S(\kappa,\omega)$ for a degenerate quantum system possesses at least one maximum as a function of $\omega>0$ for a fixed scattering vector. The simplicity of the relation (5.281) is one incentive for using $\chi''_\kappa[\omega]$ to describe the properties of degenerate quantum systems. An equally important incentive stems from the relation between $\chi''_\kappa[\omega]$ and the linear response of the target sample to a perturbation that is proportional to the variable $\hat{B}_{\boldsymbol{\kappa}}$. We will exploit this relation in a later discussion of the structure of $\chi_\kappa[\omega]$ for imperfect quantum fluids. First, we record the basic sum rules, or moment relations, for $S(\kappa,\omega)$ in terms of $\chi''_\kappa[\omega]$ using eqn (5.279)

For particle-density fluctuations, we have the fundamental sum rules,

$$\int_{-\infty}^{\infty} d\omega\, \omega S(\kappa,\omega)=\frac{1}{2\pi}\int_{-\infty}^{\infty} d\omega\, \omega\chi''_\kappa[\omega]=(\kappa^2/2M) \quad (5.282)$$

and

$$\int_{-\infty}^{\infty} d\omega \frac{S(\kappa,\omega)}{\omega}=\frac{1}{2\pi}\int_{-\infty}^{\infty} d\omega\frac{\chi''_\kappa[\omega]}{\omega}=\frac{1}{2}\chi_\kappa, \quad (5.283)$$

where $\chi_\kappa$ is the isothermal susceptibility $=\beta R(\kappa, t=0)$. In view of (5.282) and (5.283), it is usual to define a normalized frequency $\omega_0$ such that

$$\omega_0^2=(\kappa^2/M\chi_\kappa). \quad (5.284)$$

We also note the compressibility sum rule

$$\lim_{\kappa\to 0}\chi_\kappa=(1/Mv_T^2); \qquad T=0 \quad (5.285)$$

where $v_T$ is the isothermal sound velocity. Experimental results for $v_T$ for $^3$He and $^4$He are 1.80 and $2.38\times10^{12}$ Å s$^{-1}$, respectively.

The static structure factor

$$S(\kappa)=\hbar\int_{-\infty}^{\infty} d\omega\, S(\kappa,\omega)=\frac{\hbar}{\pi}\int_{0}^{\infty} d\omega\, \coth(\tfrac{1}{2}\hbar\omega\beta)\chi''_{\kappa}[\omega]. \tag{5.286}$$

In the classical limit $\hbar\to0$ we find from (5.283) and (5.286) the result $S(\kappa)\to\chi_\kappa/\beta$. At zero temperature, the Schwartz inequality provides the relation

$$\chi_\kappa \geqslant (4M/\hbar^2\kappa^2)S^2(\kappa) \tag{5.287}$$

which, combined with the compressibility sum rule, leads to $(T=0)$

$$S(\kappa)\leqslant(\hbar\kappa/2Mv_T). \tag{5.288}$$

We deduce that $S(0)=0$ at zero temperature.

The last sum rule we give is

$$\int_{-\infty}^{\infty} d\omega\,\omega^3 S(\kappa,\omega)=\frac{1}{2\pi}\int_{-\infty}^{\infty} d\omega\,\omega^3\chi''_{\kappa}[\omega]=\left(\frac{\kappa^2}{M}\right)\omega_l^2$$

where the frequency $\omega_l$ is determined by

$$\omega_l^2=\left(\frac{\hbar\kappa^2}{2M}\right)^2+\left(\frac{2\kappa^2}{M}\right)\langle\mathrm{KE}\rangle+\Omega^2(0)-\Omega^2(\kappa). \tag{5.289}$$

Here $\Omega^2(\kappa)$ is defined in (5.41), and $\langle\mathrm{KE}\rangle$ is the mean kinetic energy of a particle; for Fermi particles

$$\langle\mathrm{KE}\rangle=\sum_{\mathbf{k}}\left(\frac{\hbar^2k^2}{2M}\right)f_{\mathbf{k}}\Big/\sum_{\mathbf{k}} f_{\mathbf{k}} \tag{5.290}$$

and a similar expression for Bose particles with $f_{\mathbf{k}}$ replaced by the Bose distribution $n_{\mathbf{k}}$.

The structure factor and isothermal susceptibility have been measured for $^4$He (Cowley 1978; Price 1978). Note that in determining the susceptibility from (5.283), the integral is sensitive to values of $S(\kappa,\omega)$ at small frequencies, while the high-frequency and small-amplitude values of the response are killed off rapidly by the denominator. The susceptibility of $^3$He is not known, at present, although calculations, using measured values of $S(\kappa)$, indicate that it is similar to the $^4$He susceptibility (Yoshida and Takeno 1979). These calculations of $\chi_\kappa$ give results that are significantly different from the susceptibility of a perfect degenerate Fermi fluid, which can be calculated from the results given in § 3.6.1 and eqn (5.283). The result is

$$\chi_\kappa^{(0)}=\left(\frac{3}{4\epsilon_f}\right)\left[1+\frac{1}{y}\left\{1-(y/2)^2\right\}\ln\left|\frac{2+y}{2-y}\right|\right] \tag{5.291}$$

where the Fermi energy is given in terms of the Fermi wave vector $p_f$, by $\epsilon_f = \hbar^2 p_f^2/2M$, and $\kappa = yp_f$. The major difference between the calculated results for $\chi_\kappa$, based on a measured $S(\kappa)$, and $\chi_\kappa^{(0)}$ occur at wave vectors $\kappa < 2p_f$ where $\chi_\kappa^{(0)}$ rises to a maximum value of $(3/2\epsilon_f)$ whereas the calculated values decrease. Using the result (5.285), the experimental value of $v_T$, and $\epsilon_f = 0.428$ meV, we conclude that, as $\kappa \to 0$, $\epsilon_f \chi_\kappa$ tends to 0.42 which is more than a factor of three smaller than the result obtained using the perfect Fermi fluid result (eqn (5.291)).

We will now discuss the spectrum of density fluctuations observed in neutron scattering from liquid $^3He$. Perhaps the first point to make is that successful neutron scattering experiments with $^3He$ samples demand great ingenuity on the part of the experimentalist because of the almost daunting absorption cross-section; see Table 1.1. Notwithstanding the immense technical difficulties, several neutron scattering studies of $^3He$ have been reported (Stirling *et al.* 1976; Sköld and Pelizzari 1978; Hilton *et al.* 1980).

The response function of a perfect degenerate Fermi fluid is given in § 3.6.1. Because of quantum statistics, in the guise of the Pauli exclusion principle, the scattering is confined to a domain of $\omega, \kappa$-values. This feature of the response stems from the requirement, imposed by the Pauli exclusion principle, that the response is generated by the promotion of states across the Fermi energy surface.

The exclusion principle tends to keep fermions apart in an imperfect Fermi fluid, and the independent-particle approximation is quite successful in computing properties of normal fluids. By the same token, the Pauli principle limits fermion collisions at low temperatures, to such an extent that the time between collisions $\tau$ increases as $T^{-2}$ as $T \to 0$; the value of $\tau$ is estimated to be $\tau \sim (0.4/T^2) \times 10^{-12}$ s with $T$ in degrees K. This has a profound effect on the propagation of ordinary sound because the kinematic viscosity, and hence the damping of ordinary sound (eqn (5.160)), is proportional to $\tau$. At $T = 0$, ordinary sound ceases to propagate at any frequency, since local thermal equilibrium, which requires $\omega\tau \ll 1$, is never attained.

Density fluctuations in an imperfect Fermi fluid participate in a collective motion that is sustained by a self-consistent interaction between the particles. Called zero sound by Landau, this collective mode can be viewed as a distortion of the Fermi surface, and it occurs only in the collisionless regime where $\omega\tau \gg 1$ (Akhiezer *et al.* 1962). Hence, if we keep the frequency fixed and decrease the temperature of a Fermi fluid, ordinary sound will eventually be completely damped and zero sound will emerge at a sufficiently low temperature.

Our demonstration of the existence of a collisionless sound mode in an imperfect Fermi fluid incorporates explicitly the notion of a self-consistent interaction between the particles (Levin and Valls 1983). Let us consider the calculation of the generalized susceptibility $\chi_\kappa[\omega]$, for a Fermi fluid described by a Hamiltonian based on a two-body pair potential,

$$\hat{\mathscr{H}} = (\text{KE}) + \frac{1}{2N} \sum_{\mathbf{q}} u(\mathbf{q}) \hat{\rho}^+_{\mathbf{q}} \hat{\rho}_{\mathbf{q}}. \tag{5.292}$$

Here, the first term denotes the kinematic energy of particles of mass $M$ and $u(\mathbf{q})$ is the Fourier transform of the effective pair potential. We will assume that $u(0)$ is a constant, and therefore exclude very long-range potentials.

We choose to calculate $\chi_\kappa[\omega]$ by using linear response theory. In Appendix B it is shown that, if we apply a perturbation ($\eta > 0$)

$$\hat{\mathscr{H}}_1 = -h(t)\hat{\rho}_{\boldsymbol{\kappa}}; \qquad h(t) = h_0 \exp(\mathrm{i}\omega t + \eta t), \tag{5.293}$$

then the response of the system measured by the change induced in $\hat{\rho}^+_{\boldsymbol{\kappa}}$, denoted here by $\delta\langle\hat{\rho}^+_{\boldsymbol{\kappa}}\rangle$, is proportional to $h(t)\chi_\kappa[\omega]$. Thus we can obtain $\chi_\kappa[\omega]$, and hence $S(\kappa, \omega)$, by calculating $\delta\langle\hat{\rho}^+_{\boldsymbol{\kappa}}\rangle$ using perturbation theory. For a perfect fluid with $u(\mathbf{q}) = 0$, the result for the susceptibility, $\chi^{(0)}_\kappa[\omega]$, is

$$\chi^{(0)}_\kappa[\omega] = \frac{1}{N} \sum_q (f_{\mathbf{q}} - f_{\boldsymbol{\kappa}+\mathbf{q}})/(\hbar\omega + \epsilon_{\mathbf{q}} - \epsilon_{\boldsymbol{\kappa}+\mathbf{q}} - \mathrm{i}\eta), \tag{5.294}$$

where $\epsilon_{\mathbf{q}} = \hbar^2 q^2/2M$, the distribution function

$$f_{\mathbf{q}} = (\exp\{(\epsilon_{\mathbf{q}} - \mu)\beta\} + 1)^{-1}, \tag{5.295}$$

and $\eta$ serves to define the function at the pole $\hbar\omega = (\epsilon_{\boldsymbol{\kappa}+\mathbf{q}} - \epsilon_{\mathbf{q}})$. The imaginary part of $\chi^{(0)}_\kappa[\omega]$ is evaluated for a degenerate fluid in § 3.6.1 (eqns (3.119) and (5.281)).

Consider next the influence of the interaction term in the Hamiltonian. The total perturbation applied to the system can now be regarded as the sum of the direct term (5.293) and a term generated by $u$. The perturbing potential generated by the latter can be found from the change in the Hamiltonian caused by the change in $\hat{\rho}^+_{\boldsymbol{\kappa}}$ by an amount $\delta\langle\hat{\rho}^+_{\boldsymbol{\kappa}}\rangle$. Noting that this change will effect the Fourier component $\mathbf{q} = \boldsymbol{\kappa}$ only and keeping terms linear in $\delta\langle\hat{\rho}^+_{\boldsymbol{\kappa}}\rangle$, the change in $\hat{\mathscr{H}}$ is readily shown to be

$$\frac{1}{N} u(\boldsymbol{\kappa}) \hat{\rho}_{\boldsymbol{\kappa}} \delta\langle\hat{\rho}^+_{\boldsymbol{\kappa}}\rangle.$$

This perturbing potential acts in conjunction with the direct term (5.293). However, $\delta\langle\hat{\rho}^+_{\boldsymbol{\kappa}}\rangle$ is proportional to the susceptibility. With our definitions,

$$\delta\langle\hat{\rho}^+_{\boldsymbol{\kappa}}\rangle = -h(t)N\chi_\kappa[\omega],$$

where the factor $N$ arises because we require the spectrum of the normalized correlation function $\langle\hat{\rho}_{\boldsymbol{\kappa}}\hat{\rho}^{+}_{\boldsymbol{\kappa}}(t)\rangle/N$ (cf. eqn (5.272)). Hence, the total perturbation is now

$$\hat{\mathscr{H}}_1 = -h(t)\hat{\rho}_{\boldsymbol{\kappa}}\{1+u(\boldsymbol{\kappa})\chi_\kappa[\omega]\}. \tag{5.296}$$

We repeat the calculation of $\delta\langle\hat{\rho}^{+}_{\boldsymbol{\kappa}}\rangle$ by perturbation theory using (5.296) in place of (5.293) and find immediately

$$\begin{aligned}\frac{1}{N}\delta\langle\hat{\rho}^{+}_{\boldsymbol{\kappa}}\rangle &= -h(t)\chi^{(0)}_\kappa[\omega]\{1+u(\boldsymbol{\kappa})\chi_\kappa[\omega]\}\\ &= -h(t)\chi_\kappa[\omega],\end{aligned}$$

whereupon the susceptibility of the imperfect fluid is

$$\chi_\kappa[\omega] = \chi^{(0)}_\kappa[\omega]/\{1-u(\boldsymbol{\kappa})\chi^{(0)}_\kappa[\omega]\}. \tag{5.297}$$

The response function $S(\kappa, \omega)$ is proportional to the imaginary part of $\chi_\kappa[\omega]$ and this, in turn, is proportional to the imaginary part of the susceptibility of the perfect fluid. Thus, at first sight, $S(\kappa, \omega)$ is non-zero for the range of $\kappa, \omega$ for which $\chi^{(0)}_\kappa[\omega]$ is non-zero, and zero otherwise. However, (5.297) possesses a pole at a value of $\omega$ for fixed $\kappa$ which lies outside the range in which $\chi^{(0)}_\kappa[\omega]$ is non-zero. This pole represents the collective oscillation of the fermions and, to the approximation in which we are working, the collective oscillation has an infinite lifetime. A more sophisticated calculation would yield a finite lifetime of the collective oscillation, which we identify as zero sound, due to multi-particle collisions (Hilton *et al.* 1980).

In order to prove that (5.297) possesses a pole at long wavelengths outside the particle–hole continuum, we must solve the equation

$$1 = u(\boldsymbol{\kappa})\operatorname{Re}\chi^{(0)}_\kappa[\omega], \tag{5.298}$$

in the limit $\kappa \to 0$. The evaluation of $\operatorname{Re}\chi^{(0)}_\kappa[\omega]$ from (5.294) is straightforward using the integration variables proposed in § 3.6.1 and the result in terms of the reduced energy $x = \hbar\omega/\epsilon_{\mathrm{f}}$, and reduced wave vector $y = \kappa/p_{\mathrm{f}}$ is

$$\begin{aligned}\operatorname{Re}\chi^{(0)}_\kappa[\omega] = \left(\frac{-3}{4\epsilon_{\mathrm{f}}}\right)\Bigg\{1 &+ \frac{1}{2y}\left[1-\left(\frac{x-y^2}{2y}\right)^2\right]\ln\left|\frac{2y-(x-y^2)}{2y+(x-y^2)}\right|\\ &+ \frac{1}{2y}\left[1-\left(\frac{x+y^2}{2y}\right)^2\right]\ln\left|\frac{2y+(x+y^2)}{2y-(x+y^2)}\right|\Bigg\}.\end{aligned} \tag{5.299}$$

Note that the result (5.299) satisfies the relation $\operatorname{Re}\chi^{(0)}_\kappa[0] = -\chi^{(0)}_\kappa$ where the isothermal susceptibility $\chi^{(0)}_\kappa$ is given by (5.291). A sound mode will have a dispersion that is linear in the wave vector. We therefore seek a solution of (5.298) using (5.299) in the form $x = y\Omega$, and take the limit

$y \to 0$. A careful calculation shows that eqn (5.298) reduces to

$$1 = u(0)\frac{3}{2\epsilon_f}\left\{\frac{\Omega}{4}\ln\left(\frac{\Omega+2}{\Omega-2}\right) - 1\right\}, \tag{5.300}$$

where $\Omega > 2$ because we seek a solution outside the particle–hole continuum. Given that $u(0) > 0$, i.e. the potential is repulsive, eqn (5.300) admits a solution for all $\Omega > 2$. The strong coupling limit, $(u(0)/\epsilon_f) \gg 1$, leads to a solution

$$\Omega^2 = 2u(0)/\epsilon_f, \tag{5.301}$$

which we identify with the velocity of zero-sound within our approximate treatment of density fluctuations. If we write the dispersion relation in the form $\omega = c\kappa$, then the velocity

$$c = \{u(0)/M\}^{1/2}. \tag{5.302}$$

Hence, we find an oscillation of the density fluctuations with an infinite lifetime and a dispersion of the same form as obtained for ordinary sound. We emphasize that the zero-sound mode is physically distinct from ordinary sound, for the former originates from a coherent self-consistent interaction between the fermions in the collisionless regime whereas the latter propagates when the system is in local thermal equilibrium. Note that $c$ is a nonanalytic function of the interaction potential, and this implies that the result (5.302) cannot be obtained by a straightforward perturbative expansion in the potential.

Consider the structure in $\chi''_\kappa[\omega]$ as a function of $\omega$ for a fixed small value of $\kappa$. For frequencies within the range in which $\mathrm{Im}\,\chi^{(0)}_\kappa[\omega]$ is finite, that is $0 < x < (y^2 + 2y)$ for $y = (x/p_f) < 2$, the response $\chi''_\kappa[\omega]$ will not be too unlike $\mathrm{Im}\,\chi^{(0)}_\kappa[\omega]$ (eqn (3.119)) apart from some distortion due to the denominator in (5.297). Above this broad spectrum of response there will be a delta function contribution, at the frequency $\omega = c\kappa$, due to zero sound. The amplitude of the delta function is evaluated using the following argument. Write the denominator of (5.297) as $D_\kappa(\omega)$. The velocity $c$ is then defined by the equation $\mathrm{Re}\,D_\kappa(c\kappa) = 0$ for $\kappa \to 0$. We make a Taylor series expansion of $D_\kappa(\omega)$ about $\omega = c\kappa$ so that, in the vicinity of the pole,

$$D_\kappa(\omega) \doteq D_\kappa(c\kappa) + \dot{D}_\kappa(c\kappa)(\omega - c\kappa - i\eta)$$

where $\dot{D}_\kappa(\omega)$ is the derivative of $D_\kappa(\omega)$ with respect to $\omega$. The first term in this expansion is zero by definition, and $\dot{D}_\kappa(c\kappa)$ is purely real because the frequency $c\kappa$ lies above the frequency range in which $\mathrm{Im}\,\chi^{(0)}_\kappa$ is non-zero. Hence,

$$\chi_\kappa[\omega] \doteq \chi^{(0)}_\kappa[\omega]/\{\dot{D}_\kappa(c\kappa)(\omega - c\kappa - i\eta)\}$$

and

$$\lim_{\eta\to 0} \chi''_\kappa[\omega] = \left\{\frac{\operatorname{Re}\chi^{(0)}_\kappa[c\kappa]}{\dot{D}_\kappa(c\kappa)}\right\}\pi\,\delta(\omega - c\kappa), \tag{5.303}$$

where we have used the standard identity for the delta function,

$$\lim_{\eta\to 0} \operatorname{Im}(\omega - x - i\eta)^{-1} = \pi\delta(\omega - x).$$

Since the response must be positive, we deduce that the pole $\omega = c\kappa$ is physically significant if $\dot{D}_\kappa(c\kappa) > 0$.

The coefficient in (5.303) can be calculated easily from the result for $\chi^{(0)}_\kappa[\omega]$ in the limit of large $\omega$

$$\chi^{(0)}_\kappa[\omega] = (\kappa^2/M\omega^2) + \cdots.$$

This result can be obtained from (5.299), or the sum rules given at the beginning of the section. We find that the zero-sound contribution to the response for $\omega > 0$, in the limit of strong coupling and $\kappa \to 0$, is

$$\frac{1}{\pi}\chi''_\kappa[\omega] = \frac{\kappa}{2Mc}\,\delta(\omega - c\kappa). \tag{5.304}$$

This result satisfies the sum rule (5.282) independent of the precise value of the zero-sound velocity. However, (5.283) and the compressibility sum rule (5.285) imply that $c^2 = v_T^2$, whereas it is generally believed that $c > v_T$.

The dispersion of zero sound for larger $\kappa$ can be derived from (5.298) for a given $u(\boldsymbol{\kappa})$. For a sufficiently large $\kappa$, the dispersion enters the particle–hole continuum described by $\operatorname{Im}\chi^{(0)}_\kappa[\omega]$ and acquires a lifetime through interactions with the continuum states.

The theory described is not in accord with the experimental findings. For wave vectors $\kappa \sim 2p_f$, the width of the observed response spectrum is smaller than the theoretical prediction. However, theory and experiment are more or less in line if the $^3$He mass is increased by a factor three, i.e. the effective mass $M^* \sim 3M$. The conclusion is consistent with the results of Landau's theory, valid for small $\kappa$ and $\omega$. Hence, the neutron experiments imply that the particle interactions in $^3$He persist at wave vectors as large as $\kappa \sim 2p_f$, and free-particle scattering, centred about the recoil energy $(\hbar^2\kappa^2/2M)$, emerges at wave vectors very much greater than this.

Even with an effective mass $M^* \sim 3M$, the theory is an inadequate representation of the data, particularly in the region in which the zero-sound mode merges with the continuum. The differences between theory and experiment are greatly reduced by the introduction of a frequency-dependent potential that is chosen to match with Landau theory in the appropriate limit (Yoshida and Takeno 1979; Glyde and Khanna 1980; Levin and Valls 1983). Such a theory is a good starting point from which to explore the pressure dependence of the response because the Landau parameters as a function of pressure are well established. The zero-sound mode energy shows a marked increase with applied pressure, and the intensity of the density response decreases which exacerbates further the immense technical difficulties of the experiments (Glyde and Khanna 1980).

Thus far we have discussed only the density response. However, the measured response is the weighted sum of the density and paramagnetic scattering, cf. (5.269). The latter is predicted to be localized about a frequency which is well below the zero-sound contribution for $\kappa \sim p_{\rm f}$, and to possess an intensity that is comparable with that from zero sound. This rather substantial paramagnetic response is yet another complicating issue in performing an interpretation of the more interesting particle density response (Hilton *et al.* 1980).

We mentioned in the prologue to the section that the very different behaviour of dense $^4$He compared with $^3$He at low temperatures is due largely to the difference in the quantum statistics for boson and fermion particles. For example, boson particles tend to occupy the same single-particle state in marked contrast to particles subject to the Pauli exclusion principle. Whereas the low-lying states of a normal Fermi fluid can be expected to be similar to those of a perfect fluid, the low-lying excitations in an imperfect Bose fluid have a collective phonon character. This latter feature of a Bose fluid is believed to reflect the existence of a condensate of zero-momentum particles for temperatures below the $\lambda$-transition.

The neutron data for $^4$He below the $\lambda$-transition show that for small $\kappa$ the response is dominated by a sharp peak. This observation is consistent with an exact result for the coherent response of $^4$He in the limit $T=0$,

$$\lim_{\kappa\to 0,\,\omega\to 0} S(\kappa,\omega) = \frac{\kappa}{2Mv_T}\,\delta(\omega - v_T\kappa); \qquad \omega > 0$$
$$= 0; \qquad \omega < 0, \tag{5.305}$$

where $v_T$ is the isothermal velocity of sound. Because this result satisfies the sum rule (5.282) we are certain that the response at $\omega = v_T\kappa$ is the only contribution as $\kappa$ and $\omega$ become small.

For larger values of $\kappa$ and $\omega$ and $T \to 0$ we might expect $\chi''_\kappa[\omega]$ to be the sum of two delta functions to a good approximation. If

$$\frac{1}{\pi}\chi''_\kappa[\omega] = C\{\delta(\omega-\omega_0) - \delta(\omega+\omega_0)\}, \tag{5.306}$$

then we satisfy the sum rules (5.282) and (5.283) if $\omega_0$ is identified with (5.284) and the amplitude factor

$$C = (\omega_0\chi_\kappa/2).$$

Given the compressibility sum rule (5.285), the approximation (5.306) agrees with the exact result (5.305) in the limit $\kappa \to 0$.

The frequency $\omega_0$ shows a marked $\kappa$-dependence when the isothermal susceptibility reaches its maximum value. The dip in $\omega_0$ is usually called the roton minimum for historic reasons. Such behaviour is in accord with the experimental finding but it lacks quantitative agreement, e.g. at $\kappa = 1.9\,\text{Å}^{-1}$, which is close to the minimum in $\omega_0$, the approximate dispersion $\omega_0 = (\kappa^2/M\chi_\kappa)^{1/2} \sim 1.2\,\text{meV}$ whereas the observed value is 0.75 meV (Cowley 1978). However, the approximation is fair for smaller wave vectors, $\kappa \lesssim 0.8\,\text{Å}^{-1}$.

The experimental results contain a significant weight at high frequencies which is evidently not included in (5.306). For a wave vector $\kappa \sim 2.5\,\text{Å}^{-1}$, the phonon and high-frequency contributions to the response are, in fact, more or less equal, and increasing $\kappa$ further results in the total disappearance of the phonon contribution. Although the interpretation of the neutron measurements has attracted much attention there is no really satisfactory theoretical picture yet.

The phonon mode persists above the $\lambda$-transition at long wavelengths. It is possible to interpret this in terms of a theory which is akin to the density response theory for $^3$He (eqn (5.297)). The response function for the perfect fluid (5.294) is evaluated in § 3.6.1 for arbitrary temperatures, in terms of the chemical potential $\mu$ which is itself a function of temperature. If $\chi_\kappa^{(0)}[\omega]$ is taken to be the susceptibility of the perfect fluid, then $\mu$ is determined from the condition

$$N = \sum_{\mathbf{k}} n_{\mathbf{k}} \tag{5.307}$$

where the Bose distribution

$$n_{\mathbf{k}} = (\exp\{(\epsilon_{\mathbf{k}} - \mu)\beta\} - 1)^{-1}$$

and $\epsilon_{\mathbf{k}} = \hbar^2 k^2/2M$. The theory can be improved by incorporating the measured values of the density as a function of temperature (for a given pressure). To this end we make the replacement (Fetter and Walecka 1971)

$$(\epsilon_{\mathbf{k}} - \mu)\beta \to (M/M^*)\epsilon_{\mathbf{k}} + \theta, \tag{5.308}$$

with $\theta$ determined by (5.307) and $\theta = 0$ at the $\lambda$-transition. In (5.308) we allow for an effective mass $M^*$ which describes in part the interactions in the Bose fluid over and above what can be accounted for by a simple pair-potential. Using the saturated vapour pressure values of the particle density at the $\lambda$-transition, $M^* = 1.4\,M$, and $\theta$ increases from zero at the transition to 0.48 at 4.2 K. The theory predicts that the phonon velocity should increase as the temperature is increased from 2.2 to 4.2 K, whereas the experimental results show the opposite trend. We therefore have a similar situation regarding the confrontation of simple theories and neutron measurements as found in the previous discussion of $^3$He, in that the theories are found to be sadly lacking when a quantitative comparison is made (Mook 1983).

## REFERENCES


Akhiezer, A. I., Akhiezer, I. A., and Pomeranchuk, I. Ya. (1962). *Soviet Phys. JETP* **14,** 343.

Biggin, S., Gay, M. and Enderby, J. E. (1984). *J. Phys.* **C17,** 977.

Boon, J. P. and Yip, S. (1980). *Molecular hydrodynamics.* McGraw-Hill.

Copley, J. R. D. and Lovesey, S. W. (1975). *Rep. Prog. Phys.* **38,** 461.

Cowley, R. A. (1978). In *Quantum liquids.* J. Ruvalds and T. Regge (ed.) North Holland, Amsterdam.

Cummings, H. Z. and Levanyuk, A. P. (1983). *Light Scattering Near Phase Transitions.* North-Holland, Amsterdam.

Enderby, J. E. and Neilson, G. W. (1980). *Adv. Phys.* **29,** 323.

—— and —— (1981). *Rep. Prog. Phys.* **44,** 593.

Fetter, A. L. and Walecka, J. D. (1971). *Quantum theory of many-particle systems.* McGraw-Hill.

Glyde, H. R. and Khanna, F. C. (1980). *Can. J. Phys.* **58,** 343.

Hansen, J-P. and McDonald, I. R. (1976). *Theory of simple liquids.* Academic Press, New York.

Hilton, P. A., Cowley, R. A., Scherm, R., and Stirling, W. G. (1980). *J. Phys.* **C13,** L295.

Hoover, W. G. (1983). *Ann Rev. Phys. Chem.* **34,** 103.

Howe, R. A. (1978). In C. A. Croxton (ed.) *Progress in liquid physics.* John Wiley, New York.

Kawasaki, K. and Gunton, J. D. (1978). In C. A. Croxton (ed.) *Progress in liquid physics.* John Wiley, New York.

Levin, K. and Valls, O. T. (1983). *Phys. Rept.* **98,** 1.

Lewis, J. W. E. and Lovesey, S. W. (1977). *J. Phys.* **C10,** 3221.

—— and —— (1978). *J. Phys.* **C11,** L57.

Lovesey, S. W. (1979). *Z. Phys.* **B34,** 323.

March, N. H. and Tosi, M. P. (1977). *Atomic dynamics in liquids.* Macmillan, New York.

McGreevy, R. L., Mitchell, E. W. and Margaca, F. M. A. (1984). *J. Phys.* **C17,** 775.

Mook, H. A. (1983). *Phys. Rev. Lett.* **51,** 1454.

Mountain, R. D. (1976). *Advan. mol. relaxation Processes* **9,** 225.

—— (1982). *Phys. Rev.* **A26,** 2859.

—— Basu, P. K. (1983). *Phys. Rev.* **A28,** 370.

Powles, J. G. and Rickayzen, G. (1979). *Mol. Phys.* **38,** 1875.

Price, D. L. (1978). In *The physics of liquid and solid helium,* Part II (ed. K. H. Bennemann and J. B. Ketterson). Wiley, New York.

Résibois, P. and DeLeener, M. (1977). *Classical kinetic theory of fluids.* John Wiley, New York.

Schofield, P. (1968). *Physics of simple liquids.* North-Holland, Amsterdam.

Sears, V. F., Svensson, E. C., Martel, P., and Woods, A. D. B. (1982). *Phys. Rev. Lett.* **49,** 279.

Sjölander, A. and Turski, L. A. (1978). *J. Phys.* **C11,** 1973.

Sköld, K. and Pelizzari, C. A. (1978). *J. Phys.* **C11,** L589.
Söderstrom, O., Copley, J. R. D., Suck, J-B., and Dorner, B. (1980). *J. Phys.* **F10,** L151.
Stanley, H. E. (1971). *Introduction to phase transitions and critical phenomena.* Oxford University Press.
Stirling, W. G., Scherm, R., Hilton, P. A., and Cowley, R. A. (1976). *J. Phys.* **C9,** 1643.
van Kampen, N. G. (1981). *Stochastic processes in physics and chemistry.* North-Holland, Amsterdam.
Wahnström, G. and Sjögren, L. (1982). *J. Phys.* **C15,** 401.
Waseda, Y. (1980). *The structure of non-crystalline materials.* McGraw-Hill, New York.
Yoshida, F. and Takeno, S. (1979). *Prog. theor. Phys.* **62,** 37.

# 6
# PHYSICO-CHEMICAL APPLICATIONS

The fortuitous combination of the prominent role of protons in chemical processes and the good neutron characteristics of protons means that neutron diffraction and spectroscopy are useful for a wide range of physico-chemical studies. The proton cross-section is exceptionally large and much larger than the deuterium cross-section. The latter can be used to great advantage since the contrast pattern of hydrogenous compounds can be manipulated with dexterity in deuteration (Hayter and Penfold 1983; Furrer *et al.* 1983; Ehrhardt *et al.* 1984; McCammon 1984; Middendorf 1984). It is safe to neglect changes in the proton function that might arise from exchange with deuterium unless the function is highly mass dependent. Of course, the proton function is not always of prime interest, and selective isotope enrichment is effective with many other elements as can be seen by reference to Table 1.1.

A complete account of physico-chemical applications of neutron scattering demands a tome, and it is possibly beyond the capabilities of a single author. Here we strive to formulate some essential theory in enough detail to make modifications and extensions a feasible task. We will make extensive use of results from other chapters.

## 6.1. Diffraction by a molecular fluid

For the moment we consider diffraction by a single-component molecular fluid of density $\rho_0$. Effects of polydispersity on the diffraction pattern are discussed later in the section. We will set aside complications that can arise from the rotational and vibrational motions, and focus on the interpretation of the coherent elastic cross-section for rigid molecules.

We denote the average scattering length and position of the $j$th scattering nucleus by $\bar{b}_j$ and $\mathbf{R}_j$, respectively, and $\bar{b}_j$ is calculated using the formulae of § 1.3. With this notation, the coherent cross-section for scattering from rigid molecules is

$$\left(\frac{d\sigma}{d\Omega}\right)_{\text{coh}} = \left\langle \left| \sum_j \bar{b}_j \exp(i\boldsymbol{\kappa} \cdot \mathbf{R}_j) \right|^2 \right\rangle \tag{6.1}$$

where the average is over all possible configurations of the nuclei and, for the moment, we denote this by angular brackets. The nuclei comprise molecules and it is therefore sensible to recognize this in (6.1) by rearranging the sum into one over nuclei relative to a molecular centre of mass and then sum over all molecules.

We index the $N$ identical molecules by the label $l$. The position of the $\nu$th nucleus relative to the centre of mass of the $l$th molecule is $\mathbf{r}_{\nu(l)}$. With this notation and

$$\mathbf{R}_j - \mathbf{R}_{j'} = \mathbf{l} - \mathbf{l}' + \mathbf{r}_{\nu(l)} - \mathbf{r}_{\nu'(l')}, \tag{6.2}$$

the cross-section (6.1) reduces to

$$\left(\frac{d\sigma}{d\Omega}\right)_{coh} = \left\langle \sum_{ll'} \exp\{i\boldsymbol{\kappa}\cdot(\mathbf{l}-\mathbf{l}')\} \sum_{\nu(l)} \sum_{\nu'(l')} \bar{b}_\nu \bar{b}_{\nu'} \exp\{i\boldsymbol{\kappa}\cdot(\mathbf{r}_\nu - \mathbf{r}_{\nu'})\}\right\rangle. \tag{6.3}$$

Here it is assumed that the scattering lengths are purely real.

Let us separate the sum over molecules in (6.3) into intermolecular ($l \neq l'$) and intramolecular ($l = l'$) contributions, so that

$$\left(\frac{d\sigma}{d\Omega}\right)_{coh} = N\langle|F'(\boldsymbol{\kappa})|^2\rangle$$
$$+ \left\langle \sum_{l\neq l'} \exp\{i\boldsymbol{\kappa}\cdot(\mathbf{l}-\mathbf{l}')\} \sum_{\nu(l)} \sum_{\nu'(l')} \bar{b}_\nu \bar{b}_{\nu'} \exp\{i\boldsymbol{\kappa}\cdot(\mathbf{r}_\nu - \mathbf{r}_{\nu'})\}\right\rangle \tag{6.4}$$

where the molecular form factor is

$$F'(\boldsymbol{\kappa}) = \sum_\nu \bar{b}_\nu \exp(i\boldsymbol{\kappa}\cdot\mathbf{r}_\nu). \tag{6.5}$$

The expression (6.4) is exact given that the system is monodisperse. The second contribution contains information on the correlation between the orientations of the molecules and the intermolecular separation. The correlation is negligible for nearly spherical molecules, because a sphere has no orientation, and for this case we obtain, to a good approximation,

$$\left(\frac{d\sigma}{d\Omega}\right)_{coh} = N\langle|F'(\boldsymbol{\kappa})|^2\rangle$$
$$+ N\{S_{cm}(\kappa) - 1\}\left\langle \sum_{\nu(l)} \sum_{\nu'(l')} \bar{b}_\nu \bar{b}_{\nu'} \exp\{i\boldsymbol{\kappa}\cdot(\mathbf{r}_\nu - \mathbf{r}_{\nu'})\}\right\rangle \tag{6.6}$$

where it is understood that $l \neq l'$ in the second contribution, and $S_{cm}(\kappa)$ is the molecular structure factor (March 1984),

$$S_{cm}(\kappa) = 1 + \rho_0 \int d\mathbf{R} \exp(i\boldsymbol{\kappa}\cdot\mathbf{R})\{g_{cm}(R) - 1\}. \tag{6.7}$$

In the limit $\kappa \to 0$ we obtain for the cross-section (6.6) the result

$$\lim_{\kappa\to 0}\left(\frac{\mathrm{d}\sigma}{\mathrm{d}\Omega}\right)_{\mathrm{coh}} = N\,|F'(0)|^2 \lim_{\kappa\to 0} S_{\mathrm{cm}}(\kappa) \tag{6.8}$$

and

$$\lim_{\kappa\to 0} S_{\mathrm{cm}}(\kappa) = (\rho_0\chi_T/\beta), \tag{6.9}$$

where $\chi_T$ is the isothermal compressibility and $\beta$ is the inverse temperature.

In order to gain some insight into the full content of the cross-section we turn to a simple system of homonuclear diatomic molecules. Let the two nuclei be separated by a distance $r$ and

$$\mathbf{R}_j = \mathbf{l} + \tfrac{1}{2}(-1)^{\xi}\mathbf{r}_l \tag{6.10}$$

where $\xi = 1, 2$ labels the nuclei in a molecule. The structure factor

$$\begin{aligned} S(\kappa) &= \frac{1}{4N}\left\langle\left|\sum_{l\xi} \exp\{\mathrm{i}\boldsymbol{\kappa}\cdot(\mathbf{l}+\tfrac{1}{2}(-1)^{\xi}\mathbf{r}_l)\}\right|^2\right\rangle \\ &= \frac{1}{N}\left\langle\left|\sum_{l} \exp(\mathrm{i}\boldsymbol{\kappa}\cdot\mathbf{l})\cos(\tfrac{1}{2}\boldsymbol{\kappa}\cdot\mathbf{r}_l)\right|^2\right\rangle. \end{aligned} \tag{6.11}$$

Note that for vanishing separation of the nuclei, $r \to 0$, the structure factor (6.11) coincides with the static structure factor of a monatomic fluid. Separating the intra- and intermolecular contributions in (6.11) we find

$$\begin{aligned} NS(\kappa) &= N\langle\cos^2(\tfrac{1}{2}\boldsymbol{\kappa}\cdot\mathbf{r})\rangle \\ &\quad + \left\langle\sum_{l\neq l'} \exp\{\mathrm{i}\boldsymbol{\kappa}\cdot(\mathbf{l}-\mathbf{l}')\}\cos(\tfrac{1}{2}\boldsymbol{\kappa}\cdot\mathbf{r}_l)\cos(\tfrac{1}{2}\boldsymbol{\kappa}\cdot\mathbf{r}_{l'})\right\rangle. \end{aligned} \tag{6.12}$$

Here the dimensionless molecular form factor is

$$\begin{aligned} F(\kappa) &= \langle\cos^2(\tfrac{1}{2}\boldsymbol{\kappa}\cdot\mathbf{r})\rangle = \tfrac{1}{2}\langle\{1+\cos(\boldsymbol{\kappa}\cdot\mathbf{r})\}\rangle \\ &= \tfrac{1}{2}\{1+\sin(\kappa r)/(\kappa r)\}. \end{aligned} \tag{6.13}$$

The second contribution in (6.12) can be expressed as the sum of a term proportional to $\{S_{\mathrm{cm}}(\kappa)-1\}$ and a term that vanishes for spherical molecules. Neglecting the latter contribution it can be shown that (Hansen and McDonald 1976)

$$\begin{aligned} S(\kappa) &= \{F(\kappa)\}^2 + \{S_{\mathrm{cm}}(\kappa)-1\}\{2\sin(\tfrac{1}{2}\kappa r)/\kappa r\}^2 \\ &= \{2\sin(\tfrac{1}{2}\kappa r)/\kappa r\}^2 S_{\mathrm{cm}}(\kappa) + \left\{\frac{1}{2}+\frac{1}{(2\kappa r)}\sin(\kappa r) - [2\sin(\tfrac{1}{2}\kappa r)/\kappa r]^2\right\}. \end{aligned} \tag{6.14}$$

We will use this approximate result to study the behaviour of the structure factor $S(\kappa)$ as a function of the scattering vector. For small $\kappa$ we see that the second contribution to (6.14) vanishes, and the first contribution reduces to $S_{cm}(\kappa)$ evaluated for $\kappa \to 0$. From eqn (6.9) it follows that

$$S(\kappa) = S_{cm}(\kappa) = (\rho_0 \chi_T/\beta); \qquad \kappa \to 0. \tag{6.15}$$

The second contribution in (6.14) is very small for $(\kappa r) < 1$ since the leading-order term in its expansion is $(\kappa r)^4/720$. Thus, for long wavelengths the structure factor is the product of $S_{cm}(\kappa)$ and $(1-(\kappa r)^2/24)$ to a good approximation. In the opposite extreme, $(\kappa r) \gg 1$, the structure factor is determined by the limiting behaviour of the second contribution. Thus the structure factor $S(\kappa)$ does not resemble $S_{cm}(\kappa)$ and, for a dense fluid, the main peak in $S(\kappa)$ occurs at a $\kappa$-value that is smaller than the position of the main peak in $S_{cm}(\kappa)$.

Let us now return to the coherent cross-section (6.6) which is valid for nearly spherical molecules. Consistent with this we assume that the orientations of nuclei in different molecules are statistically independent, i.e. steric and soft forces do not induce correlations in the relative orientations of nuclei in different molecules. The coefficient of

$$\{S_{cm}(\kappa) - 1\}$$

in (6.6) is then

$$\left\langle \sum_{\nu(l)} \bar{b}_\nu \exp(i\boldsymbol{\kappa} \cdot \mathbf{r}_\nu) \right\rangle \left\langle \sum_{\nu(l')} \bar{b}_\nu \exp(-i\boldsymbol{\kappa} \cdot \mathbf{r}_\nu) \right\rangle = |\langle F'(\kappa) \rangle|^2$$

and the coherent elastic cross-section reduces to

$$\left(\frac{d\sigma}{d\Omega}\right)_{coh} = N\,|\langle F'(\kappa)\rangle|^2\, S_{cm}(\kappa) + N\{\langle |F'(\kappa)|^2\rangle - |\langle F'(\kappa)\rangle|^2\}. \tag{6.16}$$

Note that the second term vanishes for a monodisperse system of spherical molecules. The contribution of this term to the cross-section can be assessed by comparing (6.16) with (6.14) which is derived for homonuclear diatomic molecules. For the latter case the second term increases gradually with $\kappa$ from zero at $\kappa = 0$ to 0.5 in the limit of large $\kappa$ with a weak maximum at $\kappa r = 7.3$ where it achieves the value 0.54. Based on these results we can safely regard the second contribution to the coherent cross-section as a featureless, diffuse background that can be neglected for small scattering vectors.

The diffuse background does not vanish even for spherical molecules if the size distribution is polydisperse. A rigorous account of polydispersion begins with eqn (6.3) since we need to allow for a variation in molecular size (and more generally the shape) as a function of the label $l$. The size fluctuations will contribute to the diffuse background, and ideally need to be assessed for each molecular species.

In the absence of interactions between the molecules the molecular structure factor $S_{cm}(\kappa)$ is unity for all values of the scattering vector. From (6.16) we deduce the result for non-interacting molecules (cf. § 1.5)

$$\left(\frac{d\sigma}{d\Omega}\right)_{coh} = N\langle|F'(\kappa)|^2\rangle. \tag{6.17}$$

For a spherical molecule of radius $R$ with a uniform scattering length density $\rho$ (cf. § 1.5, eqn (1.56))

$$\begin{aligned} F'(\kappa) &= \rho \int_\Omega d\mathbf{R}\, \exp(i\boldsymbol{\kappa}\cdot\mathbf{R}) \\ &= \Omega\rho(3/x^3)(\sin x - x\cos x) \equiv \Omega\rho\{3j_1(x)/x\}, \end{aligned} \tag{6.18}$$

where the dimensionless variable $x = \kappa R$, $\Omega = 4\pi R^3/3$, and $j_1(x)$ is a spherical Bessel function of order 1. In the limit of small $\kappa R$ (March 1984),

$$\begin{aligned} F'(\kappa) &= \Omega\rho\{1 - \tfrac{1}{10}x^2 + \cdots\} \\ &\doteq \Omega\rho\, \exp(-\tfrac{1}{10}x^2); \qquad x = \kappa R \ll 1. \end{aligned} \tag{6.19}$$

The final expression in (6.19) can be compared with the Guinier form that is often used for very small $\kappa$. For an arbitrary molecule shape, and $\kappa \to 0$, the Guinier form is

$$\langle|F'(\kappa)|^2\rangle = (\Omega\rho)^2 \exp(-\kappa^2 R_g^2/3); \qquad \kappa \to 0, \tag{6.20}$$

where $R_g$ is the radius of gyration of the molecule. For a spherical molecule of radius $R$ we deduce from (6.19) and (6.20) the result $R_g^2 = 0.6R^2$. For an arbitrary shape molecule of volume $\Omega$,

$$\Omega R_g^2 = \int_\Omega d\mathbf{R}\, R^2. \tag{6.21}$$

The radius of gyration is often obtained from diffraction data by examining a plot of the logarithm of the cross-section as a function of $\kappa^2$. If (6.20) is valid, the data approximate a straight line whose gradient is $R_g^2/3$. A potential pitfall in the data analysis is that intermolecular interactions cause $S_{cm}(\kappa)$ to deviate from unity in the $\kappa$-region over which the data is fitted to the Guinier form, in which case the value derived for $R_g$ will not have a simple interpretation as a radius of gyration defined in (6.21) (Hayter and Penfold 1983).

The variation of the molecular form factor with $\kappa$ depends on the shape or conformation of the molecule for intermediate scattering vectors $\kappa R_g > 1$. From (6.18) we see that, for a spherical molecule, $F^2(\kappa) \propto 1/\kappa^4$ for $\kappa R_g > 1$, whereas for a rodlike molecule it can be shown that $F^2(\kappa) \propto 1/\kappa$.

For a dilute polymer solution diffraction in the Guinier region can be analysed for the variation of $R_g^2$ on the number of repeating units $Z_r$ in the monomer, while data at intermediate scattering vectors depend on the polymer conformation (Higgins 1979). If the relative orientations of the segments of the polymer chain obey a Gaussian distribution, then $R_g^2 \propto Z_r$. The corresponding coherent cross-section per monomer for a dilute monodisperse system is

$$\left(\frac{d\sigma}{d\Omega}\right)_{coh} = Z_r(\rho\Omega)^2(2/x^4)\{x^2 - 1 + \exp(-x^2)\}, \qquad (6.22)$$

where $x = \kappa R_g$ and $\rho\Omega$ is the average scattering length of one repeat unit. For $\kappa R_g < 1$, the expression (6.22) is consistent with the Guinier form, and for intermediate scattering vectors the cross-section decreases as $1/\kappa^2$. The latter result has been verified for bulk samples, and hence Gaussian conformation is more appropriate than a rodlike structure, for example.

## 6.2. Quasi-elastic scattering from single particles (Springer 1972)

The term quasi-elastic is usually applied to that part of the inelastic scattering spectrum which arises from random, or stochastic, processes that occur over a relatively long time scale. Rotational jumps of a molecule and the diffusion of a particle in a liquid or hot solid are examples of the types of motion which contribute to the quasi-elastic component of the spectrum. However, there is no precise definition of the term or the class of motions that contribute to it.

The strictly elastic contribution to the incoherent cross-section is readily defined in terms of the self-pair correlation function $G_s(\mathbf{r}, t)$ (eqn (3.23)). Recall that, when classical statistics apply, this function is the probability that, given a particle at the origin of the coordinate system at time $t = 0$, the same particle is at $\mathbf{r}$ at time $t$. The response function $S_i(\kappa, \omega)$ is

$$S_i(\kappa, \omega) = \frac{1}{2\pi\hbar}\int_{-\infty}^{\infty} dt \exp(-i\omega t) \int d\mathbf{r} \exp(i\boldsymbol{\kappa}\cdot\mathbf{r}) G_s(\mathbf{r}, t), \quad (6.23)$$

and therefore the elastic contribution to the incoherent cross-section is

$$\left(\frac{d^2\sigma}{d\Omega\, dE'}\right)^{el}_{incoh} = N\frac{\sigma_i}{4\pi}\,\delta(\hbar\omega) \int d\mathbf{r} \exp(i\boldsymbol{\kappa}\cdot\mathbf{r}) G_s(\mathbf{r}, \infty). \quad (6.24)$$

For a particle in a fluid the elastic cross-section is zero because $G_s(\mathbf{r}, \infty) = 0$, whereas it is finite, in general, for a confined particle.

In this context it is instructive to consider scattering from a particle bound in an isotropic harmonic oscillator potential. From the results given in § 3.6.2 we find that the intensity of the elastic scattering is

$$\int d\mathbf{r} \exp(i\boldsymbol{\kappa} \cdot \mathbf{r}) G_s(\mathbf{r}, \infty) = \exp\{-2W(\kappa)\} I_0(y). \tag{6.25}$$

Here

$$2W(\kappa) = \hbar\kappa^2 \coth(\tfrac{1}{2}\hbar\omega_0\beta)/2M\omega_0 \tag{6.26}$$

and $I_0(y)$ is a Bessel function of the first kind of order zero with an argument

$$y = \hbar\kappa^2/\{2M\omega_0 \sinh(\tfrac{1}{2}\hbar\omega_0\beta)\}. \tag{6.27}$$

In (6.26) and (6.27), $\omega_0$ is the frequency of vibration. We draw attention to the fact that the Debye–Waller factor, $\exp(-2W)$, on the right-hand side of (6.25) is multiplied by $I_0(y) \geqslant 1$. The latter arises from thermal fluctuations in the motion of the particle which are absent if the particle participates in a collective oscillation. In the limit of long wavelengths, $(\hbar\kappa^2/M\omega_0) \ll 1$, $I_0(y) \to 1$ and the intensity (6.25) approaches unity as $\kappa \to 0$. The confinement of the particle is reduced by decreasing the spring constant. In the limit $\omega_0 \to 0$, the right-hand side of (6.25) approaches

$$(M\beta\omega_0^2/2\pi\kappa^2)^{1/2}$$

and therefore the elastic intensity vanishes for free motion achieved with $\omega_0 = 0$.

Let us now consider a model for scattering from a particle executing harmonic oscillations about a centre which is diffusing freely in space. If there is no dynamic correlation between the two types of motion, then the response function is approximately

$$S_i(\kappa, \omega) = \frac{1}{2\pi\hbar} \int_{-\infty}^{\infty} dt \exp(-i\omega t - \kappa^2 D_s |t|) \int d\mathbf{r} \exp(i\boldsymbol{\kappa} \cdot \mathbf{r}) G_s(\mathbf{r}, t), \tag{6.28}$$

where $D_s$ is the self-diffusion constant (see § 5.3), and $G_s(\mathbf{r}, t)$ describes the oscillatory motion. The response function (6.28) comprises a quasi-elastic contribution and a sum of inelastic terms centred about $\pm\omega_0, \pm 2\omega_0, \ldots$. Each contribution has an energy spread $\sim(\hbar\kappa^2 D_s)$. The quasi-elastic contribution to the response of this model is

$$\begin{aligned} \pi\hbar S^q(\kappa, \omega) &= \left\{\frac{\kappa^2 D_s}{\omega^2 + (\kappa^2 D_s)^2}\right\} \int d\mathbf{r} \exp(i\boldsymbol{\kappa} \cdot \mathbf{r}) G_s(\mathbf{r}, \infty) \\ &= \left\{\frac{\kappa^2 D_s}{\omega^2 + (\kappa^2 D_s)^2}\right\} \exp\{-2W(\kappa)\} I_0(y), \end{aligned} \tag{6.29}$$

where the second equality follows from (6.25). We will use the notation $S^{\mathrm{q}}(\kappa, \omega)$ for the quasi-elastic response.

The contribution (6.29) is distinct from the inelastic components of the spectrum centred at $\pm\omega_0$ if $(\kappa^2 D_s) \ll \omega_0$, and this condition imposes a limitation on the value of the scattering vector. Indeed, for intermediate and large scattering vectors, the elastic response of the oscillator cannot be separated from the inelastic events at $\pm\omega_0$, $\pm 2\omega_0$, etc. and the complete incoherent response function (6.28) approaches the value

$$S_{\mathrm{i}}(\kappa, \omega) \doteq \frac{1}{2\pi\hbar} \int_{-\infty}^{\infty} \mathrm{d}t \exp(-\mathrm{i}\omega t - \kappa^2 D_s |t|) \exp(-\mathrm{i}\hbar\kappa^2 t/2M)$$

$$= \frac{(\kappa^2 D_s/\pi\hbar)}{(\omega - \hbar\kappa^2/2M)^2 + (\kappa^2 D_s)^2}; \qquad \kappa \to \infty. \tag{6.30}$$

Hence, for sufficiently large values of $\kappa$ the incoherent response approaches a broad Lorentz curve centred at the recoil energy $(\hbar^2\kappa^2/2M)$.

Two further comments about (6.29) are in order. First, the intensity

$$\hbar \int_{-\infty}^{\infty} \mathrm{d}\omega \; S^{\mathrm{q}}(\kappa, \omega)$$

is the same as that from the elastic contribution of the oscillatory motion (eqn (6.25)). Secondly, the model does not satisfy the detailed balance condition because of the approximate treatment of the diffusive motion in (6.28). We can rectify this shortcoming of the model by multiplying the expression by the detailed balance factor

$$\hbar\omega\beta\{1 - \exp(-\hbar\omega\beta)\}^{-1} = \hbar\omega\beta\{1 + n(\omega)\}. \tag{6.31}$$

A model for the response function built from the uncorrelated motions will satisfy the detailed balance condition if the individual response functions satisfy the condition independently. Let the individual response functions be denoted by $S_1(\kappa, \omega)$ and $S_2(\kappa, \omega)$. If the motions described by these functions are uncorrelated, then the model response function is

$$S_{\mathrm{i}}(\kappa, \omega) = \hbar \int_{-\infty}^{\infty} \mathrm{d}\omega' S_1(\kappa, \omega') S_2(\kappa, \omega - \omega')$$

$$= \hbar \int_{-\infty}^{\infty} \mathrm{d}\omega' S_2(\kappa, \omega') S_1(\kappa, \omega - \omega'). \tag{6.32}$$

The detailed balance condition for a system with inversion symmetry is

$$S_1(\kappa, \omega) = \exp(\hbar\omega\beta) S_1(\kappa, -\omega), \tag{6.33}$$

and there is a similar expression for $S_2(\kappa, \omega)$. Given that $S_1(\kappa, \omega)$ and $S_2(\kappa, \omega)$ satisfy the detailed balance condition, then it is straightforward to verify that $S_i(\kappa, \omega)$ (eqn (6.32)) also satisfies the condition. Returning to the example of oscillatory motion relative to a centre that is diffusing freely, we take

$$\pi S_1(\kappa, \omega) = \omega\beta\{1 + n(\omega)\}\left\{\frac{\kappa^2 D_s}{\omega^2 + (\kappa^2 D_s)^2}\right\} \tag{6.34}$$

while $S_2(\kappa, \omega)$ is the complete response of a harmonic oscillator of frequency $\omega_0$ given in § 3.6.2. If $\kappa^2 \ll (\omega_0/D_s)$, so that we can separate the quasi-elastic contribution to the response, then $S^q(\kappa, \omega)$ is obtained from (6.32) using

$$S_2(\kappa, \omega) \doteq \exp\{-2W(\kappa)\}I_0(y)\,\delta(\hbar\omega). \tag{6.35}$$

The result is the same as (6.29) multiplied by the detailed balance factor (6.31).

If the model response function (6.32) is to satisfy the sum rule (cf. (3.63))

$$\int_{-\infty}^{\infty} d\omega\, \omega S_i(\kappa, \omega) = (\kappa^2/2M), \tag{6.36}$$

then

$$\int_{-\infty}^{\infty} d\omega\, S_1(\kappa, \omega) \int_{-\infty}^{\infty} d\omega'\omega' S_2(\kappa, \omega') + \int_{-\infty}^{\infty} d\omega\, \omega S_1(\kappa, \omega) \int_{-\infty}^{\infty} d\omega' S_2(\kappa, \omega') = (\kappa^2/2\hbar M). \tag{6.37}$$

We are not able to impose (6.37) on our model of oscillatory motion relative to a centre that is diffusing because the first frequency moment of a Lorentz function does not exist. This can be overcome by demanding that $S_1(\kappa, \omega)$ is given by (6.34) for $|\omega| < \omega_c$, say, and is zero otherwise; the frequency $\omega_c$ is then related to the other parameters of the model by imposing the condition (6.37). Notice that, if the response functions $S_1(\kappa, \omega)$ and $S_2(\kappa, \omega)$ are exact models of self-motion, then $f = 1$ or 2,

$$\hbar \int_{-\infty}^{\infty} d\omega\, S_f(\kappa, \omega) = 1, \tag{6.38}$$

and

$$\int_{-\infty}^{\infty} d\omega\, \omega S_f(\kappa, \omega) = (\kappa^2/2M_f), \tag{6.39}$$

and therefore, from (6.37),

$$\sum_f (1/M_f) = (1/M). \tag{6.40}$$

The $M_f$'s are interpreted as effective masses which are related to the true mass of the scattering particle by (6.40).

The response function (6.34) describes continuous diffusion. For some applications it is appropriate to consider motion on a lattice by repeated jumps. In the limit of long wavelengths the structure of the lattice is not effective, and the motion is akin to continuous diffusion. This type of motion is observed with protons in metals (Sköld *et al.* 1979).

The stochastic motion of a particle is usually described by a rate equation for the probability $P(\mathbf{r}, t)$, that, given a particle at the origin of the coordinate system at $t = 0$, the same particle is at $\mathbf{r}$ at time $t$ (van Kampen 1981). Following the definition of $P(\mathbf{r}, t)$, we take the initial condition

$$P(\mathbf{r}, t = 0) = \delta(\mathbf{r}). \tag{6.41}$$

When classical statistics apply to the description of the target sample, $P(\mathbf{r}, t)$ and $G_s(\mathbf{r}, t)$ are identical. In view of this we shall calculate the self-response function from $P(\mathbf{r}, t)$ using the expression (cf. (6.23))

$$S_f(\kappa, \omega) = \omega\beta\{1 + n(\omega)\}\frac{1}{2\pi}\int_{-\infty}^{\infty} \mathrm{d}t \exp(-\mathrm{i}\omega t) \int \mathrm{d}\mathbf{r} \exp(\mathrm{i}\boldsymbol{\kappa}\cdot\mathbf{r})P(\mathbf{r}, t). \tag{6.42}$$

For such a model response function,

$$\int_{-\infty}^{\infty} \mathrm{d}\omega S_f(\kappa, \omega)$$

$$= (\beta/2\pi)\int_{-\infty}^{\infty} \mathrm{d}t \int_0^{\infty} \mathrm{d}\omega\; \omega \cos(\omega t)\{2n(\omega) + 1\} \int \mathrm{d}\mathbf{r} \exp(\mathrm{i}\boldsymbol{\kappa}\cdot\mathbf{r})P(\mathbf{r}, t). \tag{6.43}$$

In the classical limit $\hbar \to 0$, we can replace $\{2n(\omega) + 1\}$ by $(2/\hbar\omega\beta)$, and the integral (6.43) agrees with (6.38) in view of (6.41). However, in general, the model (6.42) will not satisfy exact sum rules; rather it is part of a model function to be used in (6.32).

Let us consider the calculation of the probability function $P(\mathbf{r}, t)$ for a particle jumping between interstitial sites of a rigid lattice. The vectors $\boldsymbol{\delta}$ define the neighbouring interstitial positions to which the particle can jump from the site $\mathbf{r}$, and the number of neighbouring positions involved is $r$. A rate equation for $P(\mathbf{r}, t)$ is

$$\tau\, \partial_t P(\mathbf{r}, t) = \frac{1}{r}\sum_{\boldsymbol{\delta}} \{P(\mathbf{r} + \boldsymbol{\delta}, t) - P(\mathbf{r}, t)\} \tag{6.44}$$

where $\tau$ is the residence time at a given site. If we multiply (6.44) by $\exp(\mathrm{i}\boldsymbol{\kappa}\cdot\mathbf{r})$ and integrate over $\mathbf{r}$ we find a solution that satisfies (6.41)

$$\int \mathrm{d}\mathbf{r} \exp(\mathrm{i}\boldsymbol{\kappa}\cdot\mathbf{r})P(\mathbf{r}, t) = \exp\{-t\gamma(\boldsymbol{\kappa})/\tau\}, \tag{6.45}$$

where the geometric factor

$$r\gamma(\boldsymbol{\kappa}) = \sum_{\boldsymbol{\delta}} \{1 - \exp(i\boldsymbol{\kappa} \cdot \boldsymbol{\delta})\}. \tag{6.46}$$

Taking $\boldsymbol{\kappa} = 0$ in (6.45) shows that the integrated probability is a constant independent of the time. If the sites of the lattice possess inversion symmetry,

$$\begin{aligned} \gamma(\kappa) &= \frac{1}{2r} \sum_{\boldsymbol{\delta}} (\boldsymbol{\delta} \cdot \boldsymbol{\kappa})^2; \qquad \kappa \to 0 \\ &= \kappa^2 l^2, \end{aligned} \tag{6.47}$$

where the equality defines the length $l$. Observe that the continuum limit of (6.44) is a standard diffusion equation. The continuum limit is reached with $\boldsymbol{\delta} \to 0$, in which case

$$P(\mathbf{r} + \boldsymbol{\delta}, t) = P(\mathbf{r}, t) + (\boldsymbol{\delta} \cdot \boldsymbol{\nabla}) P(\mathbf{r}, t) + \tfrac{1}{2} (\boldsymbol{\delta} \cdot \boldsymbol{\nabla})^2 P(\mathbf{r}, t) + \cdots.$$

Inserting this Taylor series expansion in (6.44) the term in $(\boldsymbol{\delta} \cdot \boldsymbol{\nabla})$ vanishes because of inversion symmetry, and the next term produces a diffusion equation with an effective self-diffusion constant $l^2/\tau$. The actual form of $P(\mathbf{r}, t)$ at large $\mathbf{r}$ is readily obtained from (6.45) using (6.47), and the result is

$$P(\mathbf{r}, t) = (4\pi t l^2/\tau)^{-3/2} \exp\{-r^2/(4tl^2/\tau)\}; \qquad r \to \infty.$$

Solution (6.45) is not an even function of time as it stands, although the solution of interest must have the property. The rate equation (6.44) describes the motion of the jumping particle on a time scale $t > \tau$, and the physical solution must be invariant under time reversal. Hence we compute the response function using (6.45) with $t$ replaced by $|t|$; the result cannot hold for short times because $P(\mathbf{r}, t)$ is unlikely to have a discontinuous gradient at $t = 0$. From (6.45) we obtain the response function

$$\pi S_f(\kappa, \omega) = \omega\beta\{1 + n(\omega)\} \left\{ \frac{\gamma(\boldsymbol{\kappa})/\tau}{\omega^2 + \{\gamma(\boldsymbol{\kappa})/\tau\}^2} \right\}. \tag{6.48}$$

In the long-wavelength limit we recover the result for continuous diffusion (eqn (6.34)) with an effective diffusion constant $l^2/\tau$, where $l^2$ is defined in (6.47). At larger values of $\boldsymbol{\kappa}$, the geometric function $\gamma(\boldsymbol{\kappa})$ is very anisotropic. If we neglect the contribution to the scattering which comes from the lattice vibrations, the quasi-elastic response is the product of (6.48) and the lattice Debye–Waller factor.

Scattering from lattice vibrations is the subject of Chapter 4. The elastic incoherent response is

$$\exp\{-2W(\boldsymbol{\kappa})\}\, \delta(\hbar\omega),$$

where $\exp(-2W)$ is the lattice Debye–Waller factor. For small values of $\kappa$ that are of interest in quasi-elastic scattering, the inelastic contribution to the incoherent response can be written in terms of the normalized vibrational density of states $Z(\omega) = Z(-\omega)$. The so called one-phonon approximation to the inelastic response is (cf. § 4.4)

$$\frac{\kappa^2}{2M\omega} Z(\omega)\{1 + n(\omega)\} \exp\{-2W(\boldsymbol{\kappa})\}.$$

In the limit $\omega \to 0$, the density of states for three-dimensional crystals vanishes as $\omega^2$; in this limit a useful approximation is

$$Z(\omega) = (3\omega^2/\omega_{\mathrm{D}}^3); \qquad \omega \to 0$$

where $\omega_{\mathrm{D}}$ is the Debye frequency. For $\omega \to 0$, the one-phonon approximation to the inelastic response reduces to a constant which is proportional to $\kappa^2$ and the temperature, namely,

$$\frac{\hbar^2\kappa^2}{2M} \frac{3k_{\mathrm{B}}T}{(\hbar\omega_{\mathrm{D}})^3} \exp\{-2W(\boldsymbol{\kappa})\}.$$

The contribution to the response (6.32) that arises from inelastic lattice vibrations is

$$\frac{\hbar\kappa^2}{2M} \exp\{-2W(\boldsymbol{\kappa})\} \int_{-\infty}^{\infty} \mathrm{d}\omega' \, \frac{Z(\omega')}{\omega'} \{1 + n(\omega')\} S_{\mathrm{f}}(\kappa, \omega - \omega'),$$

where $S_{\mathrm{f}}(\kappa, \omega')$ is given by (6.48) and, provided that this function is sharply peaked at $\omega' = 0$ and we consider $\omega \to 0$, the inelastic lattice contribution reduces to

$$\frac{\hbar^2\kappa^2}{2M} \cdot \frac{3k_{\mathrm{B}}T}{(\hbar\omega_{\mathrm{D}})^3} \exp\{-2W(\boldsymbol{\kappa})\}\hbar \int_{-\infty}^{\infty} \mathrm{d}\omega' S_{\mathrm{f}}(\kappa, \omega'). \tag{6.49}$$

The total incoherent response for jump diffusion on a lattice is the sum of (6.49) and the quasi-elastic contribution

$$S^{\mathrm{q}}(\kappa, \omega) = \exp\{-2W(\boldsymbol{\kappa})\} S_{\mathrm{f}}(\kappa, \omega). \tag{6.50}$$

We conclude that, if $\omega$ and $\kappa$ satisfy restrictive conditions (they must both be small), the incoherent response consists of a sharp quasi-elastic component on an $\omega$-independent background generated by the lattice vibrations.

Let us now look at a problem where the space available to the jumping particle is severely restricted (Ehrhardt *et al.* 1984). An extreme case is one in which the particle can reside at only two places apart from the origin. Initially the particle is at the origin and the probability distribution satisfies (6.41). At a later time the particle may be at the origin or either of two equivalent positions defined by the vectors $\pm\mathbf{d}$. Hence $P(\mathbf{r}, t)$ is of the form

$$P(\mathbf{r}, t) = \delta(\mathbf{r})f(t) + \{\delta(\mathbf{r}-\mathbf{d}) + \delta(\mathbf{r}+\mathbf{d})\}f'(t),$$

where $f(t)$ and $f'(t)$ are dimensionless functions that describe the relative distribution between the three available sites as a function of time. In order to satisfy (6.41), $f(0) = 1$ and $f'(0) = 0$ and, because the integrated probability is a constant independent of time, namely,

$$\int \mathrm{d}\mathbf{r} P(\mathbf{r}, t) = 1, \tag{6.51}$$

we conclude that

$$f(t) + 2f'(t) = 1.$$

From these considerations, the general form of $P(\mathbf{r}, t)$ is

$$P(\mathbf{r}, t) = \delta(\mathbf{r})f(t) + \tfrac{1}{2}\{\delta(\mathbf{r}-\mathbf{d}) + \delta(\mathbf{r}+\mathbf{d})\}\{1 - f(t)\} \tag{6.52}$$

where $f(t)$ is to be determined with the boundary condition $f(0) = 1$.

The spatial Fourier transform of $P(\mathbf{r}, t)$ required in (6.42) is

$$\int \mathrm{d}\mathbf{r} \exp(i\boldsymbol{\kappa} \cdot \mathbf{r}) P(\mathbf{r}, t) = f(t) + \cos(\boldsymbol{\kappa} \cdot \mathbf{d})\{1 - f(t)\}$$

$$= \{f(t) - f(\infty)\}\{1 - \cos(\boldsymbol{\kappa} \cdot \mathbf{d})\} + \cos(\boldsymbol{\kappa} \cdot \mathbf{d}) + f(\infty)\{1 - \cos(\boldsymbol{\kappa} \cdot \mathbf{d})\}. \tag{6.53}$$

The second equality shows that the inelastic scattering has a spatial structure factor

$$\{1 - \cos(\boldsymbol{\kappa} \cdot \mathbf{d})\}$$

which vanishes if $\boldsymbol{\kappa}$ and $\mathbf{d}$ are perpendicular. If, on the other hand, all orientations of $\mathbf{d}$ can occur with equal probability as in a polycrystal, the structure factor should be averaged over the orientation of $\boldsymbol{\kappa}$ with respect to $\mathbf{d}$ and the result is

$$\{1 - \sin(\kappa d)/(\kappa d)\}.$$

The structure factor for elastic scattering cannot be determined without a knowledge of $f(t)$ in the limit $t \to \infty$. Because all spatial configurations are equally probable in the limit of long times, $P(\mathbf{r}, t)$ must tend to a result proportional to

$$\{\delta(\mathbf{r}) + \delta(\mathbf{r}-\mathbf{d})\} + \{\delta(\mathbf{r}) + \delta(\mathbf{r}+\mathbf{d})\}.$$

The proportionality constant is determined from (6.51) to be $\frac{1}{4}$, from which we deduce that $f(\infty) = \frac{1}{2}$. The elastic structure factor is therefore

$$\tfrac{1}{2}\{1 + \cos(\boldsymbol{\kappa} \cdot \mathbf{d})\}$$

and, for random orientations of $\mathbf{d}$, this becomes

$$\tfrac{1}{2}\{1+\sin(\kappa d)/(\kappa d)\}.$$

To complete the calculation of $P(\mathbf{r}, t)$, we must find an explicit form for $f(t)$. If the time dependence of the probability distribution is obtained from a rate equation, then it will be of the form $\exp(-t/\tau)$ where $\tau$ is characteristic of the time between jumps. Given that $f(0)=1$ and $f(\infty)=\tfrac{1}{2}$, and that the passage between these limits is exponential we find

$$f(t)=\tfrac{1}{2}\{1+\exp(-|t|/\tau)\}. \qquad (6.54)$$

According to (6.53) the inelasticity in the spectrum is determined by $\{f(t)-f(\infty)\}$. Using (6.54) for $f(t)$ the inelastic component of the response is found to be a Lorentz function centred at $\omega=0$ with a half-width at half-maximum $=1/\tau$. The spatial confinement does not influence the dynamic response. This is not generally the case for scattering from particles executing jumps in a confined space, and various geometries have been studied in the context of quasi-elastic scattering from liquid crystals (Leadbetter and Richardson 1979).

Since the quasi-elastic component of the spectrum can usually be identified unambiguously in the limit of small scattering vectors, it is often appropriate to use a continuum model for $P(\mathbf{r}, t)$. For three-dimensional isotropic continuum models $P(\mathbf{r}, t)$ is a Gaussian function of $r$, and we write (cf. § 5.3.1)

$$P(\mathbf{r}, t)=\{2\pi\mu(t)\}^{-3/2}\exp\{-r^2/2\mu(t)\}; \qquad r\to\infty. \qquad (6.55)$$

The spatial Fourier transform required in (6.42) is a Gaussian function of $\kappa^2$, namely,

$$\int d\mathbf{r}\,\exp(i\boldsymbol{\kappa}\cdot\mathbf{r})P(\mathbf{r}, t)=\exp\{-\tfrac{1}{2}\kappa^2\mu(t)\}. \qquad (6.56)$$

In the case of uncorrelated jumps on a lattice we have $\mu(t)=2tl^2/\tau$ where $l^2$ is defined in (6.47). To satisfy the initial condition (6.41) it is necessary that $\mu(0)=0$. The physical interpretation of $\mu(t)$ is established by observing that, from the definition of $P(\mathbf{r}, t)$, the spatial moment

$$\int d\mathbf{r}\,r^2P(\mathbf{r}, t)=3\mu(t) \qquad (6.57)$$

is the mean-square displacement after a time $t$. The result $\mu(t)\propto t$ found for uncorrelated jumps on a lattice is characteristic of a random-walk process (van Kampen 1981).

The present formulation can be applied to quasi-elastic scattering from polymers in dilute solution (Higgins 1979). The radius of gyration of a polymer chain $R_g$ is typically $\sim 100$ Å. For $\kappa R_g \ll 1$, a continuum description is appropriate with $\mu(t) \propto tD_s$ where $D_s$ is the self-diffusion constant. Observed deviations of $D_s$ from Stokes–Einstein values are attributed to concentration-dependent effects that increase the rate of diffusive motion.

More interesting is the behaviour at intermediate scattering vectors where the internal motion of chain segments is observed. The Rouse model describes a monomer in terms of normal modes derived for a chain of balls that interact with the solvent (Bird *et al.* 1977). The solvent interaction is generated by a frictional force of strength $\xi$, and the coupling of balls is ascribed solely to an entropic force of strength $(k_B T/a^2)$ where $a \sim (R_g/10)$. For this model of polymer dynamics, $\mu(t) \propto (a^2 k_B Tt/\xi)^{1/2}$ and the width of the quasi-elastic spectrum, for intermediate scattering vectors $\kappa R_g > 1$ and $\kappa a < 1$, increases as $\kappa^4$. If coupling of the balls through the solvent is allowed, then $\mu(t) \propto (k_B Tt/\eta)^{2/3}$ where $\eta$ is the solvent viscosity, and the quasi-elastic width increases as $\kappa^3$.

## 6.3. Gas of isotropic harmonic oscillators

We calculate the cross-section for scattering from particles executing harmonic oscillations with respect to freely moving molecular centres. The model is the simplest one of interest in molecular spectroscopy. More realistic models of molecular vibrations, and the associated inelastic neutron cross-sections, are discussed in the following section. The position of the $j$th particle is

$$\mathbf{R}_j = \mathbf{l} + \mathbf{u} \tag{6.58}$$

where $\mathbf{u}$ is the displacement relative to the molecular centre. Each centre has a mass $M_{\text{mol}}$, and the scattering particles have a mass $M$ and frequency of vibration $\omega_0$. The Hamiltonian of one of the $N$ identical molecules is then

$$\mathcal{H} = \tfrac{1}{2}M_{\text{mol}}(\hat{\dot{\mathbf{l}}})^2 + \tfrac{1}{2}M(\hat{\dot{\mathbf{u}}})^2 + \tfrac{1}{2}M\omega_0^2\hat{\mathbf{u}}^2 \tag{6.59}$$

The partial differential cross-section is obtained from an expression derived in § 5.1

$$\frac{d^2\sigma}{d\Omega\, dE'} = \frac{k'}{k}\sum_{jj'}\frac{1}{2\pi\hbar}\int_{-\infty}^{\infty} dt\, \exp(-i\omega t)\overline{\{A_j A_{j'} + \tfrac{1}{4}\delta_{jj'}B_j^2 i(i+1)\}}\, Y_{jj'}(\kappa, t) \tag{6.60}$$

where the correlation function

$$Y_{jj'}(\kappa, t) = \langle \exp(-i\boldsymbol{\kappa}\cdot\hat{\mathbf{R}}_j)\exp\{i\boldsymbol{\kappa}\cdot\hat{\mathbf{R}}_{j'}(t)\}\rangle. \tag{6.61}$$

The other quantities in (6.60) relate to the nuclear scattering amplitudes which are taken to be purely real; the single-atom cross-section is

$$\sigma = 4\pi\{A^2 + \tfrac{1}{4}B^2 i(i+1)\}. \tag{6.62}$$

The horizontal bar in expression (6.60) denotes an average over the orientations of the molecules.

Since the motions of the molecular centres and the oscillating particles are taken to be independent, the correction function is

$$Y_{jj'}(\kappa, t) = \langle \exp(-i\boldsymbol{\kappa}\cdot\hat{\mathbf{l}})\exp\{i\boldsymbol{\kappa}\cdot\hat{\mathbf{l}}'(t)\}\rangle\,\langle \exp(-i\boldsymbol{\kappa}\cdot\hat{\mathbf{u}})\exp\{i\boldsymbol{\kappa}\cdot\hat{\mathbf{u}}(t)\}\rangle. \quad (6.63)$$

Given that the centres move freely and obey Boltzmann statics the first correlation function in (6.63) is obtained directly from (3.95). The second correlation function is derived in § 3.6.2. Combining the results, we obtain from (6.63) the result

$$Y_{jj'}(\kappa, t) = \exp\{i\boldsymbol{\kappa}\cdot(\mathbf{l}'-\mathbf{l})\}\exp\{\kappa^2(i\hbar t\beta - t^2)/2\beta M_{\text{mol}}\}$$
$$\times\exp\{-2W(\kappa)\}\sum_{n=-\infty}^{\infty} I_n(y)\exp\{n\omega_0(it+\tfrac{1}{2}\hbar\beta)\}, \quad (6.64)$$

where $W(\kappa)$ and $y$ are defined in (6.26) and (6.27) in terms of $\kappa^2$, $\omega_0$, $M$, and the ambient temperature $T = 1/k_B\beta$, and the integer $n$ ranges over positive and negative values.

When we substitute (6.64) into the cross-section (6.60) we must distinguish between intermolecular ($l \neq l'$) and intramolecular ($l = l'$) contributions; these are sometimes referred to as 'outer' and 'inner' terms. The outer contribution contains

$$\overline{\sum_{l\neq l'} \exp\{i\boldsymbol{\kappa}\cdot(\mathbf{l}'-\mathbf{l})\}}.$$

The quantity is evaluated by assuming the neighbours of any given molecule are smeared over a sphere of radius $p$, in which case

$$\overline{\sum_{l\neq l'} \exp\{i\boldsymbol{\kappa}\cdot(\mathbf{l}'-\mathbf{l})\}} = N\{\sin(\kappa p)/\kappa p\} = Nj_0(\kappa p). \quad (6.65)$$

From (6.64) and (6.65) and the integral (3.75), we obtain for the partial differential cross-section per molecule

$$\frac{d^2\sigma}{d\Omega\, dE'} = \frac{k'}{k}\left\{\frac{\sigma}{4\pi} + A^2 j_0(\kappa p)\right\}\exp\{-2W(\kappa)\}\left(\frac{\beta M_{\text{mol}}}{2\pi\hbar^2\kappa^2}\right)^{1/2}$$
$$\times \sum_{n=-\infty}^{\infty} I_n(y)\exp\left\{-\frac{\beta M_{\text{mol}}}{2\hbar^2\kappa^2}(\hbar\omega - E_R - n\hbar\omega_0)^2 + \tfrac{1}{2}n\hbar\omega_0\beta\right\},$$
$$(6.66)$$

where the recoil energy of a molecular centre $E_R = (\hbar^2\kappa^2/2M_{\text{mol}})$.

The integer $n$ is the number of units of energy $\hbar\omega_0$ lost ($n > 0$) or gained ($n < 0$) by the neutron in a given term in the sum. The processes in which the neutron loses energy are more strongly weighted than those in which it gains energy ($I_n(y) = I_{-n}(y)$), as one would expect, and in the limit $\beta \to \infty$ only those terms for which $n \geqslant 0$ remain. If the particle is bound very tightly to the molecule, $\omega_0$ is large. As $\omega_0$ increases, the only non-zero term is that for $n = 0$, i.e. the neutron does not have sufficient energy to excite any of the harmonic oscillator energy levels.

We now calculate the total neutron cross-section at absolute zero. It is not difficult to show that, in the limit $\beta \to \infty$, (6.66) reduces to

$$\frac{d^2\sigma}{d\Omega\, dE'} = \left\{\frac{\sigma}{4\pi} + A^2 j_0(\kappa p)\right\}\exp(-\hbar\kappa^2/2M\omega_0) \times \frac{k'}{k}\sum_{n=0}^{\infty}\frac{1}{n!}\left(\frac{\hbar\kappa^2}{2M\omega_0}\right)^n \delta(\hbar\omega - \hbar n\omega_0 - E_R). \tag{6.67}$$

If we drop the outer scattering term and integrate (6.67) over angles, we obtain

$$\frac{d\sigma}{dE'} = \sigma\frac{M_{mol}}{2\hbar^2 k^2}\sum_{n=0}^{\infty}\frac{1}{n!}\left(\frac{\omega - n\omega_0}{\gamma\omega_0}\right)^n \exp\left\{-\frac{1}{\gamma\omega_0}(\omega - n\omega_0)\right\}, \tag{6.68}$$

where $\gamma = M/M_{mol}$, and the energy satisfies the condition

$$\frac{\hbar^2}{2M_{mol}}(k-k')^2 \leqslant (\hbar\omega - n\hbar\omega_0) \leqslant \frac{\hbar^2}{2M_{mol}}(k+k')^2. \tag{6.69}$$

The integration of (6.68) over the final energy of the neutron, subject to the condition given by (6.69), is very similar to that for the scattering of a neutron by a single free nucleus, which we discussed in § 1.9. By an argument analogous to that given in § 1.9, we find that (6.69) is equivalent to

$$\frac{-\gamma'k + \surd\{k^2 - (2mn\omega_0/\hbar)(1+\gamma')\}}{1+\gamma'} \leqslant k' \leqslant \frac{\gamma'k + \surd\{k^2 - (2mn\omega_0/\hbar)(1+\gamma')\}}{1+\gamma'}, \tag{6.70}$$

where $\gamma' = m/M_{mol}$.

Using (6.70) it is straightforward to obtain the total cross-section, which can be written as

$$\sigma_t = \sigma\left(\frac{\hbar\omega_0}{4E}\right)\left(\frac{M}{m}\right)\sum_n \frac{1}{n!}\int_{x_-(n)}^{x_+(n)} dx\, x^n e^{-x}, \tag{6.71}$$

where the upper limit of the summation is that value of $n$ for which $E = n\hbar\omega_0(1+\gamma')$. The limits of integration in (6.71) are

$$x_{\pm}(n) = \frac{m}{M}\frac{1}{\hbar\omega_0(1+\gamma')^2}[E^{1/2} \pm \{E - n\hbar\omega_0(1+\gamma')\}^{1/2}]^2 \tag{6.72}$$

and $E = \hbar^2 k^2/2m$ is the energy of the incident neutron. When $m = M$ and $\gamma' = 0$, which corresponds to a proton bound harmonically to a fixed point, the cross-section (6.72) reduces to the result first derived by Fermi.

Consider the $n = 0$ term. The integral in (6.71) is easily performed, giving a total neutron cross-section

$$\sigma_t = \sigma\frac{\hbar\omega_0}{4E}\left(\frac{M}{m}\right)\left[1-\exp\left\{\frac{-4mE}{M\hbar\omega_0(1+\gamma')^2}\right\}\right]. \tag{6.73}$$

If $E/\hbar\omega_0 \ll 1$, this reduces to

$$\sigma_t \sim \frac{\sigma}{(1+\gamma')^2}, \tag{6.74}$$

which is the total cross-section for the scattering from free particles of mass $M_{mol}$. Since the limit $E/\hbar\omega_0 \ll 1$ corresponds to particles that are very tightly bound to their corresponding molecules we should expect the mass of the molecules to enter the cross-section formula rather than that of the scattering particles.

In the opposite limit of $E \gg \hbar\omega_0$, the total cross-section has the asymptotic form

$$\begin{aligned}\sigma_t &= \frac{\sigma}{4}\frac{\hbar\omega_0}{E}\left(\frac{M}{m}\right)\sum_{n=0}^{E/\hbar\omega_0(1+\gamma')}\frac{1}{n!}\int_0^\infty dx\, x^n e^{-x}\\ &= \frac{\sigma}{4}\left(\frac{M}{m}\right)\left(\frac{1}{1+\gamma'}\right).\end{aligned} \tag{6.75}$$

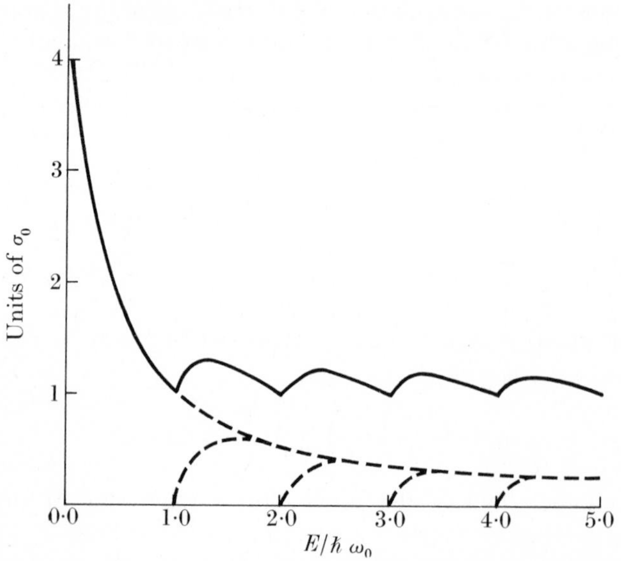


FIG. 6.1. Cross-section for a proton bound harmonically in a crystal (in units of the free single-nucleus cross-section $\sigma_0 = \sigma/4$) as a function of incident neutron energy $E$. The broken lines are the individual contributions from the terms in the summation.

Figure 6.1 shows the cross-section for a proton ($M=m$) bound harmonically in a crystal ($M_{\mathrm{mol}}=\infty$), in units of $\sigma/4$, as a function of the incident energy $E$. The broken lines in Fig. 6.1 are the individual contributions from the terms in the summation. This cross-section corresponds to that expected for hydrogen atoms harmonically bound in crystals, the so-called Einstein model in solid state theory, and is realized in practice by very few solids (Sköld *et al.* 1979; Boland *et al.* 1978).

## 6.4. Molecular spectroscopy

The vibrational spectra of nuclei in molecules can be determined from inelastic incoherent neutron scattering experiments. A very simple example of single-particle motion in a molecule is considered in the preceding section. For a single particle executing harmonic oscillations relative to a free molecular centre, scattering from the fundamental mode is proportional to the response function derived from (6.64) with $n=\pm 1$, namely,

$$S_f(\kappa,\omega)=\exp\{-2W(\kappa)\}\mathrm{I}_1(y)\frac{1}{2\pi\hbar}\int_{-\infty}^{\infty}\mathrm{d}t\,\exp(-\mathrm{i}\omega t)$$

$$\times\exp\left\{\frac{\kappa^2}{2\beta M_{\mathrm{mol}}}(\mathrm{i}\hbar t\beta-t^2)\right\}[\exp\{\omega_0(\mathrm{i}t+\tfrac{1}{2}\hbar\beta)\}+\exp\{-\omega_0(\mathrm{i}t+\tfrac{1}{2}\hbar\beta)\}]$$

$$=(\beta/4\pi E_{\mathrm{R}})^{1/2}\exp\{-2W(\kappa)\}\mathrm{I}_1(y)$$

$$\times\left\{\exp\left[-\frac{\beta}{4E_{\mathrm{R}}}(\hbar\omega-E_{\mathrm{R}}-\hbar\omega_0)^2+\tfrac{1}{2}\hbar\omega_0\beta\right]\right.$$

$$\left.+\exp\left[-\frac{\beta}{4E_{\mathrm{R}}}(\hbar\omega-E_{\mathrm{R}}+\hbar\omega_0)^2-\tfrac{1}{2}\hbar\omega_0\beta\right]\right\} \qquad (6.76)$$

where the recoil energy of the molecule $E_{\mathrm{R}}=(\hbar^2\kappa^2/2M_{\mathrm{mol}})$ and the second equality follows from the integral identity (3.75). The vibrational frequency of the particle is $\omega_0$, and $W(\kappa)$ and the argument of the Bessel function are given in (6.26) and (6.27), respectively.

The response (6.76) is seen to consist of two Gaussian peaks centred at $(E_{\mathrm{R}}\pm\hbar\omega_0)$, and the width of each peak is proportional to $(k_{\mathrm{B}}TE_{\mathrm{R}})^{1/2}$. In the limit $E_{\mathrm{R}}\to 0$, achieved by taking $M_{\mathrm{mol}}\to\infty$, say, the response reduces to a pair of delta functions with arguments $(\omega\pm\omega_0)$, as we shall see. The limit $M_{\mathrm{mol}}\to\infty$ is appropriate for a particle embedded in a crystal. However, the result (6.76) in this limit is not the same as the one-phonon incoherent cross-section derived in § 4.4 unless we replace the Bessel function $\mathrm{I}_1(y)$ with its small argument value

$$\mathrm{I}_1(y)=(y/2);\qquad y\to 0. \qquad (6.77)$$

For $E_R \to 0$ and $y \ll 1$, so that we can use (6.77), the response (6.76) can be written in the quintessential form (Furrer *et al.* 1983)

$$S_f(\kappa, \omega) = \frac{\kappa^2}{2M\omega} Z(\omega)\{1+n(\omega)\}\exp\{-2W(\kappa)\}, \tag{6.78}$$

where $Z(\omega) = Z(-\omega)$ is the normalized vibrational density of states and the detailed balance factor is defined in (6.31). It is instructive to consider the steps involved in reducing (6.76) to (6.78).

First, a delta function can be defined as

$$\delta(x) = \lim_{\epsilon \to 0^+} (\pi\epsilon)^{-1/2} \exp(-x^2/\epsilon). \tag{6.79}$$

In (6.76) we identify $\epsilon = (4E_R/\beta)$ and find, for $\epsilon \to 0$,

$$\begin{aligned} S_f(\kappa, \omega) &\\ &= \exp\{-2W(\kappa)\}\mathrm{I}_1(y)\,\{\delta(\hbar\omega - \hbar\omega_0)\exp(\tfrac{1}{2}\hbar\omega_0\beta) + \delta(\hbar\omega + \hbar\omega_0)\exp(-\tfrac{1}{2}\hbar\omega_0\beta)\} \\ &= \exp\{-2W(\kappa) + \tfrac{1}{2}\hbar\omega\beta\}\mathrm{I}_1(y)\,\{\delta(\hbar\omega - \hbar\omega_0) + \delta(\hbar\omega + \hbar\omega_0)\}. \end{aligned}$$

The second equality follows using the identity

$$\exp(-\tfrac{1}{2}\hbar\omega_0\beta)\,\delta(\hbar\omega + \hbar\omega_0) = \exp(\tfrac{1}{2}\hbar\omega\beta)\,\delta(\hbar\omega + \hbar\omega_0).$$

Next we invoke (6.77) given that $y$ (eqn (6.27)) is small compared with unity. In the definition of $y$,

$$\{\sinh(\tfrac{1}{2}\hbar\omega_0\beta)\}^{-1} = 2\exp(\tfrac{1}{2}\hbar\omega_0\beta)n(\omega_0),$$

where $n(\omega_0)$ is the Bose occupation function that occurs in the detailed balance factor (6.31). Combining these results and using the identities

$$\hbar\,\delta(\hbar\omega) = \delta(\omega)$$

and

$$\exp(\hbar\omega\beta)n(\omega) = 1 + n(\omega) = -n(-\omega),$$

we arrive at the result, valid for $E_R \to 0$ and $y \ll 1$,

$$S_f(\kappa, \omega) = \frac{\kappa^2}{2M\omega}\{1+n(\omega)\}\{\delta(\omega-\omega_0)+\delta(\omega+\omega_0)\}\exp\{-2W(\kappa)\}. \tag{6.80}$$

Results (6.78) and (6.80) are identical if we identify

$$Z(\omega) = \delta(\omega - \omega_0) + \delta(\omega + \omega_0), \tag{6.81}$$

which is the correct expression for the density of states of a single harmonic oscillator of frequency $\omega_0$.

We find that an expression similar to (6.78) holds for more realistic models of molecular vibrations. Before we proceed to these models we emphasize that the result (6.78) is valid when the scattering vector is not large. This restriction on $\kappa$ is implied by the use of (6.77). If

$$(\hbar\kappa^2/M\omega_0) \gg 1,$$

it is not even possible to select the fundamental mode in the response for a multitude of modes with equal amplitude are engaged at large scattering vectors. To see this we refer to expression (6.64). For $y \ll 1$ and $n > 0$,

$$\mathrm{I}_n(y) \propto y^n; \qquad y \to 0$$

so that the fundamental mode, with $n = 1$, is the largest contribution to the inelastic spectrum. On the other hand, for $y \gg 1$, the Bessel function is independent of $n$, and all modes contribute with equal amplitude. In this instance the representation of the oscillator correlation function used in (6.64) is no longer useful, and we must seek the appropriate expression beginning with (6.63). The analysis for $y \gg 1$ is given in §§ 3.6.2 and 4.7.2. Using the result given in (4.176), the correlation function (6.64) in the limit $\kappa \to \infty$ is found to be

$$Y_{jj'}(\kappa, t) = \exp\{\mathrm{i}\boldsymbol{\kappa} \cdot (\mathbf{l}' - \mathbf{l})\}\exp\left\{\frac{\kappa^2}{2\beta M_{\mathrm{mol}}}(\mathrm{i}\hbar t\beta - t^2)\right\}$$

$$\times \exp\left\{\frac{\mathrm{i}\hbar t\kappa^2}{2M} - \frac{1}{2}t^2\langle(\boldsymbol{\kappa}\cdot\hat{\mathbf{v}})^2\rangle\right\}; \qquad \kappa \to \infty \tag{6.82}$$

where $\mathbf{v}$ is the velocity of the scattering particle. The corresponding incoherent response function is

$$S_{\mathrm{i}}(\kappa, \omega) = \{2\pi\hbar^2\Delta^2\}^{-1/2} \exp\left\{-\left(\omega - \frac{\hbar\kappa^2}{2M} - E_{\mathrm{R}}/\hbar\right)^2 \Big/ 2\Delta^2\right\} \tag{6.83}$$

where the width function

$$\Delta^2 = \langle(\boldsymbol{\kappa}\cdot\hat{\mathbf{v}})^2\rangle + (2E_{\mathrm{R}}/\hbar^2\beta).$$

We conclude from this result that, in the limit of large scattering vectors, the incoherent cross-section approaches a Gaussian function of $\omega$, centred at the frequency $(\hbar\kappa^2/2M + E_{\mathrm{R}}/\hbar)$, whose width increases linearly with $\kappa$.

The discussion thus far has assumed that the motions of the oscillating particle and the molecular centre are uncorrelated for (6.63) is the product of correlation functions for the individual motions. Some of the limitations that stem from this assumption are discussed in § 6.2. For polyatomic molecules we will assume that the translational, rotational, and vibrational motions are all uncorrelated. Here we are primarily concerned with the vibrational spectra of nuclei in polyatomic molecules although, as we have seen, the spectra are not in general independent of the other motions (Califano *et al.* 1981).

For a polyatomic molecule containing $N_0$ nuclei we are concerned with the rotational and vibrational motions of a system with $3N_0$ degrees of freedom. Of these, three correspond to translational and three to rotational motion of the system as a whole. (If the molecule is linear, the number of vibrational degrees of freedom is $3N_0-5$ since, in this case, it is not meaningful to speak of a rotation of a linear molecule about its axis.) The energy of a system of particles undergoing small displacements is

$$E=\sum_{\xi,\alpha}\tfrac{1}{2}M_\xi(\dot{u}_\xi^{(\alpha)})^2+\sum_{\substack{\xi,\xi'\\ \alpha,\beta}}\tfrac{1}{2}u_\xi^{(\alpha)}k_{\xi,\xi'}^{(\alpha,\beta)}u_{\xi'}^{(\beta)}, \qquad (6.84)$$

where $u_\xi^{(\alpha)}$ is the $\alpha$ component of the displacement of the $\xi$th nucleus from its equilibrium position, $M_\xi$ its mass, and $k_{\xi,\xi'}^{(\alpha,\beta)}$ a force constant. We can now make a linear transformation of the displacements that will eliminate the coordinates corresponding to the translational motion and rotation of the system in (6.84), and reduce the second term to the sum of squares. We will not pause here to show this for the classical case but we quantize the system and show how the Hamiltonian corresponding to (6.84) can be expressed as the sum of $3N-6$ independent harmonic oscillators (Weissbluth 1978).

Let $\phi_\xi^{(\alpha)}(\lambda)$ satisfy

$$\sum_{\xi,\alpha}M_\xi^{-1/2}k_{\xi,\xi'}^{(\alpha,\beta)}M_{\xi'}^{-1/2}\phi_\xi^{(\alpha)}(\lambda)=\omega_\lambda^2\phi_{\xi'}^{(\beta)}(\lambda), \qquad (6.85)$$

where, for a unitary transformation,

$$\sum_{\xi,\alpha}[\phi_\xi^{(\alpha)}(\lambda)]^*\phi_\xi^{(\alpha)}(\lambda')=\delta_{\lambda,\lambda'}, \qquad (6.86)$$

where $\lambda$ labels the $3N_0-6$ eigenmodes. We now define Bose operators $\hat{a}_\lambda$ and $\hat{a}_\lambda^+$ such that

$$\hat{u}_\xi^{(\alpha)}=\sum_\lambda\left(\frac{\hbar}{2M_\xi\omega_\lambda}\right)^{1/2}\phi_\xi^{(\alpha)}(\lambda)[\hat{a}_\lambda\exp(-\mathrm{i}\omega_\lambda t)+\hat{a}_{-\lambda}^+\exp(\mathrm{i}\omega_\lambda t)], \quad (6.87)$$

so $\hat{u}_\xi^{(\alpha)}$ is real if $[\phi_\xi^{(\alpha)}(\lambda)]^*=\phi_\xi^{(\alpha)}(-\lambda)$. In terms of the Bose operators the Hamiltonian corresponding to (6.84) is simply

$$\mathcal{H}=\sum_\lambda\hbar\omega_\lambda(\hat{a}_\lambda^+\hat{a}_\lambda+\tfrac{1}{2}). \qquad (6.88)$$

The eigenmodes can, of course, be classified according to the symmetry elements of the equilibrium configuration of nuclei. For example, the spherical top molecule $CH_4$ has $3N_0-6=9$ normal vibrations: a non-degenerate one (belonging to the representation $\Gamma_1$) that involves only the hydrogen nuclei (along the C—H bond) with $\hbar\omega_1=0.37$ eV; a twofold degenerate one ($\Gamma_3$) involving only the hydrogen nuclei (perpendicular to the C—H bond), $\hbar\omega_2=0.172$ eV; and two threefold degenerate ones ($\Gamma_4$), $\hbar\omega_3=0.168$ eV and $\hbar\omega_4=0.391$ eV.

The contribution to (6.63) from the vibrational motion of the nuclei can be calculated by essentially the same method as that adopted in § 3.6.2. From eqn (6.87) we can write

$$\hat{\mathbf{u}}_\nu(t)=\sum_\lambda \mathbf{c}_\nu^{(\lambda)}\hat{Q}^{(\lambda)}(t), \tag{6.89}$$

where $\hat{Q}^{(\lambda)}(t)=(\hbar/2\omega_\lambda)^{1/2}(\hat{a}_\lambda+\hat{a}^+_{-\lambda})$, so that

$$\exp\{i\boldsymbol{\kappa}\cdot\hat{\mathbf{u}}_\nu(t)\}=\prod_\lambda \exp\{i\boldsymbol{\kappa}\cdot\mathbf{c}_\nu^{(\lambda)}\hat{Q}^{(\lambda)}(t)\}. \tag{6.90}$$

The formal development now parallels that given in § 3.6.2, the only difference being that there is a product over all allowed eigenfrequencies $\omega_\lambda$; we find

$$\begin{aligned}
&\langle\exp(-i\boldsymbol{\kappa}\cdot\hat{\mathbf{u}}_\nu)\exp\{i\boldsymbol{\kappa}\cdot\hat{\mathbf{u}}_{\nu'}(t)\}\rangle\\
&=\exp\{-W_\nu(\boldsymbol{\kappa})-W_{\nu'}(\boldsymbol{\kappa})+\langle\boldsymbol{\kappa}\cdot\hat{\mathbf{u}}_\nu\boldsymbol{\kappa}\cdot\hat{\mathbf{u}}_{\nu'}(t)\rangle\}\\
&=\exp\left\{-W_\nu(\boldsymbol{\kappa})-W_{\nu'}(\boldsymbol{\kappa})+\sum_\lambda\frac{\hbar(\boldsymbol{\kappa}\cdot\mathbf{c}_\nu^{(\lambda)})(\boldsymbol{\kappa}\cdot\mathbf{c}_{\nu'}^{(\lambda)})^*}{2\omega_\lambda\sinh(\frac{1}{2}\hbar\omega_\lambda\beta)}\cosh[\omega_\lambda(it+\tfrac{1}{2}\hbar\beta)]\right\}\\
&=\exp\{-W_\nu(\boldsymbol{\kappa})-W_{\nu'}(\boldsymbol{\kappa})\}\prod_\lambda\exp\left\{\frac{\hbar(\boldsymbol{\kappa}\cdot\mathbf{c}_\nu^{(\lambda)})(\boldsymbol{\kappa}\cdot\mathbf{c}_{\nu'}^{(\lambda)})^*}{2\omega_\lambda\sinh(\frac{1}{2}\hbar\omega_\lambda\beta)}\cosh[\omega_\lambda(it+\tfrac{1}{2}\hbar\beta)]\right.\\
&=\exp\{-W_\nu(\boldsymbol{\kappa})-W_{\nu'}(\boldsymbol{\kappa})\}\\
&\quad\times\prod_\lambda\left(\sum_{n_\lambda}\exp\{n_\lambda\omega_\lambda(it+\tfrac{1}{2}\hbar\beta)\}\,\mathrm{I}_{n_\lambda}\{(\hbar/2\omega_\lambda)(\boldsymbol{\kappa}\cdot\mathbf{c}_\nu^{(\lambda)})(\boldsymbol{\kappa}\cdot\mathbf{c}_{\nu'}^{(\lambda)})^*/\sinh(\tfrac{1}{2}\hbar\omega_\lambda\beta)\}\right).
\end{aligned} \tag{6.91}$$

Here, $n_\lambda$ is an integer and

$$\begin{aligned}
2W_\nu(\boldsymbol{\kappa})&=\langle(\boldsymbol{\kappa}\cdot\hat{\mathbf{u}}_\nu)^2\rangle\\
&=\sum_\lambda|\boldsymbol{\kappa}\cdot\mathbf{c}_\nu^{(\lambda)}|^2\,(\hbar/2\omega_\lambda)\coth(\tfrac{1}{2}\hbar\omega_\lambda\beta).
\end{aligned}$$

The representation of the correlation function in terms of contributions from individual modes is useful for small scattering vectors. For large scattering vectors many modes contribute to the correlation function and hence an alternative representation must be sought as shown in the introduction to this section. Result (6.91) applies for harmonic vibrations of an oriented assembly of nuclei. In comparing (6.91) with (6.64), say, for a single harmonic oscillator remember that the vectors $\mathbf{c}_\nu^{(\lambda)}$ in (6.89) are proportional to $M_\nu^{-1/2}$.

The incoherent vibrational cross-section for scattering from the assembly of nuclei is

$$\left(\frac{d^2\sigma}{d\Omega\, dE'}\right)_{\text{incoh}} = \frac{k'}{k}\frac{1}{4\pi}\sum_\nu \sigma_i^{(\nu)} \frac{1}{2\pi\hbar}\int_{-\infty}^{\infty} dt\, \exp(-i\omega t) \times\langle \exp(-i\boldsymbol{\kappa}\cdot\hat{\mathbf{u}}_\nu)\exp\{i\boldsymbol{\kappa}\cdot\hat{\mathbf{u}}_\nu(t)\}\rangle, \tag{6.93}$$

where $\sigma_i^{(\nu)}$ is the incoherent cross-section for the $\nu$th nucleus. From (6.91) we obtain the desired result

$$\left(\frac{d^2\sigma}{d\Omega\, dE'}\right)_{\text{incoh}} = \frac{k'}{k}\frac{1}{4\pi\hbar}\sum_\nu \sigma_i^{(\nu)} \exp\{-2W_\nu(\boldsymbol{\kappa}) + \tfrac{1}{2}\hbar\omega\beta\}$$

$$\times\prod_\lambda \left(\sum_{n_\lambda} I_{n_\lambda}\{(\hbar/2\omega_\lambda)\,|\boldsymbol{\kappa}\cdot\mathbf{c}_\nu^{(\lambda)}|^2/\sinh(\tfrac{1}{2}\hbar\omega_\lambda\beta)\}\,\delta\left(\omega - \sum_\lambda n_\lambda\omega_\lambda\right)\right). \tag{6.94}$$

The reader might find it useful to compare and contrast this result with the corresponding result for a single harmonic oscillator given in § 3.6.2.

The elastic incoherent cross-section is the contribution to (6.94) that is proportional to $\delta(\omega)$, namely,

$$\left(\frac{d^2\sigma}{d\Omega\, dE'}\right)^{\text{el}}_{\text{incoh}}$$

$$= \delta(\omega)\frac{1}{4\pi\hbar}\sum_\nu \sigma_i^{(\nu)} \exp\{-2W_\nu(\boldsymbol{\kappa})\}\prod_\lambda I_0\{(\hbar/2\omega_\lambda)\,|\boldsymbol{\kappa}\cdot\mathbf{c}_\nu^{(\lambda)}|^2/\sinh(\tfrac{1}{2}\hbar\omega_\lambda\beta)\}. \tag{6.94}$$

If the argument of the Bessel function $I_0(y)$ is small compared with unity, then we can make use of the limiting form,

$$I_0(y) = 1 + (y/2)^2; \qquad y \to 0.$$

Given that $y \ll 1$ for all modes, the elastic cross-section reduces to

$$\left(\frac{d\sigma}{d\Omega}\right)^{\text{el}}_{\text{incoh}} = \frac{1}{4\pi}\sum_\nu \sigma_i^{(\nu)} \exp\{-2W_\nu(\boldsymbol{\kappa})\}. \tag{6.96}$$

Note that the elastic scattering is not isotropic in general because the Debye–Waller factor depends on the orientation of $\boldsymbol{\kappa}$ relative to the displacements of the nuclei. Debye–Waller factors are the subject of § 4.3.

The cross-section for scattering from a fundamental mode is obtained from the general expression (6.94) by selecting the terms $n_\lambda = \pm 1$ for the mode of interest and $n_\lambda = 0$ for the remaining modes. If the arguments of the Bessel functions are small, then we can take $I_0(y)$ to be unity. The fundamental mode of interest will contribute a term that is proportional to $I_1(y) = (y/2)$ for $y \ll 1$. Hence, the cross-section for scattering from a fundamental mode labelled $\lambda$ is

$$\left(\frac{d^2\sigma}{d\Omega\, dE'}\right)^{f}_{\text{incoh}} = \frac{k'}{k}\frac{1}{4\pi}\sum_{\nu} \sigma_i^{(\nu)} \exp\{-2W_\nu(\boldsymbol{\kappa}) + \tfrac{1}{2}\hbar\omega\beta\}$$
$$\times \frac{|\boldsymbol{\kappa}\cdot\mathbf{c}_\nu^{(\lambda)}|^2}{4\omega_\lambda \sinh(\frac{1}{2}\hbar\omega_\lambda\beta)}\{\delta(\omega-\omega_\lambda)+\delta(\omega+\omega_\lambda)\}. \quad (6.97)$$

This expression is the analogue of (6.80) for scattering from the fundamental mode of a single particle executing small-amplitude harmonic oscillations. Applying the results of the discussion that precedes (6.80) to (6.97) we obtain an alternative form for incoherent scattering by a fundamental mode,

$$\left(\frac{d^2\sigma}{d\Omega\, dE'}\right)^{f}_{\text{incoh}}$$
$$= \frac{k'}{k}\frac{\{1+n(\omega)\}}{8\pi}\sum_{\nu} \sigma_i^{(\nu)} \exp\{-2W_\nu(\boldsymbol{\kappa})\}\frac{1}{\omega_\lambda}|\boldsymbol{\kappa}\cdot\mathbf{c}_\nu^{(\lambda)}|^2\{\delta(\omega-\omega_\lambda)+\delta(\omega+\omega_\lambda)\} \quad (6.98)$$

where the detailed balance factor is defined in (6.31).

The scattering will be most intense for those nuclei which have large incoherent cross-sections and displacement vectors, e.g. protons. Expressions (6.97) and (6.98) are based on the assumption that the arguments of the Bessel functions in (6.94) are small. This will not be valid for sufficiently large values of the scattering vector, and in this instance the representation (6.91) of the correlation function is no longer useful because a multitude of modes contribute to the response (Howard and Waddington 1980). The behaviour of the correlation function in the limit $\kappa \to \infty$ is discussed in the introduction to this section and in §§ 3.6.2 and 4.8. For intermediate values of the scattering vector, overtones lead to a rich structure in the cross-section that cannot be described adequately by simple mathematical analysis (Warner *et al.* 1983).

If it is necessary to include other motions of the nuclei in the cross-section and these motions can be taken to be independent of the vibrational motion, then the incoherent cross-section is obtained from

(6.93) with the vibrational correlation function multiplied by the appropriate correlation functions for the other motions of interest. An alternative form is based on the convolution theorem for Fourier transforms which is illustrated in (6.32). The effect of translational motion on scattering from the fundamental mode is contained in (6.76), and it produces a Gaussian line whose width increases linearly with the scattering vector and the square root of the ambient temperature. The correlation functions for free rotations of polyatomic molecules are calculated in § 6.5.2.

## 6.5. Scattering by molecular gases

We assume that the molecules do not possess a magnetic moment and we will only concern ourselves with the scattering from gases consisting of diatomic, spherical, and symmetric top polyatomic molecules. Because the neutrons are principally scattered by the nuclei and since, furthermore, the energy required to excite a molecule out of its normal electron configuration is large compared to the energy of a thermal neutron, the atomic electrons do not usually give a direct contribution to the scattering (Lovesey *et al.* 1982). However, they are involved in so far as the dynamical states of the nuclei depend on the total symmetry of the molecule and, therefore, on the symmetry of the normal electron configuration. It is found that in the great majority of cases the wave function of the electrons in this state is completely symmetrical, i.e. the wave function is invariant with respect to all the elements of the symmetry group of the molecule. This empirical rule is found to apply to nearly all diatomic and polyatomic molecules (Weissbluth 1978).

Because the separation of the vibrational and rotational energy states of molecules can be of the same order of magnitude as the energy associated with thermal and cold neutrons, we can meaningfully discuss the scattering from molecules only if we have some understanding of these dynamical states. Thus we will briefly discuss the molecular terms, first for diatomic molecules and then for polyatomic molecules, before considering the calculation of the neutron cross-sections.

The energy levels of a diatomic molecule can be shown to be represented by

$$E = U_e + B_e K(K+1) + \hbar\omega_e(\nu + \tfrac{1}{2}). \tag{6.99}$$

In (6.99) $U_e$ is the electron energy, including the energy of the Coulomb interaction of the nuclei at their equilibrium separation, the second term is the rotational energy from the rotation of the molecule, and the third the energy of the vibrations of the nuclei within the molecule. The parameters in the second two terms of (6.99) are defined as follows (for molecules with zero electron spin; it is found that the normal state of nearly all molecules is a singlet): $\hat{\mathbf{K}}$ is the sum of the orbital angular momentum of the electrons and the angular momentum of the nuclei, so that $\hat{\mathbf{K}}^2 = K(K+1)$ is a conserved quantity, $B_e = \hbar^2/(2M_{mol}r_e^2)$ ($r_e$ is the equilibrium separation of the nuclei and $M_{mol}$ the mass of the molecule) is called the rotational constant, $\omega_e$ is the frequency of the vibrations of the nuclei, and the number $\nu = 0, 1, 2, \ldots$, called the vibrational quantum number, denumerates the levels with a given $K$ in order of increasing energy. This expression is a first approximation to the energy levels and is valid when the nuclei undergo small displacements from their equilibrium separation. In the next approximation the rotational and vibrational degrees of freedom are coupled. In all but one of the examples considered here we shall assume that the rotational and vibrational motions are independent of each other.

If the intervals between the vibrational, rotational, and electron levels are denoted by $\Delta E_\nu$, $\Delta E_r$, and $\Delta E_{el}$, respectively, then, quite generally,

$$\Delta E_{el} \gg \Delta E_\nu \gg \Delta E_r. \tag{6.100}$$

Hence, as is clear from the following table, the rotational levels must dominate the structure of the slow neutron cross-section.

| (eV) | $H_2$ | $N_2$ |
|---|---|---|
| $-U_e$ | 4.7 | 7.5 |
| $\hbar\omega_e$ | 0.54 | 0.29 |
| $10^3 \times B_e$ | 7.6 | 0.25 |
| $r_e$ (Å) | 0.74 | 1.09 |

Let us now examine the restrictions imposed on the rotational levels of a *diatomic molecule* by the spatial symmetry of the nuclei. We assume the molecule has two identical nuclei at positions $\mathbf{R}_1$, $\mathbf{R}_2$ so that the total wave functions can be represented as

$$\Psi = \Phi_{Elec}\Phi_{Vib}(|\mathbf{R}_1 - \mathbf{R}_2|)\Phi_{Rot}(\mathbf{R}_1 - \mathbf{R}_2)\chi_{Nuc}, \tag{6.101}$$

where $\Phi_{Elec}$ is the electron wave function involving all the electron coordinates $\mathbf{r}_1, \mathbf{r}_2, \ldots, \mathbf{r}_n$ and their spins $s_1, s_2, \ldots, s_n$ with the nuclear positions $\mathbf{R}_1$ and $\mathbf{R}_2$ appearing as parameters. $\Phi_{Vib}$ is the vibrational wave function of the two nuclei and is even in their separation $\mathbf{R}_1 - \mathbf{R}_2$. $\Phi_{Rot}(\mathbf{R}_1 - \mathbf{R}_2)$ is the rotational wave function of the two nuclei and $\chi_{Nuc}$ is the total nuclear spin wave function of the molecule. We know that $\Psi$ must be antisymmetric in the exchange of any pair of electrons, so that commonly $\Phi_{\text{Elec}}$ can in a first approximation be represented as a determinant wave function. Because the nuclei are identical, $\Psi$ must also be either antisymmetric or symmetric under interchange of the nuclei, and this introduces an important restriction on the allowed rotational wave functions. To understand this we must examine in outline the classification of molecular eigenstates.

We first consider the electronic wave function of a diatomic molecule with identical nuclei. Obviously the molecular axis is a symmetry axis and in almost every molecule (NO is a well-known exception) the total electronic wave function has the maximum possible symmetry in the ground state and is non-degenerate. It is customary to label such states as $\Sigma$-type. In addition we specify three other electronic quantum numbers; the spin multiplicity, $2S+1$; the symmetry on reflection $(R)$ in a plane containing the molecular axis, even $(+)$ or odd $(-)$; the symmetry on inversion of coordinates of all the electrons $(\mathscr{I})$ about the middle of the molecular axis, even (*gerade*) or odd (*ungerade*).

To illustrate this notation consider Fig. 6.2, which shows a diatomic system and, schematically, an $s$ orbital and two $p$ orbitals on each atom. Then, for example, a single electron in the *s-bonding orbital*

$$S_b \simeq \{s_1(\mathbf{r}) + s_2(\mathbf{r})\}$$

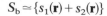

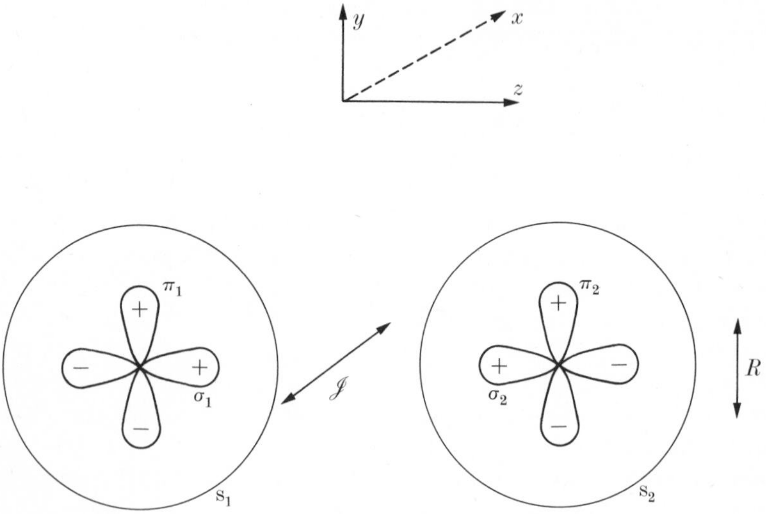


FIG. 6.2. Schematic diagrams of $s$ and $p$ orbitals in a diatomic molecule. The other $p$ orbitals, $\mu_1$ and $\mu_2$, are degenerate with $\pi_1$ and $\pi_2$ and point into the page, i.e. along the $x$ axes.

gives a wave function which is non-degenerate, therefore labelled $\Sigma$; has spin $\frac{1}{2}$ and therefore spin multiplicity 2, therefore labelled $^2\Sigma$; is even on reflection in the $x$–$z$ plane, therefore labelled $^2\Sigma^+$; and even on inversion $\mathscr{I}$, therefore labelled $^2\Sigma_g^+$. A single electron in the *s-antibonding orbital*

$$S_{\mathrm{ab}} \simeq \{s_1(\mathbf{r}) - s_2(\mathbf{r})\}$$

would give a state labelled $^2\Sigma_u^+$. Two electrons in the $s$-bonding orbital would give an electronic wave function corresponding to the $H_2$ molecule and then

$$\Phi_{\mathrm{Elec}} = S_{\mathrm{b}}(\mathbf{r}_1)S_{\mathrm{b}}(\mathbf{r}_2)\frac{1}{\sqrt{2}}(\alpha_1\beta_2 - \beta_1\alpha_2)$$

is labelled $^1\Sigma_g^+$.

As a final example, the ground state of $O_2$ has, apart from closed electronic shells, eight electrons to fit in the 12 possible $p$ states. In a simple molecular orbital theory,

$$\Phi_{\mathrm{Elec}} = \mathrm{Det}\{\sigma_{\mathrm{b}}^{\alpha}(1)\sigma_{\mathrm{b}}^{\beta}(2)\pi_{\mathrm{b}}^{\alpha}(3)\pi_{\mathrm{b}}^{\beta}(4)\mu_{\mathrm{b}}^{\alpha}(5)\mu_{\mathrm{b}}^{\beta}(6)\pi_{\mathrm{ab}}^{\alpha}(7)\mu_{\mathrm{ab}}^{\alpha}(8)\},$$

where Det{ } stands for the full determinant of the eight electrons. This state has $S = 1$, is odd because an odd number of $\pi$ orbitals is involved, and is *gerade* because an even number of anti-bonding orbitals are occupied. Therefore the ground state of $O_2$ is labelled $^3\Sigma_g^-$.

Consider now the total wave function of the diatomic molecule. The Hamiltonian of the molecule is invariant with respect to a simultaneous change in sign of the coordinates of all the *electrons* and the *nuclei*. Hence such an operation either leaves the total wave function unchanged (*positive*) or multiplies it by $-1$ (*negative*). For $\Sigma$ terms the sign depends only on $K$ and is given by $(-1)^K$ for $\Sigma^+$ and $(-1)^{K+1}$ for $\Sigma^-$ terms. These results can be understood as follows. From (6.99) the motion of the nuclei is seen to be equivalent to that of a single particle, of orbital angular momentum $K$, in a centrally symmetric field. Hence, when the signs of all the coordinates are changed the nuclear wave function is multiplied by $(-1)^K$. For the electron $\Sigma$ term, inversion is equivalent to reflection in a plane passing through the molecular axis and we thus arrive at the above results for the signs of $\Sigma^{\pm}$ terms under inversion.

Lastly, consider the wave function of the two identical nuclei. If the nuclei have integral spin $i$ the coordinate nuclear wave function must be *symmetric* under inversion whereas if $i$ is half-integral the coordinate wave function must be *antisymmetric*. It follows that if the total nuclear spin $I$ is even, then the coordinate nuclear wave function must be symmetric whereas if the total nuclear spin $I$ is odd then the coordinate nuclear wave function must be antisymmetric.

We have now determined both how the total wave function of the electrons and nuclei transforms under inversion and also how the separate wave functions of the electrons and nuclei transform under inversion.

Since the transformation property of the nuclear wave function under inversion depends on the total nuclear spin $I$ and the transformation properties of the total wave function of the electrons and nuclei are determined by $K$, it follows that $I$ and $K$ are correlated for a given electron $\Sigma$-type level. As an example we examine a $\Sigma_g^{\pm}$ term. The sign of the total wave function of a $\Sigma^+$ term is given by $(-1)^K$; hence for $\Sigma_g^+$ the total wave function is positive for even $K$, and therefore symmetric in the nuclear wave function, while for odd $K$ the total wave function is negative and consequently the nuclear wave function is antisymmetric. If $i=0$, there are no antisymmetric states and thus no rotational states with odd $K$. An important example is that of the hydrogen molecule, where $i=\frac{1}{2}$. For $I=0$ (para-hydrogen) there are no rotational states with odd $K$, and states with odd values of $K$ only for $I=1$ (ortho-hydrogen).

At high temperatures, when many rotational levels are excited, these rules are not of importance, but for low excitation numbers observed in supersonic gas jets, for example, the restrictions are quite necessary.

If we ignore the vibrational motion of the nuclei, the rotational states of a polyatomic molecule are given by the eigenstates of the Hamiltonian of a rigid body, namely

$$\hat{\mathcal{H}}=\tfrac{1}{2}\hbar^2\left[\frac{\hat{J}_\xi^2}{I_A}+\frac{\hat{J}_\eta^2}{I_B}+\frac{\hat{J}_\zeta^2}{I_C}\right], \tag{6.102}$$

where $\hat{J}_\xi$, $\hat{J}_\eta$, and $\hat{J}_\zeta$ are the components of the angular momentum for the three principal axes $\xi$, $\eta$, $\zeta$ of the molecule and $I_A$, $I_B$, and $I_C$ the principal moments of inertia.

The simplest case of (6.102) is when all three principal moments of inertia are equal, $I_A=I_B=I_C=I$. For this case,

$$\hat{\mathcal{H}}=\frac{\hbar^2}{2I}\hat{J}^2, \tag{6.103}$$

with eigenvalues

$$E_J=\frac{\hbar^2}{2I}J(J+1). \tag{6.104}$$

Each of these energy levels is degenerate with respect to the $2J+1$ directions of the angular momentum relative to the body itself, i.e. with respect to the values of $J_\zeta=K$. (We use $K$ to denote both the total orbital angular momentum of a diatomic molecule and also the eigenvalue of the $\zeta$ component of angular momentum for polyatomic molecules. This should not lead to any confusion.) The value of the rotational constant $\hbar^2/2I$ for methane is ~0.65 meV.

The symmetric top has two principal moments equal, $I_A$ and $I_B$, say, and a corresponding Hamiltonian

$$\hat{\mathcal{H}}=\frac{\hbar^2}{2I_A}\hat{J}^2+\tfrac{1}{2}\hbar^2\left(\frac{1}{I_C}-\frac{1}{I_A}\right)\hat{J}_\zeta^2. \tag{6.105}$$

For a given $J$ and $K$, the eigenvalues are

$$E_{JK} = \hbar^2 J(J+1)/2I_A + \tfrac{1}{2}\hbar^2\left(\frac{1}{I_C} - \frac{1}{I_A}\right)K^2. \tag{6.106}$$

The degeneracy with respect to values of $K$ which occurred for a spherical top are here partly removed. The values of the energy are the same only for values of $K$ differing in sign alone. It can be shown that the eigenfunctions of (6.105) can be expressed in terms of the rotation matrix (Weissbluth 1978; Edmonds 1960), $\mathcal{D}^{(J)}_{MK}(\alpha\beta\gamma)$, where the Euler angles $\alpha$, $\beta$, $\gamma$ define the orientation of the axes fixed in the molecule with respect to the laboratory frame of reference. The eigenfunctions of (6.105), properly normalized, are

$$|JMK\rangle = \left(\frac{8\pi^2}{2J+1}\right)^{1/2} \mathcal{D}^{(J)}_{MK}(\alpha\beta\gamma). \tag{6.107}$$

$\mathcal{D}^{(J)}_{MK}$ is the eigenfunction of $\hat{J}^2$ with eigenvalue $\hbar^2 J(J+1)$ and the eigenfunction of $\hat{J}_z$, say, with eigenvalue $\hbar M$. It is simultaneously the eigenfunction of $\hat{J}_\zeta$ with an eigenvalue $\hbar K$.

We cannot give a discussion of the classification of the molecular terms for a polyatomic molecule without the aid of group theory. Since this would lead us too far afield we shall not pursue this topic any further but merely note that the wave function of the normal electron state is completely symmetrical in nearly all stable polyatomic molecules (Califano *et al.* 1981).

### 6.5.1. *Diatomic molecules*

The form of the cross-section for scattering by a diatomic molecule is discussed in § 1.7. Here we present details of the calculation of the matrix elements involved in the cross-section.

From the discussion given in the introduction to this section it follows that in calculating the cross-section we need to distinguish between the two cases where the changes in the total spin of the nuclei, $\Delta I = I' - I$, and orbital angular momentum, $\Delta K = K' - K$, are even and odd. Following the notation used in § 1.7 we label the two cases (a) and (b)

$$\Delta I = 0, \qquad |\Delta K| = 0, 2, \ldots; \quad \text{case (a)}$$
$$|\Delta I| = 1, \qquad |\Delta K| = 1, 3, \ldots; \quad \text{case (b).}$$

The corresponding partial cross-sections are given in eqns (1.69) and (1.72). From the latter we obtain the partial differential cross-sections by averaging over all initial states and summing over all final states with the conservation of energy included by a delta function as in (1.23).

We denote the rotational energy

$$E_K = B_e K(K+1)$$

and

$$p_K = \exp(-\beta E_K) \Big/ \sum_K \exp(-\beta E_K). \tag{6.108}$$

In averaging over the initial states we must allow for the fact that for each $K$ there are $(2K+1)$ values of $M_K$. Hence, the thermal average for the rotational states

$$\langle(\cdots)\rangle = \sum_K \{p_K/(2K+1)\} \sum_{M_K} \langle KM_K |(\cdots)| KM_K\rangle. \tag{6.109}$$

We assume that the molecule is initially in its vibrational ground state. The partial differential cross-section for case (a) is, from eqn (1.69),

$$\begin{aligned}\left(\frac{d^2\sigma}{d\Omega\, dE'}\right)_a &= \frac{k'}{k}\{4A^2+\tfrac{1}{4}B^2I(I+1)\}\sum_K\{p_K/(2K+1)\} \\ &\quad\times\sum_{M_K}\sum_{K',M_{K'}}\sum_{\nu'}|\langle K'M_{K'}\nu'|\cos\tfrac{1}{2}\boldsymbol{\kappa}\cdot\hat{\mathbf{r}}\,|KM_K\nu=0\rangle|^2\,\delta(\hbar\omega+E_K-E_{K'}+\nu'\hbar\omega_e) \\ &= \frac{k'}{k}\{4A^2+\tfrac{1}{4}B^2I(I+1)\}\frac{1}{2\pi\hbar}\int_{-\infty}^{\infty} dt\,\exp(-i\omega t) \\ &\quad\times\sum_K\{p_K/(2K+1)\}\sum_{M_K}\sum_{K',M_{K'}}\sum_{\nu'}\exp\left\{\frac{it}{\hbar}(E_{K'}-E_K)+i\omega_e\nu' t\right\} \\ &\quad\times|\langle K'M_{K'}\omega'|\cos\tfrac{1}{2}\boldsymbol{\kappa}\cdot\hat{\mathbf{r}}\,|KM_K\nu=0\rangle|^2\end{aligned} \tag{6.110}$$

where we have used the integral representation of the delta function to obtain the final expression. A similar expression can be derived for case (b) starting from (1.72), and the only differences are that it involves the matrix element of $\sin(\tfrac{1}{2}\boldsymbol{\kappa}\cdot\hat{\mathbf{r}})$ and a different combination of scattering lengths.

Expression (6.110) applies to a diatomic molecule with a fixed centre of mass. For a gas it is more appropriate to consider the translational motion of the centre of mass to be free. We incorporate this feature in the cross-section by assuming that the translation motion is independent of the other motions, and therefore include the following factor in the integrand of (6.110)

$$\exp\left\{\frac{\kappa^2}{4M\beta}(i\hbar t\beta - t^2)\right\}. \tag{6.111}$$

This prescription for the motion of the centre of mass is used also in (6.64) for a model of scattering from a particle executing harmonic oscillations relative to a freely moving centre.

To calculate the vibrational contribution to the matrix elements in (6.110), let

$$\hat{\mathbf{r}} = \mathbf{d} + \hat{\mathbf{u}}, \tag{6.112}$$

where $|\mathbf{d}| = r_e$ and $\mathbf{u}$ is the displacement of the nuclei from their equilibrium separation. The motion of the two nuclei relative to each other is equivalent to a single, harmonically bound particle of mass $M/2$. Hence, from eqn (6.87),

$$\hat{u} = \left(\frac{\hbar}{M\omega_e}\right)^{1/2} (\hat{a} + \hat{a}^+)$$

and we require the matrix elements

$$\langle \nu' | \exp\left\{ \pm \frac{i\kappa}{2} \eta \left(\frac{\hbar}{M\omega_e}\right)^{1/2} (\hat{a} + \hat{a}^+) \right\} | \nu = 0 \rangle, \tag{6.113}$$

where $\eta$ is the cosine of the angle between $\boldsymbol{\kappa}$ and $\mathbf{r}$. Since

$$\exp \hat{A} \exp \hat{B} = \exp(\hat{A} + \hat{B} + \tfrac{1}{2}[\hat{A}, \hat{B}]), \tag{6.114}$$

if $[\hat{A}, \hat{B}]$ is a c-number, and

$$\exp(\pm \alpha \hat{a}^+) | \nu \rangle = \sum_{h=0}^{\infty} \left\{ \frac{(\nu + h)!}{\nu!} \right\}^{1/2} \frac{(\pm\alpha)^h}{h!} | \nu + h \rangle,$$

we have

$$\begin{aligned} &\langle \nu' | \exp\{\pm \alpha (\hat{a} + \hat{a}^+)\} | \nu = 0 \rangle \\ &\qquad = e^{-\alpha^2/2} \langle \nu' | \exp(\pm\alpha\hat{a}) \exp(\pm\alpha\hat{a}^+) | \nu = 0 \rangle \\ &\qquad = e^{-\alpha^2/2} \left(\frac{1}{\nu'!}\right)^{1/2} \sum_{h=0}^{\infty} \frac{(\pm\alpha)^{\nu'+2h}}{h!} = \frac{(\pm\alpha)^{\nu'}}{(\nu'!)^{1/2}} e^{\alpha^2/2}. \end{aligned}$$

So that, with $\alpha = \frac{1}{2} i \kappa \eta (\hbar/\omega_e M)^{1/2}$,

$$\begin{aligned} &|\langle K' M_{K'} \nu' | \cos \tfrac{1}{2} \boldsymbol{\kappa} \cdot \hat{\mathbf{r}} | K M_K \nu = 0 \rangle|^2 \\ &\quad = \frac{1}{\nu'} |\langle K' M_{K'} | \exp(-\kappa^2 \eta^2 \hbar / 8\omega_e M) \\ &\quad \times \tfrac{1}{2} \{ \exp(\tfrac{1}{2} i \kappa \eta r_e) \alpha^{\nu'} + \exp(-\tfrac{1}{2} i \kappa \eta r_e)(-\alpha)^{\nu'} \} | K M_K \rangle|^2 \\ &\quad = \frac{1}{\nu'!} \left(\frac{\kappa^2 \hbar}{4\omega_e M}\right)^{\nu'} |\langle K' M_{K'} | \eta^{\nu'} \exp\left(-\frac{\kappa^2 \eta^2 \hbar}{8 M \omega_e} + \frac{i}{2} \kappa \eta r_e\right) | K M_K \rangle|^2. \end{aligned} \tag{6.115}$$

The last line follows because $\Delta K$ is an even integer. The result for the corresponding matrix element of $\sin \frac{1}{2} \boldsymbol{\kappa} \cdot \hat{\mathbf{r}}$ is the same.

The rotational wave functions of a diatomic molecule are spherical harmonics $Y^K_{M_K}$ (Weissbluth 1978; Judd 1975). Thus

$$\begin{aligned}
&\langle K'M_{K'}|\,\eta^{\nu'}\exp\left(-\frac{\kappa^2\eta^2\hbar}{8M\omega_e}+\frac{i}{2}\kappa\eta r_e\right)|KM_K\rangle\\
&=\int_0^{2\pi}d\phi\int_{-1}^{1}d\eta\, Y^{*K'}_{M_{K'}}\eta^{\nu'}\exp\left(-\frac{\kappa^2\eta^2\hbar}{8M\omega_e}+\frac{i}{2}\kappa\eta r_e\right)Y^K_{M_K}\\
&=(-)^{M_{K'}}\sum_{l,m}\left\{\frac{(2K+1)(2K'+1)(2l+1)}{4\pi}\right\}^{1/2}\begin{pmatrix}K' & K & l\\ -M_{K'} & M_K & m\end{pmatrix}\\
&\quad\times\begin{pmatrix}K' & K & l\\ 0 & 0 & 0\end{pmatrix}\int_0^{2\pi}d\phi\int_{-1}^{1}d\eta\, Y^{*l}_m\eta^{\nu'}\exp\left(-\frac{\kappa^2\eta^2\hbar}{8M\omega_e}+\frac{i}{2}\kappa\eta r_e\right)\\
&=(-)^{M_{K'}}\{(2K'+1)(2K+1)\}^{1/2}\sum_l(2l+1)\\
&\quad\times\begin{pmatrix}K' & K & l\\ -M_{K'} & M_K & 0\end{pmatrix}\begin{pmatrix}K' & K & l\\ 0 & 0 & 0\end{pmatrix}\mathscr{A}_{l\nu'},
\end{aligned}\tag{6.116}$$

where

$$\mathscr{A}_{l\nu'}=\tfrac{1}{2}\int_{-1}^{1}d\eta\, P_l(\eta)\eta^{\nu'}\exp\left(-\frac{\kappa^2\eta^2\hbar}{8M\omega_e}+\frac{i}{2}\kappa\eta r_e\right)\tag{6.117}$$

and $P_l$ is the Legendre polynomial or order $l$. In deriving (6.116) we have used an expansion of the product of two spherical harmonics in terms of $3j$ symbols namely

$$\begin{aligned}
Y^{*l_1}_{m_1}Y^{l_2}_{m_2}&=(-)^{m_1}\sum_{l,m}\left\{\frac{(2l_1+1)(2l_2+1)(2l+1)}{4\pi}\right\}^{1/2}\\
&\quad\times\begin{pmatrix}l_1 & l_2 & l\\ -m_1 & m_2 & m\end{pmatrix}Y^{*l}_m\begin{pmatrix}l_1 & l_2 & l\\ 0 & 0 & 0\end{pmatrix}
\end{aligned}$$

and the second $3j$ symbol is non-zero only if $l_1+l_2+l$ is an even integer. The integral over $\phi$ in the second line is non-zero if $m=0$ and

$$Y^l_0=\left(\frac{2l+1}{4\pi}\right)^{1/2}P_l.$$

From (6.116),

$$\begin{aligned}
&\frac{1}{2K+1}\sum_{M_K,M_{K'}}\left|\langle K'M_{K'}|\,\eta^{\nu'}\exp\left(-\frac{\eta^2\kappa^2\hbar}{8M\omega_e}+\frac{i}{2}\kappa\eta r_e\right)|KM_K\rangle\right|^2\\
&=(2K'+1)\sum_{M_K,M_{K'}}\sum_{l,l'}(2l+1)(2l'+1)\begin{pmatrix}K' & K & l\\ -M_{K'} & M_K & 0\end{pmatrix}\\
&\quad\times\begin{pmatrix}K' & K & l\\ 0 & 0 & 0\end{pmatrix}\begin{pmatrix}K' & K & l'\\ -M_{K'} & M_K & 0\end{pmatrix}\begin{pmatrix}K' & K & l'\\ 0 & 0 & 0\end{pmatrix}\mathscr{A}_{l\nu'}\mathscr{A}^*_{l'\nu'}\\
&=(2K'+1)\sum_{l=|K'-K|}^{K'+K}(2l+1)\begin{pmatrix}K' & K & l\\ 0 & 0 & 0\end{pmatrix}^2|\mathscr{A}_{l\nu'}|^2
\end{aligned}\tag{6.118}$$

because of the orthogonality relation

$$\sum_{m_1,m_2} \begin{pmatrix} j_1 & j_2 & j_3 \\ m_1 & m_2 & m_3 \end{pmatrix} \begin{pmatrix} j_1 & j_2 & j_3' \\ m_1 & m_2 & m_3' \end{pmatrix} = \frac{1}{2j_3+1}\,\delta_{j_3j_3'}\delta_{m_3,m_3'}\Delta(j_1j_2j_3), \tag{6.119}$$

where $\Delta(j_1j_2j_3)=1$ if $j_1$, $j_2$, $j_3$ satisfy the triangular condition, and is zero otherwise.

The contribution to the cross-section from the translation motion is given in (6.111), and the resulting time integral in (6.110) from this term and (6.118) is of the same form as for the scattering from free nuclei. The final expressions for the partial differential cross-sections for the two cases in which $\Delta K$ is an even integer ($\Delta I=0$) and when $\Delta K$ is an odd integer ($\Delta I=\pm1$) can be written

$$\frac{\mathrm{d}^2\sigma}{\mathrm{d}\Omega\,\mathrm{d}E'} = \left\{ \begin{matrix} [4A^2+B^2\tfrac{1}{4}I(I+1)], & \text{(a)} \\ \tfrac{1}{4}B^2[4i(i+1)-I(I+1)], & \text{(b)} \end{matrix} \right\} \frac{k'}{k}\left(\frac{\beta M}{\pi\hbar^2\kappa^2}\right)^{1/2}$$

$$\times \sum_{K',K} p_K(2K'+1)\sum_{\nu'}\left(\frac{\hbar\kappa^2}{4M\omega_e}\right)^{\nu'}\frac{1}{\nu'!}$$

$$\times \exp\left\{-\frac{\beta M}{\hbar^2\kappa^2}\left(E_{K'}-E_K+\nu'\hbar\omega_e+\frac{\hbar^2\kappa^2}{4M}-\hbar\omega\right)^2\right\}$$

$$\times \sum_{l=|K'-K|}^{K'+K}(2l+1)\begin{pmatrix} K' & K & l \\ 0 & 0 & 0 \end{pmatrix}^2 |\mathscr{A}_{l\nu'}|^2. \tag{6.120}$$

If, as may well be the case, the incident neutron energy is less than $\hbar\omega_e$, the vibrational states need not be considered; i.e. we can put $\nu'=0$ in (6.120) and use

$$\mathscr{A}_{0l} = \tfrac{1}{2}\int_{-1}^{1} \mathrm{d}\eta\, \exp(\tfrac{1}{2}\mathrm{i}\kappa r_e\eta)P_l(\eta) = \mathrm{i}^l j_l(\tfrac{1}{2}\kappa r_e) \simeq \mathrm{i}^l\frac{(\kappa r_e/2)^l}{(2l+1)!!}, \tag{6.121}$$

if in addition $\tfrac{1}{2}\kappa r_e \ll 1$. In the latter case only the first few terms in the summation over $l$ need be retained. The $l=0$ term is non-zero only if $K+K'$ is an even integer, which requires $\Delta K$ to be even.

As a particular example of (6.120) let us consider the scattering from hydrogen. Here $\hbar\omega_e = 0.54$ eV so that for slow neutrons we can usually ignore the vibrational levels. For the scattering from para-hydrogen

$(I=0)$ we then have

$$\frac{d^2\sigma}{d\Omega\, dE'}=\frac{1}{4}\frac{k'}{k}\left(\frac{\beta m}{\pi\hbar^2\kappa^2}\right)^{1/2}\sum_{K=0,2,4,\ldots} p_K\left[(3b^{(+)}+b^{(-)})^2\sum_{K'=0,2,4,\ldots}(2K'+1)\right.$$
$$\times\exp\left\{-\frac{\beta m}{\hbar^2\kappa^2}\left(E_{K'}-E_K+\frac{\hbar^2\kappa^2}{4m}-\hbar\omega\right)^2\right\}$$
$$\times\sum_{l=|K'-K|}^{K'+K}(2l+1)\begin{pmatrix}K' & K & l\\ 0 & 0 & 0\end{pmatrix}^2 j_l^2(\kappa r_e/2)+3(b^{(+)}-b^{(-)})^2$$
$$\times\sum_{K'=1,3,5,\ldots}(2K'+1)\exp\left\{-\frac{\beta m}{\hbar^2\kappa^2}\left(E_{K'}-E_K+\frac{\hbar^2\kappa^2}{4m}-\hbar\omega\right)^2\right\}$$
$$\left.\times\sum_{l=|K'-K|}^{K'+K}(2l+1)\begin{pmatrix}K' & K & l\\ 0 & 0 & 0\end{pmatrix}^2 j_l^2(\kappa r_e/2)\right], \qquad (6.122)$$

and for the scattering from ortho-hydrogen $(I=1)$,

$$\frac{d^2\sigma}{d\Omega\, dE'}=\frac{1}{4}\frac{k'}{k}\left(\frac{\beta m}{\pi\hbar^2\kappa^2}\right)^{1/2}$$
$$\times\sum_{K=1,3,5,\ldots} p_K\left[\{(3b^{(+)}+b^{(-)})^2+2(b^{(+)}-b^{(-)})^2\}\sum_{K'=1,3,5,\ldots}(2K+1)\right.$$
$$\times\exp\left\{-\frac{\beta m}{\hbar^2\kappa^2}\left(E_{K'}-E_K+\frac{\hbar^2\kappa^2}{4m}-\hbar\omega\right)^2\right\}$$
$$\times\sum_{l=|K'-K|}^{K'+K}(2l+1)\begin{pmatrix}K' & K & l\\ 0 & 0 & 0\end{pmatrix}^2 j_l^2(\kappa r_e/2)$$
$$+(b^{(+)}-b^{(-)})^2\sum_{K'=0,2,4,\ldots}(2K'+1)\exp\left\{-\frac{\beta m}{\hbar^2\kappa^2}\left(E_{K'}-E_K+\frac{\hbar^2\kappa^2}{4m}-\hbar\omega\right)^2\right\}$$
$$\left.\times\sum_{l=|K'-K|}^{K'+K}(2l+1)\begin{pmatrix}K' & K & l\\ 0 & 0 & 0\end{pmatrix}^2 j_l^2(\kappa r_e/2)\right]. \qquad (6.123)$$

We have written these expressions in terms of the scattering lengths $b^{(\pm)}$ which are given in § 1.4.

### 6.5.2. *Polyatomic molecules*

In our discussions of the diatomic molecule in §§ 1.7 and 6.5.1, we found that the nuclear spin and coordinates were not independent. The same is true, of course, for any homonuclear molecule and strictly speaking the wave functions used to evaluate the matrix elements must be symmetrized, i.e. for polyatomic molecules that contain two or more identical nuclei their spatial wave functions must be either symmetric or antisymmetric with respect to the interchange of two of the identical nuclei, depending on whether these have integral or half-integral spins. However, we know on quite general grounds that the purely quantum mechanical effect of spin correlation is negligible at high temperatures and for nuclei of mass $M \gg m$; we shall, therefore, assume that the matrix elements can be evaluated with unsymmetrized wave functions (Hama and Miyagi 1973). An exception would be molecules in a supersonic gas jet whose internal states are characterized by a local temperature of a few degrees K.

The appropriate form of the cross-section is discussed in §§ 1.7 and 5.1. In terms of correlation functions, the partial differential cross-section is given by (6.60) and (6.61) with the position vector

$$\mathbf{R}_j = \mathbf{l} + \mathbf{d}_\nu + \mathbf{u}_\nu \tag{6.124}$$

where $\mathbf{d}_\nu$ denotes equilibrium positions of the nuclei relative to the centre of mass of the molecule and $\mathbf{u}_\nu$ are their displacements from the equilibrium configuration. The total Hamiltonian of the target system is then the sum of $N$ single-molecule Hamiltonians, each of which we take to be

$$\hat{\mathcal{H}}_\mathrm{t} + \hat{\mathcal{H}}_\mathrm{r} + \hat{\mathcal{H}}_\mathrm{v}, \tag{6.125}$$

where $\hat{\mathcal{H}}_\mathrm{t}$ is the Hamiltonian of the translational motion of the centre of mass, and $\hat{\mathcal{H}}_\mathrm{r}$ and $\hat{\mathcal{H}}_\mathrm{v}$ the Hamiltonians that describe the rotational and vibrational motions of the nuclei. Obviously $\hat{\mathbf{l}}$, $\hat{\mathbf{d}}_\nu$, and $\hat{\mathbf{u}}_\nu$ do not commute with $\hat{\mathcal{H}}_\mathrm{t}$, $\hat{\mathcal{H}}_\mathrm{r}$, and $\hat{\mathcal{H}}_\mathrm{v}$, respectively, but in addition $\hat{\mathbf{u}}_\nu$ does not commute with $\hat{\mathcal{H}}_\mathrm{r}$ because the vectors $\mathbf{u}_\nu$ rotate with the molecule.

Let us assume that we can separate the translational, rotational, and vibrational contributions to the correlation function (6.61) and hence write

$$Y_{jj'}(\boldsymbol{\kappa}, t) \doteq \langle \exp(-\mathrm{i}\boldsymbol{\kappa}\cdot\hat{\mathbf{l}})\exp\{\mathrm{i}\boldsymbol{\kappa}\cdot\hat{\mathbf{l}}'(t)\}\rangle \\ \times \langle \exp(-\mathrm{i}\boldsymbol{\kappa}\cdot\hat{\mathbf{d}}_\nu)\exp\{\mathrm{i}\boldsymbol{\kappa}\cdot\hat{\mathbf{d}}_{\nu'}(t)\}\rangle\langle \exp(-\mathrm{i}\boldsymbol{\kappa}\cdot\hat{\mathbf{u}}_\nu)\exp\{\mathrm{i}\boldsymbol{\kappa}\cdot\hat{\mathbf{u}}_{\nu'}(t)\}\rangle. \tag{6.126}$$

The first of these correlation functions we approximate by the result for free motion, namely,

$$\langle \exp(-\mathrm{i}\boldsymbol{\kappa}\cdot\hat{\mathbf{l}})\exp\{\mathrm{i}\boldsymbol{\kappa}\cdot\hat{\mathbf{l}}'(t)\}\rangle \\ = \exp\{\mathrm{i}\boldsymbol{\kappa}\cdot(\mathbf{l}'-\mathbf{l})\}\exp\left\{\frac{\kappa^2}{2\beta M_\mathrm{mol}}(\mathrm{i}\hbar t\beta - t^2)\right\}, \tag{6.127}$$

where $M_\mathrm{mol}$ is the mass of the molecule. The vibrational correlation function in (6.126) is given explicitly by (6.91).

Because the energy of the first vibrational level is in general much larger than the separation of the rotational levels and also larger than the energy associated with thermal neutrons (for methane the lowest vibrational energy level is at 0.168 eV) it is clear from the outset that the cross-section for the scattering slow neutrons will depend strongly on the rotational states of the nuclei and not on their vibrational states. It is usually sufficient, in fact, to assume that the nuclei are in their vibrational ground state. The scattering of epithermal neutrons, with energies larger than $\hbar\omega_e$, will involve many rotational and vibrational states. For sufficiently energetic neutrons, it is a good approximation to sum over the rotational states using closure (cf. § 1.8).

We calculate the rotational term in (6.127), restricting ourselves to spherical and symmetric-top molecules. For a symmetric molecule

$$\langle \exp\{-\mathrm{i}\boldsymbol{\kappa}\cdot\hat{\mathbf{d}}_v\}\exp\{\mathrm{i}\boldsymbol{\kappa}\cdot\hat{\mathbf{d}}_{v'}(t)\}\rangle = \sum_{JK} P_{JK}\frac{1}{2J+1}\sum_{M}\langle JMK|\exp\{-\mathrm{i}\boldsymbol{\kappa}\cdot\hat{\mathbf{d}}_v\}\exp\{\mathrm{i}\boldsymbol{\kappa}\cdot\hat{\mathbf{d}}_{v'}(t)\}|JMK\rangle. \tag{6.128}$$

where $|JMK\rangle$ are the normalized eigenfunctions of the symmetric-top molecule, eqn (6.107), and

$$P_{JK} = \frac{\exp(-\beta E_{JK})}{\sum \exp(-\beta E_{JK})} \tag{6.129}$$

is the Boltzmann factor. We have taken account of the $(2J+1)$-fold degeneracy of the energy levels $E_{JK}$, eqn (6.106), with respect to the quantum number $M$ (this degeneracy is merely a consequence of the molecule having no preferred orientation). If we introduce a complete set of eigenfunctions $|J'M'K'\rangle$ in (6.128), then we require to calculate the matrix elements

$$\langle J'M'K'|\exp\{\mathrm{i}\boldsymbol{\kappa}\cdot\hat{\mathbf{d}}_v(t)\}|JMK\rangle = \exp\{\mathrm{i}t(E_{J'K'}-E_{JK})/\hbar\}\langle J'M'K'|\exp(\mathrm{i}\boldsymbol{\kappa}\cdot\hat{\mathbf{d}}_v)|JMK\rangle. \tag{6.130}$$

If $\tilde{\mathbf{d}}'_v$ and $\tilde{\boldsymbol{\kappa}}'$ are unit vectors along $\mathbf{d}_v$ and $\boldsymbol{\kappa}$ referred to the rotating coordinate frame of reference in the molecule, then the exponential factor has the standard expansion (Edmonds 1960)

$$\exp(\mathrm{i}\boldsymbol{\kappa}\cdot\mathbf{d}_v) = 4\pi\sum_{l=0}^{\infty}\sum_{n=-l}^{l}\mathrm{i}^l j_l(\kappa d_v)\,Y_n^{*l}(\tilde{\mathbf{d}}'_v)\,Y_n^l(\tilde{\boldsymbol{\kappa}}'). \tag{6.131}$$

However, the scattering vector $\boldsymbol{\kappa}$ is required referred to the laboratory frame of reference and not the frame of reference rotating with the molecule. If the rotation of the axes fixed in the molecule with respect to those in the laboratory is specified by the Euler angles $\{\alpha\beta\gamma\}$, then

$$Y_n^l(\tilde{\boldsymbol{\kappa}}') = \sum_{n'=-l}^{l} Y_{n'}^l(\tilde{\boldsymbol{\kappa}})\mathscr{D}_{n'n}^{(l)}(\alpha\beta\gamma), \tag{6.132}$$

where $\tilde{\boldsymbol{\kappa}}$ is the corresponding unit vector in the laboratory frame of reference. Thus,

$$\langle J'M'K'|\exp(\mathrm{i}\boldsymbol{\kappa}\cdot\hat{\mathbf{d}}_v)\,|JMK\rangle = \int \mathrm{d}\Omega \left(\frac{2J'+1}{8\pi^2}\right)^{1/2} \{\mathscr{D}^{(J')}_{M'K'}(\alpha\beta\gamma)\}^*$$

$$\times 4\pi \sum_{l=0}^{\infty}\sum_{n=-l}^{l} \mathrm{i}^l j_l(\kappa d_v)\,\mathrm{Y}^{*l}_n(\tilde{\mathbf{d}}'_v) \sum_{n'=-l}^{l} \mathrm{Y}^l_{n'}(\tilde{\boldsymbol{\kappa}})\mathscr{D}^{(l)}_{n'n}(\alpha\beta\gamma)\left(\frac{2J+1}{8\pi^2}\right)^{1/2}\mathscr{D}^{(J)}_{MK}(\alpha\beta\gamma)$$

$$=(-1)^{M'-K'}\{(2J+1)(2J'+1)\}^{1/2}4\pi\sum_{lnn'}\mathrm{i}^l j_l(\kappa d_v)\,\mathrm{Y}^{*l}_n(\tilde{\mathbf{d}}'_v)\,\mathrm{Y}^l_{n'}(\tilde{\boldsymbol{\kappa}})$$

$$\times\begin{pmatrix} J & J' & l\\ M & -M' & n'\end{pmatrix}\begin{pmatrix} J & J' & l\\ K & -K' & n\end{pmatrix}. \qquad (6.133)$$

In obtaining the second line we have used the relationship

$$\{\mathscr{D}^{(J)}_{MK}(\alpha\beta\gamma)\}^* = (-1)^{M-K}\mathscr{D}^{(J)}_{-M,-K}(\alpha\beta\gamma)$$

together with the following identity for the integral of a triple product or rotation matrices in terms of $3j$ symbols,

$$\frac{1}{8\pi^2}\int \mathrm{d}\Omega\,\mathscr{D}^{(j_1)}_{m_1'm_1}(\alpha\beta\gamma)\mathscr{D}^{(j_2)}_{m_2'm_2}(\alpha\beta\gamma)\mathscr{D}^{(j_3)}_{m_3'm_3}(\alpha\beta\gamma)$$

$$=\begin{pmatrix} j_1 & j_2 & j_3\\ m_1' & m_2' & m_3'\end{pmatrix}\begin{pmatrix} j_1 & j_2 & j_3\\ m_1 & m_2 & m_3\end{pmatrix}.$$

We can choose the $z$-axis to coincide with the direction defined by $\tilde{\boldsymbol{\kappa}}$, i.e. we can put

$$\mathrm{Y}^l_n(\tilde{\boldsymbol{\kappa}}) = \left(\frac{2l+1}{4\pi}\right)^{1/2}. \qquad (6.134)$$

Thus, in conjunction with (6.133),

$$\frac{1}{2J+1}\sum_{M}\sum_{J'M'K'}\langle JMK|\exp(-\mathrm{i}\boldsymbol{\kappa}\cdot\hat{\mathbf{d}}_v)\,|J'M'K'\rangle\langle J'M'K'|\exp\{\mathrm{i}\boldsymbol{\kappa}\cdot\hat{\mathbf{d}}_{v'}(t)\}\,|JMK\rangle$$

$$=\frac{1}{2J+1}\sum_{M}\sum_{J'M'K'}\exp\{\mathrm{i}t(E_{J'K'}-E_{JK})/\hbar\}4\pi(2J+1)(2J'+1)$$

$$\times\sum_{lnn'}\sum_{l_1n_1n_1'}\{(2l+1)(2l_1+1)\}^{1/2}\,\mathrm{i}^l(-\mathrm{i})^{l_1}j_{l_1}(\kappa d_v)j_l(\kappa d_{v'})\,\mathrm{Y}^{*l}_n(\tilde{\mathbf{d}}'_v)\,\mathrm{Y}^{l_1}_{n_1}(\tilde{\mathbf{d}}'_{v'})$$

$$\times\begin{pmatrix} J & J' & l\\ M & -M' & n'\end{pmatrix}\begin{pmatrix} J & J' & l\\ K & -K' & n\end{pmatrix}\begin{pmatrix} J & J' & l_1\\ M & -M' & n_1'\end{pmatrix}\begin{pmatrix} J & J' & l_1\\ K & -K' & n_1\end{pmatrix}.$$

Because of the orthogonality relation for the $3j$ symbols given by eqn (6.119), this is

$$\sum_{J'K'} \exp\{it(E_{J'K'}-E_{JK})/\hbar\}4\pi(2J'+1) \sum_{l=|J-J'|}^{J+J'} j_l(\kappa d_v)j_l(\kappa d_{v'})$$
$$\times Y_n^{*l}(\tilde{\mathbf{d}}'_v)Y_n^l(\tilde{\mathbf{d}}'_{v'})\begin{pmatrix} J & J' & l \\ K & -K' & n \end{pmatrix}^2$$
$$= \sum_{J'K'} \exp\{it(E_{J'K'}-E_{JK})/\hbar\}\mathcal{L}_{vv'}(JJ';KK'). \quad (6.135)$$

We have therefore shown that, for a symmetric-top molecule,

$$\langle \exp\{-i\boldsymbol{\kappa}\cdot\hat{\mathbf{d}}_v\}\exp\{i\boldsymbol{\kappa}\cdot\hat{\mathbf{d}}_{v'}(t)\}\rangle$$
$$= \sum_{JJ'}\sum_{KK'} P_{JK}\exp\{it(E_{J'K'}-E_{JK})/\hbar\}\mathcal{L}_{vv'}(JJ';KK'), \quad (6.136)$$

where $\mathcal{L}$ is defined by eqn (6.135). The corresponding expression for the spherical-top molecule is readily obtained from eqn (6.136), for in this case the eigenstates are independent of the quantum number $K$ (eqn (6.104)) so that

$$P_J = \frac{\exp(-\beta E_J)}{\sum_J (2J+1)\exp(-\beta E_J)} \quad (6.137)$$

and

$$\sum_{KK'} \mathcal{L}_{vv'}(JJ';KK') = (2J'+1)\sum_{l=|J-J'|}^{J+J'} j_l(\kappa d_v)j_l(\kappa d_{v'})P_l(\cos\theta_{vv'}), \quad (6.138)$$

where $P_l$ is the Legendre polynomial of order $l$ and $\theta_{vv'}$ is the angle between $\tilde{\mathbf{d}}'_v$ and $\tilde{\mathbf{d}}'_{v'}$.

The actual calculation of $\mathcal{L}$ for a given molecule is, clearly, a laborious task. In the limit $\kappa d_v \ll 1$, however, which corresponds to very low-energy incident neutrons, only the first few terms in the summation over $l$ need be taken on account of the well-known expansion of spherical Bessel functions,

$$j_l(x) = \frac{x^l}{(2l+1)!!}\left\{1 - \frac{x^2}{2(2l+3)} + \cdots\right\}. \quad (6.139)$$

As we have already mentioned, for slow neutron scattering the molecule may usually be regarded as being in its vibrational ground state. We will, in addition, introduce an approximate method of averaging the vibrational contribution over the orientations of the molecule, namely

$$\exp\left[-\sum_\lambda (\hbar/4\omega_\lambda)\overline{\{|\boldsymbol{\kappa}\cdot\mathbf{c}_v^{(\lambda)}|^2 + |\boldsymbol{\kappa}\cdot\mathbf{c}_{v'}^{(\lambda)}|^2\}}\right] = \exp(-\kappa^2\gamma_{vv'}),$$

where

$$\gamma_{vv'} = \sum_{\lambda} (\hbar/12\omega_\lambda)\{|\mathbf{c}_v^{(\lambda)}|^2 + |\mathbf{c}_{v'}^{(\lambda)}|^2\}. \tag{6.140}$$

Within the various approximations that have been introduced, the neutron cross-section for the scattering from a symmetric-top molecule is,

$$\begin{aligned}\frac{\mathrm{d}^2\sigma}{\mathrm{d}\Omega\,\mathrm{d}E'} = \frac{k'}{k}\left(\frac{\beta M_{\mathrm{mol}}}{2\pi\hbar^2\kappa^2}\right)^{1/2} &\sum_{v,v'} \{A_v A_{v'} + \tfrac{1}{4} i_v(i_v+1)\,\delta_{vv'} B_v^2\} \\ &\times \exp(-\kappa^2\gamma_{vv'}) \sum_{JJ'}\sum_{KK'} P_{JK}\mathscr{L}_{vv'}(JJ';KK') \\ &\times \exp\left\{-\frac{\beta M_{\mathrm{mol}}}{2\hbar^2\kappa^2}\left(E_{J'K'} - E_{JK} - \hbar\omega + \frac{\hbar^2\kappa^2}{2M_{\mathrm{mol}}}\right)^2\right\},\end{aligned} \tag{6.141}$$

where we have dropped the 'outer' scattering term. The corresponding cross-section for the scattering from spherical molecules has an analogous form to (6.141), it being only necessary to take account of eqns (6.137)–(6.138).

### 6.5.3. *Quasi-classical approximation (Krieger and Nelkin 1957; Turchin 1965; Parks et al. 1970; Williams 1966)*

The approximation to the cross-section that we discuss in this-subsection is valid when the incident neutron energy is large compared to the level spacing of the rotational states and when many of these levels are thermally excited. Under these conditions it is not necessary to consider the contribution to the scattering from individual levels but only the average of a small number; this is what we mean by a quasi-classical treatment of the scattering process. Mathematically this means that the operator nature of the variables associated with the rotational motion is ignored and the thermal averaging is carried through in a straightforward classical manner. Krieger and Nelkin (1957) found that such a procedure gave the result

$$\begin{aligned}&\langle \exp(-i\boldsymbol{\kappa}\cdot\hat{\mathbf{l}})\exp\{i\boldsymbol{\kappa}\cdot\hat{\mathbf{l}}(t)\}\rangle \langle \exp(-i\boldsymbol{\kappa}\cdot\hat{\mathbf{d}}_v)\exp\{i\boldsymbol{\kappa}\cdot\hat{\mathbf{d}}_{v'}(t)\}\rangle \\ &\qquad \simeq \exp[-i\boldsymbol{\kappa}\cdot(\hat{\mathbf{d}}_v - \hat{\mathbf{d}}_{v'})]\exp\left\{-\frac{1}{2\beta}\boldsymbol{\kappa}\cdot\mathbf{M}_v^{-1}\cdot\boldsymbol{\kappa}(t^2 - i\hbar t\beta)\right\},\end{aligned} \tag{6.142}$$

where $\mathbf{M}_v^{-1}$ is the *mass tensor*. Its components are

$$\{M_v^{-1}\}_{ii} = \frac{d_{vj}^2}{I_{vk}} + \frac{d_{vk}^2}{I_{vj}} + \frac{1}{M_{\mathrm{mol}}}$$

and

$$\{M_v^{-1}\}_{ij} = -\frac{d_{vi}d_{vj}}{I_{vk}} \quad (i \neq j), \tag{6.143}$$

where $i$, $j$, $k$ are referred to the directions of the principal axes of inertia and $d_\xi$, $d_\eta$, $d_\zeta$ are the coordinates of the $v$th nucleus under consideration with respect to the centre of mass of the molecule.

If we assume, as above, that the molecule can be regarded to be in its vibrational ground state, which requires, in particular, the incident neutron energy to be less than the energy needed to excite the first vibrational level, the cross-section is proportional to the integral over the time $t$ of the product of the vibrational term

$$\exp\left[-\sum_\lambda \frac{\hbar}{4\omega_\lambda}\{|\boldsymbol{\kappa}\cdot\mathbf{c}_v^{(\lambda)}|^2+|\boldsymbol{\kappa}\cdot\mathbf{c}_{v'}^{(\lambda)}|^2\}\right] \tag{6.144}$$

with the two factors given by the right-hand side of (6.142). To average this product over the orientations of the molecule, we replace it by the average of the products and take the average of the arguments of the exponential in (6.144) and the second factor on the right-hand side of (6.142). The vibrational contribution is then identical with that given in he previous section, eqn. (6.140), while the translational and rotational motions contribute

$$\overline{\exp\{-i\boldsymbol{\kappa}\cdot(\mathbf{d}_v-\mathbf{d}_{v'})\}\exp\left\{-\frac{1}{2\beta}\boldsymbol{\kappa}\cdot\mathbf{M}_v^{-1}\cdot\boldsymbol{\kappa}(t^2-i\hbar t\beta)\right\}}$$
$$\simeq\overline{\exp\{-i\boldsymbol{\kappa}\cdot(\mathbf{d}_v-\mathbf{d}_{v'})\}}\exp\left\{-\frac{1}{2\beta}\overline{\boldsymbol{\kappa}\cdot\mathbf{M}_v^{-1}\cdot\boldsymbol{\kappa}}(t^2-i\hbar t\beta)\right\}$$
$$=j_0(\kappa\,|\mathbf{d}_v-\mathbf{d}_{v'}|)\exp\left\{-\frac{\kappa^2}{2\beta M_v^{(0)}}(t^2-i\hbar t\beta)\right\}, \tag{6.145}$$

where $3/M_v^{(0)}=\mathrm{Tr}\,M_v^{-1}$.

With these approximations the neutron cross-section is

$$\frac{\mathrm{d}^2\sigma}{\mathrm{d}\Omega\,\mathrm{d}E'}=\frac{k'}{k}\sum_{v,v'}\{A_vA_{v'}+\tfrac{1}{4}i_v(i_v+1)\delta_{v,v'}B_v^2\}\frac{1}{2\pi\hbar}\int_{-\infty}^{\infty}\mathrm{d}t$$
$$\times\exp(-i\omega t)\exp(-\kappa^2\gamma_{vv'})j_0(\kappa\,|\mathbf{d}_v-\mathbf{d}_{v'}|)\exp\left\{-\frac{\kappa^2}{2\beta M_v^{(0)}}(t^2-i\hbar t\beta)\right\}$$
$$=\frac{k'}{k}\sum_{vv'}\{A_vA_{v'}+\tfrac{1}{4}i_v(i_v+1)\delta_{v,v'}B_v^2\}j_0(\kappa\,|\mathbf{d}_v-\mathbf{d}_{v'}|)$$
$$\times\exp(-\kappa^2\gamma_{vv'})\left(\frac{\beta M_v^{(0)}}{2\pi\hbar^2\kappa^2}\right)^{1/2}\exp\left\{-\frac{\beta M_v^{(0)}}{2\hbar^2\kappa^2}\left(\hbar\omega-\frac{\hbar^2\kappa^2}{2M_v^{(0)}}\right)^2\right\}. \tag{6.146}$$

As a function of the scattering angle, this expression will be most reliable at large scattering angles because here the large recoil energy means that an envelope of rotational states is involved.

The total direct scattering cross-section in the Krieger–Nelkin approximation is (cf. § 3.7)

$$\sigma_{(v)} = \{A_v^2 + \tfrac{1}{4}i_v(i_v+1)B_v^2\}\frac{\pi}{2m\gamma_{vv}E}$$
$$\times[\Phi\{\surd(M_v^{(0)}\beta E/m)\} - (1-c)^{1/2}\exp(-M_v^{(0)}\beta Ec/m)\Phi\{\surd(M_v^{(0)}\beta E(1-c)/m)\}], \tag{6.147}$$

where

$$c = 1 + \frac{\beta(1+M_v^{(0)}/m)^2}{8\gamma_{vv}M_v^{(0)}}.$$

Because the total cross-section is insensitive to the contribution of individual dynamical states of the nuclei, this expression should give a good representation of the total cross-section, provided that the energy of the incident neutrons is large compared to the separation of the rotational levels.

## REFERENCES


Bird, R. B. *et al.* (1977). *Dynamics of polymeric liquids.* Vol. 2. John Wiley, New York.

Boland, B. C., Mildner, D. F. R., Stirling, G. C., Bunce, L. J., Sinclair, R. N., and Windsor, C. G. (1978). *Nucl. Inst. Meth.* **154,** 349.

Califano, S., Schettino, V., and Neto, N. (1981). *Lecture notes in chemistry,* Vol. 26. Springer-Verlag, Berlin.

Edmonds, A. R. (1960). *Angular momentum in quantum mechanics.* Princeton University Press, Princeton, New Jersey.

Ehrhardt, K.-D., Buchenau, U., Samuelsen, E. J., and Maier, H. D. (1984). *Phys. Rev.* **B29,** 996.

Furrer, A., Stöckli, A., Hälg, W., Kühlbrandt, W., Mühlethaler, K., and Wehrli, E. (1983). *Helv. Phys. Acta.* **56,** 655.

Hama, J. and Miyagi, H. (1973). *Prog. theoret. Phys.* **50,** 1142.

Hansen, J-P. and McDonald, I. R. (1976). *Theory of simple liquids.* Academic Press, New York.

Hayter, J. B. and Penfold, J. (1983). *Colloid and Polymer Sci.* **261,** 1022.

Higgins, J. S. (1979). In G. Kostorz (ed.) *Treatise on materials science and technology,* Vol. 15. Academic Press, New York.

Howard, J. and Waddington, T. C. (1980). *Advances in infra-red Raman spectroscopy,* Vol. 7. Heydin.

Judd, B. R. (1975). *Angular momentum theory for diatomic molecules.* Academic Press, New York.

Krieger, T. J. and Nelkin, M. S. (1957). *Phys. Rev.* **106,** 290.

Leadbetter, A. J. and Richardson, R. M. (1979). *The molecular physics of liquid crystals.* Academic Press, New York.

Lovesey, S. W., Bowman, C. D., and Johnson, R. G. (1982). *Z. Phys.* **B47,** 137.

March, N. H. (1984). *Phys. Chem. Liquids* **13,** 163.

McCammon, J. A. (1984). *Rep. Prog. Phys.* **47,** 1.

Middendorf, H. D. (1984). *Ann. Rev. Biophys.* **13.**

Parks, D. E., Nelkin, M. S., Beyster, J. R., and Wikner, N. F. (1970). *Slow neutron scattering and thermalization.* Benjamin.
Sköld, K., Mueller, M. H., and Brun, T. O. (1979). In G. Kostorz (ed.) *Treatise on materials science and technology*, Vol. 15. Academic Press, New York.
Springer, T. (1972). *Springer tracts in modern physics*, Vol. 64. Springer-Verlag, Berlin.
Turchin, V. F. (1965). *Slow neutrons.* Israel Program for Scientific Translations.
van Kampen, N. G. (1981). *Stochastic processes in physics and chemistry.* North-Holland, Amsterdam.
Warner, M., Lovesey, S. W., and Smith, J. (1983). *Z. Phys.* **B51** 109.
Weissbluth, M. (1978). *Atoms and molecules.* Academic Press, New York.
Williams, M. M. R. (1966). *The slowing down and thermalization of neutrons.* North-Holland, Amsterdam.

# APPENDIX A

THIS appendix is devoted to (1) a more detailed derivation of the cross-section formula (1.21). In particular we justify the choice of phase for the scattering amplitude $f$, eqn (1.13b), and clarify the relationship with the formal theory of scattering. Also (2) we discuss the concept of a complex scattering length in relation to the absorption of thermal neutrons by nuclei.

## A.1. Scattering amplitude $f$

First we discuss the scattering of a neutron with momentum $\mathbf{p} = \hbar\mathbf{k}$ by a target system consisting of a single particle moving in a potential well. This system is described by the Hamiltonian $\hat{\mathscr{H}}_{\mathrm{T}}$. We choose an arbitrary origin close to the target system with respect to which the position vectors of the neutron and target particle are $\mathbf{r}$ and $\mathbf{R}$ respectively. If the interaction potential between the neutron and the target particle is $\hat{V}(\mathbf{r}, \mathbf{R})$ the Schrödinger equation for the system that consists of the incident neutron and the target system is

$$\left[\hat{\mathscr{H}}_{\mathrm{T}} + \frac{1}{2m}\hat{\mathbf{p}}^2 + \hat{V}(\mathbf{r}, \mathbf{R}) - \mathscr{E}\right]\Psi(\mathbf{r}, \mathbf{R}) = 0. \tag{A.1}$$

In (A.1), $\mathscr{E}$ is the total energy, i.e. the sum of the energy of the incident neutron $E$ and the initial energy of the target.

The total wave function $\Psi$ can be expanded in terms of the complete set of eigenfunctions of the target Hamiltonian $\phi_\lambda(\mathbf{R})$, which satisfy

$$(\hat{\mathscr{H}}_{\mathrm{T}} - E_\lambda)\phi_\lambda(\mathbf{R}) = 0.$$

If

$$\Psi(\mathbf{r}, \mathbf{R}) = \sum_{\lambda'} A_{\lambda'}(\mathbf{r})\phi_{\lambda'}(\mathbf{R}), \tag{A.2}$$

the coefficients $A_{\lambda'}$ are found to satisfy the following inhomogeneous differential equation

$$(\nabla^2 + k_{\lambda'}^2)A_{\lambda'}(\mathbf{r}) = \frac{2m}{\hbar^2}\int \mathrm{d}\mathbf{R}\,\phi_{\lambda'}^*(\mathbf{R})\,\hat{V}(\mathbf{r}, \mathbf{R})\Psi(\mathbf{r}, \mathbf{R}), \tag{A.3}$$

in which

$$k_{\lambda'}^2 = \frac{2m}{\hbar^2}(\mathscr{E} - E_{\lambda}').$$

Equation (A.3) can be recast as an integral equation with the aid of the Green function $\mathscr{G}_\lambda(\mathbf{r}-\mathbf{r}')$, which satisfies (A.3) when the source term is replaced by $\delta(\mathbf{r}-\mathbf{r}')$. Explicitly,

$$\mathscr{G}_\lambda(\mathbf{r}-\mathbf{r}') = \frac{-1}{4\pi\,|\mathbf{r}-\mathbf{r}'|}\exp\{ik_\lambda\,|\mathbf{r}-\mathbf{r}'|\},$$

and eqn (A.3) can be written

$$A_{\lambda'}(\mathbf{r}) = A^{(0)}_{\lambda'}(\mathbf{r}) + \frac{2m}{\hbar^2}\int \mathrm{d}\mathbf{r}'\mathscr{G}_{\lambda'}(\mathbf{r}-\mathbf{r}')\int \mathrm{d}\mathbf{R}\phi^*_{\lambda'}(\mathbf{R})\,\hat{V}(\mathbf{r}',\mathbf{R})\Psi(\mathbf{r}',\mathbf{R}). \tag{A.4}$$

Here $A^{(0)}_{\lambda'}(\mathbf{r})$ is a solution of the homogeneous equation and is like $\exp(i\mathbf{k}_{\lambda'}\cdot\mathbf{r})$, i.e. like $\exp(ik_{\lambda'}r\tilde{\mathbf{n}}\cdot\tilde{\mathbf{n}}')$ where $\tilde{\mathbf{n}}$ is a unit vector along the direction of the incident plane wave and $\tilde{\mathbf{n}}'$ a unit vector along $\mathbf{r}$.

Before we can proceed any further we must establish the boundary conditions on the coefficients $A_\lambda$. If the target is in the state $\phi_\lambda$ before the scattering process, then, clearly, $A_\lambda$ must have an asymptotic form that represents the sum of the incident and scattered wave; thus $A_\lambda$ must behave like

$$A_\lambda \sim \exp(ik_\lambda r\tilde{\mathbf{n}}\cdot\tilde{\mathbf{n}}') + f_{\lambda\lambda}(\tilde{\mathbf{n}},\tilde{\mathbf{n}}')\frac{1}{r}\exp(ik_\lambda r)$$

for $r=|\mathbf{r}|$ much greater than the dimensions of the target. The functions $A_{\lambda'\neq\lambda}$ must represent scattered waves only, and so have the asymptotic form

$$A_{\lambda'\neq\lambda} \sim f_{\lambda\lambda'}(\tilde{\mathbf{n}},\tilde{\mathbf{n}}')\frac{1}{r}\exp(ik_{\lambda'}r).$$

From these expressions we deduce that $r^{-2}\,|f_{\lambda\lambda'}(\tilde{\mathbf{n}},\tilde{\mathbf{n}}')|^2$ is the number of neutrons per unit volume at a distance $\mathbf{r}$ from the target. Of these, the number crossing unit area per unit time is proportional to $k_{\lambda'}r^{-2}\,|f_{\lambda\lambda'}|^2$, whereas in the incident beam the number crossing unit area per unit time is proportional to $k_\lambda$. Hence, by definition,

$$\left(\frac{\mathrm{d}\sigma}{\mathrm{d}\Omega}\right)^{\lambda}_{\lambda'} = \frac{k_{\lambda'}}{k_\lambda}\,|f_{\lambda\lambda'}(\tilde{\mathbf{n}},\tilde{\mathbf{n}}')|^2. \tag{A.5}$$

Thus, in view of this discussion, we require solutions of (A.4) for $|\mathbf{r}'| \ll r$. Since $|\mathbf{r}-\mathbf{r}'| \sim r - \tilde{\mathbf{n}}'\cdot\mathbf{r}'$ the scattering amplitude $f_{\lambda\lambda'}$ is

$$f_{\lambda\lambda'}(\tilde{\mathbf{n}},\tilde{\mathbf{n}}') = -\frac{m}{2\pi\hbar^2}\int \mathrm{d}\mathbf{r}'\exp(-ik_{\lambda'}\tilde{\mathbf{n}}'\cdot\mathbf{r}')\int \mathrm{d}\mathbf{R}\phi^*_{\lambda'}(\mathbf{R})\,\hat{V}(\mathbf{r}',\mathbf{R})\Psi(\mathbf{r}',\mathbf{R}). \tag{A.6}$$

If the perturbation on the incident wave due to scattering is small, then to a good approximation $\Psi(\mathbf{r}, \mathbf{R})$ can be replaced by the initial wave function of the total system, i.e. $\Psi(\mathbf{r}, \mathbf{R}) \simeq \phi_\lambda(\mathbf{R})\exp(ik_\lambda \tilde{\mathbf{n}} \cdot \mathbf{r})$. This approximation is called the Born approximation. Inserting it in the above equation for $f_{\lambda\lambda'}(\tilde{\mathbf{n}}, \tilde{\mathbf{n}}')$, we have

$$f_{\lambda\lambda'}(\tilde{\mathbf{n}}, \tilde{\mathbf{n}}') = -\frac{m}{2\pi\hbar^2}\int \mathrm{d}\mathbf{r}' \exp(-ik_{\lambda'}\, \tilde{\mathbf{n}}' \cdot \mathbf{r}') \times \int \mathrm{d}\mathbf{R}\, \phi^*_{\lambda'}(\mathbf{R})\, \hat{V}(\mathbf{r}', \mathbf{R})\phi_\lambda(\mathbf{R})\exp(ik_\lambda \tilde{\mathbf{n}} \cdot \mathbf{r}'). \quad \text{(A.7)}$$

This expression verifies the choice of phase adopted in Chapter 1.

It is worthwhile to mention that in the formal theory of scattering (Newton 1982) it is usual to define an operator $\hat{T}$, the $t$-matrix operator, such that when it operates on the initial state of the total system it is equal to $\hat{V}\Psi$. An equation for $\hat{T}$ is easily found which can, in principle, be solved to any order in the interaction potential $\hat{V}$. The first approximation is $\hat{T} \simeq \hat{V}$ and this leads immediately to eqn (A.7) for the scattering amplitude.

The expression given in (A.7) for $f_{\lambda\lambda'}$ has been derived for the case of a single target particle in a potential well, whereas we are for the most part concerned with scattering from an array of such target systems. The amplitude for the latter problem is obtained to an approximation consistent with that used to derive (A.7) if $\hat{V}$ is replaced by the sum of the potentials between the incident neutron and the individual target systems. Clearly this would be an invalid procedure if there was an appreciable probability that, before escaping to a large distance from the target, the incident neutron makes a collision with more than one of the individual target systems in the array. This probability will be very small if the amplitude scattered by each target system is small, which is the condition under which the use of the Born approximation is justified.

If we build the condition of conservation of energy into (A.5), as in Chapter 1, then the partial differential cross-section is, from (A.7),

$$\left(\frac{\mathrm{d}^2\sigma}{\mathrm{d}\Omega\, \mathrm{d}E'}\right)_{\lambda'}^{\lambda} = \frac{k'}{k}\left(\frac{m}{2\pi\hbar^2}\right)^2 \left| \langle \lambda' | \int \mathrm{d}\mathbf{r} \exp(-\mathrm{i}\mathbf{k}' \cdot \mathbf{r})\, \hat{V} \exp(\mathrm{i}\mathbf{k} \cdot \mathbf{r})\, |\lambda\rangle \right|^2 \delta(\hbar\omega + E_\lambda - E_{\lambda'}) \quad \text{(A.8)}$$

in agreement with the result quoted in Chapter 1. Because of the delta function in (A.8) we are able to drop the suffix $\lambda'$ on $k_{\lambda'}$ and simply denote the final energy of the neutron by $E' = \hbar^2 k'^2/2m$.

## A.2. Complex and energy-dependent scattering lengths

In the text we have for the most part assumed the scattering length $b$ to be independent of the incident neutron energy $E=\hbar^2k^2/2m$. On the whole this is a valid assumption for slow neutrons, but in some instances the scattering displays a very pronounced energy dependence as is evident from Fig. A.1, which shows the *total* cross-section for lutetium in the range $0.01\ \text{eV} \leqslant E \leqslant 1.0\ \text{eV}$. The peak in the cross-section at $E=0.14$ eV is due to resonant absorption of neutrons by the Lu nuclei. Our understanding of this and related processes is based upon the compound theory of nuclei (Mughabghab *et al.* 1981).

The scattering process we consider is more general than that discussed elsewhere, for we must allow the target nuclei to possess internal structure. Elastic scattering thus refers to processes in which the state of

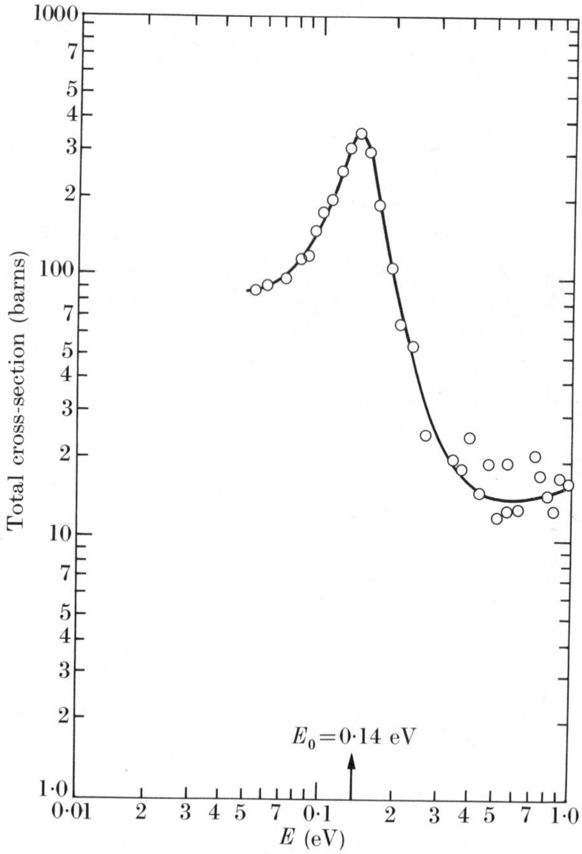


FIG. A.1. Total cross-section for lutetium. The parameters characterizing the resonance are: $E_0=0.142\pm0.001$ eV; $\Gamma_n=0.009\pm0.093$ meV; $\Gamma_\gamma=63\pm5$ meV.

the scattered neutron is identical to that of the incident neutron. Inelastic scattering includes processes in which the neutron is scattered leaving the nucleus in an excited state or when it is captured by the nucleus. For thermal neutrons the dominant inelastic process is radiative capture mentioned above in which neutrons are captured by nuclei which subsequently emit $\gamma$-rays.

The ingoing neutron shares its energy among all the nucleons. For a large number of nucleons the probability that the additional energy is concentrated in any one is small. Furthermore any nucleon must overcome reflection at the surface of the nucleus before it can be emitted and for protons there is in addition to this the Coulomb potential barrier. Hence the compound system consisting of the incident neutron plus target nucleus exists for a long period of time and, in consequence, the decay of the compound nucleus is independent of its formation. Also, it follows from the uncertainty principle that the energy width of the state is very small, particularly for large $Z$ nuclei, and hence neutron cross-sections can show resonance behaviour. Measurements show that in general both the width and spacing of resonances decrease as the mass number increases. Magic number nuclei form an exception to this trend. As we have already mentioned, the commonest reaction in the thermal energy range is radiative capture and in fact this is observed to be present to some extent for all nuclei.

It is not appropriate for us to give here anything bordering on a detailed discussion of resonance reactions. Our aim is to demonstrate the connection between a complex scattering amplitude and absorption of neutrons and the form of the elastic and reaction cross-sections in the vicinity of a resonance.

We represent the nuclei by a spherical scattering potential. The standard analysis of scattering by such a potential in terms of partial waves of angular momentum $l$ gives the elastic $\sigma_e$ and reaction $\sigma_r$ cross-sections as the sum of the components

$$\sigma_e^{(l)} = \frac{\pi}{k^2}(2l+1)\,|1-\alpha_l|^2 \tag{A.9a}$$

and

$$\sigma_r^{(l)} = \frac{\pi}{k^2}(2l+1)(1-|\alpha_l|^2), \tag{A.9b}$$

where $\alpha_l$ is related to the (complex) phase shift $\eta_l$ by

$$\alpha_l = \exp(2i\eta_l). \tag{A.10}$$

The wavelength of the incident thermal neutrons is much greater than both the mean radius of the nucleus and the range of interaction of the nucleon–nucleon force. This permits us to restrict attention to the $s$-wave ($l=0$) components of the partial wave expansion. For very small $k$

$$\alpha_0 \simeq 1+2ikf, \tag{A.11}$$

where the scattering amplitude $f$ is complex and independent of $k$. Thus as $k \to 0$

$$\sigma_e^{(0)} \simeq 4\pi\,|f|^2 \tag{A.12}$$

and

$$\sigma_r^{(0)} \simeq \frac{2\pi i}{k}(f^*-f) = \frac{4\pi}{k}\,\mathrm{Im}\,f. \tag{A.13}$$

Equation (A.12) agrees with (1.33) if $f$ is replaced by $b$. It is straightforward to show that the reaction cross-section (A.13) corresponds to absorption of the incident neutrons.

The s-wave component of the net outward radial flux of neutrons from the spherical scattering potential is

$$\frac{\hbar}{4mr^2k}(|\alpha_0|^2-1) \tag{A.14}$$

and this has the value

$$-\frac{\hbar}{mr^2}\,\mathrm{Im}\,f \tag{A.15}$$

in the limit $k \to 0$. Thus the net outward flux is proportional to Im $b$ and is negative, as it must be if absorption occurs.

In the presence of resonance reactions we can no longer take the scattering amplitude to be independent of $k$. We denote the energy at which the resonance occurs by $E_0$ and assume the nucleus to have a well-defined radius $R$. For zero-spin nuclei, $f$ is found to have the form

$$f = -R - \frac{\Gamma_n(E-E_0)}{2k\{(E-E_0)^2+\frac{1}{4}\Gamma^2\}} + \frac{i\Gamma\Gamma_n}{4k\{(E-E_0)^2+\frac{1}{4}\Gamma^2\}}. \tag{A.16}$$

Here $\Gamma_n \propto \sqrt{E}$ is the neutron width and $\Gamma$ the total width of the resonance. The first term on the right-hand side is simply the amplitude for scattering by an impenetrable sphere of radius $R$. Note that the two real terms can interfere with one another; their sum for $E<E_0$ may be positive.

The elastic scattering cross-section is given by substituting (A.16) into (A.9a) to give

$$\sigma_e^{(0)} = 4\pi R^2 + \frac{\pi\Gamma_n^2 + 4\pi kR\Gamma_n(E-E_0)}{k^2\{(E-E_0)^2+\frac{1}{4}\Gamma^2\}} \tag{A.17}$$

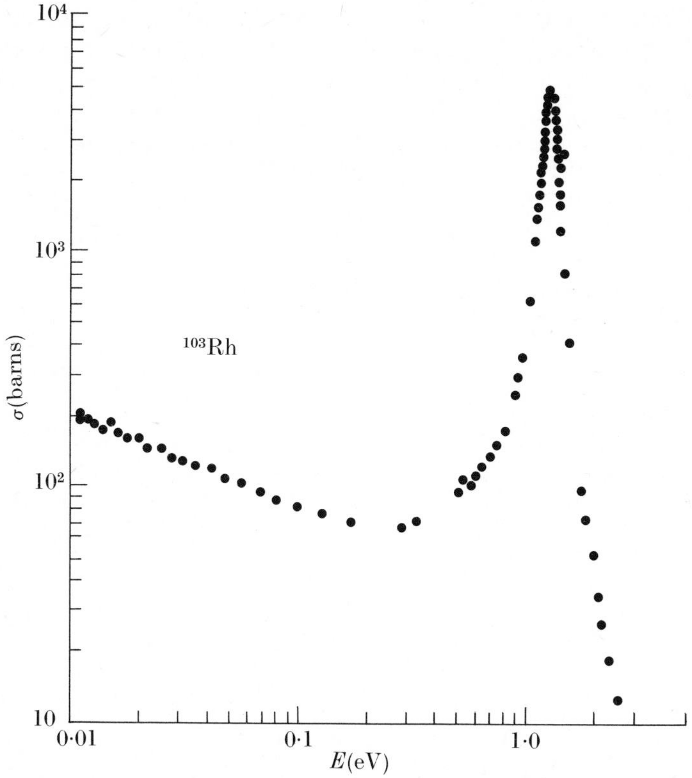


FIG. A.2. Capture cross-section of $^{103}$Rh showing the '$1/v$' rise due to the resonance towards low energies.

and near resonance the radiative capture cross-section has the form

$$\sigma_{\mathrm{r}}^{(0)} = \frac{\pi\Gamma_n\Gamma_\gamma}{k^2\{(E-E_0)^2+\frac{1}{4}\Gamma^2\}} \tag{A.18}$$

with $\Gamma = \Gamma_n + \Gamma_\gamma$. This is obtained by substituting (A.16) into (A.9b).

From the latter and the knowledge that $\Gamma_n \propto k$, it follows that, at energies much below the resonance, $\sigma_{\mathrm{r}}^{(0)} \propto E^{-1/2}$, i.e. the capture cross-section increases as $k \to 0$ as the inverse of the neutron velocity (cf. A.13). This '$1/v$' rise in the capture cross-section is naturally most pronounced when there is a resonance at low energies and is well illustrated by $^{103}$Rh shown in Fig. A.2.

The resonant scattering of neutrons has been formulated with a view to exploring its potential advantages for studying the properties of materials (Word and Trammell 1981). The cross-sections can be expressed in terms of correlation functions from which it is evident that, in most cases, the experiments are not likely to be readily interpreted in terms of material properties. However, the information provided by the experiments cannot be obtained from non-resonant scattering.

## REFERENCES

Mughabghab, S. F., Divadeenam, M., and Holden, N. E. (1981). *Neutron cross-sections*, Vol. 1. Academic Press, New York.

Newton, R. G. (1982). *Scattering theory of waves and particles.* Texts and monographs in physics. Springer-Verlag, Berlin.

Word, R. E. and Trammell, G. T. (1981). *Phys. Rev.* **B24,** 2430.

# APPENDIX B LINEAR RESPONSE THEORY

The spectrum of spontaneous fluctuations in a variable, the density of a sample say, is proportional to the response of the sample to a weak external perturbation that couples to the same variable. Students of neutron scattering are unlikely to find this statement of the so-called fluctuation-dissipation theorem a revelation or awesome (Reichl 1980). For the neutron scattering cross-section is clearly a measure of the sample's response to the neutrons and, following Van Hove, the cross-section is readily expressed in terms of the spectrum of spontaneous fluctuations in the variable that describes the neutron–matter interaction. The Van Hove representation of the cross-section in terms of a correlation function, which is usually called a scattering law or dynamic structure factor $S(\boldsymbol{\kappa}, \omega)$, is thus an example of the fluctuation-dissipation theorem. Indeed, in this book we frequently describe $S(\boldsymbol{\kappa}, \omega)$ as a response function, because it determines the cross-section, although the explicit form of $S(\boldsymbol{\kappa}, \omega)$ is a correlation function that is independent of the neutron probe.†

A prime function of this appendix is to explore the full import of the fluctuation-dissipation theorem. This is achieved by formulating linear response theory for systems that are close to thermal equilibrium; the linear approximation employed is equivalent to the Born approximation for neutron scattering, as we shall see. Linear response theory can be used to obtain expressions for $S(\boldsymbol{\kappa}, \omega)$ by exploiting the fluctuation-dissipation theorem. An example of such a calculation is given in § B.4, and several more complicated examples occur in the text.

Looking at the literature or this book the reader might feel that there is a plethora of seemingly dissimilar functions which are used to calculate $S(\boldsymbol{\kappa}, \omega)$. The functions include the generalized susceptibility $\chi_{\boldsymbol{\kappa}}[\omega]$, the relaxation function $R(\kappa, t)$, and Green functions. Authors purport to select the most convenient or natural function in which to couch their discussion. In fact the choice is to a large extent a matter of personal taste. A second aim of this appendix is to establish the link between the various functions using the framework of linear response theory. We will also give some reasons why certain types of calculation are best handled in terms of a particular function. The key point to note is that $S(\boldsymbol{\kappa}, \omega)$ itself is not very often a convenient function to employ in deriving

† In general the cross-section is the sum of different scattering laws weighted by various combinations of scattering lengths.

**Table B.1**

*Summary of principal response functions*

| | |
|---|---|
| Linear response of a system close to thermal equilibrium | $\left.\begin{array}{l}\text{Change in } \langle\hat{A}\rangle \text{ at time } t \text{ due}\\ \text{to a perturbation} = -\hat{B}h(t)\end{array}\right\} = \int_{-\infty}^{t} \mathrm{d}t'\phi_{\mathrm{AB}}(t-t')h(t')$ |
| | $\phi_{\mathrm{AB}}(t) = \frac{\mathrm{i}}{\hbar}\langle[\hat{A}(t), \hat{B}]\rangle$ |
| Generalized susceptibility | $\chi_{\mathrm{AB}}[\omega] = -\lim_{\epsilon\to 0^+}\int_0^{\infty} \mathrm{d}t \exp(-\mathrm{i}\omega t - \epsilon t)\phi_{\mathrm{AB}}(t)$ |
| | $= \chi'_{\mathrm{AB}}[\omega] + \mathrm{i}\chi''_{\mathrm{AB}}[\omega]$ |
| Relaxation function | $R_{\mathrm{AB}}(t) = \int_0^{\beta} \mathrm{d}\mu\langle\{\hat{B} - \langle\hat{B}\rangle\}\{\hat{A}(t+\mathrm{i}\hbar\mu) - \langle\hat{A}\rangle\}\rangle = R_{\mathrm{BA}}(-t)$ |
| or | $\phi_{\mathrm{AB}}(t) = -\partial_t R_{\mathrm{AB}}(t)$ |
| | $R_{\mathrm{AB}}(\omega) = \frac{1}{2\pi}\int_{-\infty}^{\infty} \mathrm{d}t \exp(-\mathrm{i}\omega t)R_{\mathrm{AB}}(t) = R_{\mathrm{BA}}(-\omega)$ |
| Green function | $G(t) = -\mathrm{i}\theta(t)\langle[\hat{B}(t), \hat{B}^+]\rangle = -\theta(t)\hbar\phi_{\mathrm{BB}^+}(t)$ |
| | $G(\omega) = \int_{-\infty}^{\infty} \mathrm{d}t \exp(\mathrm{i}\omega t)G(t) \equiv \langle\langle\hat{B}; \hat{B}^+\rangle\rangle = G'(\omega) + \mathrm{i}G''(\omega)$ |
| Static isothermal susceptibility | $\chi_{\mathrm{AB}} = R_{\mathrm{AB}}(t=0) = -\chi'_{\mathrm{AB}}[0]$ |
| | $S(\omega) = \frac{1}{2\pi\hbar}\int_{-\infty}^{\infty} \mathrm{d}t \exp(-\mathrm{i}\omega t)\langle\hat{B}\hat{B}^+(t)\rangle$ |
| | $= \frac{1}{2\pi\hbar}\int_{-\infty}^{\infty} \mathrm{d}t \exp(\mathrm{i}\omega t)\langle\hat{B}(t)\hat{B}^+\rangle = \{1+n(\omega)\}\frac{1}{\pi}\chi''_{\mathrm{B^+B}}[\omega]$ |
| | $= \{1+n(\omega)\}\omega R_{\mathrm{B^+B}}(\omega) = -\{1+n(\omega)\}\frac{1}{\pi\hbar}G''(\omega)$ |
| where the detailed balance factor is | $\{1+n(\omega)\} = \{1-\exp(-\hbar\omega\beta)\}^{-1} = -n(-\omega)$ |

approximate theories of the neutron cross-section although, of course, it is always the ultimate goal of theory. The principal response functions are listed in Table B.1.

We will use the model of a damped harmonic oscillator, described in § B.3, to illustrate various results derived from linear response theory. The model is defined in terms of a phenomenological equation for particle displacements subject to harmonic restoring forces and frictional and random forces coming from the ambient medium (van Kampen 1981). In § B.3 we calculate the change in the particle displacement induced by an impulsive perturbation, and from this we derive the generalized susceptibility for the model. The spectrum of spontaneous fluctuations in the displacements is deduced in § B.4 by applying the fluctuation-dissipation theorem.

## B.1. Generalized susceptibility, $\chi_{AB}[\omega]$

We consider the behaviour of an isolated system initially in equilibrium at a temperature $T = 1/(k_B\beta)$, when a small time-dependent perturbation is allowed to act on it. At the time $t = -\infty$ the system is described by the Hamiltonian $\hat{\mathcal{H}}_0$; at the later time $t$ the total Hamiltonian is

$$\hat{\mathcal{H}} = \hat{\mathcal{H}}_0 - \hat{\mathcal{H}}_1 = \hat{\mathcal{H}}_0 - \hat{B}h(t). \tag{B.1}$$

The response of the system to the external perturbation $\hat{\mathcal{H}}_1$ is observed in the change of a variable A, which itself does not depend explicitly on time. Quite generally we can write for a *linear* response,

$$\overline{A(t)} = \langle \hat{A} \rangle + \int_{-\infty}^{t} dt'\,\phi_{AB}(t-t')h(t'), \tag{B.2}$$

where $\hat{A}$ is the operator corresponding to the variable A and the equilibrium average

$$\langle \hat{A} \rangle = \mathrm{Tr}\,\hat{\rho}_0\hat{A} = \frac{1}{Z}\mathrm{Tr}\exp(-\beta\hat{\mathcal{H}}_0)\hat{A}. \tag{B.3}$$

$\phi_{AB}(t)$ is called a *response function*. It measures the change in A due to a perturbation localized at an earlier time $t = 0$, i.e. an impulsive perturbation $h(t) \propto \delta(t)$. Clearly $\phi_{AB}(t)$ is purely real for a real perturbation.

Any time-dependent perturbation can, by Fourier transformation, be represented as a set of monochromatic components whose time dependence is given through $\exp(i\omega t)$. In such cases $h(t) \propto \exp(i\omega t)$. When we derive an explicit expression for $\phi_{AB}(t)$ in terms of the operators $\hat{A}$ and $\hat{B}$ (cf. § B.4) we assume that $\hat{\mathcal{H}}_1$ describes an adiabatic process, i.e. a process that at every instant of time leaves the system in equilibrium. The condition on $\hat{\mathcal{H}}_1$ is fulfilled if we make $h(t)$ increase very slowly with increasing $t$, and this is achieved by replacing $\exp(i\omega t)$ by $\exp(i\omega t + \epsilon t)$, where $\epsilon$ is a small positive quantity. For in this instance the perturbation is zero for $t = -\infty$ and it can be made to increase as slowly as required through the choice of $\epsilon$; we shall take the limit of $\epsilon \to 0^+$. Thus, taking $h(t) = h\exp(\epsilon t)\cos\omega t$, where $h$ is a real constant, we obtain from (B.2)

$$\overline{A(t)} = \langle \hat{A} \rangle - h\,\mathrm{Re}\{\exp(i\omega t)\chi_{AB}[\omega]\}, \tag{B.4}$$

where

$$-\chi_{AB}[\omega] = \lim_{\epsilon\to 0^+}\int_0^{\infty} dt\,\phi_{AB}(t)\exp(-i\omega t - \epsilon t) \tag{B.5}$$

defines the *generalized susceptibility function* $\chi_{AB}[\omega]$.

Note that we define the one-sided Fourier transform in (B.5) by $\chi_{AB}[\omega]$. We reserve the notation of round brackets about the argument $\omega$ for a conventional Fourier transform.

$\chi_{AB}[\omega]$ embodies the details of the dynamical properties of the system described by $\hat{\mathcal{H}}_0$, since if $\hbar\omega$ coincides with the energy of an excitation in the system $\chi_{AB}$ must peak at this energy. Thus, in principle, the energy spectrum of the excitations can be obtained by examining the poles of the generalized susceptibility function. A simple example is discussed in § B.3.

Let us consider more closely the structure of the response function $\phi_{AB}(t)$. From both (B.2) and (B.5) it is observed that $\phi_{AB}(t)$ is required only for $t>0$. This is simply a consequence of the fact that a response to a perturbation cannot be observed before the latter has been applied. It is therefore usual to call $\phi_{AB}(t)$ a *causal* function. Nonetheless it proves convenient in subsequent discussions to have $\phi_{AB}(t)$ defined for all values of $t$. This is most naturally achieved by using the explicit expression for $\phi_{AB}(t)$ derived in § B.4.

## B.2. Analytic properties of $\chi_{AB}[\omega]$

$\chi_{AB}[\omega]$ is, in general, a complex function, which we write

$$\chi_{AB} = \chi'_{AB} + i\chi''_{AB}. \tag{B.6}$$

Since $\phi_{AB}(t)$ must be real for a real external perturbation it follows from (B.5) that

$$\chi'[\omega] = \chi'[-\omega], \qquad \chi''[\omega] = -\chi''[-\omega]. \tag{B.7}$$

For the present discussion we omit the suffixes $AB$ on both the generalized susceptibility and the response function.

$\chi'$ and $\chi''$ satisfy certain relationships: in particular they are conjugate functions, which can be derived as follows. We rewrite (B.5) as a conventional Fourier transform by using the unit step function

$$\begin{aligned} \theta(t) &= 1; \qquad t>0 \\ &= 0; \qquad t<0 \end{aligned} \tag{B.8}$$

and the result is

$$\chi[\omega] = -\int_{-\infty}^{\infty} dt\ \phi(t)\theta(t)\exp(-i\omega t). \tag{B.9}$$

An integral representation of $\theta(t)$ can be derived by considering the integral

$$I(t) = \int_{-\infty}^{\infty} d\omega\ \exp(i\omega t)/(\omega - i\epsilon), \tag{B.10}$$

where $\epsilon$ is real and positive. This integral is evaluated by using the theorem of residues from the theory of functions of a complex variable.

The result is

$$I(t) = 2\pi \mathrm{i} \exp(-\epsilon t); \qquad t > 0$$
$$= 0; \qquad t < 0. \tag{B.11}$$

From (B.11) we deduce that

$$2\pi \mathrm{i}\theta(t) = \lim_{\epsilon \to 0^+} \int_{-\infty}^{\infty} \mathrm{d}u \exp(\mathrm{i}ut)/(u - \mathrm{i}\epsilon), \tag{B.12}$$

whereupon, from (B.9),

$$\chi[\omega] = -\lim_{\epsilon \to 0^+} \int_{-\infty}^{\infty} \frac{\mathrm{d}t}{2\pi \mathrm{i}} \int_{-\infty}^{\infty} \frac{\mathrm{d}u \exp\{\mathrm{i}ut - \mathrm{i}\omega t\}\phi(t)}{(u - \mathrm{i}\epsilon)}$$
$$= -\lim_{\epsilon \to 0^+} \int_{-\infty}^{\infty} \frac{\mathrm{d}t}{2\pi \mathrm{i}} \int_{-\infty}^{\infty} \frac{\mathrm{d}u \exp(-\mathrm{i}ut)\phi(t)}{(\omega - u - \mathrm{i}\epsilon)} \tag{B.13}$$

where the second equality is obtained by a change of integration variable.

If $\phi(t)$ is *an odd function of t*, as is usually the case,

$$\int_{-\infty}^{\infty} \mathrm{d}t \exp(-\mathrm{i}ut)\phi(t) = -2\mathrm{i} \int_{0}^{\infty} \mathrm{d}t \sin(ut)\phi(t)$$
$$= -2\mathrm{i}\chi''[u]$$

and (B.13) reduces to

$$\chi[\omega] = \frac{1}{\pi} \lim_{\epsilon \to 0^+} \int_{-\infty}^{\infty} \frac{\mathrm{d}u\, \chi''[u]}{(\omega - u - \mathrm{i}\epsilon)}. \tag{B.14}$$

We now use the integral identity

$$\lim_{\epsilon \to 0^+} \int_{-\infty}^{\infty} \frac{\mathrm{d}u}{(\omega - u - \mathrm{i}\epsilon)} = \mathrm{i}\pi \int_{-\infty}^{\infty} \mathrm{d}u\, \delta(\omega - u) + P \int_{-\infty}^{\infty} \frac{\mathrm{d}u}{(\omega - u)} \tag{B.15}$$

where the second term on the right-hand side is the principal part of the integral. Applied to (B.14) we deduce that

$$\chi[\omega] = \mathrm{i}\chi''[\omega] + \frac{1}{\pi} P \int_{-\infty}^{\infty} \frac{\mathrm{d}u\, \chi''[u]}{(\omega - u)}$$

and, since $\chi = \chi' + \mathrm{i}\chi''$,

$$\chi'[\omega] = \frac{1}{\pi} P \int_{-\infty}^{\infty} \frac{\mathrm{d}u\, \chi''[u]}{(\omega - u)}. \tag{B.16}$$

Result (B.16) is usually called a dispersion relation and is a consequence of causality. Given $\chi''[\omega]$, which is often called the dissipative part of the susceptibility, $\chi'[\omega]$ can be calculated from (B.16). Hence, the real and imaginary parts of a causal function are not independent functions.

Because $\chi''[\omega]$ is an odd function of the frequency, (B.16) can be reduced to

$$\chi'[\omega]=\frac{2}{\pi}P\int_0^\infty \frac{\mathrm{d}u\, u\chi''[u]}{(\omega^2-u^2)} \tag{B.17}$$

and a case of particular interest is the static isothermal susceptibility (cf. (B.62))

$$\chi=-\chi'[0]=\frac{2}{\pi}\int_0^\infty \mathrm{d}u\{\chi''[u]/u\}. \tag{B.18}$$

When $\phi(t)$ is an odd function,

$$\phi(t)=\frac{1}{\pi \mathrm{i}}\int_{-\infty}^{\infty} \mathrm{d}\omega\, \exp(\mathrm{i}\omega t)\chi''[\omega]$$

and it follows that the frequency moments of $\chi''[\omega]$ are related to derivatives of $\phi(t)$ evaluated at $t=0$, namely,

$$\frac{1}{\pi}\int_{-\infty}^{\infty} \mathrm{d}\omega\, \omega^{2n+1}\chi''[\omega]=(-1)^n \partial_t^{2n+1}\phi(t)\big|_{t=0} \tag{B.19}$$

where $n$ is a positive integer.

A useful inequality for the susceptibility can be derived by using the standard argument for the Schwartz inequality. The result is

$$\int_0^\infty \mathrm{d}\omega\left(\frac{\chi''[\omega]}{\omega}\right)\int_0^\infty \mathrm{d}\omega\, \omega\, \chi''[\omega]\geqslant\left\{\int_0^\infty \mathrm{d}\omega\chi''[\omega]\right\}^2. \tag{B.20}$$

## B.3. Damped harmonic oscillator

We consider the motion of a damped harmonic oscillator in order to illustrate how some of the previous formulae and relations work out in practice. The results are also of use in the text, but our primary objective here is to illustrate the formalism of linear response theory.

Consider the motion of a particle, of mass $M$, in one dimension subject to a harmonic restoring force of strength $M\omega_0^2$. If the displacement at time $t$ is $x(t)$, the force

$$M\ddot{x}(t)+M\omega_0^2 x(t)$$

is balanced by the sum of the applied force and the force exerted by the ambient medium. We assume that the latter is comprised of a frictional part, whose average value is proportional to the average velocity of the particle, and a random force. This amounts to separating the medium force into components distinguished by their characteristic time scales as in the theory of Brownian motion. If the random force varies rapidly on the time scale of interest, we can set its average value to zero. Hence, in the presence of an applied force $h(t)$, when the particle is driven away from its equilibrium position and $\langle x(t)\rangle \neq 0$, the average displacement satisfies the equation of motion,

$$M\langle \ddot{x}(t)\rangle + M\omega_0^2\langle x(t)\rangle = h(t) - M\gamma\langle \dot{x}(t)\rangle. \qquad \text{(B.21)}$$

Here $\gamma$ is the friction coefficient.

From the definition of the linear response function (B.2) we have

$$\langle x(t)\rangle = \int_{-\infty}^{t} \mathrm{d}t'\phi(t-t')h(t'). \qquad \text{(B.22)}$$

Let us consider an impulsive applied force

$$h(t) = h\,\delta(t),$$

where $h$ is a constant. In this case (B.22) reduces to

$$\langle x(t)\rangle = h\phi(t)$$

and it follows that, for a Laplace variable $s = (\mathrm{i}\omega + \epsilon)$ with $\epsilon \to 0^+$,

$$\begin{aligned}\tilde{x}(s) &= \int_0^\infty \mathrm{d}t\,\exp(-st)\langle x(t)\rangle \\ &= \int_0^\infty \mathrm{d}t\,\exp(-st)h\phi(t) \\ &= -h\chi[\omega].\end{aligned} \qquad \text{(B.23)}$$

We will therefore solve (B.21) for $\tilde{x}(s)$. Using the result

$$\int_0^\infty \mathrm{d}t\,\exp(-st)\langle \dot{x}(t)\rangle = -\langle x(0)\rangle + s\tilde{x}(s)$$

together with the initial conditions $\langle x(0)\rangle = \langle \dot{x}(0)\rangle = 0$, we find from (B.21) that $\tilde{x}(\mathrm{i}\omega)$ satisfies

$$M(-\omega^2 + \omega_0^2 + \mathrm{i}\gamma\omega)\tilde{x}(\mathrm{i}\omega) = h.$$

Thus, from (B.23),

$$\chi[\omega] = -\tilde{x}\,(\mathrm{i}\omega)/h = \{M(\omega^2 - \omega_0^2 - \mathrm{i}\gamma\omega)\}^{-1} \qquad \text{(B.24)}$$

and $\chi[\omega]$ has the dimension of $(\text{length})^2/\text{energy}$.

From (B.24) we deduce that the static susceptibility

$$\chi = -\chi[0] = (1/M\omega_0^2) \qquad \text{(B.25)}$$

and the dissipative component

$$M\chi''[\omega] = \left\{\frac{\gamma\omega}{(\omega^2 - \omega_0^2)^2 + (\gamma\omega)^2}\right\}. \qquad \text{(B.26)}$$

Note that $\chi''[\omega]$ is an odd function of $\omega$. To obtain the behaviour in the limit $\gamma \to 0$, we rewrite $\chi[\omega]$ in the form

$$M\chi[\omega] = \frac{1}{\epsilon}\left\{\frac{(\gamma/2)}{\omega - \frac{i}{2}\gamma - \frac{1}{2}\epsilon} - \frac{(\gamma/2)}{\omega - \frac{i}{2}\gamma + \frac{1}{2}\epsilon}\right\} \tag{B.27}$$

where

$$\epsilon^2 = 4\omega_0^2 - \gamma^2.$$

We now utilize the standard representation of the delta function

$$\delta(x) = \frac{1}{\pi}\,\mathrm{Im}\lim_{\eta \to 0^+}(x - i\eta)^{-1}$$

to obtain the result

$$M\lim_{\gamma \to 0}\chi''[\omega] = \frac{\pi}{2\omega_0}\{\delta(\omega - \omega_0) - \delta(\omega + \omega_0)\}. \tag{B.28}$$

Thus, for vanishing frictional damping, the dissipative part of the susceptibility reduces to the sum of a pair of delta functions at the frequencies $\pm\omega_0$.

The response function $\phi(t)$ can be calculated by inverting the Fourier transform (B.9). The Fourier integral can be expressed in terms of the integral $I(t)$ (eqn (B.10)) by using the representation of $\chi[\omega]$ given in (B.27). From (B.11) we find, for $t > 0$, the result

$$\begin{aligned}\phi(t) &= -\frac{1}{2\pi}\int_{-\infty}^{\infty} dt\, \exp(i\omega t)\chi[\omega] \\ &= \frac{2}{M\epsilon}\exp\{-\tfrac{1}{2}\gamma t\}\sin(\tfrac{1}{2}\epsilon t); \qquad t > 0.\end{aligned} \tag{B.29}$$

For small damping, $\phi(t)$ oscillates in time with a frequency $\sim\omega_0$. If the damping is sufficiently large, $\epsilon^2 < 0$ and then $\phi(t)$ does not oscillate; we call this an overdamped harmonic oscillator.

The dispersion relation (B.16) can be verified using (B.26). The integrals involved in the calculation can be performed using the theorem of residues from the theory of functions of a complex variable. We will not pause to give the mathematical details of the calculation.

The model discussed in this section is an example of a system whose dynamics is specified by a stochastic equation of motion (van Kampen 1981). This arises when the medium force is separated into a frictional damping and a random component. Perhaps this feature of the model is most apparent when we set the spring constant to zero ($\omega_0 = 0$) for then $\chi''[\omega]$ (eqn (B.26)) reduces to a Lorentz function that describes a simple diffusive motion.

For most systems of interest to us, a Hamiltonian description exists. We can then obtain an explicit expression for the response function $\phi(t)$ by using first-order perturbation theory. The perturbative calculation is described in the following section. Given an operational form for $\phi(t)$, we establish the fluctuation dissipation theorem. The theorem enables us to calculate the spectrum of spontaneous fluctuations from the dissipative susceptibility.

## B.4. Calculation of the response function $\phi_{AB}(t)$ and the fluctuation-dissipation theorem

The external perturbation $\hat{\mathscr{H}}_1$ is defined such that that its average value is very small compared to that of $\hat{\mathscr{H}}_0$ and such that the density matrix $\hat{\rho}$ at time $t$ is determined by

$$i\hbar\dot{\hat{\rho}}(t)=[\hat{\mathscr{H}},\hat{\rho}] \tag{B.30}$$

with the boundary condition $\hat{\rho}(-\infty)=\hat{\rho}_0$.

The condition that $\hat{\mathscr{H}}_1$ must satisfy in order that the equation of motion (B.30) can be used to calculate $\hat{\rho}$ is that at every instant $-\infty<t$ the system is in equilibrium at time $t$. This (adiabatic) process is usually achieved by taking $\hat{\mathscr{H}}_1$ to have an overriding time dependence of the form $\exp(\epsilon t)$, $\epsilon>0$, for $t<0$, but we can also admit a perturbation that is applied instantaneously, provided it is such that the system as a whole is in an equilibrium state immediately following its application.

With $\hat{\rho}(t)=\hat{\rho}_0+\Delta\hat{\rho}(t)$, and neglecting second-order quantities, we obtain from (B.30)

$$i\hbar\Delta\dot{\hat{\rho}}=[-\hat{\mathscr{H}}_1,\hat{\rho}_0]-[\Delta\hat{\rho},\hat{\mathscr{H}}_0].$$

If $\Delta\hat{\rho}(t)=\exp(-it\hat{\mathscr{H}}_0/\hbar)\,\delta\hat{\rho}(t)\exp(it\hat{\mathscr{H}}_0/\hbar)$, then $\delta\hat{\rho}(t)$ satisfies

$$i\hbar\,\delta\dot{\hat{\rho}}=\exp(it\hat{\mathscr{H}}_0/\hbar)[-\hat{\mathscr{H}}_1,\hat{\rho}_0]\exp(-it\hat{\mathscr{H}}_0/\hbar),$$

and hence,

$$\Delta\hat{\rho}(t)=\frac{1}{i\hbar}\int_{-\infty}^{t}\mathrm{d}t'h(t')[\hat{\rho}_0,\hat{B}(t'-t)]. \tag{B.31}$$

We then have, from (B.31),

$$\overline{A(t)}-\langle\hat{A}\rangle=\frac{1}{i\hbar}\int_{-\infty}^{t}\mathrm{d}t'h(t')\mathrm{Tr}\{[\hat{\rho}_0,\hat{B}(0)]\hat{A}(t-t')\},$$

so that, from the definition of $\phi_{AB}(t-t')$ (eqn (B.2)),

$$\begin{aligned}\phi_{AB}(t-t')&=\frac{1}{i\hbar}\mathrm{Tr}\{[\hat{\rho}_0,\hat{B}(0)]\hat{A}(t-t')\}\\&=\frac{i}{\hbar}\langle[\hat{A}(t),\hat{B}(t')]\rangle.\end{aligned} \tag{B.32}$$

This is a key result, for we can immediately obtain the relation between the spectrum of spontaneous fluctuations and the linear response.

The function that describes the spectrum of scattered neutrons in the Van Hove formulation is of the form (we omit the wave-vector dependence of the variables for the moment)

$$S(\omega)=\frac{1}{2\pi\hbar}\int_{-\infty}^{\infty} \mathrm{d}t \exp(-\mathrm{i}\omega t)\langle\hat{B}\hat{B}^{+}(t)\rangle. \quad \text{(B.33)}$$

This function is purely real, as expected. In order to relate $S(\omega)$ to the response function

$$\phi(t)\equiv\phi_{B^+B}(t)=\frac{\mathrm{i}}{\hbar}\langle[\hat{B}^{+}(t),\hat{B}]\rangle, \quad \text{(B.34)}$$

we examine the Fourier transform of $\phi(t)$. Now,

$$\frac{1}{2\pi}\int_{-\infty}^{\infty} \mathrm{d}t \exp(-\mathrm{i}\omega t)\phi(t)=\frac{\mathrm{i}}{2\pi\hbar}\int_{-\infty}^{\infty} \mathrm{d}t \exp(-\mathrm{i}\omega t)\langle\hat{B}^{+}(t)\hat{B}-\hat{B}\hat{B}^{+}(t)\rangle$$

$$=-\mathrm{i}S(\omega)+\frac{\mathrm{i}}{2\pi\hbar}\int_{-\infty}^{\infty} \mathrm{d}t \exp(-\mathrm{i}\omega t)\langle\hat{B}^{+}(t)\hat{B}\rangle, \quad \text{(B.35)}$$

where we have used the definition of $S(\omega)$ in arriving at the last equality. To proceed we employ the identity

$$\langle\hat{B}^{+}(t)\hat{B}\rangle=\langle\hat{B}\hat{B}^{+}(t+\mathrm{i}\hbar\beta)\rangle. \quad \text{(B.36)}$$

To prove (B.36) one can either use the explicit representation of the correlation function in terms of the eigenstates of $\mathscr{H}_0$, or write the thermal average as a trace and use the invariance of the trace to a cyclic permutation of the operators. Applied to the last term in (B.35), the identity leads to the result

$$\int_{-\infty}^{\infty} \mathrm{d}t \exp(-\mathrm{i}\omega t)\langle\hat{B}^{+}(t)\hat{B}\rangle=\int_{-\infty}^{\infty} \mathrm{d}t \exp(-\mathrm{i}\omega t)\langle\hat{B}\hat{B}^{+}(t+\mathrm{i}\hbar\beta)\rangle$$

$$=\exp(-\hbar\omega\beta)2\pi\hbar S(\omega). \quad \text{(B.37)}$$

Combining this last result with (B.35) we obtain the desired relation

$$\frac{1}{2\pi}\int_{-\infty}^{\infty} \mathrm{d}t \exp(-\mathrm{i}\omega t)\phi(t)=-\mathrm{i}\{1-\exp(-\hbar\omega\beta)\}S(\omega) \quad \text{(B.38)}$$

or

$$S(\omega)=\{1+n(\omega)\}\frac{\mathrm{i}}{2\pi}\int_{-\infty}^{\infty} \mathrm{d}t \exp(-\mathrm{i}\omega t)\phi(t),$$

where the detailed balance factor,

$$1+n(\omega)=\{1-\exp(-\hbar\omega\beta)\}^{-1}=-n(-\omega). \quad \text{(B.39)}$$

Relation (B.38) is the so-called fluctuation dissipation theorem, for $S(\omega)$ is the spectrum of spontaneous fluctuation in the variable $B$, and the right-hand side is the response, or dissipation, of the target sample. We will explicitly verify that the right-hand side is the energy dissipation in a later development. More general forms of the theorem are given in § B.7.

If $\phi(t)$ is an odd function of $t$, the fluctuation dissipation theorem reduces to

$$S(\omega)=\{1+n(\omega)\}\frac{1}{2\pi}\int_{-\infty}^{\infty}\mathrm{d}t\,\sin(\omega t)\phi(t)$$
$$=\{1+n(\omega)\}\frac{1}{\pi}\chi''[\omega]. \tag{B.40}$$

The last form follows from the definition of the generalized susceptibility (B.5), and it is the form that is most useful in the interpretation of neutron scattering data. We now verify that $\phi(t)$ is an odd function of the time. Consider, for example, the coherent scattering in terms of the particle density fluctuations. In this instance

$$\hat{B}\equiv\hat{\rho}_{\boldsymbol{\kappa}}=\sum_j \exp(-\mathrm{i}\boldsymbol{\kappa}\cdot\hat{\mathbf{R}}_j), \tag{B.41}$$

where $\hat{\mathbf{R}}_j$ is the position operator of the $j$th particle. The correlation function of interest is

$$\langle\hat{\rho}_{\boldsymbol{\kappa}}\hat{\rho}^{+}_{\boldsymbol{\kappa}}(t)\rangle\equiv\langle\hat{\rho}_{\boldsymbol{\kappa}}\hat{\rho}_{-\boldsymbol{\kappa}}(t)\rangle.$$

For target samples that possess inversion symmetry, the correlation function depends on the magnitude of $\boldsymbol{\kappa}$ and not its direction, so that

$$\langle\hat{\rho}_{\boldsymbol{\kappa}}\hat{\rho}_{-\boldsymbol{\kappa}}(t)\rangle=\langle\hat{\rho}_{-\boldsymbol{\kappa}}\hat{\rho}_{\boldsymbol{\kappa}}(t)\rangle.$$

These results are sufficient to prove that the corresponding response function is an odd function of time.

The fluctuation dissipation theorem can be used to obtain $S(\omega)$ for a given model of $\chi''[\omega]$. For the damped harmonic oscillator model discussed in § B.3,

$$S(\omega)=\frac{1}{2\pi\hbar}\int_{-\infty}^{\infty}\mathrm{d}t\,\exp(-\mathrm{i}\omega t)\langle\hat{x}\hat{x}(t)\rangle$$
$$=\{1+n(\omega)\}\frac{1}{\pi M}\frac{\gamma\omega}{\{(\omega^2-\omega_0^2)^2+(\gamma\omega)^2\}}. \tag{B.42}$$

In the limit of vanishing frictional damping,

$$S(\omega)=\{1+n(\omega)\}\frac{1}{2M\omega_0}\{\delta(\omega-\omega_0)-\delta(\omega+\omega_0)\}$$
$$=\frac{1}{2M\omega_0}\{[1+n(\omega_0)]\,\delta(\omega-\omega_0)+n(\omega_0)\,\delta(\omega+\omega_0)\}, \tag{B.43}$$

where the final form results from the identity

$$1+n(\omega)+n(-\omega)=0. \tag{B.44}$$

Result (B.43) can be interpreted as the sum of annihilation and creation of single quanta events. At high temperatures, $\hbar\omega_0\beta \ll 1$, the population factors in (B.43) are essentially equal so that the annihilation and creation processes contribute equally to the spectra. In the opposite limiting case $\hbar\omega_0\beta \gg 1$, the population factor $n(\omega_0) \to 0$ and $S(\omega)$ contains only a creation process. This result for $T \to 0$ is a consequence of the fact that at low temperatures there are no thermally excited states of the harmonic oscillator to participate in the response.

Let us now use (B.42) to explore more features of the damped harmonic oscillator model. Inverting the Fourier transform gives the particle displacement auto-correlation function

$$\langle \hat{x}\hat{x}(t)\rangle = \hbar \int_{-\infty}^{\infty} \mathrm{d}\omega \, \exp(\mathrm{i}\omega t) S(\omega)$$

and, since

$$\partial_t^2 \langle \hat{x}\hat{x}(t)\rangle = -\langle \dot{\hat{x}}\dot{\hat{x}}(t)\rangle = -\langle \hat{v}\hat{v}(t)\rangle,$$

where $\hat{v}$ is the particle velocity, we find that

$$\begin{aligned}\langle \hat{v}\hat{v}(t)\rangle &= \hbar \int_{-\infty}^{\infty} \mathrm{d}\omega \, \omega^2 \exp(\mathrm{i}\omega t) S(\omega) \\ &\equiv \frac{\hbar}{\pi} \int_{-\infty}^{\infty} \mathrm{d}\omega \, \omega^2 \, \{1+n(\omega)\} \exp(\mathrm{i}\omega t) \chi''[\omega].\end{aligned} \tag{B.45}$$

The classical limit of this result is of interest. Taking $\hbar \to 0$ on the right-hand side leads to

$$\frac{\hbar}{\pi} \int_{-\infty}^{\infty} \mathrm{d}\omega \left(\frac{\omega^2}{\hbar\omega\beta}\right) \exp(\mathrm{i}\omega t) \chi''[\omega]$$

and, for $t=0$, this is simply

$$\frac{1}{\pi\beta} \int_{-\infty}^{\infty} \mathrm{d}\omega \, \omega \, \chi''[\omega] = \frac{1}{\beta} \dot{\phi}(0),$$

where the final equality follows from (B.19). Assembling the results we have, for classical systems,

$$\langle v^2 \rangle = \frac{1}{\pi\beta} \int_{-\infty}^{\infty} \mathrm{d}\omega \, \omega \chi''[\omega] = \frac{1}{\beta} \dot{\phi}(0). \tag{B.46}$$

Applying this to the damped harmonic oscillator model, with $\phi(t)$ given by (B.29), we recover the equipartition result $\langle v^2 \rangle = 1/M\beta$. We may also derive the result

$$\int_{-\infty}^{\infty} \mathrm{d}t \langle vv(t)\rangle = \frac{2}{\beta} \lim_{\omega\to 0} \omega\chi''[\omega].$$

The left-hand side is proportional to the self-diffusion constant $D_s$ (cf. Chapter 5). The final result is a Green–Kubo formula for the diffusion constant in terms of the displacement susceptibility,

$$\beta D_s = \lim_{\omega\to 0} \omega\chi''[\omega]. \tag{B.47}$$

For $\chi''[\omega]$ given in (B.26), the result is $D_s = 0$ for $\omega_0 \neq 0$, i.e. the diffusion constant of a bound particle is zero as should be expected. On the other hand, for $\omega_0 = 0$, (B.26) and (B.47) yield the relation

$$D_s = (1/M\beta\gamma)$$

which is the familiar Einstein relation between the friction coefficient $\gamma$ and the self-diffusion constant.

We finish this subsection by noting that the dispersion relation (B.17) can be written in terms of $S(\omega)$ by using (B.40). The result is

$$\chi'[\omega] = 2P\int_0^{\infty} \frac{\mathrm{d}u\, u\, S(u)\{1-\exp(-\hbar u\beta)\}}{(\omega^2 - u^2)} \tag{B.49}$$

whereupon the static susceptibility satisfies

$$\chi = -\chi'[0] = 2\int_0^{\infty} \mathrm{d}u \frac{S(u)}{u}\{1-\exp(-\hbar u\beta)\}. \tag{B.50}$$

The relations (B.49)–(B.50) take a very simple form for degenerate quantum systems achieved in the limit of absolute zero.

## B.5. Relaxation function, $R_{AB}(t)$

The relaxation function describes the behaviour of the change in a variable after the external perturbation has been switched off. An explicit expression for the function is obtained from (B.2) by taking $h(t)$ to be $h\exp(\epsilon t)$ for $t \leq 0$ and zero for $t > 0$. The change in the average value of A induced by this discontinuous perturbation is,

$$\begin{aligned}\overline{A(t)} - \langle\hat{A}\rangle &= h\int_{-\infty}^{0} \mathrm{d}t' \exp(\epsilon t')\phi_{AB}(t-t') \\ &= h\int_{t}^{\infty} \mathrm{d}t' \exp\{\epsilon(t-t')\}\phi_{AB}(t'). \end{aligned} \tag{B.51}$$

The relaxation function

$$R_{AB}(t) = \int_t^\infty dt' \phi_{AB}(t'). \tag{B.52}$$

A more useful form is

$$\partial_t R_{AB}(t) = -\phi_{AB}(t), \tag{B.53}$$

which is used to define $R_{AB}(t)$ for positive and negative times. The change in a variable induced by a discontinuous perturbation will vanish after a sufficiently long time elapse; this will occur for bulk samples because relaxation processes always exist in many-particle systems in thermal contact with an ambient medium. In consequence, we demand that $R_{AB}(t) \to 0$ as $t \to \infty$.

If the response $\phi_{AB}(t)$ is an odd function, then $R_{AB}(t)$ is an even function of $t$. Let $\tilde{R}_{AB}(s)$ be the Laplace transform of $R_{AB}(t)$, namely,

$$\tilde{R}_{AB}(s) = \int_0^\infty dt \exp(-st) R_{AB}(t), \tag{B.54}$$

and from (B.53) we find, for $s = (i\omega + \epsilon)$ and $\epsilon \to 0^+$,

$$-R_{AB}(0) + s\tilde{R}_{AB}(s) = \chi_{AB}[\omega]. \tag{B.55}$$

An immediate consequence of (B.55) is a relation between the dissipative susceptibility $\chi''_{AB}[\omega]$ and the real part of the Laplace transform of the relaxation function

$$\omega \tilde{R}'_{AB}(i\omega) = \chi''_{AB}[\omega]. \tag{B.56}$$

The spectrum of spontaneous fluctuations can be expressed in terms of $\tilde{R}'_{AB}(i\omega)$ using the fluctuation-dissipation theorem.

A very useful representation of $R_{AB}(t)$ is

$$R_{AB}(t) = \int_0^\beta d\mu \langle \hat{B}\hat{A}(t + i\hbar\mu)\rangle - \beta\langle \hat{A}\rangle\langle \hat{B}\rangle. \tag{B.57}$$

Perhaps the simplest way to verify (B.57) is to prove that it satisfies (B.53) with $\phi_{AB}(t)$ given by (B.32). The proof follows by applying the Heisenberg equation of motion

$$i\hbar\partial_t \hat{A}(t) = [\hat{A}(t), \hat{\mathscr{H}}]. \tag{B.58}$$

The constant of integration in (B.57) is chosen so as to make $R_{AB}(t) \to 0$ for $t \to \infty$. The law of increase in entropy, or loss of information, requires that there is no correlation between processes in a bulk sample that are well separated in time; thus,

$$\lim_{t \to \infty} \langle \hat{B}\hat{A}(t)\rangle = \langle \hat{B}\rangle\langle \hat{A}\rangle. \tag{B.59}$$

This result would not hold, for example, if there were a persistent mode in the sample to which either $\hat{A}$ or $\hat{B}$ coupled. However, in reality such modes do not exist in bulk samples since there is an almost infinite number of degrees of freedom and the mode will, eventually, decay.

In the next section we prove that $R_{AB}(t=0)$ is identical with the isothermal susceptibility.

## B.6. Isothermal susceptibility, $\chi_{AB}$

We consider now the change in the average value of an observable $A$ due to a small time-dependent perturbation $\mathscr{H}_1$. As before we write the total Hamiltonian

$$\hat{\mathscr{H}} = \hat{\mathscr{H}}_0 - \hat{\mathscr{H}}_1$$

and seek

$$\frac{1}{Z'} \operatorname{Tr} \exp(-\beta\hat{\mathscr{H}})\hat{A},$$

where

$$Z' = \operatorname{Tr} \exp(-\beta\hat{\mathscr{H}}).$$

A general solution to the problem can be obtained by defining $\hat{S}(\beta)$ by the equation

$$\exp\{-\beta(\hat{\mathscr{H}}_0 - \hat{\mathscr{H}}_1)\} = \exp(-\beta\hat{\mathscr{H}}_0)\hat{S}(\beta),$$

for then it satisfies

$$\frac{\partial \hat{S}}{\partial \beta} = \hat{\mathscr{H}}_1(-i\hbar\beta)\hat{S}(\beta),$$

with the boundary condtion $\hat{S}(0) = 1$. In principle $\hat{S}(\beta)$ can be obtained to any order in the perturbation by iterating the integral equation

$$\hat{S}(\beta) = 1 + \int_0^\beta \mathrm{d}\mu \hat{\mathscr{H}}_1(-i\hbar\mu)\hat{S}(\mu).$$

For our purposes, however, it is sufficient to take just the first-order solution because we have specified that the perturbation is small. Thus

$$\hat{S}(\beta) = 1 + \int_0^\beta \mathrm{d}\mu\, \hat{\mathscr{H}}_1(-i\hbar\mu),$$

$$Z' = Z(1 + \beta\langle\hat{\mathscr{H}}_1\rangle),$$

and

$$\frac{1}{Z'} \operatorname{Tr} \exp(-\beta\hat{\mathscr{H}})\hat{A} = \left\{\langle\hat{A}\rangle + \int_0^\beta \mathrm{d}\mu \langle\hat{\mathscr{H}}_1(-i\hbar\mu)\hat{A}\rangle\right\}(1 + \beta\langle\hat{\mathscr{H}}_1\rangle)^{-1},$$

and so the change in the average value of $A$ is

$$\int_0^\beta \mathrm{d}\mu \langle \hat{\mathscr{H}}_1(-\mathrm{i}\hbar\mu)\hat{A}\rangle - \beta\langle\hat{A}\rangle\langle\hat{\mathscr{H}}_1\rangle$$

to first order in the perturbation.

If we write $\hat{\mathscr{H}}_1 = h\hat{B}$, where $h$ is the relevant perturbation parameter, the *isothermal susceptibility* is, from the last expression,

$$\chi_{AB} = \int_0^\beta \mathrm{d}\mu \langle \hat{B}(-\mathrm{i}\hbar\mu)\hat{A}\rangle - \beta\langle\hat{A}\rangle\langle\hat{B}\rangle.$$

$$= \int_0^\beta \mathrm{d}\mu \langle \hat{B}\hat{A}(\mathrm{i}\hbar\mu)\rangle - \beta\langle\hat{A}\rangle\langle\hat{B}\rangle \tag{B.60}$$

Comparing (B.60) with $R_{AB}(t=0)$ obtained from (B.57), we find

$$R_{AB}(0) = \chi_{AB} \tag{B.61}$$

and, from (B.55),

$$\chi_{AB}[0] = -\chi_{AB}. \tag{B.62}$$

## B.7. Correlation functions and symmetry relationships

Consider two arbitrary operators $\hat{O}$ and $\hat{P}$ and define the *spectral functions* $G^{<}_{OP}(\omega)$ and $G^{>}_{OP}(\omega)$ such that

$$\langle \hat{P}(0)\hat{O}(t)\rangle = \int_{-\infty}^{\infty} \mathrm{d}\omega\, G^{<}_{OP}(\omega)\exp(\mathrm{i}\omega t) \tag{B.63}$$

and

$$\langle \hat{O}(t)\hat{P}(0)\rangle = \int_{-\infty}^{\infty} \mathrm{d}\omega\, G^{>}_{OP}(\omega)\exp(\mathrm{i}\omega t). \tag{B.64}$$

Since

$$\langle \hat{O}(t)\hat{P}(0)\rangle = \langle \hat{O}(0)\hat{P}(-t)\rangle = \langle \hat{P}(0)\hat{O}(t+\mathrm{i}\hbar\beta)\rangle,$$

it follows that

$$G^{>}_{OP}(\omega) = \exp(-\hbar\omega\beta)G^{<}_{OP}(\omega). \tag{B.65}$$

We now derive very general symmetry relations for $G^{<}_{OP}$ and $G^{>}_{OP}$. If $(\mathscr{H}_0 - E_m)\psi_m = 0$, then

$$[G^{<}_{OP}(\omega)]^* = \frac{\hbar}{Z}\sum_{mn}(P_{mn})^*(O_{nm})^*\exp(-\beta E_m)\delta(\hbar\omega - E_n + E_m)$$

$$= \frac{\hbar}{Z}\sum_{mn}(\hat{P}^+)_{nm}(\hat{O}^+)_{mn}\exp(-\beta E_m)\delta(\hbar\omega - E_n + E_m)$$

$$= G^{<}_{P^+O^+}(\omega) \tag{B.66}$$

and, also,

$$G^{<}_{OP}(-\omega) = \exp(-\hbar\omega\beta)G^{<}_{PO}(\omega) = G^{>}_{PO}(\omega). \tag{B.67}$$

Thus, from (B.66),

$$\{G^{<}_{OP}(-\omega)\}^{*} = G^{>}_{O^{+}P^{+}}(\omega), \tag{B.68}$$

which could have alternatively been derived from the identity

$$\langle \hat{P}(0)\hat{O}(t)\rangle^{*} = \langle \hat{O}^{+}(t)\hat{P}^{+}(0)\rangle. \tag{B.69}$$

So far we have not imposed any restrictions on the form of the two Hermitian operators $\hat{A}$ and $\hat{B}$. Let $\hat{A}$ and $\hat{B}$ be unchanged under time reversal; the operators $\hat{A}$ and $\hat{B}$ are then purely real. In addition we assume the body to have no magnetic structure and not to be subject to an external magnetic field, so the wave functions of the stationary states are also real. Hence

$$(\psi_n^{*}, \hat{A}\psi_m) = (\psi_n^{*}, \hat{A}^{+}\psi_m) = (\psi_m, \hat{A}^{*}\psi_n^{*}) = (\psi_m, \hat{A}\psi_n^{*}), \tag{B.70}$$

so

$$\langle \hat{B}(0)\hat{A}(t)\rangle = \langle \hat{A}(0)\hat{B}(t)\rangle \tag{B.71}$$

and

$$G^{<}_{AB}(\omega) = G^{<}_{BA}(\omega), \tag{B.72}$$

from which it follows that

$$\langle \hat{A}(0)\hat{B}(t) - \hat{B}(t)\hat{A}(0)\rangle = \langle \hat{B}(0)\hat{A}(t) - \hat{A}(t)\hat{B}(0)\rangle$$

or

$$\phi_{AB}(t) = -\phi_{AB}(-t). \tag{B.73}$$

In the presence of a magnetic field $\mathbf{H}$ the matrix elements of $\hat{A}$, say, satisfy

$$\{\hat{A}(\mathbf{H})\}_{mn} = \{\hat{A}(-\mathbf{H})\}_{nm},$$

so in place of (B.71) we have

$$\langle \hat{B}(0)\hat{A}(t)\rangle_{\mathbf{H}} = \langle \hat{A}(0)\hat{B}(t)\rangle_{-\mathbf{H}}, \tag{B.74}$$

and an analogous relationship holds for the transformation properties of the response function.

Finally, we assume that $\hat{A}$, say, changes sign under time reversal; then $\hat{A}$ must be purely imaginary. Hence

$$(\psi_n^{*}, \hat{A}\psi_m) = (\psi_n^{*}, \hat{A}^{+}\psi_m) = (\psi_m, \hat{A}^{*}\psi_n^{*}) = -(\psi_m, \hat{A}\psi_n^{*}),$$

i.e.

$$A_{nm} = -A_{mn},$$

and it follows that

$$\langle \hat{B}(0)\hat{A}(t)\rangle = -\langle \hat{A}(0)\hat{B}(t)\rangle \tag{B.75}$$

and

$$G^{<}_{AB}(\omega) = -G^{<}_{BA}(\omega), \tag{B.76}$$

with the result that $\phi_{AB}(t)$ must be an even function of $t$.

If both $\hat{A}$ and $\hat{B}$ change sign under time reversal, then the results (B.71), (B.72) and (B.73) apply.

The extension of (B.75) to the case where an external magnetic field is present is straightforward since, from (B.74), we have

$$\langle \hat{B}(0)\hat{A}(t)\rangle_{\mathbf{H}} = -\langle \hat{A}(0)\hat{B}(t)\rangle_{-\mathbf{H}}. \tag{B.77}$$

From the representation of $R_{AB}(t)$ given by eqn (B.57) we can show that

$$R_{AB}(t) = R_{BA}(-t). \tag{B.78}$$

A proof of (B.78) follows from the use of the identity (B.36), and the result $\langle \hat{A}(t)\hat{B}\rangle = \langle \hat{A}\hat{B}(-t)\rangle$. Consider then

$$R_{BA}(t) = \int_0^\beta \mathrm{d}\mu \langle \hat{A}\hat{B}(t + \mathrm{i}\hbar\mu)\rangle - \beta\langle \hat{A}\rangle\langle \hat{B}\rangle.$$

The correlation function that appears in $R_{BA}(t)$ is

$$\langle \hat{B}(t - \mathrm{i}\hbar\beta + \mathrm{i}\hbar\mu)\hat{A}\rangle = \langle \hat{B}\hat{A}(-t + \mathrm{i}\hbar\beta - \mathrm{i}\hbar\mu)\rangle$$

and, changing the integration variable $\mu \to \beta - \mu$, leads immediately to result (B.78).

It follows from (B.53) that if the response function $\phi_{AB}(t)$ is an odd function of $t$ then $R_{AB}(t)$ is an even function of $t$ and vice versa. Hence (B.78) tells us that when $\phi_{AB}(t)$ is an odd function of $t$, $R_{AB}(t) = R_{BA}(t) = R_{AB}(-t) = R_{BA}(-t)$.

Let,

$$R_{AB}(\omega) = \frac{1}{2\pi}\int_{-\infty}^{\infty} \mathrm{d}t \exp(-\mathrm{i}\omega t) R_{AB}(t). \tag{B.79}$$

Then from (B.78) we have the general relationship

$$R_{AB}(\omega) = R_{BA}(-\omega) \tag{B.80}$$

and

$$R^*_{AB}(\omega) = R_{BA}(\omega). \tag{B.81}$$

Also, we know that when $\phi_{AB}(t)$ is an odd function of $t$, $R_{AB}(t)$ is an even function of $t$. Whence when (B.73) is valid we have from (B.78)

$$R_{AB}(t) = R_{BA}(t) \tag{B.82}$$

and

$$R_{AB}(\omega) = R_{AB}(-\omega), \tag{B.83}$$

with both $R_{AB}(t)$ and $R_{AB}(\omega)$ purely real.

From the preceding section it is easy to derive the relation

$$\pi R_{AB}(\omega)\hbar\omega = \tanh(\tfrac{1}{2}\hbar\omega\beta) \int_{-\infty}^{\infty} dt \exp(-i\omega t)\langle\{\hat{A}(t)\hat{B}(0)\}\rangle, \tag{B.84}$$

where the integrand on the right-hand side contains the symmetrized correlation function

$$\langle\{\hat{A}(t)\hat{B}(0)\}\rangle = \tfrac{1}{2}\langle\hat{A}(t)\hat{B}(0) + \hat{B}(0)\hat{A}(t)\rangle. \tag{B.85}$$

If $\phi_{AB}$ is an even function of $t$,

$$\langle\{\hat{A}(t)\hat{B}(0)\}\rangle = \frac{\hbar}{2\pi i}\int_{-\infty}^{\infty} d\omega \exp(i\omega t) \coth(\tfrac{1}{2}\hbar\omega\beta)\chi'_{AB}[\omega], \tag{B.86}$$

while, if $\phi_{AB}(t)$ is an odd function of $t$(cf.(B.38)),

$$\langle\{\hat{A}(t)\hat{B}(0)\}\rangle = \frac{\hbar}{2\pi}\int_{-\infty}^{\infty} d\omega \exp(i\omega t) \coth(\tfrac{1}{2}\hbar\omega\beta)\chi''_{AB}[\omega]. \tag{B.87}$$

The fluctuation-dissipation theorem as expressed by (B.86) or (B.87) is abstract and therefore its physical content a little obscured. From (B.87) we can, in fact, show that the power spectrum of a fluctuating variable $A$, say, is related to the rate of energy dissipation of an external field coupled to the system by its corresponding operator $\hat{A}$. It is worthwhile to indicate how we can arrive at this result by a more physical argument than that used to derive the expressions given above.

For this discussion it is convenient to assume that $\langle\hat{A}\rangle$ is zero. Alternatively, we could employ an operator $\hat{\mathscr{A}}$ defined as

$$\hat{\mathscr{A}} = \hat{A} - \langle\hat{A}\rangle$$

so that $\langle(\hat{\mathscr{A}})^2\rangle$ is the mean square fluctuation of the variable $A$.

As a result of the action of the external field, the system described by the Hamiltonain $\hat{\mathscr{H}}_0$ must change its state and, in doing so, absorb energy from the field. The rate of absorption of energy from the field can be calculated as follows. If $E$ is the mean energy of the complete system then, denoting the average of the total system by a bar, and setting $\hat{B} \equiv \hat{A}$,

$$\frac{dE}{dt} = \overline{\frac{\partial\hat{\mathscr{H}}}{\partial t}} = -\bar{A}(t)\frac{dh}{dt}.$$

This is the rate at which the external field feeds energy into the system, i.e. the rate at which energy is dissipated from the external field,

$Q$. We now put

$$h(t)=\tfrac{1}{2}\{h_0 \exp(\mathrm{i}\omega t)+h_0^* \exp(-\mathrm{i}\omega t)\},$$

and from (B.4) and (B.5) we have

$$-\bar{A}(t)=\tfrac{1}{2}h_0\chi_{AA}[\omega]\exp(\mathrm{i}\omega t)+\tfrac{1}{2}h_0^*\chi_{AA}^*[\omega]\exp(-\mathrm{i}\omega t),$$

and hence

$$Q=-\bar{A}(t)\frac{\mathrm{d}h}{\mathrm{d}t}$$

$$=\frac{-\mathrm{i}\omega}{4}[h_0h_0^*\{\chi_{AA}[\omega]-\chi_{AA}^*[\omega]\}-h_0^2\chi_{AA}[\omega]\exp(2\mathrm{i}\omega t)$$
$$-h_0^{*2}\chi_{AA}^*[\omega]\exp(-2\mathrm{i}\omega t)].$$

The last two terms oscillate with time and give no mean dissipation rate. Hence only the first term survives to give the average dissipation rate as

$$Q=\tfrac{1}{2}\omega\,|h_0|^2\chi''_{AA}[\omega]. \tag{B.88}$$

(B.88) shows that the energy dissipation $Q$ is determined by the imaginary part of the generalized susceptibility.

If $W_{mn}$ is the probability per unit time for the transition from the state $m$ of $\hat{\mathscr{H}}_0$ to the state $n$ as a consequence of the external perturbation, then it follows that the energy dissipated per second is

$$Q_m=\sum_n(E_n-E_m)W_{mn}.$$

The total rate of dissipation of energy is obtained from this by averaging over all the states $m$, i.e.

$$Q=\frac{1}{Z}\sum_{mn}\exp(-\beta E_m)(E_n-E_m)W_{mn}. \tag{B.89}$$

From perturbation theory we know that

$$W_{mn}=\frac{\pi}{2\hbar}|h_0|^2\,|A_{mn}|^2\{\delta(\hbar\omega+E_n-E_m)+\delta(\hbar\omega+E_m-E_n)\},$$

so that

$$Q=\frac{\pi\omega}{2\hbar}|h_0|^2G_{AA}^<(\omega)\{1-\exp(-\hbar\omega\beta)\}. \tag{B.90}$$

The *power spectrum* of $A$ is defined by

$$\mathscr{G}_A(\omega)=\frac{1}{2\pi}\int_{-\infty}^{\infty}\mathrm{d}t\exp(-\mathrm{i}\omega t)\langle\{\hat{A}(t)\hat{A}(0)\}\rangle=\tfrac{1}{2}\{\exp(-\hbar\omega\beta)+1\}G_{AA}^<(\omega). \tag{B.91}$$

Thus, on combining (B.88) with (B.90) and (B.91), we have

$$\mathscr{G}_A(\omega) = \frac{\hbar}{2\pi}\chi''_{AA}[\omega]\coth(\tfrac{1}{2}\hbar\omega\beta), \tag{B.92}$$

which is the desired relationship.

It follows immediately from (B.92) that

$$\langle(\hat{A})^2\rangle = \frac{\hbar}{\pi}\int_0^\infty d\omega\,\chi''_{AA}[\omega]\coth(\tfrac{1}{2}\hbar\omega\beta), \tag{B.93}$$

which relates the mean square fluctuation of $\hat{A}$ to the response of the system to an external field coupled to the system by $\hat{A}$. This could, as mentioned above, have been obtained directly from (B.87), since $\phi_{AA}(t)$ must be an odd function of $t$. However, the foregoing derivation of (B.93) demonstrates more clearly the physical ideas involved.

Furthermore, we see why the theorem arises in the study of the Born approximation to the differential scattering cross-section, since, for an interaction of the form considered above, the cross-section for the process in which an incident wave $\exp(i\mathbf{k}\cdot\mathbf{r})$ is scattered into the state $\exp(i\mathbf{k}'\cdot\mathbf{r})$, the target state changing from $m$ to $n$, is proportional to

$$\frac{k'}{k}\left|\int d\mathbf{r}\exp(-i\mathbf{k}'\cdot\mathbf{r})W_{mn}\exp(i\mathbf{k}\cdot\mathbf{r})\right|.$$

Therefore $Q$ is proportional to the negative of the mean energy lost by the incident particle, i.e. the differential retardation.

## B.8. Green functions (Rickayzen 1980; Inkson 1984)

When an approximate calculation of the cross-section is best made directly from the equation of motion for the operator that describes the neutron–matter interaction, then it is usually very convenient to couch the necessary development in terms of Green functions. The Green function

$$G(t) = -i\theta(t)\langle[\hat{B}(t), \hat{B}^+]\rangle \tag{B.94}$$

satisfies the equation of motion

$$i\hbar\partial_t G(t) = \hbar\,\delta(t)\langle[\hat{B}, \hat{B}^+]\rangle + (-i)\theta(t)\langle[[\hat{B}(t), \mathscr{H}], \hat{B}^+]\rangle. \tag{B.95}$$

In deriving this result we have used the following relation between the derivative of the step function and the delta function,

$$\partial_t\theta(t) = \delta(t), \tag{B.96}$$

and the equation of motion for a Heisenberg operator, (B.58). The inhomogeneous term on the right-hand side of (B.95) gives the boundary condition on the solution in terms of the Fourier transform of $G(t)$.

We define the Fourier transform

$$G(\omega)=\int_{-\infty}^{\infty} \mathrm{d}t\, \exp(\mathrm{i}\omega t)G(t) \tag{B.97}$$

and also introduce a notation that proves to be very convenient,

$$G(\omega)=\langle\langle \hat{B}; \hat{B}^{+}\rangle\rangle \tag{B.98}$$

Using the standard integral representation for the delta function in (B.95), the equation of motion becomes

$$\hbar\omega G(\omega)=\hbar\langle[\hat{B}, \hat{B}^{+}]\rangle+\langle\langle[\hat{B}, \mathcal{H}]; \hat{B}^{+}\rangle\rangle. \tag{B.99}$$

Here we have used a mixed notation, and written the primary Green function as in (B.97) and the more complicated Green function, which arises from the time development of $\hat{B}(t)$ in (B.94), in the notation specified in (B.98).

Before discussing equation (B.99) any further we establish the relation between $G(\omega)$ and the function of main interest

$$\begin{aligned} S(\omega)&=\frac{1}{2\pi\hbar}\int_{-\infty}^{\infty} \mathrm{d}t\, \exp(-\mathrm{i}\omega t)\langle\hat{B}\hat{B}^{+}(t)\rangle \\ &=\frac{1}{2\pi\hbar}\int_{-\infty}^{\infty} \mathrm{d}t\, \exp(\mathrm{i}\omega t)\langle\hat{B}(t)\hat{B}^{+}\rangle, \end{aligned} \tag{B.100}$$

where the second form follows from the identity $\langle\hat{B}\hat{B}^{+}(t)\rangle=\langle\hat{B}(-t)\hat{B}^{+}\rangle$. We insert the definition of $G(t)$ in (B.97) and use the integral representation of the step function

$$\theta(t)=\frac{\mathrm{i}}{2\pi}\lim_{\epsilon\to 0^{+}}\int_{-\infty}^{\infty} \frac{\mathrm{d}u\, \exp(-\mathrm{i}ut)}{(u+\mathrm{i}\epsilon)} \tag{B.101}$$

which follows from (B.12). The result is, with $\epsilon\to 0^{+}$,

$$\begin{aligned} G(\omega)&=\frac{1}{2\pi}\int_{-\infty}^{\infty} \frac{\mathrm{d}u}{(\omega-u+\mathrm{i}\epsilon)}\int_{-\infty}^{\infty} \mathrm{d}t\, \exp(\mathrm{i}tu)\langle\hat{B}(t)\hat{B}^{+}-\hat{B}^{+}\hat{B}(t)\rangle \\ &=\hbar\int_{-\infty}^{\infty} \frac{\mathrm{d}u}{(\omega-u+\mathrm{i}\epsilon)}\{1-\exp(-\hbar u\beta)\}S(u) \end{aligned} \tag{B.102}$$

and the final form is achieved with the aid of (B.36). Because $S(\omega)$ is purely real we deduce from (B.102), using (B.15), the desired relation between the imaginary part of $G(\omega)$ and $S(\omega)$

$$G''(\omega)=-\pi\hbar\{1-\exp(-\hbar\omega\beta)\}S(\omega)$$

or

$$S(\omega)=-\{1+n(\omega)\}\frac{1}{\pi\hbar}G''(\omega). \tag{B.103}$$

By comparing (B.103) with (B.40) and (B.56) we find the following relations between the generalized susceptibility, relaxation function, and Green function,

$$\chi''[\omega] = -\frac{1}{\hbar} G''(\omega) = \omega \tilde{R}'(\mathrm{i}\omega). \tag{B.104}$$

Note the dimensionality of the various functions; if the variable $\hat{B}$ is dimensionless, $\chi[\omega]$, $G(\omega)$, and $\tilde{R}(\mathrm{i}\omega)$ have dimensions (energy)$^{-1}$, (frequency)$^{-1}$ and (energy$\times$frequency)$^{-1}$, respectively.

We consider the calculation of the displacement Green function for a harmonic oscillator model as an illustration of the use of the equation of motion (B.99). The Hamiltonian is

$$\mathscr{H} = \frac{1}{2M}\hat{\mathbf{p}}^2 + \tfrac{1}{2}M\omega_0^2\hat{\mathbf{u}}^2, \tag{B.105}$$

where $\hat{\mathbf{p}}$ is the momentum conjugate to the displacement $\hat{\mathbf{u}}$, i.e.

$$[\hat{u}_\alpha, \hat{p}_\beta] = \mathrm{i}\hbar\delta_{\alpha\beta}. \tag{B.106}$$

Let

$$G_{\alpha\beta}(\omega) = \langle\langle \hat{u}_\alpha ; \hat{u}_\beta \rangle\rangle, \tag{B.107}$$

and this Green function satisfies the equation,

$$\hbar\omega G_{\alpha\beta}(\omega) = \langle\langle [\hat{u}_\alpha, \mathscr{H}]; \hat{u}_\beta \rangle\rangle. \tag{B.108}$$

The inhomogeneous term in (B.108) is zero because the displacements commute. The Green function on the right-hand side of (B.108) is calculated with the aid of the relation

$$[\hat{u}_\alpha, \mathscr{H}] = \frac{\mathrm{i}\hbar}{M}\hat{p}_\alpha \tag{B.109}$$

and then

$$\omega G_{\alpha\beta}(\omega) = \frac{\mathrm{i}}{M}\langle\langle \hat{p}_\alpha ; \hat{u}_\beta \rangle\rangle. \tag{B.110}$$

We now form an equation for $\langle\langle \hat{p}_\alpha ; \hat{u}_\beta \rangle\rangle$ using (B.99),

$$\begin{aligned} \hbar\omega\langle\langle \hat{p}_\alpha ; \hat{u}_\beta \rangle\rangle &= -\mathrm{i}\hbar^2\delta_{\alpha\beta} + \tfrac{1}{2}M\omega_0^2\langle\langle [\hat{p}_\alpha, \hat{\mathbf{u}}^2]; \hat{u}_\beta \rangle\rangle \\ &= -\mathrm{i}\hbar^2\delta_{\alpha\beta} - \mathrm{i}\hbar M\omega_0^2 G_{\alpha\beta}(\omega). \end{aligned} \tag{B.111}$$

Combining (B.110) and (B.111) we arrive at the final result

$$(\omega^2 - \omega_0^2)G_{\alpha\beta}(\omega) = \frac{\hbar}{M}\delta_{\alpha\beta} \tag{B.112}$$

or

$$G_{\alpha\beta}(\omega) = \frac{\hbar\delta_{\alpha\beta}}{2M\omega_0}\{(\omega-\omega_0)^{-1} - (\omega+\omega_0)^{-1}\}$$

from which we obtain

$$G''_{\alpha\beta}(\omega)=\frac{-\pi\hbar\delta_{\alpha\beta}}{2M\omega_0}\{\delta(\omega-\omega_0)-\delta(\omega+\omega_0)\}. \tag{B.113}$$

We compare this result with (B.28) for the susceptibility of a damped harmonic oscillator in the limit of vanishing damping; the results (B.28) and (B.113) satisfy the general relation (B.104) between the dissipative susceptibility and the imaginary part of the Green function.

## B.9. Generalized Langevin equation (Lovesey 1980; Grabert 1982)

The generalized Langevin equation is an *exact* representation of the equation of motion in which the force is separated into components that have distinctly different time scales; the result is given in eqn (B.136). Applied to the conserved variables of a system the representation reproduces the equations of linearized hydrodynamics in the limit of long wavelengths and times. In consequence, it is an ideal starting point for the construction of a generalized hydrodynamic theory, in which the transport coefficients depend on the wave vector and frequency. The generalized Langevin equation can also be used to generate a systematic perturbation expansion in terms of coupling constants, as we will show later in this subsection.

### *B.9.1. Definitions*

We will not derive the generalized Langevin equation, for although the derivation is relatively straightforward it is lengthy to present precisely and the strengths and weaknesses of the representation can be fully appreciated through examples. One strength of the representation is that it is readily applied to multidimensional variables, e.g. the three conserved variables of a monatomic liquid. In view of this we review the properties of the generalized Langevin equation for a variable $\hat{A}$ which is, in general, a column matrix; $\hat{A}^+$ is its Hermitian row matrix. The notation $(\hat{B}, \hat{A}^+)$ denotes a matrix, the elements of which we take to be relaxation functions as defined in § B.5. In other words $(\hat{B}, \hat{A}^+)$ denotes a scalar product, and, if $\hat{B}=\hat{A}(t)$ is an element of the product,

$$R(t)=(\hat{A}(t), \hat{A}^+)=\int_0^\beta \mathrm{d}\mu\langle\hat{A}^+\hat{A}(t+\mathrm{i}\hbar\mu)\rangle-\beta\,|\langle\hat{A}\rangle|^2$$

$$\equiv\int_0^\beta \mathrm{d}\mu\langle\{\hat{A}^+-\langle\hat{A}^+\rangle\}\{\hat{A}(t+\mathrm{i}\hbar\mu)-\langle\hat{A}\rangle\}\rangle. \tag{B.114}$$

We recall that $R_{AB}(t)=R_{BA}(-t)$ (eqn (B.78)) so that, for $t=0$,

$$(\hat{A}, \hat{B})=(\hat{B}, \hat{A}). \tag{B.115}$$

Also,

$$(\hat{A}, \hat{B}) = (\hat{B}^+, \hat{A}^+)^*, \tag{B.116}$$

$$(\hat{A}, \hat{A}^+) \geqslant 0, \tag{B.117}$$

and the Schwartz inequality is

$$(\hat{A}, \hat{A}^+)(\hat{B}, \hat{B}^+) \geqslant |(\hat{A}, \hat{B}^+)|^2. \tag{B.118}$$

If we are dealing with Fourier components of local Hermitian variables, denoted by $\hat{A}_{\mathbf{k}}$ and $\hat{B}_{\mathbf{q}}$, say, then translational invariance requires

$$(\hat{A}_{\mathbf{k}}(t), \hat{B}^+_{\mathbf{q}}) = \delta_{\mathbf{k},\mathbf{q}}(\hat{A}_{\mathbf{k}}(t), \hat{B}^+_{\mathbf{k}}). \tag{B.119}$$

Inversion symmetry requires

$$(\hat{A}_{\mathbf{k}}(t), \hat{B}^+_{\mathbf{k}}) = (\hat{A}_{-\mathbf{k}}(t), \hat{B}^+_{-\mathbf{k}}), \tag{B.120}$$

and, when $\hat{A}^+_{\mathbf{k}} = \hat{A}_{-\mathbf{k}}$, the scalar product $(\hat{A}_{\mathbf{k}}(t), \hat{A}^+_{\mathbf{k}})$ is purely real.

If $\hat{B} = \hat{A}$ and $\hat{A}$ is Hermitian, then, in zero magnetic field,

$$(\hat{A}(t), \hat{A}) = (\hat{A}(-t), \hat{A}) = (\hat{A}, \hat{A}(t)). \tag{B.121}$$

Taking $\hat{A}$ to be a single variable, $\hat{A}$ and $\dot{\hat{A}}$ have opposite time reversal signature and

$$(\dot{\hat{A}}, \hat{A}) = 0. \tag{B.122}$$

Notice that for a quantum system the last result follows directly from the general relation, eqn (B.53),

$$(\dot{\hat{A}}(t), \hat{B}) = \frac{-\mathrm{i}}{\hbar}([\hat{A}(t), \mathscr{H}], \hat{B}) = \frac{-\mathrm{i}}{\hbar}\langle[\hat{A}(t), \hat{B}]\rangle. \tag{B.123}$$

From (B.123) we can deduce the identity $(\dot{\hat{A}}, \hat{B}) = -(\hat{A}, \dot{\hat{B}})$.

Following these preliminary definitions, and various other pertinent results, we turn to the generalized Langevin equation itself. Let the normalized relaxation function (matrix) be

$$\mathbf{F}(t) = (\hat{A}(t), \hat{A}^+) \cdot (\hat{A}, \hat{A}^+)^{-1}, \tag{B.124}$$

where the dot denotes matrix multiplication. The Laplace transform satisfies the equation

$$\begin{aligned}\tilde{\mathbf{F}}(s) &= \int_0^\infty \mathrm{d}t \exp(-st)\mathbf{F}(t) \\ &= \{s\mathscr{I} - \mathrm{i}\boldsymbol{\Omega} + \tilde{\mathbf{M}}(s)\}^{-1},\end{aligned} \tag{B.125}$$

where $\mathscr{I}$ is the unit matrix, and the matrices $\boldsymbol{\Omega}$ and $\tilde{\mathbf{M}}(s)$ are to be defined. First, the frequency matrix

$$\mathrm{i}\boldsymbol{\Omega} = \dot{\mathbf{F}}(0) = (\dot{\hat{A}}, \hat{A}^+) \cdot (\hat{A}, \hat{A}^+)^{-1}, \tag{B.126}$$

and this can have nonzero matrix elements when $\hat{A}$ contains two or more components.

In order to define $\tilde{\mathbf{M}}(s)$, which is usually referred to as a memory function matrix for reasons that will become apparent, we must introduce a projection operator $P$ that is defined by

$$P\hat{B} = (\hat{B}, \hat{A}^{+}) \cdot (\hat{A}, \hat{A}^{+})^{-1} \cdot \hat{A}. \tag{B.127}$$

From (B.127) it is evident that $P\hat{B}$ is the projection of $\hat{B}$ on to $\hat{A}$. The projection operator is linear and Hermitian

$$(P\hat{A}, \hat{B}^{+}) = (\hat{A}, (P\hat{B})^{+}) \tag{B.128}$$

and idempotent

$$P^2 = P. \tag{B.129}$$

We shall also write the equation of motion for $\hat{A}(t)$ in terms of the Liouville operator $\hat{\mathscr{L}}$,

$$\partial_t \hat{A}(t) = \dot{\hat{A}}(t) = \mathrm{i}\hat{\mathscr{L}}\hat{A}(t). \tag{B.130}$$

For a quantal system,

$$\hbar\hat{\mathscr{L}}\hat{A} = [\mathscr{H}, \hat{A}], \tag{B.131}$$

and the appropriate Poisson bracket is used for a classical system. Using this notation, we define the column matrix

$$\hat{f} = (1-P)\dot{\hat{A}} = \dot{\hat{A}} - \mathrm{i}\mathbf{\Omega} \cdot \hat{A}, \tag{B.132}$$

which evolves in time through the operator

$$\exp\{\mathrm{i}t(1-P)\hat{\mathscr{L}}\}$$

which differs from the usual time development by the presence of the projection operator. Thus, with $\hat{f} \equiv \hat{f}(0)$ defined in (B.132),

$$\hat{f}(t) = \exp\{\mathrm{i}t(1-P)\hat{\mathscr{L}}\}\hat{f} \tag{B.133}$$

and the memory function matrix

$$\mathbf{M}(t) = (\hat{f}(t), \hat{f}^{+}) \cdot (\hat{A}, \hat{A}^{+})^{-1} \tag{B.134}$$

with the Laplace transform

$$\tilde{\mathbf{M}}(s) = \int_0^{\infty} \mathrm{d}t \exp(-st)\mathbf{M}(t). \tag{B.135}$$

The complications that arise from the presence of the projection operator in $\hat{f}(t)$ is the price paid for the construction of an exact generalized Langevin equation.

Result (B.125) is equivalent to the equation of motion

$$\partial_t \hat{A}(t) = \mathrm{i}\mathbf{\Omega} \cdot \hat{A}(t) - \int_0^t \mathrm{d}\bar{t}\, \mathbf{M}(t-\bar{t}) \cdot \hat{A}(\bar{t}) + f(t), \tag{B.136}$$

given that the random force (eqn (B.132)) satisfies

$$(\hat{f}(t), \hat{A}^+) = 0. \quad \text{(B.137)}$$

The structure of the memory function, which contains the nonlinear aspects of the equation of motion, is revealed in part by expressing it as a continued fraction expansion. To this end consider the case when the column matrix $\hat{A}$ contains two elements, the variable $\hat{Q}$ and its time derivative $\dot{\hat{Q}}$. The random force has just one component in this case

$$\hat{f}_2 = \ddot{\hat{Q}} + \theta\hat{Q} \quad \text{(B.138)}$$

where $\theta = (\hat{Q}, \hat{Q}^+)^{-1}$ and the memory function matrix has one element $M_{22}(t)$. We then find from (B.125)

$$\theta \int_0^\infty \mathrm{d}t \exp(-st)(\hat{Q}(t), \hat{Q}^+) = \{s + \tilde{M}(s)\}^{-1}$$
$$= \{s + \theta(\dot{\hat{Q}}, \dot{\hat{Q}}^+)/[s + \tilde{M}_{22}(s)]\}^{-1}. \quad \text{(B.139)}$$

It is easy to show from (B.139) that, for short times,

$$\theta(\hat{Q}(t), \hat{Q}^+) = 1 - \tfrac{1}{2}t^2 M(t=0) + \cdots \quad \text{(B.140)}$$

so that

$$M(0) = -(\ddot{\hat{Q}}, \hat{Q}^+)\theta = (\dot{\hat{Q}}, \dot{\hat{Q}}^+)\theta. \quad \text{(B.141)}$$

In view of this result it follows from (B.139) that $M(t)$ satisfies an equation

$$\partial_t M(t) = -\int_0^t \mathrm{d}\bar{t}\, M_{22}(t-\bar{t})M(\bar{t}), \quad \text{(B.142)}$$

from which we conclude that $M_{22}(t)$ is the memory function of the memory function $M(t)$.

A little thought shows that if we use a variable $\hat{A}$ that comprises $\hat{Q}$ and all its derivatives then we generate an infinite chain of equations of which (B.139) and

$$\tilde{M}(s) = \theta(\dot{\hat{Q}}, \dot{\hat{Q}}^+)/\{s + \tilde{M}_{22}(s)\}$$

are the first two members; to be precise we generate a continued fraction expansion for the memory function $\tilde{M}(s)$.

The properties of the continued fraction expansion are readily summarized. For this purpose let $\hat{A}$ denote a single variable and

$$F(t) = (\hat{A}(t), \hat{A}^+)/(\hat{A}, \hat{A}^+).$$

Then,

$$\tilde{F}(s) = \int_0^\infty \mathrm{d}t \exp(-st)F(t)$$
$$= \{s + \tilde{K}^{(1)}(s)\}^{-1}, \quad \text{(B.143)}$$

where, for $n \geqslant 1$,

$$\tilde{K}^{(n)}(s) = \delta_n / \{s + \tilde{K}^{(n+1)}(s)\}. \qquad \text{(B.144)}$$

The coefficient $\delta_n$ in (B.144) is the initial value of $K^{(n)}(t)$ and it can be expressed in terms of the frequency moments of the Fourier transform

$$F(\omega) = \frac{1}{\pi} \operatorname{Re} \tilde{F}(\mathrm{i}\omega). \qquad \text{(B.145)}$$

The $n$th moment is defined to be

$$\langle \omega^n \rangle = \int_{-\infty}^{\infty} \mathrm{d}\omega \, \omega^n F(\omega) \qquad \text{(B.146)}$$

and, since $F(\omega)$ is an even function of $\omega$, the nonzero moments have $n$ equal to an even integer. Moreover, the short-time expansion of $F(t)$ is

$$F(t) = 1 - \frac{t^2}{2!} \langle \omega^2 \rangle + \frac{t^4}{4!} \langle \omega^4 \rangle - \cdots \qquad \text{(B.147)}$$

and

$$s\tilde{F}(s) = 1 - \langle \omega^2 \rangle / s^2 + \langle \omega^4 \rangle / s^4 - \cdots . \qquad \text{(B.148)}$$

With the notation (B.146), the first three $\delta$s are

$$\delta_1 = \langle \omega^2 \rangle,$$
$$\delta_2 = (\langle \omega^4 \rangle / \langle \omega^2 \rangle) - \langle \omega^2 \rangle, \qquad \text{(B.149)}$$

and

$$\delta_2 \delta_3 = (\langle \omega^6 \rangle / \langle \omega^2 \rangle) - (\langle \omega^4 \rangle / \langle \omega^2 \rangle)^2.$$

The result obtained in § B.3 for the susceptibility of a damped harmonic oscillator is equivalent to a simple termination of the continued fraction expansion. To see how this comes about, let $\hat{A} = \hat{x}$ the displacement operator, and

$$F(t) = (\hat{x}(t), \hat{x}) / (\hat{x}, \hat{x}).$$

The approximation amounts to terminating the expansion (B.144) by setting $\tilde{K}^{(2)}(s) = \gamma$, where $\gamma$ is a constant; thus

$$\tilde{F}(s) \doteq \{s + \delta_1 / (s + \gamma)\}^{-1}, \qquad \text{(B.150)}$$

and the corresponding susceptibility is, from (B.55),

$$\chi[\omega] \doteq (\hat{x}, \hat{x}) \{s\tilde{F}(s) - 1\}$$
$$= -(\hat{x}, \hat{x}) \delta_1 / (s^2 + s\gamma + \delta_1), \qquad \text{(B.151)}$$

where $s = \mathrm{i}\omega$. We now recognize $\delta_1$ to be $\omega_0^2$ and, for a harmonic oscillator of natural frequency $\omega_0$,

$$\hat{x}(t) = \left(\frac{\hbar}{2M\omega_0}\right)^{1/2} \{\hat{a}\exp(-\mathrm{i}\omega_0 t) + \hat{a}^+\exp(\mathrm{i}\omega_0 t)\}, \tag{B.152}$$

where $\hat{a}$ and $\hat{a}^+$ are the usual Bose operators. Result (B.152) enables us to calculate the relaxation function $(\hat{x}, \hat{x})$ from the definition (B.114), with the result

$$(\hat{x}, \hat{x}) = 1/M\omega_0^2. \tag{B.153}$$

On combining (B.151) and (B.153), we obtain (B.24) and identify $\gamma$ with the frictional damping.

We turn now to the special time development of the fluctuating force. The significance of the projection operator in the time development is extracted by considering the correlation function of the operator

$$\hat{g}(t) = \exp(\mathrm{i}t\hat{\mathscr{L}})\hat{f} \tag{B.154}$$

which differs from $\hat{f}(t)$ (eqn (B.133)) only in the form of the time development operator; in particular $\hat{g}(0) = \hat{f}(0)$. It can be shown that (Lovesey 1980)

$$\int_0^\infty \mathrm{d}t \exp(-st)(\hat{g}(t), \hat{g}^+) \cdot (\hat{A}, \hat{A}^+)^{-1}$$
$$= (s\mathscr{I} - \mathrm{i}\mathbf{\Omega}) \cdot \{s\mathscr{I} - \mathrm{i}\mathbf{\Omega} + \tilde{\mathbf{M}}(s)\}^{-1} \cdot \tilde{\mathbf{M}}(s), \tag{B.155}$$

where $\tilde{\mathbf{M}}(s)$ is the Laplace transform of the random-force auto-correlation function (eqn (B.134)). The result (B.155) shows that $(\hat{g}(t), \hat{g}^+)$ and $(\hat{f}(t), \hat{f}^+)$ are not simply related, i.e. the projection operator in the time development of $\hat{f}(t)$ appears to have a nontrivial effect.

There is one important case when the two auto-correlation functions are the same, and it arises in the proof that we obtain the standard Green–Kubo formulae for transport coefficients in the hydrodynamic limit. The generic form a conservation equation for a classical variable is

$$\dot{\varphi}_\mathbf{k} + \mathrm{i}kJ_\mathbf{k} = 0, \tag{B.156}$$

where $J_\mathbf{k}$ is the flux associated with the variable $\varphi_\mathbf{k}$. From the definition (B.126),

$$\mathrm{i}\Omega = (\dot{\varphi}_\mathbf{k}, \varphi_\mathbf{k}^*)(\varphi_\mathbf{k}, \varphi_\mathbf{k}^*)^{-1} = -\mathrm{i}k(J_\mathbf{k}, \varphi_\mathbf{k}^*)(\varphi_\mathbf{k}, \varphi_\mathbf{k}^*)^{-1}, \tag{B.157}$$

and therefore

$$\hat{f} = \hat{g} = \dot{\varphi}_\mathbf{k} - \mathrm{i}\Omega\varphi_\mathbf{k} = -\mathrm{i}kJ_\mathbf{k} + \mathrm{i}k(J_\mathbf{k}, \varphi_\mathbf{k}^*)(\varphi_\mathbf{k}, \varphi_\mathbf{k}^*)^{-1}\varphi_\mathbf{k}$$
$$= -\mathrm{i}k\mathring{J}_\mathbf{k} \tag{B.158}$$

where the last equality defines $\mathring{J}_{\mathbf{k}}$; note that $\mathring{J}_{\mathbf{k}}$ is orthogonal to $\varphi_{\mathbf{k}}$. We observe that $\mathrm{i}\Omega$ and the memory matrix $M_k(t) \propto k^2(\mathring{J}_{\mathbf{k}}(t), \mathring{J}^*_{\mathbf{k}})$ vanish in the limit $k \to 0$. From this it follows that eqn (B.155) leads to the important result

$$\lim_{k\to 0} \int_0^\infty \mathrm{d}t \exp(-st)(\mathring{J}_{\mathbf{k}}(t), \mathring{J}^*_{\mathbf{k}})(\varphi_{\mathbf{k}}, \varphi^*_{\mathbf{k}})^{-1} = \lim_{k\to 0} \{\tilde{M}_k(s)/k^2\}. \quad \text{(B.159)}$$

The left-hand side is precisely the form of a Green–Kubo formula and, in the limit of long wavelengths, this is seen to be equal to the memory function whose time development is given by the complicated operator in (B.133). We conclude that, for conserved variables, the complication in the time development of the random forces vanishes in the long-wavelength limit (Mountain 1976).

### *B.9.2. Anharmonic lattice vibrations*

An important aspect of the phonon cross-section is the influence of anharmonic interactions which always exist, to some extent, in real crystals. If the anharmonic interactions are weak, then we can study their influence on the cross-section by perturbation theory. Such a calculation provides a useful guide to the behaviour of the cross-section in the vicinity of a structural phase transition, but, since a transition is driven by anharmonic terms, a simple perturbative calculation can be no more than a guide to the full behaviour.

Here we study the influence of weak cubic and quartic interactions on the phonon cross-section by calculating the lattice displacement susceptibility. The anharmonic interactions are included to leading order in a systematic expansion. Our primary objective in presenting the development at this juncture is to illustrate the use of the generalized Langevin equation reviewed in the preceding section. The method is well suited to the problem for two reasons. First, the Hamiltonian is conveniently expressed in terms of the displacement and momentum normal coordinates so it is natural to work with these conjugate variables. Hence we develop the perturbation theory in terms of a two-component column-matrix variable whose elements are proportional to the displacement and momentum normal coordinates. A second complicating feature of the perturbation expansion is that the perturbation is the sum of cubic and quartic interactions which must be treated in a consistent manner. It is relatively easy to keep track of the various terms in the generalized Langevin representation because it affords a very compact formalism. In fact we will derive an exact expression for the displacement susceptibility as a starting point for our discussion. A trivial approximation to the exact expression leads to the result for a damped harmonic oscillator derived in

§ B.3, and then we proceed to a systematic perturbative study of the influence of cubic and quartic interactions.

The natural variables are the normal coordinate displacement $\hat{Q}_{\mathbf{k}}$ and the momentum $\hat{P}_{\mathbf{k}}$. These operators satisfy the commutation relation

$$[\hat{Q}_{\mathbf{k}}, \hat{P}^{+}_{\mathbf{q}}] = \mathrm{i}\hbar\delta_{\mathbf{k},\mathbf{q}} \tag{B.160}$$

and we require $\hat{Q}_{\mathbf{k}} = \hat{Q}^{+}_{-\mathbf{k}}$ and $\hat{P}_{\mathbf{k}} = \hat{P}^{+}_{-\mathbf{k}}$ in order that the lattice displacement and momentum are real. The harmonic Hamiltonian for particles of mass $M$ is

$$\mathscr{H}_0 = \frac{1}{2}\sum_{\mathbf{k}}\left\{\frac{1}{M}\hat{P}^{+}_{\mathbf{k}}\hat{P}_{\mathbf{k}} + M\omega_k^2\hat{Q}^{+}_{\mathbf{k}}\hat{Q}_{\mathbf{k}}\right\} \tag{B.161}$$

where $\omega_k$ is the lattice dispersion frequency. The anharmonic contribution to the total Hamiltonian is

$$\hat{\mathscr{H}}_1 = -\sum_{123} J(1, 2, 3)\hat{Q}_1\hat{Q}_2\hat{Q}_3 - \sum_{1234} K(1, 2, 3, 4)\hat{Q}_1\hat{Q}_2\hat{Q}_3\hat{Q}_4. \tag{B.162}$$

Here we use numbers to label wave vectors, i.e. $\mathbf{k}_1 \equiv 1$, etc. Each interaction term in (B.162) conserves momentum, so that $J(1, 2, 3)$ is zero unless the wave vectors satisfy $1+2+3=0$ and $K(1, 2, 3, 4)$ is zero unless $1+2+3+4=0$. The interaction coefficients are unchanged by a permutation of the wave vectors since the normal coordinate operators commute with each other, i.e. they are completely symmetric in the wave-vector indices. Because $\hat{Q}_{\mathbf{k}}$ and $\hat{P}_{\mathbf{k}}$ appear in $\mathscr{H}_0$, we will work with a column-vector variable. Using (B.160), we can verify that the time derivative of $\hat{Q}_{\mathbf{k}}$ is

$$\begin{aligned}\dot{\hat{Q}}_{\mathbf{k}} &= -\frac{\mathrm{i}}{\hbar}[\hat{Q}_{\mathbf{k}}, (\mathscr{H}_0 + \mathscr{H}_1)] \\ &= -\frac{\mathrm{i}}{\hbar}[\hat{Q}_{\mathbf{k}}, \mathscr{H}_0] = P_{\mathbf{k}}/M.\end{aligned} \tag{B.163}$$

In view of this result we choose our variable to be the column vector

$$\hat{A} = \mathrm{column}\{\hat{Q}_{\mathbf{k}}, \hat{P}_{\mathbf{k}}/M\}. \tag{B.164}$$

The variables in $\hat{A}$ are orthogonal and thus $(\hat{A}, \hat{A}^{+})$ is a diagonal matrix. We will find that only one element of the random force is nonzero, and this further simplifies the structure of the calculation.

We use the notation

$$(\hat{Q}_{\mathbf{k}}, \hat{Q}^{+}_{\mathbf{k}}) = \chi_k/M, \tag{B.165}$$

where the susceptibility has dimension (frequency)$^{-2}$. Observe that, from (B.163),

$$(\hat{P}_{\mathbf{k}}, \hat{P}^{+}_{\mathbf{k}}) = M(\dot{\hat{Q}}_{\mathbf{k}}, \hat{P}^{+}_{\mathbf{k}}) = -\frac{\mathrm{i}M}{\hbar}\langle[\hat{Q}_{\mathbf{k}}, \hat{P}^{+}_{\mathbf{k}}]\rangle = M, \tag{B.166}$$

and, since $(\hat{\dot{A}}, \hat{B}) = -(\hat{A}, \hat{\dot{B}})$,

$$(\hat{\dot{P}}_{\mathbf{k}}, \hat{Q}^+_{\mathbf{k}}) = -(\hat{P}_{\mathbf{k}}, \hat{\dot{Q}}^+_{\mathbf{k}}) = -\frac{1}{M}(\hat{P}_{\mathbf{k}}, \hat{P}^+_{\mathbf{k}}) = -1. \tag{B.167}$$

Using these results, we find the frequency matrix

$$\begin{aligned} \mathrm{i}\boldsymbol{\Omega} &= \text{column}\{\hat{\dot{Q}}_{\mathbf{k}}, \hat{\dot{P}}_{\mathbf{k}}/M\} \cdot \text{row}\{\hat{Q}^+_{\mathbf{k}}, \hat{P}^+_{\mathbf{k}}/M\} \cdot (\hat{A}, \hat{A}^+)^{-1} \\ &= \begin{pmatrix} 0 & 1 \\ -1/\chi_k & 0 \end{pmatrix} \end{aligned} \tag{B.168}$$

and the random-force column vector

$$\hat{f} = \hat{\dot{A}} - \mathrm{i}\boldsymbol{\Omega} \cdot \hat{A} = \text{column}\{0, \hat{f}_2\}$$

where

$$\hat{f}_2 = \frac{1}{M}\hat{\dot{P}}_{\mathbf{k}} + \frac{1}{\chi_k}\hat{Q}_{\mathbf{k}}. \tag{B.169}$$

An equation for $\hat{\dot{P}}_{\mathbf{k}}$ is readily obtained,

$$\begin{aligned} \hat{\dot{P}}_{\mathbf{k}} &= \frac{-\mathrm{i}}{\hbar}[\hat{P}_{\mathbf{k}}, (\mathcal{H}_0 + \mathcal{H}_1)] \\ &= -M\omega_k^2 \hat{Q}_{\mathbf{k}} + \hat{Z}_{\mathbf{k}} \end{aligned} \tag{B.170}$$

where the operator $\hat{Z}_{\mathbf{k}}$ arises from the anharmonic terms, namely,

$$\begin{aligned} \hat{Z}_{\mathbf{k}} &= \frac{-\mathrm{i}}{\hbar}[\hat{P}_{\mathbf{k}}, \mathcal{H}_1] \\ &= 3\sum_{23} J(-k, 2, 3)\hat{Q}_2\hat{Q}_3 + 4\sum_{234} K(-k, 2, 3, 4)\hat{Q}_2\hat{Q}_3\hat{Q}_4. \end{aligned} \tag{B.171}$$

We now construct a formally exact result for the relaxation function

$$\int_0^\infty \mathrm{d}t \exp(-st)(\hat{Q}_{\mathbf{k}}(t), \hat{Q}^+_{\mathbf{k}}) \tag{B.172}$$

which is, given (B.164), the $(1, 1)$ matrix element of $\tilde{\mathbf{F}}(s)$ defined in (B.125). Let the one nonzero element of the memory function matrix be denoted by $(s = (\mathrm{i}\omega + \epsilon)$ with $\epsilon \to 0^+)$

$$\int_0^\infty \mathrm{d}t \exp(-st)(\hat{f}_2(t), \hat{f}_2^+) = \tilde{\theta}_k(s)/M; \tag{B.173}$$

then the relaxation function (B.172) is readily found after some simple matrix manipulation to be

$$(\chi_k/M)\frac{\{s + \tilde{\theta}_k(s)\}}{s\{s + \tilde{\theta}_k(s)\} + (1/\chi_k)} \tag{B.174}$$

and the displacement susceptibility of interest is, from (B.55),

$$M\chi_k[\omega]=\{\omega^2-(1/\chi_k)-\mathrm{i}\omega\tilde{\theta}_k(\mathrm{i}\omega)\}^{-1}. \tag{B.175}$$

If the memory function $\tilde{\theta}_k(\mathrm{i}\omega)$ is replaced by a constant, result (B.175) has the same structure as the susceptibility of a damped harmonic oscillator (cf. (B.24)).

The calculation of $\chi_k[\omega]$ requires expressions for both $\chi_k$ and $\tilde{\theta}_k(\mathrm{i}\omega)$. For the former we combine the definitions (B.165) and (B.170) and find, on using (B.167), the result

$$\omega_k^2\chi_k=1+(\hat{Z}_{\mathbf{k}},\hat{Q}_{\mathbf{k}}^+).$$

Applying (B.170) a second time leads to our starting point for the calculation of $\chi_k$, namely,

$$\omega_k^2\chi_k=1-\frac{1}{M\omega_k^2}(\hat{P}_{\mathbf{k}}^+,\hat{Z}_{\mathbf{k}})+\frac{1}{M\omega_k^2}(\hat{Z}_{\mathbf{k}},\hat{Z}_{\mathbf{k}}^+). \tag{B.176}$$

Because $\hat{Z}_{\mathbf{k}}$ is derived from the anharmonic terms, (B.176) can be expanded order-by-order in the interaction coefficients. Given that $\langle\hat{Q}_{\mathbf{k}}\rangle=0$ for all wave vectors, a simple calculation shows that

$$(\hat{P}_{\mathbf{k}},\hat{Z}_{\mathbf{k}})=-12\sum_q K(-k,k,q,-q)\langle|\hat{Q}_{\mathbf{q}}|^2\rangle. \tag{B.177}$$

Note that this expression is exact.

To proceed further we must introduce some approximations. The leading-order terms in $(\hat{Z}_{\mathbf{k}},\hat{Z}_{\mathbf{k}}^+)$ are of order $J^2$. We find

$$(\hat{Z}_{\mathbf{k}},\hat{Z}_{\mathbf{k}}^+)\doteq 9\sum_{23}\sum_{2'3'}J(-k,2,3)J^*(-k,2',3')(\hat{Q}_2\hat{Q}_3,\hat{Q}_{2'}^+\hat{Q}_{3'}^+). \tag{B.178}$$

The calculation of $\tilde{\theta}_k(s)$ starts from an appropriate approximation for $\hat{f}_2$, eqn (B.169). The leading-order term arises when we use the approximation $\chi_k\doteq 1/\omega_k^2$,

$$\hat{f}_2\doteq(3/M)\sum_{23}J(-k,2,3)\hat{Q}_2\hat{Q}_3, \tag{B.179}$$

and the corresponding result for $\tilde{\theta}_k(s)$ is

$$\tilde{\theta}_k(s)\doteq(9/M)\sum_{23}\sum_{2'3'}J(-k,2,3)J^*(-k,2',3')$$
$$\times\int_0^\infty \mathrm{d}t\,\exp(-st)(\hat{Q}_2(t)\hat{Q}_3(t),\hat{Q}_{2'}^+\hat{Q}_{3'}^+). \tag{B.180}$$

In view of the fact that the random force is proportional to the strength of the cubic term, and therefore a small quantity, we neglect the projection operator in the time evolution of the random force. Thus, the relaxation function in (B.180) is evaluated for harmonic vibrations described by $\hat{\mathscr{H}}_0$, and the time dependence of the $\hat{Q}_{\mathbf{k}}$'s will be combinations of $\exp(\pm it\omega_k)$. This scheme is entirely consistent with our approximation of keeping leading-order terms from $\hat{\mathscr{H}}_1$.

Before we make the calculation of the relaxation function in (B.180) we take advantage of a formal rearrangement. Consider the quantity in the denominator of $\chi_k[\omega]$ (eqn (B.175))

$$(1/\chi_k)+s\tilde{\theta}_k(s)$$

and expand $\chi_k^{-1}$ using (B.176) and use (B.180) for $\tilde{\theta}_k(s)$. The approximate result can be written in a compact form with the definition

$$\omega_k^2(T)=\omega_k^2-(12/M)\sum_q K(-k,k,q,-q)\langle|\hat{Q}_{\mathbf{q}}|^2\rangle, \tag{B.181}$$

and the final form, correct to order $K$ and $J^2$, is

$$(1/\chi_k)+s\tilde{\theta}_k(s)\doteq\omega_k^2(T)+(9/M)\sum_{23}\sum_{2'3'}J(-k,2,3)J^*(-k,2',3')$$
$$\times\int_0^\infty \mathrm{d}t\,\exp(-st)\{\partial_t(\hat{Q}_2(t)\hat{Q}_3(t),\,\hat{Q}_{2'}^+\hat{Q}_{3'}^+)\}. \tag{B.182}$$

The Laplace transform now involves the derivative of the relaxation function which is particularly easy to evaluate in view of the identity (B.123). To make the calculation we need the following results. For harmonic motion,

$$[\hat{Q}_{\mathbf{k}}(t),\,\hat{Q}_{\mathbf{k}}^+]=\frac{-i\hbar}{M\omega_k}\sin(t\omega_k) \tag{B.183}$$

and

$$\langle\hat{Q}_{\mathbf{k}}(t)\hat{Q}_{\mathbf{k}}^+\rangle=(\hbar/2M\omega_k)\{(1+n_k)\exp(-it\omega_k)+n_k\exp(it\omega_k)\}, \tag{B.184}$$

where $n_k$ is the Bose occupation factor for the frequency $\omega_k$. With the aid of these results, a straightforward and tedious calculation leads to

$$\partial_t(\hat{Q}_2(t)\hat{Q}_3(t),\,\hat{Q}_{2'}^+\hat{Q}_{3'}^+)=\frac{-i}{\hbar}\langle[\hat{Q}_2(t)\hat{Q}_3(t),\,\hat{Q}_{2'}^+\hat{Q}_{3'}^+]\rangle$$
$$=(-\hbar/2M^2\omega_2\omega_3)\{(1+n_2+n_3)\sin\{t(\omega_2+\omega_3)\}$$
$$+(n_3-n_2)\sin\{t(\omega_2-\omega_3)\}\}(\delta_{22'}\delta_{33'}+\delta_{23'}\delta_{32'}). \tag{B.185}$$

The Laplace transform of this result is inserted in (B.182), and the final form for the displacement susceptibility is, from (B.175),

$$(M\chi_k[\omega])^{-1}=\omega^2-(1/\chi_k)-i\omega\tilde{\theta}_k(i\omega)$$
$$=\omega^2-\omega_k^2(T)-\Sigma_k(\omega) \tag{B.186}$$

and our approximation for the self-energy is

$$\Sigma_k(\omega) = (9\hbar/2M^3) \sum_{pq} \frac{|J(-k, p, q)|^2}{\omega_p \omega_q}$$
$$\times \{(1 + n_p + n_q)[(\omega - \Omega_+ - i\epsilon)^{-1} - (\omega + \Omega_+ - i\epsilon)^{-1}]$$
$$+ (n_q - n_p)[(\omega - \Omega_- - i\epsilon)^{-1} - (\omega + \Omega_- - i\epsilon)^{-1}]\}, \qquad \text{(B.187)}$$

where the frequencies

$$\Omega_\pm = \omega_p \pm \omega_q. \qquad \text{(B.188)}$$

Because the interaction coefficient $J$ conserves the wave vector in the interaction process, the wave vectors in (B.187) satisfy $\mathbf{k} = \mathbf{p} + \mathbf{q}$, and therefore one sum is trivial.

A few comments about the results are in order. First, we note from the definition of $\omega_k^2(T)$ that, for $K$ positive, it is possible to have $\omega_k^2(T) = 0$ for some value of $\mathbf{k}$ and the temperature. The occurrence of a soft mode is a signature for a structural phase transition. A study of the conditions for which $\omega_k^2(T) = 0$ demands a complete knowledge of the quartic interaction coefficient. If we employ the approximation (B.184) for the mean-square of the displacement normal coordinate and consider the high-temperature limit in which $(1 + n_k) \doteq n_k = (k_B T/\hbar\omega_k)$, then the equation $\omega_k^2(T) = 0$ reduces to

$$\omega_k^2 - (12 k_B T/M^2) \sum_q \{K(-k, k, q, -q)/\omega_q^2\} = 0. \qquad \text{(B.189)}$$

A more complete study of the soft-mode condition recognizes that the normal-mode frequency contains a contribution from the self-energy. Given that the self-energy is a correction to the normal-mode frequency it is sufficient to evaluate it at the 'bare' frequency $\omega_k$. The 'renormalized' normal-mode frequency is then $\{\omega_k^2(T) + \Sigma_k'(\omega_k)\}^{1/2}$ where $\Sigma_k'$ is the real part of the self-energy. A self-consistent equation for the soft-mode condition is obtained by using (B.184) for $\langle|\hat{Q}_\mathbf{q}|^2\rangle$ in $\omega_k(T)$ with $\omega_q$ replaced by the renormalized normal-mode frequency. In the limit of high temperatures, we recover eqn (B.189) with $\omega_q^2$ replaced by $\{\omega_q^2(T) + \Sigma_q'(\omega_q)\}$, and the solution depends on the cubic and quartic interaction coefficients in a nontrivial manner.

Let us turn now to a consideration of the self-energy (B.187). We are not able to evaluate this expression without a complete prescription for the cubic interaction coefficient and the lattice dispersion frequency. Even so we can reach a few useful conclusions. We focus attention on the damping of the collective mode which arises from the imaginary part of the self-energy. From (B.187) we deduce that, for $\epsilon \to 0^+$,

$$\Sigma_k''(\omega) = (9\pi\hbar/2M^3) \sum_{pq} \frac{|J(-k, p, q)|^2}{\omega_p \omega_q}$$
$$\times \{(1+n_p+n_q)[\delta(\omega-\Omega_+)-\delta(\omega+\Omega_+)]+(n_q-n_p)[\delta(\omega-\Omega_-)-\delta(\omega+\Omega_-)]\}. \quad \text{(B.190)}$$

Note that $\Sigma_k''(\omega)$ is an odd function of $\omega$. The damping is finite for $T \to 0$ due to zero-point fluctuations. In the opposite limiting case we find

$$\Sigma_k''(\omega) = (9\pi k_B T\omega/2M^3) \sum_{pq} \frac{|J(-k, p, q)|^2}{(\omega_p \omega_q)^2}$$
$$\times \{\delta(\omega-\Omega_+)+\delta(\omega+\Omega_+)+\delta(\omega-\Omega_-)+\delta(\omega+\Omega_-)\} \quad \text{(B.191)}$$

and, by analogy with the damped harmonic oscillator model of § B.3, we can identify the quantity $\{\Sigma_k''(\omega)/\omega\}$ evaluated at $\omega = \omega_k$ as a wave-vector-dependent friction coefficient.

## REFERENCES

Grabert, H. (1982). *Springer tracts in modern physics*, Vol. 95. Springer-Verlag, Berlin.

Inkson, J. C. (1984). *Many-Body Theory of Solids*. Plenum Press, New York.

Lovesey, S. W. (1980). *Condensed matter physics, dynamic correlations*. Frontiers in physics, Vol. 49. Benjamin/Cummings, Reading, MA.

Mountain, R. D. (1976). *Advan. mol. relaxation Processes* **9,** 225.

Reichl, L. E. (1980). *A modern course in statistical physics*. Edward Arnold, London.

Rickayzen, G. (1980). *Green's functions and condensed matter*. Academic Press, New York.

van Kampen, N. G. (1981). *Stochastic processes in physics and chemistry*. North-Holland, Amsterdam.

# INDEX


absorption of neutrons §§ 1.3, A.2
  cross-sections Table 1.1
alloys
  diffuse elastic scattering § 2.5
  inelastic, mass-defect § 4.7.3
anharmonicity
  in phonon theory §§ 4.5, B.9.2
  role in structural phase transitions § 4.6

Bloch's identity § 3.8
Boltzmann
  distribution § 3.6.1
  fluid, total cross-section § 3.7
Born approximation §§ 1.2, A.1
Bose
  distribution function eqn (3,101)
  fluid §§ 3.6.1, 5.6
    perfect, response function eqn (3.111)
Bose–Einstein condensation § 5.6
Bragg
  cut-off Fig. 2.9
  scattering § 2.4
Bragg's law 44
Bravais lattices § 2.1, Fig. 2.1, Table 2.1
Brillouin
  doublet § 5.4, eqn (5.161)
  zone § 4.1, Fig. 4.1
Brownian motion eqn (5.63)

charge density fluctuations § 5.5
coherent cross-sections
  classical fluid
    monatomic § 5.4
    multicomponent § 5.5
  elastic § 2.2
  one-phonon § 4.4
  quantum fluid § 5.6
contrast factors § 1.5
correlation functions, temporal
  classical approximation §§ 3.2, 3.5
  coherent § 3.2
  complex conjugate eqn (3.44)
  incoherent § 3.2, eqn (3.92)
  nuclear scattering eqn (3.6)
    impulse approximation § 3.4
  stationary condition eqn (3.42)
  velocity eqns (5.62), (5.67), (5.72)
critical scattering § 5.4.1, eqn (5.162)
cross-section
  Boltzmann particle eqn (3.98)
    total § 3.7
  bound nuclei § 1.3
  Bragg § 2.4
  classical approximation §§ 3.5, 5.1
  diatomic molecule § 1.7, 6.1, 6.5.1
  differential eqn (1.12)
  fluids
    classical
      binary eqn (5.233)
      coherent eqns (5.38), (5.97)
      single particle eqns (5.29), (5.58)
    quantum
      Bose eqn (3.111), § 5.6
      Fermi eqn (3.119)
  gases
    diatomic § 6.5.1
    perfect § 3.7
    polyatomic § 6.5.2
  general expression eqn (1.23)
  harmonic oscillator §§ 3.6.2, 6.3
  impulse approximation § 3.4
  lattice vibrations
    anharmonic §§ 4.5, 4.6
    multiphonon § 4.8
    one-phonon § 4.4
      coherent eqn (4.89)
      incoherent eqn (4.92)
  Laue monotonic 56
  molecular
    diatomic §§ 1.7, 6.5.1
    polyatomic § 6.5.2
    quasi-classical § 6.5.3
  mass defect § 4.7.2
  partial differential eqn (1.23)
  single free particle § 1.9
  static approximation §§ 1.8, 5.4.1
  total, single atom § 1.4
  unpolarized neutrons, general expression eqn (5.5)
  Van Hove formulation § 3.1
crystal lattice
  dynamics §§ 4.1, 4.5
  normal vibrations § 4.1
  mass defects §§ 4.7.2, 4.7.3
  vectors § 2.1, Table 2.1
  vectors, reciprocal § 2.1, Table 2.2
  vibrational frequency spectrum eqn (4.39), Fig. 4.3
    Debye model eqn (4.42)

Debye model density of phonon states eqn (4.42)
Debye–Scherrer cones Fig. 2.8
Debye–Waller factor § 4.3
detailed balance
  condition eqn (3.49)
  factor eqn (5.17), Table B.1
diffraction
  Bragg § 2.4
  diffuse § 2.5
  macromolecules §§ 1.5, 6.1
  multicomponent classical fluid §§ 5.5, 6.1
diffuse elastic scattering from alloys § 2.5
diffusion
  constant eqns (5.47), (5.62)
  continuous § 5.3, eqn (5.49)
  jump model eqn (6.44)
  spherical particle § 5.3.2
dynamic structure factor, response function § 3.1

Einstein relation eqn (5.66)
elastic scattering
  Bragg §§ 1.3, 2.4, 4.2
  diffuse §§ 1.3, 2.5
  total, powder eqn (4.47)
electron–phonon interaction § 4.6
energy conversion data § 1
entropy fluctuations, Rayleigh peak § 5.4.2
exchange interaction §§ 1.7, 3.6.1, 5.1, 6.5

Fermi
  distribution function eqn (3.102)
  pseudo-potential eqn (1.30)
Fermi's Golden rule eqn (1.8)
fluctuation–dissipation theorem eqn (3.54), § B.4
fluids
  classical
    binary § 5.5
    Boltzmann, perfect § 3.6.1
      total § 3.7
    monatomic
      coherent § 5.4
      single-particle § 5.3
  quantum
    Bose §§ 3.6.1, 5.6
    Fermi §§ 3.6.1, 5.6
  supercooled §§ 5.4.2, 5.4.3
force constants §§ 4.1, B.9.2
form factor
  diatomic molecule eqn (6.13)
  Gaussian polymer eqn (6.22)
  Guinier form eqn (6.20)
  molecular eqn (1.55)
  spherical molecule eqns (1.56), (6.18)

gases
  diatomic molecules § 1.7
  free particles § 3.6.1
  harmonic oscillators § 6.3
  molecular § 6.5
Gaussian approximation:
  to single-particle motion in fluids § 5.3.1, eqn (5.50)
  to multiphonon cross-section § 4.8
  to quasi-elastic scattering § 6.2
Green functions
  definition § B.8, Table B.1
  lattice displacement § 4.7
  mass defect § 4.7.2
  oscillator, hybridized with phonon enviroment § 4.7.1
Green–Kubo formulae
  diffusion constant eqn (5.62)
  thermal conductivity eqn (5.176)
  viscosity, kinematic eqn (5.175)
Grüneisen parameter § 4.5

harmonic oscillator § 3.6.2
  damped §§ 4.7.1, B.3
  gas of § 6.3, Fig. 6.1
Heisenberg operator eqn (3.3)
helium
  $^3$He § 5.6
  $^4$He § 5.6
hydrodynamics
  fluid equations § 5.4.2
  single particle motion § 5.3.2
hydrogen
  ortho- § 1.7
  para- § 1.7

impulse approximation § 3.4
incoherent approximation, multiphonon cross-section § 4.8
isotopic incoherence §§ 1.3, 2.3

Kohn anomaly § 4.6
Krieger–Nelkin approximation § 6.5.3

Langevin equation, Brownian motion eqn (5.63)
  generalized, Mori formulation § B.9
lattice
  defects § 2.5
  mixed, harmonic § 4.7
  vectors § 2.1
  vectors, reciprocal § 2.1
  vibrations
    anharmonic § 4.5
    harmonic § 4.1
Laue method § 2.4, Figs. 2.5, 2.6
  monotonic scattering eqn (2.55)

linear response theory
  applied to Fermi fluid § 5.6
  review of Appendix B

mass defect § 4.7.2
mass-tensor approximation § 6.5.3
Maxwell relaxation time § 5.4.2
mean field theory
  liquid–vapour transition § 5.4.1, eqn (5.162)
  structural phase transition § 4.6
molecular spectroscopy § 6.4
molecules
  diatomic §§ 1.7, 6.5.1
  polyatomic §§ 6.1, 6.5.2
moments § 3.3
  classical fluid § 5.2
  quantum fluid § 5.6
  *see also* sum rules
multiphonon cross-section
  mass expansion § 4.8
  Gaussian approximation § 4.8

Navier–Stokes equation eqn (5.142)
nuclear
  forces § 1.3
  spin orientations §§ 1.7, 5.1, 5.6

pair distribution function eqn (3.28)
  classical monatomic fluid eqn (5.102), Fig. 5.1
partial differential cross-section eqn (1.23)
phase transitions
  liquid–vapour § 5.4
  structural § 4.6
phonon
  damping §§ 4.5, 4.7.3, B.9.2
    interference effects § 4.5
  energy § 4.1, Figs. 4.2, 4.3
  polarization vector § 4.1
    phase convention eqn (4.30)
  resonance modes § 4.7
  scattering
    coherent eqn (4.89)
    incoherent eqn (4.92)
    multiphonon § 4.8
  wave vector, sum eqn (4.13)
polarization, in cross-section § 1.2

quantum fluids
  cross-section eqn (5.269)
  Bose §§ 3.6.1, 5.6
  Fermi §§ 3.6.1, 5.6
quasi-elastic scattering §§ 5.4.3, 6.2

radiative capture §§ 1.3, 4.1
radius of gyration eqn (6.21)
rate equation for jump diffusion eqn (6.44)
Rayleigh peak § 5.3.2
relaxation function eqn (5.16), § B.5, Table B.1
response function §§ 3.1, B.4, Table B.1
rocking curve § 2.4
rotational energy states 264
roton minimum 236

Scattering
  amplitude eqn (1.13b), § A.1
    operator § 1.6
  length § 1.3
    bound values Table 1.1
    energy dependent § A.2
scattering law, response function § 3.1
self-correlation function, single-particle response §§ 3.2, 3.4
  classical monatomic fluid § 5.3
  gaussian approximation § 5.3.1
  phonon § 4.4
  quasi-elastic contribution § 6.2
scattering length, bound § 1.3, Table 1.1
small-angle scattering §§ 1.5, 6.1
soft mode theory
  of structural phase transition 129–30
  of liquid–glass transition § 5.4.3
static approximation § 1.8, 1.9
  classical monatomic fluid § 5.4.1
structure factor
  classical fluid
    monatomic eqn (5.33), Fig. 5.1
    multicomponent § 5.5, eqn (5.264)
  quantum fluid eqn (5.286)
    Schwartz inequality eqn (5.287)
  unit cell eqn (2.28)
sum rules, monatomic fluid:
  classical § 5.2
  compressibility eqn (5.285)
  quantum § 5.6
susceptibility:
  isothermal § B.6, Table B.1
  generalized, wave vector and frequency dependent § B.1, Table B.1

t-matrix for mass defects § 4.7.3
thermal conduction in fluids § 5.4.2

unit-cell § 2.1
  structure factor eqn (2.28)

Van Hove formulation of the cross-section § 3.1
viscoelastic theory, of classical monatomic fluid § 5.4.3

Zero sound 230